CW00870776

Progress in Mathematical Physics
Volume 27

Floyd Williams

Topics in Quantum Mechanics

Birkhäuser
Boston • Basel • Berlin

Floyd Williams
University of Massachusetts
Department of Mathematics
Amherst, MA 01003
U.S.A.

Library of Congress Cataloging-in-Publication Data

Williams, Floyd L.
Topics in quantum mechanics / Floyd Williams.
p. cm.– (Progress in mathematical physics ; v. 27)
Includes bibliographical references and index.
ISBN 0-8176-4311-7 (alk. paper) — ISBN 3-7643-4311-7 (alk. paper)
1. Quantum theory. 2. Mathematical physics. I. Title. II. Series.

QC174.12 .W538 2003
530.12–dc21

2002038317
CIP

AMS Subject Classifications: 20C30, 20C35, 22E40, 33C45, 33C55, 34A05, 34L20, 34L40, 35J10, 35Q40, 42A38, 43A85, 70H03, 70H05, 70H25, 70S15, 81P05, 81Q05, 81Q15, 81Q30, 81R05, 81T13, 81T20

Printed on acid-free paper.

Birkhäuser

ISBN 0-8176-4311-7 SPIN 10890279
ISBN 3-7643-4311-7

Reformatted from the author's files by John Spiegelman, Philadelphia, PA.
Printed in the United States of America.

9 8 7 6 5 4 3 2 1

Birkhäuser Boston • Basel • Berlin
A member of BertelsmannSpringer Science+Business Media GmbH

To My Mother,
Mrs. Lee Edna Rollins

Contents

Preface xiii

Acknowledgments xv

I Introductory Concepts in Quantum Theory 1

0 Units of Measurement 3

1 Quantum Mechanics: Some Remarks and Themes 7

1.1 Planck Quantization 7
1.2 The Bohr Frequency Rule and Emission Spectra 9
1.3 Schrödinger's Equation 14
1.4 The de Broglie Wavelength 17
1.5 Heisenberg's Uncertainty Principle 20
1.6 Bohr's Complementarity Principle 22

2 Equations of Motion in Classical Mechanics 25

2.1 Introduction 25
2.2 Motion in One Dimension 26
2.3 Motion in Three Dimensions 31
2.4 Many Particles in Three Dimensions 32
2.5 Coulomb and Gravitational Potentials 34
2.6 Hamilton's Equations of Motion 36
2.7 Lagrange's Equations of Motion 40
2.8 Center of Mass Coordinates 44

3 Quantization and the Schrödinger Equation **51**
3.1 The One-Dimensional Schrödinger Equation 51
3.2 The Quantized Harmonic Oscillator 53
3.3 A Finite Potential Well 53
3.4 Schrödinger's Equation for a System 58
3.5 Schrödinger's Equation for an Atom 60
3.6 Non-Quantization of the Free Particle 63
3.7 Quantization, Large Eigenvalues, and Spectral Zeta Functions . . 65
3.8 The Diffeomorphism Φ Again 72
3.9 Some Brief Remarks on Axioms 77

4 Hypergeometric Equations and Special Functions **81**
4.1 Initial Remarks 81
4.2 Differential Equations of Hypergeometric Type 82
4.3 Reduction to Canonical Form 83
4.4 Solutions of $\sigma y'' + \tau y' + \lambda y = 0$ 87
4.5 Examples of the Reduction Technique 91
4.6 Examples: Hermite, Legendre, Laguerre, Jacobi 95
4.7 Orthogonality 98
4.8 Normalized Wave Functions for the Harmonic Oscillator 109
4.9 Further Examples of the Reduction Technique: The Pöschl–Teller Potential 110
Appendix 4A: Spherical Harmonics 115
Appendix 4B: The Polynomials L_n 121

5 Hydrogen-like Atoms **123**
5.1 Solutions of Schrödinger's Equation for Hydrogenic Atoms 124
5.2 Normalized Wave Functions 130
5.3 The Wave Functions ψ_{nlm} for Small n 132
5.4 Quantization of Angular Momentum 138
5.5 The Average Distance of the Electron from the Nucleus 148
5.6 The Ritz–Rydberg Formula: Further Examples 150
5.7 Wave Function Orthogonality 152
5.8 Ground State Energy of Hydrogen-like Atoms 154
Appendix 5A: Commutation Relations for Angular Momentum Operators 154

6 Heisenberg's Uncertainty Principle **157**
6.1 Heisenberg's Inequality 157
6.2 Heisenberg's Uncertainty Principle 162
6.3 Example 169

7 Group Representations and Selection Rules **171**
7.1 Group Representations 171
7.2 Contragredient and Tensor Product Representations 178
7.3 Subrepresentations 181

7.4 Group Actions 182
7.5 An Abstract Selection Rule 183
7.6 Some Preliminary Applications of Theorem 7.2 188
7.7 Time-Dependent Perturbations and Spectroscopic Selection Rules 190
7.8 Spectroscopic Selection Rules for the Hydrogenic Atom 201
7.9 A General Quantum Mechanical Selection Rule 202
7.10 Group Invariance of Schrödinger's Equation 203
7.11 The Representations $1^{\pm}$ of S_n, and the Irreducible Representations of S_3 204
Appendix 7A: Proof of Theorems 7.2 and 7.6 208
Appendix 7B: Proof of Proposition 7.2 213
Appendix 7C: Proof of Equation (7.1.10) 214
Appendix 7D: Remarks on Equation (7.7.54) 215

8 The Quantized Hamiltonian for a Charged Particle in an Electromagnetic Field 217
8.1 The Motion of a Charged Particle in an Electromagnetic Field . . 217
8.2 Classical and Quantized Hamiltonian for a Charged Particle 223
Appendix 8A: The Electric Field Due to a Collection of Point Charges, Electric Potential, and Voltage 225
Appendix 8B: Electric Dipole Moments 230

9 Spin Wave Functions 233
9.1 Spin Operators 233
9.2 Pauli Spin Wave Functions 236
9.3 The Zeeman Effect for Weak Magnetic Fields 240
Appendix 9A: Magnetic Dipole Moments 245
Appendix 9B: Quantization of the Magnetic Dipole Moment $\vec{\mu}$ 248
Appendix 9C: Construction of Higher Spin Operators 249

10 Introduction to Multi-Electron Atoms 253
10.1 First-Order Correction to E_o 253
10.2 Helium-Like Atoms: Their Ground-State Energies 256
10.3 Pauli's Exclusion Principle and Electron Configurations 260
Appendix 10A: On the Integral in (10.2.19) 267

II Some Selected Topics 269

11 Fresnel Integrals and Feynman Integrals 271
11.1 Fresnel Integrals 271
11.2 Feynman Path Integrals 280
Appendix 11A: Gaussian Integrals 288

12 Path Integral for the Harmonic Oscillator 291

13 Euclidean Path Integrals 299
13.1 Euclidean Path Integrals 299

13.2 Wave Function Expansion of the Euclidean Path Integral 303

14 The Density Matrix and Partition Function in Quantum Statistical Mechanics **307**
14.1 Helmholtz Free Energy, Entropy and Internal Energy 307
14.2 A Zeta Function Representation of the Free Energy 310
Appendix 14A: Jacobi Inversion 314

15 Zeta Regularization **317**

16 Helmholtz Free Energy for Certain Negatively Curved Space-Times, and the Selberg Trace Formula **321**
16.1 A Zeta Function for Kaluza–Klein Space-Times Modeled on $SO_1(d,1)/SO(d)$, and a Trace Formula for Compact Quotients of $SO_1(d,1)$ 321
16.2 Temperature Asymptotics 331

17 The Zeta Function of a Product of Laplace Operators and the Multiplicative Anomaly for X_Γ^d **333**
17.1 $\det L_1 L_2 = \det L_1 \det L_2$? 333
17.2 Explicit Meromorphic Structure of $\zeta_\Gamma(s; b_1, b_2)$ 335

18 Schrödinger's Equation and Gauge Theory **341**
Patrick Shanahan 341
18.1 A Brief Introduction to Gauge Theory 341
18.2 343
18.3 346
18.4 349
18.5 352
About the Author 357

General Appendices 359

Appendix A: Some Further Electron Configurations **361**

Appendix B: Mendeléev Periodic Table **365**

Appendix C: Determinants for String World-Sheets That Are Tori: Another Example of Zeta Regularization **369**

Appendix D: Evaluation of the Integral $I_n = \int_{\mathbb{R}^n}\int_{\mathbb{R}^n} \frac{e^{-\alpha(|x|+|y|)}}{|x-y|^{n-\beta}}\,dxdy$ **373**

Appendix E: Some Informal Comments on QFT **379**

References **385**

Index **393**

List of Figures

1.1 Newton's prism experiment 10
1.2 Spectral lines 12
1.3 Photographic plate for the Balmer series (experimental observation) 13
1.4 Circular motion 18
1.5 Traveling wave 19
1.6 Single slit diffraction of electrons 21
1.7 Computing Δp 23

2.1 Simple harmonic motion (cf. Figure 1.4) 29
2.2 A well of depth d 30
2.3 Particle on a cylinder 40

3.1 Finite potential well 54
3.2 Tangent intersections 56
3.3 Cotangent intersections 57
3.4 Turning points for the classical region R_E 68

4.1 Spherical coordinates 116

5.1 Electrostatic coulomb force on an electron 124
5.2 Graph of $\psi_{100}(x,0,0)$ 133
5.3 Angular momentum 138
5.4 Quantum angular momentum 139
5.5 Spatial quantization of angular momentum 140

7.1 An electromagnetic wave propagated in the y-direction. The electric and magnetic field vectors are perpendicular to each other and to the direction in which the wave is propagated. 199

8.1 Direction of E_B 225
8.2 Example 226
8.3 Multiple charges 227
8.4 Two charges in oil 229
8.5 Dipole 230
8.6 Dipole in a uniform field 231
8.7 Circular displacement 231

9.1 Spatial quantization of S 235
9.2 Zeeman splitting in a magnetic field 242
9.3 Right-hand rule 246
9.4 Magnetic dipole moment 247

10.1 Shell diagram for Na 262
10.2 Shell diagram for Cl 263

11.1 Closed contour 272
11.2 A lower bound for the sine 273
11.3 Path of particle 280
11.4 A polygonal path 282

18.1 Basic set-up for a fiber bundle 343
18.2 Transition mappings 344
18.3 Isotopic spin 354

A.1 Electron configurations 363

B.1 Metals 366
B.2 Nonmetals 367
B.3 Rare earth elements 368

Preface

Quantum mechanics and quantum field theory are highly successful physical theories that have numerous practical applications. Largely mathematical in character, these theories continue to stimulate the imaginations of applied mathematicians and purists as well. In recent years, in particular, as a new array of tools have emerged, including a representative amount from the domain of so-called pure mathematics, interest in both the conceptual and physical aspects of these beautiful subjects has especially blossomed. Given the emergence of newer and often spectacular applications of mathematics to quantum theory, and to theoretical physics in general, one notes that certain communication gaps between physicists and mathematicians continue to be bridged.

This text on quantum mechanics, designed primarily for mathematics students and researchers, is an attempt to bridge further gaps. Although the mathematical style presented is generally precise, it is counterbalanced at some points by a relaxation of precision, as our overall purpose is to capture the basic flavor of the subject both formally and intuitively. The approach is one in which we attempt to maintain sensitivity with respect to diverse backgrounds of the readers, including those with modest backgrounds in physics. Thus we have included several concrete computational examples to fortify stated principles, several appendices, and certain basic physical concepts that help to provide for a reasonably self-contained account of the material, especially in the first 11 chapters.

Part I, which consists of the first ten chapters, is an introduction to the basic classical themes and ideas of quantum mechanics. Part II, Chapters 11–18, consists of selected special topics with some emphasis on path integrals and applications of zeta functions and the Selberg trace formula. The material is somewhat more technical in contrast to that of Part I. The last chapter is a further discussion of special topics by guest contributor Patrick Shanahan.

Further remarks concerning the content and scope of this work now follow. In Chapter 1 we provide a brief overview of the physical aspects of the subject, with a historic discussion of some basic themes: Planck's quantization, Bohr's frequency rule, Schrödinger wave functions, deBroglie wavelength, Heisenberg's uncertainty principle, emission spectra, Balmer's formula, and so on—themes which set the tone of the work and which are subject to further analysis in Chapters 3, 5, 6, and 7. A discussion of matters such as spin, Pauli wave functions, Pauli's exclusion principle, and the Zeeman effect, for example, are deferred to Chapters 9 and 10.

Chapter 2 covers background material on classical mechanics, which is needed for the transition to quantum mechanics. One could employ symplectic machinery here in a modern approach, but we have chosen a more modest route—one suitable for our purposes. No effort is made, moreover, to consider geometric quantization—say in the context of a line bundle with a connection whose curvature is the symplectic form on a base manifold. Material from Chapter 2 is needed also for Chapters 8 and 18 where the motion of a charged particle in an electromagnetic field is considered.

A study of hydrogen-like atoms occupies most of Chapter 5. The discussion there is extended in Chapter 7 which takes up spectroscopic selection rules for such atoms. Since these rules follow from a general abstract selection rule, which in fact is a pure group-theoretic result, we develop the group theory necessary for a proof of the latter. The statement presented clarifies some of those found in the physical literature which seem to rest on hypotheses not as well articulated. We also discuss, among other things, the group invariance of Schrödinger's equation, and we provide some remarks on representations of the symmetric group for purposes of Chapter 10. It is observed, for example, that the latter group has exactly two irreducible representation of degree 1, up to equivalence—a mathematical fact which can be viewed as corresponding in some sense to the somewhat mysterious fact that nature allows only for symmetric or antisymmetric wave functions in the description of indistinguishable particles. More extended and systematic applications of group theory are found in other excellent texts, particularly in the three references listed at the conclusion of Chapter 7.

The Schrödinger equation in the context of gauge theory is considered in Chapter 18. That chapter, contributed by Pat Shanahan, takes as a starting point material from Chapter 8. An introduction to gauge theory is presented, with a discussion of the Maxwell and Yang–Mills equations.

Appendix E in the general appendices at the end of the text is presented to help ease the transition from Part I to Part II, by providing a few minor comments on quantum field theory and statistics. Some of the material in Part II requires of the reader some elementary knowledge of Riemannian geometry.

One of the special features of this book, not found in physics texts, is the systematic use of the theory of hypergeometric equations, by A. Nikiforov and V. Uvarov, to develop uniformly the theory of special functions and to solve

Schrödinger equations without the usual recourse to power series methods. By this approach, one obtains a priori quantization of energy conditions from a single point of view. This theory, developed in the fourth chapter, has many applications apart from quantum theory. We also indicate in Part II the important role, which is ever increasing, that automorphic forms play in physical theories. In particular, we consider some critical applications of Selberg's trace formula.

Acknowledgments

The shape and development of the text derives much from many stimulating and informative conversations the author has had in past years with various persons at the Universities of Massachusetts, Missouri, and Vermont, and at Johnson C. Smith University, Hong Kong University of Science and Technology, the National University of Singapore, the State University of Economics and Finances of Saint Petersburg, Russia, the Pulvoko Astronomical Observatory in Saint Petersburg, Russia, and the State University of Londrina, Brazil. The author expresses herewith considerable gratitude for the kind hospitality accorded him at these distinguished universities and at the Pulvoko Observatory.

The author also extends a very great measure of thanks to Volker Ecke and to Hala Allouba of the University of Massachusetts for the many long hours of labor they invested in the preparation of the manuscript. Their dedication and quality of work throughout have simply been superb. I am much appreciative of and indebted to these two young scholars in mathematics and computer science for assistance in many directions.

This work is dedicated to my dear mom, Mrs. Lee Edna Rollins, with lots of love and lifelong admiration. Her many years of love and wisdom have always kept me in good stead.

Floyd Williams

Amherst, Massachusetts

July, 2001

Part I

Introductory Concepts in Quantum Theory

0

Units of Measurement

This book is largely targeted toward a mathematical audience, whose grasp and recollection of principles of physics may vary from small to great. It seems prudent therefore that we should adopt a somewhat self-contained approach and proceed along elementary lines. In many cases we shall attempt to complement stated principles with concrete computational examples. Towards this end, one necessity is a review of some of the basic units of measurement. Also towards the end of providing a largely self-contained account we will supply generous amounts of detail at times in the discussion of much of the material.

It will be convenient to mention terms like "energy," or "radiation," in this chapter and the next without providing some precision of meaning. Key terms and notions will be defined and developed as we move along. Some readers may choose to browse lightly through the remainder of this chapter and to return to it later as necessity dictates.

Classical physics involves the notion of fundamental quantities such as length, time, mass, force, and energy. Various systems have been devised for measuring these quantities. Length (the distance between two points) can be measured in miles, feet, or meters, for example—the conversions being that 1 mile = $5,280$ feet and 1 meter = 3.28 feet. Time (a measure of duration) will usually be expressed in seconds. Mass is a measure of inertia—i.e., of the resistance an object offers to a change in its motion as some force is applied to it. The notions of mass and force are not independent ones as we may regard force, defined intuitively, as a push or pull which tends to produce change of motion (or position) of an object. If one takes the notion of force as fundamental, for example, then one can *define* mass via Newton's second law of motion. Namely, the acceleration a experienced by an object (i.e., the rate of change of its velocity) due to a force F acting on it is directly proportional to F: $a = \kappa F$ for some $\kappa > 0$. Then we can define the *mass* m of the object by $m = \frac{1}{\kappa}$; thus $F = ma$. Energy is a measure of the capacity for work to be done, work being the product of the magnitude of force and the displacement it causes. Energy may be *potential* (due to position) or *kinetic* (due to motion). A more careful discussion of these matters will be given in Chapter 2.

In the English system mass is measured in *pounds* and force is measured in *poundals*: one poundal of force is that amount which will impart to a mass of one pound an acceleration of one foot per second in one second. We shall be interested in the French system where mass is measured in *grams* and force is measured in *dynes*: one dyne of force is that amount which will impart to a mass of one gram an acceleration of one centimeter per second in one second. Here a mass of one pound = 453.6 grams, a centimeter $\stackrel{\text{def}}{=}$ meter/100 = 0.0328 feet = 0.3936 inches. Therefore one poundal of force = 1 pound of mass $\times 1$ ft./sec^2 = 453.6 grams $\times \frac{\text{centimeter}}{0.0328\ \text{sec}^2} = 13829.27 \frac{\text{g cm}}{\text{sec}^2}$ (where we also write "g" for grams, "cm" for centimeters) $\Longrightarrow$ 1 dyne = $\frac{1}{13829.27}$ poundal. One also refers to the c.g.s. system where length, mass, time are measured in centimeters, grams, seconds. For example, work (= force × distance) in the c.g.s. system is measured in dyne-centimeters. One dyne-centimeter is called an *erg* ($= 1 \frac{\text{g cm}^2}{\text{sec}^2}$).

Energy can be measured in erg units. For example, the kinetic energy T of an object of mass m with velocity of magnitude v is given by $T = \frac{1}{2}mv^2$; we shall see in Chapter 2 *why* the $\frac{1}{2}$ appears in this definition. Thus in the c.g.s. system the units of T are grams $\times \frac{(\text{cm})^2}{\text{sec}^2}$ = ergs. One can also measure calories in terms of ergs: 1 calorie (the amount of heat required to raise the temperature of one gram of water one degree centigrade (or Celsius)) = 4.187×10^7 ergs.

The unit of *action* (= work × times) is the erg-second. The fundamental constant that we consider is *Planck's constant* h given by

$$h = 6.62608 \times 10^{-27} \text{ erg-seconds.} \tag{0.0.1}$$

We also write $\hbar$ for $h/2\pi$ (following Dirac):

$$\hbar \stackrel{\text{def}}{=} \frac{h}{2\pi} = 1.05457 \times 10^{-27} \text{erg-seconds.} \tag{0.0.2}$$

There will be occasions to consider periodic motion—for example the motion of a fixed point on the circumference of a circle. In this case the *frequency* ν of the motion is the number of revolutions per second of the point, and the *period* T is the amount of time required for one revolution; thus $\nu = \frac{1}{T}$ has the unit (second)$^{-1}$. The unit of frequency is the *hertz* (Hz), in honor of Heinrich Hertz who discovered radio waves in 1887. One hertz = (second)$^{-1}$, or one cycle per second. $h\times$ frequency is therefore measured in ergs, and therefore should represent some energy quantity. This indeed is the case, as expressed by the *Bohr frequency rule*, discussed in Chapter 1, Section 1.2. Also note formula (1.1.1).

Recall next that two electrically charged objects attract or repel each other according as the charges are opposite or like, respectively. In the c.g.s. *Gaussian* system charge is measured in *electrostatic units* (esu), defined as follows: One esu is that amount of charge on a test object for which an equal object of like charge is repelled by a force of 1 dyne if the two objects are placed 1 cm apart in a vacuum. Experiments in 1909 by Ehrenhaft and in 1910 by Millikan especially have determined that the charge $-e$ on the electron is given by

$$e = 4.80321 \times 10^{-10}\text{esu}. \tag{0.0.3}$$

The esu is also called a *stat-Coulomb*. One has Coulomb's law (due to Charles Coulomb (1736-1806))

$$F = \kappa \frac{q_1 q_2}{r^2}, \tag{0.0.4}$$

which we shall discuss further in Chapter 2, for the magnitude F of the force of attraction or repulsion of two objects with charges q_1, q_2 separated a distance r apart; $\kappa > 0$ is a suitable proportionality constant.

Example

Two objects of like charge experience a repulsive force of 9 dynes. They are separated a distance of 0.20 cm apart. What is the modulus $|q|$ of the charge q of these objects?

Solution: By Coulomb's law $\kappa q^2 = r^2 F = (0.20)^2 \cdot 9 \text{ cm}^2 \text{g} \frac{\text{cm}}{\text{sec}^2} = 0.36 \frac{\text{g cm}^3}{\text{sec}^2} \implies |q| = \frac{0.6}{\sqrt{\kappa}}$ esu.

In addition to the electron charge $-e$ given in (0.0.3) we shall need the electron, proton, and neutron masses m_e, m_p, m_n:

$$m_e = 9.10939 \times 10^{-28} \text{ grams}, \tag{0.0.5}$$

$$m_p = 1.67263 \times 10^{-24} \text{ grams} \tag{0.0.6}$$

$$= 1836.16\ m_e, \tag{0.0.7}$$

$$m_n = 1.67493 \times 10^{-24} \text{ grams} \tag{0.0.8}$$

$$= 1838.69\ m_e. \tag{0.0.9}$$

The hydrogen atom consists of an electron (with negative charge $-e$) and a proton with positive charge e. The *reduced mass* μ of hydrogen is defined by $\mu = \frac{m_e m_p}{m_e + m_p}$. Thus by (0.0.5) and (0.0.7),

$$\mu = \frac{1836.16\ m_e^2}{1837.16\ m_e} = 0.999\ m_e = 9.1044 \times 10^{-28} \text{ grams}. \tag{0.0.10}$$

Given this value of μ we can compute *Bohr's radius* a_o defined by

$$a_o \stackrel{\text{def}}{=} \frac{h^2}{4\pi^2 \mu e^2}. \tag{0.0.11}$$

First note indeed that a_o has the unit of length: By Coulomb's law (0.0.4) e^2 is measured in units $(\text{cm})^2$ dynes $= \text{g} \frac{\text{cm}^3}{\text{sec}^2}$. Then the quotient in (0.0.11) has units

$$\frac{\left[\frac{\text{g}\times(\text{cm})^2}{\text{sec}}\right]^2}{\text{g}^2 \frac{\text{cm}^3}{\text{sec}^2}} = \text{centimeters}.$$

By equations (0.0.1), (0.0.3) and (0.0.10),

$$a_o = \frac{43.904936 \times 10^{-54}}{8292.3234 \times 10^{-48}} \text{ cm } = 0.52946 \times 10^{-8} \text{ cm} . \tag{0.0.12}$$

As we shall show in Chapter 5, contrary to Niels Bohr's earlier version of the hydrogen atom which proposed that the electron assumes a circular orbit about its nucleus, the electron (in its lowest energy state) maintains an average distance of $\frac{3}{2}a_o$ cm from the nucleus. We will calculate, moreover, that the probability of finding the electron within a sphere about the origin of radius a_o is 0.3235. When working with small atomic distances it is convenient at times to use *Ångström* units (Å): 1 Å $\stackrel{\text{def}}{=}$ 10^{-8} cm. Thus we may write $a_o = 0.529$ Å.

The quantity

$$\mu_B \stackrel{\text{def}}{=} \frac{e\hbar}{2m_e c}, \tag{0.0.13}$$

where

$$c = 2.99792458 \times 10^{10} \frac{\text{cm}}{\text{sec}} \tag{0.0.14}$$

is the speed of light in a vacuum, will be encountered in Chapter 9. μ_B is called the *Bohr magneton*. By (0.0.2), (0.0.3), (0.0.5), and (0.0.14),

$$\mu_B = 9.27401 \times 10^{-21} \frac{\text{esu-sec}^2 \times \text{erg}}{\text{gm} \times \text{cm}}. \tag{0.0.15}$$

We have enough information on hand to compute, for the record, the famous *fine structure constant* $\alpha \stackrel{\text{def}}{=} e^2/\hbar c$, a dimensionless quantity which controls minute spin and magnetic effects of electrons in atoms. By (0.0.2), (0.0.3), (0.0.14), and the remarks following (0.0.11),

$$\begin{aligned} \alpha &= \frac{23.070826 \times 10^{-20} \text{ cm}^2 \text{ dynes}}{3.1615212 \times 10^{-17} \text{ erg-sec cm sec}^{-1}} \\ &= 7.29738 \times 10^{-3} = \frac{1}{137.036}. \end{aligned} \tag{0.0.16}$$

For a particle of charge q and mass m the quantity $\omega = \frac{qB}{mc}$ will appear in Chapters 8 and 9, where B is a real number representing a certain magnetic field strength. The latter has units $\frac{\text{erg-sec}}{\text{esu-cm}^2}$. Thus ω has units $\frac{\text{esu erg sec}}{\text{g} \frac{\text{cm}}{\text{sec}} \text{ esu cm}^2} = \frac{\text{esu sec}^2 \text{ gcm}^2}{\text{gcm}^3 \text{ esu sec}^2} = \frac{1}{\text{sec}}$. Indeed $\omega/2\pi$ will be the frequency of a harmonic oscillator. A common measure of magnetic field strength is the *Tesla*, after Nikola Tesla (1856–1943), a leading figure in the development of alternating current. The latter measure, however, is based on SI (Système International) units where, for example, length, mass, and force are measured in meters, kilograms, and Newtons, respectively.

1

Quantum Mechanics: Some Remarks and Themes

1.1 Planck Quantization

Quantum mechanics to a large extent is a subject which conceptually is abstract and very mathematical in nature. Despite its abstract formalism it has proved to be a storehouse of ideas of enormous practical value. These radically new ideas, boldly proposed and strikingly uncommon, have helped to unlock deep secrets of the world, especially at the microscopic level where the breakdown of classical physics becomes rather apparent. There flows indeed from this subject a vast stream of concrete, everyday life applications. The phenomenon of transistors, television, semiconductors, computers, lasers, to name a few, instruments of undebated importance for communication, education, energy requirements, medicine, etc.—all involve quantum mechanical effects. The work of P. Lenard and especially of A. Einstein on the photoelectric effect, for example, helped toward the development of television. Our world has been positively changed by the fruits of this elegant subject and its forward thinking founders.

The present chapter provides for casual, sometimes heuristic and incomplete introductory remarks on some of the themes of these founders—remarks which thus serve also as a perspective for this book. We introduce, for example, Planck's quantization, Bohr's frequency rule, Schrödinger's equation, de Broglie's wavelength, and Heisenberg's uncertainty principle. These themes are subject to further detailed analysis and development (in some cases) in ensuing chapters.

December 14th, 1900 is considered to be the birthday of quantum mechanics. On that day Max Karl Ernst Ludwig Planck presented to the German Physical Society a paper entitled "On the Theory of the Energy Distribution Law of the

Normal Spectrum." It is appropriate to formally begin this book with some remarks on Planck's quantization—his radical idea on which the aforementioned paper was based.

The first sentence in Chapter 2 of Herman Weyl's wonderful, scholarly book, *The Theory of Groups and Quantum Mechanics* [87], is well worth quoting here: "The magic formula

$$E = h\nu \tag{1.1.1}$$

from which the whole of quantum theory is developed, establishes a universal relationship between the frequency ν of an oscillatory process and the energy E associated with such a process." Here h is a universal constant of nature discovered in October, 1900 by Planck. Planck was led to formula (1.1.1) in his study of radiation emitted by a solid hot body. The problem was to find theoretically a formula for the energy density of the radiation as a function of its frequency ν. Such a function provides for a *spectral emission curve*. Such curves were obtained from experiments prior to 1900, but physicists were unable to account for them on the basis of the classical theory of electrodynamics and statistical mechanics. After some desperate attempts, Planck found that a satisfactory solution of the problem could be obtained if one made the following radical assumption: the energy of emitted (or absorbed) radiation cannot take on an arbitrary value but only integral multiples $0, h\nu, 2h\nu, 3h\nu, \ldots$ of the basic energy value $E = h\nu$. Thus the energy is not arbitrary and "continuous" in value, but is *quantized* (restricted). This *quantization of energy* would be further looked into by Einstein.

Five years later Einstein would explain that a beam of light was like a hail of particles, called *quanta*. Later the name *photon* (due to G. N. Lewis (1926)) would come into vogue. Einstein's proposal was that all electromagnetic radiation of frequency ν (compare Table 1.1 in the next section) consists of discrete particle-like "bundles" or "packets," each of energy $= h\nu$. If an electromagnetic wave consists of 3000 such packets (photons), for example, then it has an energy of $3000h\nu$ units. The idea of quantization of energy has of course profound consequences concerning the structure of matter.

Example

Suppose a photon has a frequency of 6.1×10^{12} Hz. Using the value $h = 6.626 \times 10^{-27}$ erg-sec (see equation (0.0.1)) we see that the photon has energy $E = h\nu = 6.626 \times 10^{-27}\text{erg-sec} \times 6.1 \times 10^{12}\text{sec}^{-1} = 4.042 \times 10^{-14}$ ergs. We will discuss the relationship between the frequency ν of a wave and its *wavelength* λ in Section 1.4 of this chapter. As we shall see $\nu\lambda =$ the speed of the wave. By this fact, we have for a photon $\nu\lambda = c$ where c is the speed of light in a vacuum. By (0.0.14)

$$c = 2.99792458 \times 10^{10}\text{cm/sec}. \tag{1.1.2}$$

The wavelength of the preceding photon is $\lambda = \frac{c}{\nu} = \frac{3\times 10^{10}\,\frac{\text{cm}}{\text{sec}}}{6.1\times 10^{12}\,\text{sec}^{-1}} = 0.492 \times 10^{-2}$ cm, which is in the *infrared* range; compare Table 1.1 in the next section. Consider a second photon with wavelength $\lambda = 5 \times 10^{-9}$ cm in the x-*ray* range. We claim that this x-ray photon is a million times more energetic than the infrared photon. Its energy is $E = h\nu = \frac{hc}{\lambda} = \frac{6.626\times10^{-27}\times3\times10^{10}}{5\times10^{-9}}$ ergs $= 3.976 \times 10^{-8}$ ergs $= 0.98 \times 10^{6}(4.042 \times 10^{-14})$ ergs, which verifies the claim.

1.2 The Bohr Frequency Rule and Emission Spectra

In 1913, Niels Bohr, inspired by Planck's quantization, formulated a *frequency rule* which is described as follows. Suppose an atom (or some quantum system) undergoes a transition from one state of energy E to a second state of energy E'. Then radiation will be emitted or absorbed. If $E > E'$ (i.e., the system proceeds from a state of higher energy to a state of lower energy) then radiation will be *emitted*. The frequency ν of the emitted radiation will be given by $E - E' = h\nu$ (which compares with Planck's equation (1.1.1)). Similarly, if the system passes from a lower energy state E to a higher energy state E' ($E < E'$), then it will *absorb* radiation of a frequency ν given by $E' - E = h\nu$. Thus, in general

$$|E - E'| = h\nu. \tag{1.2.1}$$

The frequency rule plays a decisive role in the theoretical derivation of formulas for frequencies (or wavelengths) observed in certain spectroscopic lines—in particular in the derivation of *Balmer's formula*. This application is one of the beautiful, spectacular achievements of the Bohr theory. Before considering such formulas and launching directly into a general discussion of spectroscopic lines, it will be useful to recall first the elementary notion of the spectrum of ordinary light.

During the latter part of the 17th century Christian Huygens and Isaac Newton conducted major, initial investigations into the nature of light. In his celebrated prism experiment Newton demonstrated that ordinary "white light" was really a mixture of light of various colors ranging from red to violet. Namely, suppose a beam of light (sunlight, for example) is allowed through a narrow slit to enter a dark room and pass through a prism as indicated in Figure 1.1. As the beam strikes the prism it is *refracted* (i.e., it is bent due to a change in its speed). The refracted light leaving the prism is allowed to strike a screen in the dark room. Observing the screen, one will note a band of colors (red, orange, yellow, green, blue, indigo, violet) formed thereon, called the *spectrum* of visible light.

The term *dispersion* refers to this spreading out or separation of white light into its constituent colors. Each color corresponds to a certain wavelength λ, or frequency $\nu = \frac{\text{speed of wave}}{\lambda}$ (as noted earlier), with $\lambda \approx 4 \times 10^{-5}$ cm for violet (with short wavelength and high frequency) and $\lambda \approx 7.5 \times 10^{-5}$ cm for red (with long wavelength and low frequency).

The spectrum of light is part of the general *electromagnetic spectrum* of a range of waves with wavelengths greater than or less than those of visible light, given in the following table.

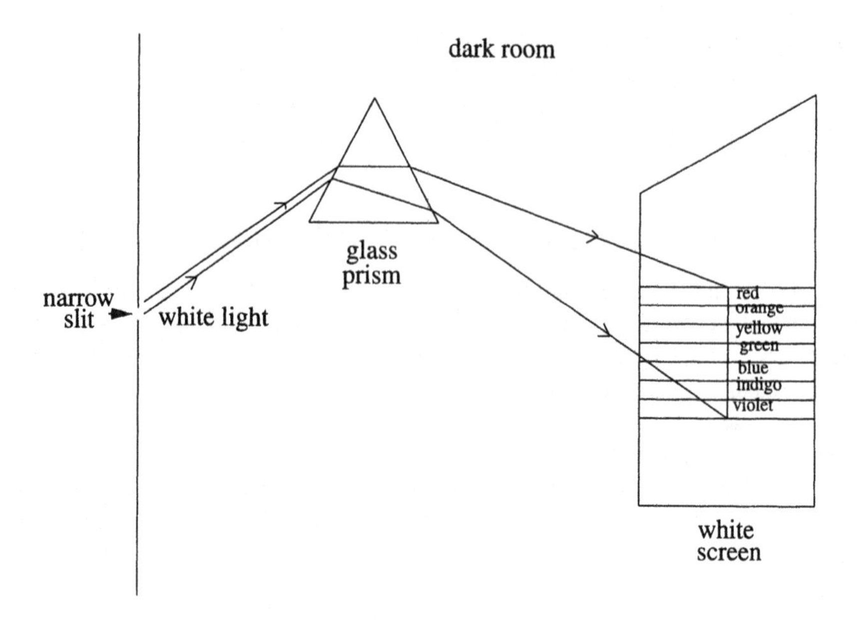

Figure 1.1: Newton's prism experiment

In other words, light is simply electromagnetic radiation in a certain frequency range detectable by the human eye (as discovered by James Maxwell and Heinrich Hertz).

The values given in Table 1.1 are approximate and they vary from one text to another.

Newton's prism experiment concerning the spectrum of light eventually prompted the study of *emission spectra* of various atoms and molecules. The latter study provides deep insight into atomic and molecular structure, and provides in particular experimental proof that only certain specific energy states can be realized. Just as each of us as individuals has a distinguished set of identifying fingerprints, atoms, molecules, elements, etc. have a distinguished emission spectrum, which might be *continuous*, or might consist of *lines*. Similar to Figure 1.1 one has the following typical set-up in diagram form for observing a line spectrum, say of a luminous gas for example.

In the discharge tube (containing the gas) electrons flowing from the negative electrode to the positive electrode collide with the gas. From these collisions there results the emission of light or radiation by the atoms of the gas. As in Figure 1.1 this light when passed through a prism and allowed to collect on a plate appears as a set of lines, colored images corresponding to some definite wavelength. The fact that these lines are discrete (each is separated by a particular distance) corresponds to the fact that the atoms in their excited state emit their radiation only quantum mechanically—i.e., at a discrete set of specific wavelengths. In this example no continuous distribution of wavelengths is observed.

Table 1.1: Electromagnetic spectrum

	λ wavelength in cm	$\nu = \frac{c}{\lambda}$ frequency in Hz
long radio waves	$10^5 - 10^9$	$3 \times 10^5 - 30$
AM radio waves	$10^3 - 10^4$	$3 \times 10^7 - 3 \times 10^6$
FM radio and TV waves	$1.5 \times 10^2 - 7.5 \times 10^2$	$2 \times 10^8 - 4 \times 10^7$
short radio waves	$10^{-1} - 1$	$3 \times 10^{11} - 3 \times 10^{10}$
microwaves	$10^{-2} - 10$	$3 \times 10^{12} - 3 \times 10^9$
infrared waves	$10^{-3} - 10^{-2}$	$3 \times 10^{13} - 3 \times 10^{12}$
visible light	$7.5 \times 10^{-5} - 4 \times 10^{-5}$	$4 \times 10^{14} - 7.5 \times 10^{14}$
ultraviolet rays	$10^{-7} - 10^{-4}$	$3 \times 10^{17} - 3 \times 10^{14}$
x-rays	$10^{-10} - 10^{-6}$	$3 \times 10^{20} - 3 \times 10^{16}$
gamma rays	$10^{-12} - 10^{-8}$	$3 \times 10^{22} - 3 \times 10^{18}$

In the important case of hydrogen, for example, for each integer $n \geq 1$ there is a corresponding (infinite) set or series of spectral lines $\{L_j^{(n)}\}_{j=n+1}^{\infty}$. For $n = 1, 2, 3, 4, 5$ for example we obtain the following five series, each named after its discoverer: the *Lyman* series $\{L_j^{(1)}\}_{j=2}^{\infty}$, the *Balmer* series $\{L_j^{(2)}\}_{j=3}^{\infty}$, the *Paschen* series $\{L_j^{(3)}\}_{j=4}^{\infty}$, the *Brackett* series $\{L_j^{(4)}\}_{j=5}^{\infty}$, and the *Pfund* series $\{L_j^{(5)}\}_{j=6}^{\infty}$, respectively. For the Balmer series we have the following illustration of its spectrum (in part). Recall that 1 Ångström $= 10^{-8}$ cm.

In 1885 Johann Balmer offered a simple formula to account for the empirical wavelength values appearing in Figure 1.3. His formula was

$$\frac{1}{\lambda} = K\left(\frac{1}{2^2} - \frac{1}{j^2}\right), \qquad j = 3, 4, 5, \ldots \tag{1.2.2}$$

for the wavelength λ of the line $L_j^{(2)}$, where K is some constant. On the basis of experimental evidence he deduced that by taking $\frac{1}{K} = 911.76$ Å formula (1.2.2) found agreement with the known data. Thus

$$\lambda = 911.76\left(\frac{4j^2}{j^2 - 4}\right) \text{Å}. \tag{1.2.3}$$

Take $j = 8$ for example. Then formula (1.2.3) gives $\lambda = 911.76(\frac{4\times 64}{64-4})$Å $=$ 3890.176 Å for the wavelength corresponding to the line $L_8^{(2)}$—which is the value given in Figure 1.3. If we let $j \to \infty$ in equation (1.2.3) we obtain the wavelength $\lambda_\infty = 3647.04$ Å which the limiting line $L_\infty^{(2)}$ in Figure 1.3 corresponds to.

Although Balmer's formula provides an accurate description of the wavelengths occurring in his series in Figure 1.3, the formula does not provide an *explanation* of spectral discreteness. A theoretical derivation of Balmer's formula, and of the

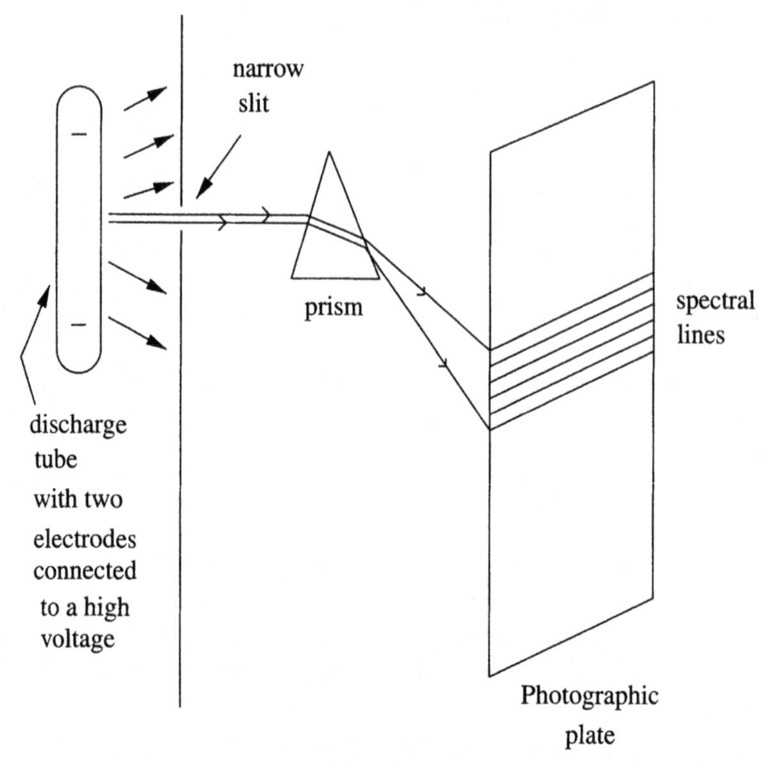

Figure 1.2: Spectral lines

more general Ritz–Rydberg type formulas for wavelengths in other series, is possible as we shall see now. In Chapter 5 it will be shown in detail that the solution of Schrödinger's equation (discussed in the next section) leads to the energy states of hydrogen, which are given by a discrete sequence $\{E^{(n)}\}_{n=1}^{\infty}$. Thus Schrödinger's equation carries within itself (and supersedes) Planck's quantization, and it carries in particular (in conjunction with the Bohr frequency rule) the explanation of spectral discreteness. Namely, $E^{(n)}$ is given by

$$E^{(n)} = -\frac{2\mu e^4 \pi^2}{n^2 h^2} \tag{1.2.4}$$

where $-e$ is the electron charge and μ is the reduced mass of hydrogen; see (0.0.3), (0.0.10). Define

$$R = \frac{2\pi^2 \mu e^4}{h^3 c} \tag{1.2.5}$$

so that by (1.2.4)

$$E^{(n)} = -\frac{Rhc}{n^2}. \tag{1.2.6}$$

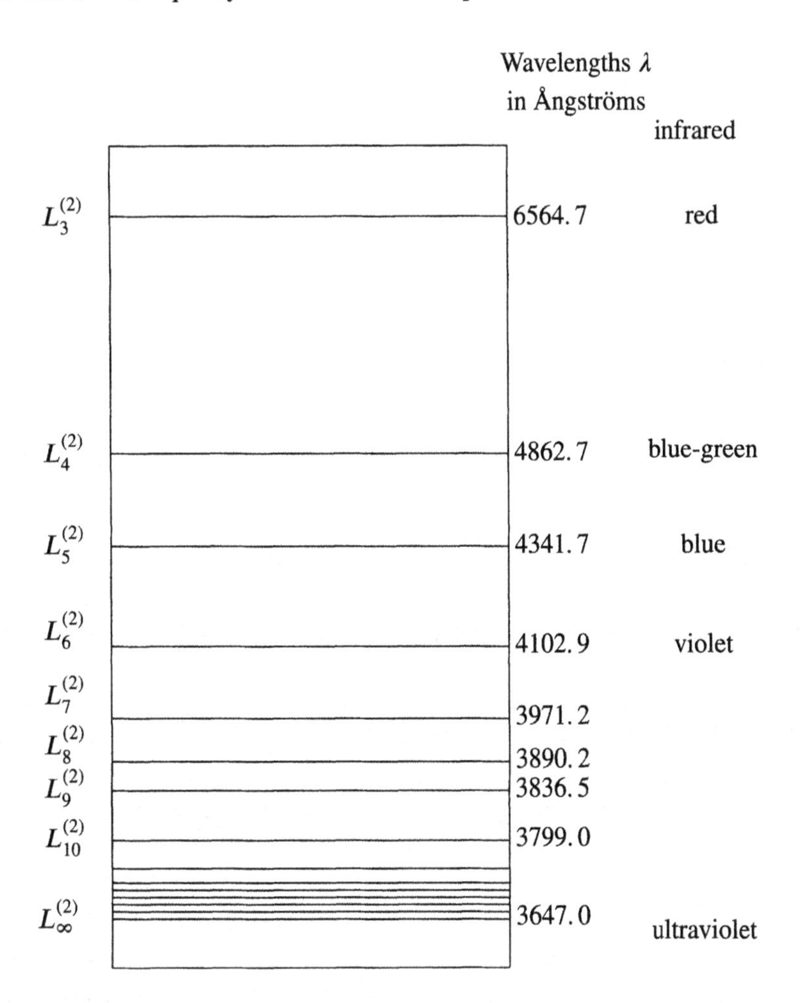

Figure 1.3: Photographic plate for the Balmer series (experimental observation)

Here c is the speed of light; see (1.1.2). Equation (1.2.6) can also be obtained by the Bohr theory. R is called the *Rydberg constant* (after the Swedish physicist Johannes Rydberg). Now by the Bohr frequency rule, a transition $E^{(n)} \to E^{(n')}$ from one energy state to another leads to radiation of frequency ν given by equation (1.2.1):

$$h\nu = |E^{(n)} - E^{(n')}| = Rhc \left| \frac{1}{(n')^2} - \frac{1}{n^2} \right| . \tag{1.2.7}$$

If $n_2 > n_1$ (i.e., $E^{(n_2)} > E^{(n_1)}$) then in the transition $E^{(n_2)} \to E^{(n_1)}$ radiation will be *emitted* (as we have seen), and since the frequency ν and wavelength λ of this radiation are related by $c = \lambda \nu$ we may express equation (1.2.7) as

$$\frac{1}{\lambda} = R\left(\frac{1}{n_1^2} - \frac{1}{n_2^2}\right), \quad \text{for } E^{(n_2)} \to E^{(n_1)}, n_2 > n_1, \tag{1.2.8}$$

which gives Balmer's formula (1.2.2): take $n_1 = 2$. More generally (1.2.8) is the Ritz–Rydberg formula. The experimental value of R is

$$R_e = 109678 \text{ cm}^{-1} = 0.00109678(\text{Å})^{-1}. \tag{1.2.9}$$

The theoretical value R_t of R is given by definition (1.2.5). By (0.0.1), (0.0.3), (0.0.10), (1.1.2)

$$R_t = 109677.78 \text{ cm}^{-1} = 0.0010967778(\text{Å})^{-1} \tag{1.2.10}$$

which agrees with (1.2.9). Here recall that the units of e^2 are gram $\frac{\text{cm}^3}{\text{sec}^2}$ (by Coulomb's law) and therefore the units of e^4 are gram$^2 \frac{\text{cm}^6}{\text{sec}^4}$, which implies that the units of $R = R_t$ are indeed cm^{-1}. Note moreover that $\frac{1}{R_t} = 911.74$ Å so that R_t is in close agreement with Balmer value $\frac{1}{K} = 911.76$ Å given earlier; i.e., $R_t \sim K$.

The lines in Figure 1.3 which we have labeled $L_3^{(2)}, L_4^{(2)}, L_5^{(2)}, \ldots$ are usually labeled $H_\alpha, H_\beta, H_\gamma, \ldots$ in spectroscopy. We now see that for the Balmer series (where $n_1 = 2$ in (1.2.8)) the red line $L_3^{(2)} = H_\alpha$ corresponds to the transition $E^{(3)} \to E^{(2)}$—a higher energy state to a lower in which radiation of wavelength λ is emitted, where λ is given theoretically by equation (1.2.8), where we use $\frac{1}{R_t} = 911.74$ Å: $\lambda = 911.74\left(\frac{1}{4} - \frac{1}{9}\right)^{-1}$ Å$= 6564.53$ Å, which compares with the value $\lambda = 6564.7$ Å of Figure 1.3. Similarly the blue-green line $L_4^{(2)} = H_\beta$ corresponds to the transition $E^{(4)} \to E^{(2)}$, the blue line $L_5^{(2)} = H_\gamma$ corresponds to the transition $E^{(5)} \to E^{(2)}$, etc. The wavelengths of the emitted radiation are given (theoretically) by $\lambda = 911.74\left(\frac{1}{4} - \frac{1}{16}\right)^{-1}$ Å, $911.74\left(\frac{1}{4} - \frac{1}{25}\right)^{-1}$ Å, $\cdots =$ 4862.61 Å, 4341.62 Å, ... (for the transitions $E^{(4)} \to E^{(2)}$, $E^{(5)} \to E^{(2)}, \ldots$) respectively—which differ just slightly from the values given in Figure 1.3. Further computational examples for other spectral series will be given later. As we have indicated a careful derivation of formula (1.2.4) (on which the preceding analysis was based) will be given in Chapter 5. A "theoretical" account of the Bohr frequency rule will appear in Chapter 7; see remarks in **(i)** following equation (7.7.58).

1.3 Schrödinger's Equation

Bohr's theory was remarkably accurate in accounting for the spectrum of hydrogen, as we have noted. For atoms with two electrons however, such as Helium, predictions of the theory conflicted with experimental results. Corrections to and generalizations of the theory were provided by the gifted teacher Arnold Sommerfeld, about 1916–1919. Despite this, these earlier versions of quantum mechanics, with their limited applicability and lack of a firm conceptual foundation, would have to give way to a more substantial and comprehensive theory. Credits would be due to great theorists such as L. de Broglie, E. Schrödinger, W. Heisenberg, M. Born, P. Jordan, P. Dirac, W. Pauli, and to others.

In 1924 de Broglie assigned a wavelength λ to an electron (or to an arbitrary particle) moving with a speed v. Thus he associated a wave character or behavior to a particle of mass m, a character which Einstein had noted in his work (on the photoelectric effect, for example). De Broglie's precise mathematical formula was

$$\lambda = \frac{h}{mv} = \frac{h}{p}, \tag{1.3.1}$$

where $p \overset{\text{def}}{=} mv$ is the *momentum* of the particle. In the next section we will discuss de Broglie's bold discovery in more detail.

The task of describing mathematically a de Broglie matter wave (or of setting up a *wave function* ψ for a particle) was taken up by Erwin Schrödinger in a series of classic papers [75, 76] in 1926, wherein he announced the renowned *wave equation*, which in one dimension reads

$$\frac{\hbar^2}{2m}\frac{d^2\psi}{dx^2} + (E - V(x))\psi(x) = 0. \tag{1.3.2}$$

Here $E \in \mathbb{R}$ is a fixed real number (which will represent the quantum energy of the particle), and $V(x)$ is the potential energy of the particle; potential and kinetic energy will be defined carefully in the next chapter. Schrödinger's equation (in one dimension or in higher dimensions) is one of the most important equations of modern physics. It is the key equation that we will consider, of course. Schrödinger showed that his new *wave mechanics* was equivalent to the *matrix mechanics* version of quantum theory by Heisenberg, Born, and Jordan, and moreover that his equation, in his words, "carries within itself the quantum conditions"—conditions noted by Bohr and Sommerfeld, which were also discovered by W. Wilson and J. Ishiwara in 1915; compare Section 3.7, Chapter 3; in particular see equation (3.7.8) there. Namely, generally speaking, energies of a quantum system cannot assume arbitrarily prescribed values, but are given by a discrete set of eigenvalues E of a second-order differential operator H, say

$$H = -\frac{\hbar^2}{2m}\frac{d^2}{dx^2} + M_V \tag{1.3.3}$$

in the one-dimensional case, where M_V is multiplication by the function $V(x)$, since equation (1.3.2) can be written as

$$H\psi = E\psi, \tag{1.3.4}$$

where one requires a non-zero square-integrable solution ψ over $\mathbb{R}$:

$$||\psi||^2 \overset{\text{def}}{=} \int_{\mathbb{R}} |\psi(x)|^2 \, dx < \infty \tag{1.3.5}$$

for dx = Lebesgue measure on $\mathbb{R}$. From a normalized wave function ψ (i.e., ψ such that $||\psi||^2 = 1$) one extracts the following physical information—Max Born's profound probabilistic interpretation (1926):

$$\int_a^b |\psi(x)|^2\, dx \text{ is the probability that the particle will be located within the interval } [a, b]. \tag{1.3.6}$$

Or in three dimensions, for example,

$$\iiint_{\mathfrak{R}} |\psi(x, y, z)|^2\, dxdydz \text{ is the probability of finding the particle in a region } \mathfrak{R} \text{ in space} \tag{1.3.7}$$

(compare, for example, Section 5.5 of Chapter 5), where equation (1.3.2) is replaced by the equation

$$\frac{\hbar^2}{2m}\left(\frac{\partial^2\psi}{\partial x^2} + \frac{\partial^2\psi}{\partial y^2} + \frac{\partial^2\psi}{\partial z^2}\right) + (E - V(x, y, z))\psi(x, y, z) = 0. \tag{1.3.8}$$

Equations (1.3.2), (1.3.8) are *time independent* Schrödinger equations, compared to the more general *time dependent* Schrödinger equations

$$i\hbar\frac{\partial\psi}{\partial t}(x, t) = -\frac{\hbar^2}{2m}\frac{\partial^2\psi}{\partial x^2}(x, t) + V(x)\psi(x, t) \stackrel{\text{def}}{=} H\psi(x, t), \tag{1.3.9}$$

$$i\hbar\frac{\partial\psi}{\partial t}(x, y, z, t) = -\frac{\hbar^2}{2m}\left(\frac{\partial^2\psi}{\partial x^2} + \frac{\partial^2\psi}{\partial y^2} + \frac{\partial^2\psi}{\partial z^2}\right) + V(x, y, z)\psi(x, y, z, t). \tag{1.3.10}$$

If $\psi(x)$ is a solution of the time independent equation (1.3.2), for some (energy level) E, then we easily obtain a time dependent solution $\psi(x, t)$ of (1.3.9) by setting

$$\psi(x, t) = \psi(x)\exp(Et/i\hbar). \tag{1.3.11}$$

Similarly, a solution $\psi(x, y, z)$ of (1.3.8) leads immediately to a solution

$$\psi(x, y, z, t) \stackrel{\text{def}}{=} \psi(x, y, z)\exp(Et/i\hbar) \tag{1.3.12}$$

of equation (1.3.10). Thus we may focus on the time independent equation.

There is a heuristic procedure for obtaining the right-hand side of the time dependent equations (1.3.9), (1.3.10). For example we obtain the differential operator H in (1.3.3) (or in (1.3.9)) by employing the "quantization rules" $p = mv \to \frac{\hbar}{i}\frac{\partial}{\partial x}$, $V \to M_V$ by which the operators $\frac{\hbar}{i}\frac{\partial}{\partial x}$, M_V correspond to the "classical observables" p, V. Here $p^2 \to \left(\frac{\hbar}{i}\frac{\partial}{\partial x}\right)^2 = -\hbar^2\frac{\partial^2}{\partial x^2}$ so that the kinetic energy $T \stackrel{\text{def}}{=} \frac{1}{2}mv^2 = \frac{p^2}{2m} \to -\frac{\hbar^2}{2m}\frac{\partial^2}{\partial x^2}$, and the *total energy* $T + V$, or *Hamiltonian*,

$\to -\frac{\hbar^2}{2m}\frac{\partial^2}{\partial x^2} + M_V \stackrel{\text{def}}{=} H$, as desired (which is the reason for the symbol H). Note that the operators $D_x \stackrel{\text{def}}{=} \frac{\hbar}{i}\frac{d}{dx}$, M_x satisfy the Heisenberg–Born–Jordan commutation relations

$$[D_x, M_x] \stackrel{\text{def}}{=} D_x \circ M_x - M_x \circ D_x = \frac{\hbar}{i} \cdot 1. \tag{1.3.13}$$

In later chapters a good deal more will be said about Schrödinger's equation and solutions of it in particular cases.

1.4 The de Broglie Wavelength

In his famous 1924 Paris doctoral thesis, Louis Victor de Broglie boldly proclaimed that to a particle of mass m moving with a velocity v is associated a wavelength λ given by $\lambda = \frac{h}{mv}$. Strictly speaking this equations holds for non-relativistic particles where v is much less than the velocity of light. h is expressed in units of $\frac{\text{mass}\times(\text{length})^2}{\text{time}}$ so that indeed $\frac{h}{mv}$ is a length quantity. Notice that from time to time we will, carelessly, fail to distinguish between velocity v and *speed*–the magnitude $|v|$ of v. Thus $\lambda = \frac{h}{mv}$ really means $\lambda = \frac{h}{m|v|}$. De Broglie's claim was verified by experiments of C. Davisson and L. Germer, and by G. Thompson, O. Stern, F. Rupp and others. These interesting experiments demonstrated for example that beams of electrons could indeed be diffracted. Interference phenomena in such beams were also noted. Even before the birth of quantum mechanics Einstein noted wave-like properties of atoms in a monotonic gas. The wave-like character of a moving particle is now very well established. We will indicate one route by which one might arrive at the fundamental assertion $\lambda = \frac{h}{mv}$. For this we pause for a moment to understand the notion of a traveling wave and that of a wavelength.

Consider first a point P which travels along the circumference of a circle Γ, uniformly in a counterclockwise direction with angular speed ω measured in radians per second. Call R the radius of Γ, and call $\theta = \theta(t)$ the angular position of P at time t as indicated in Figure 1.4.

$\omega \stackrel{\text{def}}{=} \frac{d\theta}{dt} \implies \theta(t) = t\omega + \theta(0)$ which implies that the vertical displacement y is given by $y = R\sin\theta = R\sin(t\omega + \theta(0))$. y is plotted as a function of t on the right side of the diagram in Figure 1.4. The amount of time required for P to complete one revolution about the center of Γ is the *period* T of the motion: For any instant t, $\theta(t+T) = \theta(t) + 2\pi$; i.e., $(t+T)\omega + \theta(0) = t\omega + \theta(0) + 2\pi \implies T = \frac{2\pi}{\omega}$. The *frequency* ν of the motion is the number of revolutions per second: $\nu = \frac{1}{T} \implies y = R\sin(t\omega + \theta(0)) = R\sin(\frac{2\pi}{T}t + \theta(0)) = R\sin(2\pi\nu t + \theta(0))$, which describes the variation of the displacement y with respect to time t. Note that ν need not be a whole number.

For the concept of a traveling wave one considers however displacement with respect to time and distance. Thus given a fixed constant C (which will have a physical meaning presently) consider the function f of two variables t, x given by

$$f(t,x) = y\left(\frac{x}{C} - t\right) \stackrel{\text{i.e.,}}{=} R\sin\left[2\pi\nu\left(\frac{x}{C} - t\right) + \theta(0)\right]. \tag{1.4.1}$$

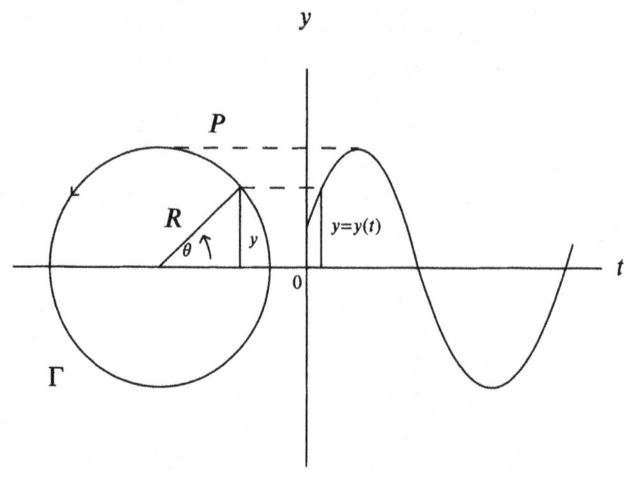

Figure 1.4: Circular motion

One checks immediately that f is a solution of the *wave equation*

$$\frac{\partial^2 u}{\partial t^2} = C^2 \frac{\partial^2 u}{\partial x^2}. \tag{1.4.2}$$

The plots of $x \to f(0,x), f(t,x)$ (for t fixed) represent, intuitively, a wave that has traveled to the right a distance $x'_o - x_o$ in t seconds, the dashed curve in Figure 1.5 being $x \to f(t,x)$. The speed of the wave is given by the distance traveled divided by the time. Now $x_o = -\frac{C\theta(0)}{2\pi v}$ and $x'_o = x_o + Ct \implies x'_o - x_o = Ct \implies$ the speed of the wave is the constant C. The solid curve $(x \to f(0,x))$ has a maximum at the indicated points m_1, m_2, m_3, where $m_1 = \frac{C}{4v} - \frac{C\theta(0)}{2\pi v} = \frac{C}{4v} + x_o$, $m_2 = \frac{5C}{4v} + x_o$. By definition, the *wavelength* λ is the distance between two successive crests:

$$\lambda \overset{\text{def}}{=} m_2 - m_1 = \frac{C}{v} \tag{1.4.3}$$

$$\implies f(t,x) = R \sin\left[2\pi\left(\frac{x}{\lambda} - vt\right) + \theta(0)\right] \tag{1.4.4}$$

$$= R \sin\left[\frac{2\pi x}{\lambda} - \omega t + \theta(0)\right] \tag{1.4.5}$$

in equation (1.4.1). Thus, abstractly, we regard a *traveling wave* as a function f of two variables ("time and position") given by

$$f(t,x) = A \sin(kx \mp \omega t + \delta) \tag{1.4.6}$$

$$= A \cos(kx \mp \omega t + \delta_1) \qquad \text{for } \delta_1 = \delta - \frac{\pi}{2} \tag{1.4.7}$$

for real constants A, k, ω, δ, called the *harmonic amplitude*, *wave number* (or *propagation constant*), *angular speed* (or *angular frequency*), and *phase constant*, respectively, of f. f "moves to the right or left" according to the choice

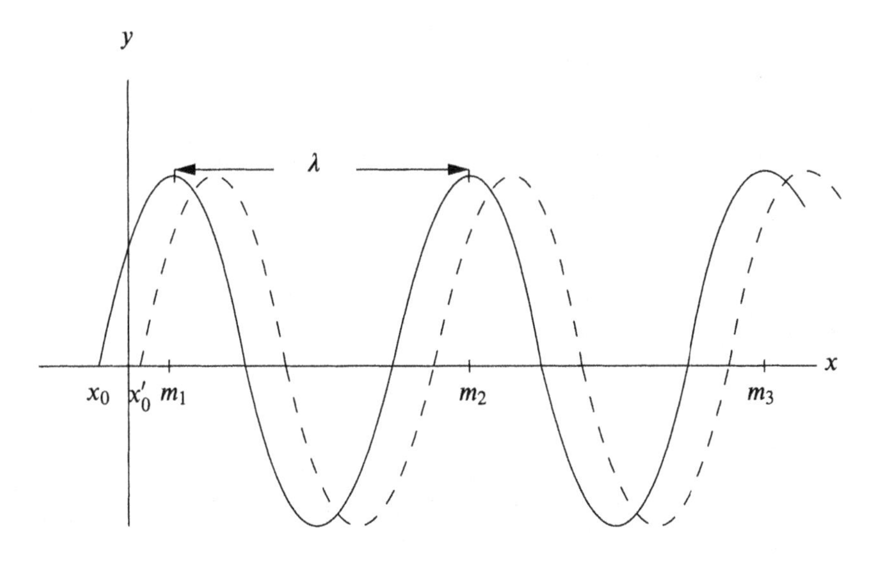

Figure 1.5: Traveling wave

of $-$ or $+$ respectively in (1.4.6). The *frequency* ν of f is $\frac{\omega}{2\pi}$; the *wavelength* λ is given by $k = \frac{2\pi}{\lambda}$, and the *speed* or *phase velocity* C of f is given by $C = \nu\lambda = \frac{\omega}{k}$; compare (1.4.3), (1.4.4), (1.4.5). f satisfies the wave equation (1.4.2). Using the identity $\sin(a \mp b) = \sin a \cos b \mp \cos a \sin b$ we can write

$$f(t,x) = f_1(t,x) \mp f_2(t,x) \qquad \text{for} \tag{1.4.8}$$

$$f_1(t,x) = A\sin(kx+\delta)\cos\omega t, \tag{1.4.9}$$

$$f_2(t,x) = A\cos(kx+\delta)\sin\omega t, \tag{1.4.10}$$

where in f_1, f_2 the variables t, x are *separated.* Moreover f_1, f_2 are also solutions of the wave equation (1.4.2) (since $C^2k^2 = \omega^2$).

Based on an analogy with Schrödinger's equation we now sketch why it is reasonable that one should have the de Broglie relation $\lambda = \frac{h}{mv}$. Our argument of course does not constitute a "proof" since in fact we consider the statement $\lambda = \frac{h}{mv}$ as axiomatic. Let $\psi_1(x) = A\sin(kx+\delta)$, $\psi_2(x) = A\cos(kx+\delta)$ so that in (1.4.9), (1.4.10) $f_1(t,x) = \psi_1(x)\cos(\omega t)$, $f_2(x,t) = \psi_2(x)\sin(\omega t)$. If $\psi(x) =$ either $\psi_1(x)$ or $\psi_2(x)$, then ψ satisfies

$$\psi''(x) + k^2\psi(x) = 0. \tag{1.4.11}$$

On the other hand, in the time-independent Schrödinger equation (equation (1.3.2)) regard $E - V(x)$ as the classical kinetic energy $T = \frac{p^2}{2m}$ where $p = mv$ is the momentum of our particle of mass m with velocity v. That is, regard the total energy E as $V + T$, as in the discussion following (1.3.12). Then, if we (incorrectly) write equation (1.3.2) as

$$\psi''(x) + \frac{8\pi^2 m}{h^2}\frac{p^2}{2m}\psi(x) = 0, \tag{1.4.12}$$

using (0.0.2), and compare (1.4.12) with equation (1.4.11), we see that the square k^2 of the wave number of the traveling wave f in (1.4.8) corresponds to $\frac{8\pi^2 m}{h^2}\frac{p^2}{2m} = \frac{4\pi^2 p^2}{h^2}$, or that k corresponds to $\frac{2\pi p}{h}$. But $k = \frac{2\pi}{\lambda}$ where λ is the wavelength of the traveling wave. Thus $\frac{2\pi}{\lambda}$ corresponds to $\frac{2\pi p}{h}$, or λ corresponds to $\frac{h}{p} = \frac{h}{mv}$, which suggests exactly the de Broglie relation $\lambda = \frac{h}{mv} = \frac{h}{p}$.

As P travels the circumference of Γ in Figure 1.4 its projection onto the diameter of Γ (on the t-axis) travels back and forth along this diameter. This back and forth motion is called *simple harmonic motion* (or simple harmonic oscillation), which we shall look at further in Chapter 2. simple harmonic oscillator.

Examples

1. Let f be given by

$$f(t,x) = 6.7\sin(4\pi x + 7t) \tag{1.4.13}$$

where x is measured in centimeters and t is measured in seconds. Then f represents a wave traveling along the x-axis in the negative direction (because of the plus sign $+$) with harmonic amplitude 6.7 centimeters, wave number $k = 4\pi$, angular frequency $\omega = 7$ radians/sec, and phase constant $\delta = 0$. The frequency of the wave is $\nu = \frac{\omega}{2\pi} = \frac{3.5}{\pi}$ Hz. The speed of the wave is $C = \frac{\omega}{k} = \frac{7}{4\pi} = \frac{1.75}{\pi}$ cm/sec, and the wavelength is $\lambda = \frac{2\pi}{k} = \frac{1}{2}$ centimeter.

2. The speed of a wave of frequency $\nu = 18.3$ Hz and wavelength $\lambda = 30.9$ cm is given by $C = \nu\lambda = (18.3\ \frac{1}{\text{sec}})(30.9\text{ cm}) = 565.5\ \frac{\text{cm}}{\text{sec}}$.

3. A golf ball of mass 50 grams moves at a speed of 60 feet/sec $= \frac{60}{3.28}$ meter/sec $= 1829.3$ cm/sec. Its de Broglie wavelength is given by $\lambda = \frac{h}{mv} = \frac{6.63\times10^{-27}}{50\times1829.3}$ cm $= 7.24\times10^{-32}$ cm, which is too small to be detected.

4. An electron moves with a kinetic energy of one *electron volt*, i.e., 1.60×10^{-12} ergs. What is its de Broglie wavelength?

Solution: The electron mass is $m_e = 9.11\times10^{-28}$ grams; see (0.0.5). The kinetic energy $T = \frac{1}{2}m_e v^2 \implies v = \sqrt{\frac{2T}{m_e}} \implies m_e v = \sqrt{2Tm_e} = \sqrt{2\times1.6\times10^{-12}\times9.11\times10^{-28}}\text{ g}\ \frac{\text{cm}}{\text{sec}} = \sqrt{29.192}\times10^{-20}\text{ g}\ \frac{\text{cm}}{\text{sec}} = 5.4\times10^{-20}$ g $\frac{\text{cm}}{\text{sec}}$. Therefore $\lambda = \frac{h}{m_e v} = \frac{6.63\times10^{-27}}{5.4\times10^{-20}}$ cm $= 1.23\times10^{-7}$ cm. We note that by Table 1.1 this wavelength is in the x-ray range, which is considerably less than the range of visible light $4\times10^{-5} - 7\times10^{-5}$ cm (violet to red).

1.5 Heisenberg's Uncertainty Principle

Among the foundational principles of quantum theory one of the most celebrated is the *Heisenberg uncertainty principle*, discovered by Werner Heisenberg in 1927. The principle states that it is impossible to measure simultaneously and precisely the position and momentum of a particle. One can measure very precisely the position x, or the velocity v (and thus the momentum $p =$ mass $\times\, v$) separately, but there is a natural limit to knowing both the values of position and velocity simultaneously and as accurately as one would desire, especially at the microscopic level. The very act of measuring accurately the position of an electron, for example, interferes with the measuring of its momentum. The limitation in simultaneity and precision of measurement of x and p can be ascribed to the wave nature of the electron, as we shall see.

If $\Delta x, \Delta p$ denote the *uncertainty* in the position x and the momentum p, respectively, of a particle, then Heisenberg's principle is usually expressed analytically as

$$(\Delta x)(\Delta p) \geq \frac{\hbar}{2}. \tag{1.5.1}$$

Thus as the uncertainty Δx in x decreases (i.e., as the position of the particle becomes more accurately known), the uncertainty Δp in p increases. Similarly as Δp decreases, Δx increases. So far $\Delta x, \Delta p$ have not been carefully defined. This will be done in Chapter 6, where a proof of (1.5.1) will be taken up. In fact, there we will provide two proofs of the fundamental inequality (1.5.1). For now, let us consider how one might arrive at (1.5.1) by proceeding heuristically.

As we considered a beam of light entering a narrow slit in Figure 1.1 (also cf. Figure 1.2), consider now a beam of electrons entering a narrow slit as indicated in Figure 1.6.

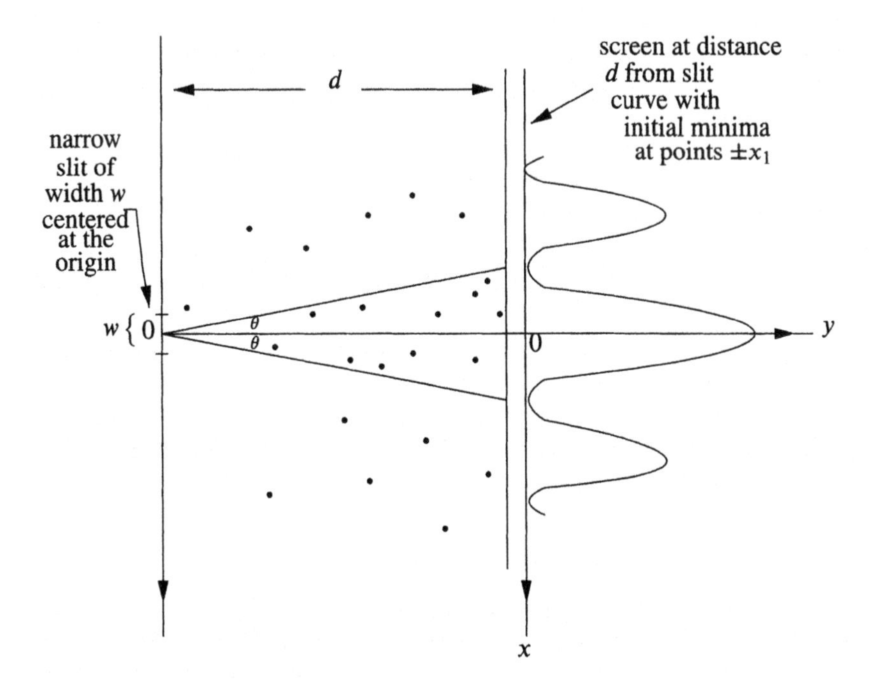

Figure 1.6: Single slit diffraction of electrons

After passing through the slit the field of electrons is diffracted (i.e., is "spread out") and eventually impacts the fluorescent screen throughout a range of x-values. It is because of the *wave* nature (not the particle nature) of electrons that this diffraction phenomenon is exhibited. Most of the electrons will land on the screen within the symmetric interval $[-x_1, x_1]$ about the origin 0, where $\pm x_1$ are initial minimal points of an intensity distribution of illumination curve for electrons impacting the screen, as indicated in Figure 1.6.

Suppose we follow the path of a single electron e entering the slit. As it enters the uncertainty Δx in its position x will be chosen to correspond to the width w of the slit ($\Delta x \approx w$), which we could make arbitrarily small. As e passes through the slit and is diffracted in an arbitrary and unpredictable direction, an uncertainty in its velocity and momentum is introduced, and it is impossible to determine in advance at what point e will strike the screen. It has the best chance however of striking within the interval $[-x_1, x_1]$. Now for light of wavelength λ passing through the slit one has the following simple diffraction theory result:

$$\sin\theta = \frac{\lambda}{w} \tag{1.5.2}$$

for θ in Figure 1.6. We consider θ very small so that $\sin\theta \approx \tan\theta = \frac{x_1}{d} \implies x_1 \approx \frac{d\lambda}{w}$. Also now choose λ to correspond to the de Broglie wavelength $\lambda = \frac{h}{p}$ of e. After leaving the slit the momentum of e acquires an x-component p_x (because of diffraction) whose magnitude we take as a measure of the uncertainty of e's momentum: $\Delta p \approx p_x$.

As Figure 1.7 indicates $\Delta p \approx p_x = p\sin\theta \approx \frac{h}{\lambda}\frac{\lambda}{w}$ (by (1.5.2)) $= \frac{h}{w}$. Recalling that $w \approx \Delta x$ we see that $(\Delta x)(\Delta p) \approx h = 2\pi\hbar \geq \frac{\hbar}{2}$, which somewhat crudely establishes equation (1.5.1) for now, until a clean, rigorous argument in Chapter 6 is presented.

The uncertainty principle was stated here for the variables position and momentum x, p, but it extends to other "conjugate" variables as well. For time and energy t, E, for example, one has

$$(\Delta t)(\Delta E) \geq \frac{\hbar}{2} \tag{1.5.3}$$

for suitable definitions of Δt, ΔE. A physical meaning of (1.5.3) for example is that the energy of a molecule in an excited state for some time period cannot be exactly determined. Heisenberg's principles spell the downfall of *classical determinism*: the future is not necessarily determined by initial data of the past—a matter which we return to later.

1.6 Bohr's Complementarity Principle

The wave-like character of matter has been noted in this chapter—in discussions ranging from de Broglie's wavelength to electron diffraction, for example, as illustrated in Figure 1.6. The dual character of an electron for example as wave-like or particle-like, as revealed in numerous experiments, or the dual character

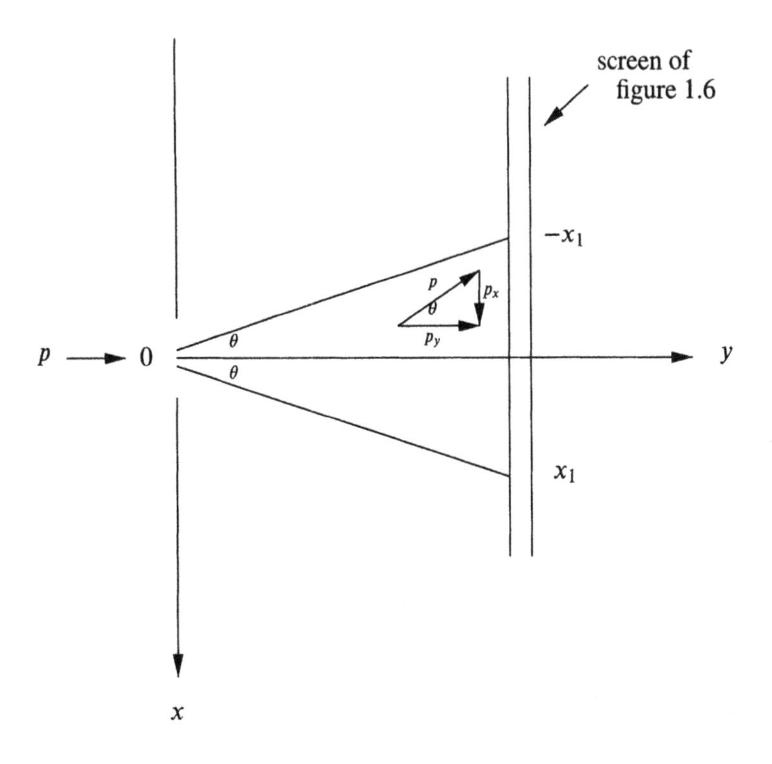

Figure 1.7: Computing Δp

of other physical entities, eventually led to the enunciation of a *complementarity principle* by Niels Bohr in 1928—a year after Heisenberg uncertainty was discovered. Bohr's principle, like Heisenberg's principle, forbids a certain simultaneity. Namely, it states the impossibility of one observing simultaneously both the wave and particle-like nature of matter, and that both aspects must be taken into account in any attempt to completely describe a physical entity.

Thus the wave and particle nature of matter *complement* each other. Whether one or the other of these natures is observed in some experiment depends on the type of experiment being performed. Any experiment designed to manifest a wave-like attribute can provide no manifestation of particle-like attributes—the converse being true as well.

The profound principles of uncertainty and complementarity of Heisenberg and Bohr, with their weighty philosophic implications, are empirically rooted. They express fundamental limitations on human measurements and observations—limitations which reflect deep intrinsic realities of nature, and which cannot be surmounted by the construction of "better mouse traps."

2

Equations of Motion in Classical Mechanics

2.1 Introduction

In classical mechanics one studies the motions of objects on a macroscopic level. These motions, of a particle or a system of particles, of the planets around the sun, or of a pendulum, for example, are governed by differential equations. These equations, or their solutions, are called *equations of motion*, which in earliest form go back to Newton. Other forms are due to Lagrange, Hamilton, Jacobi, and others.

Newton's basic law, his second law of motion, states that a force F acting on an object of mass m produces an acceleration a of that object (in the direction of F) of the magnitude $|F|/m$ where $|F|$ is the magnitude of F. Taking the notion of a force as fundamental or axiomatic one can use this law in fact to *define* mass : $m = |F|/|a|$. For an object moving along the x-axis, say, under the influence of F the acceleration is given in terms of the second derivative of its position $x(t)$ with respect to time t. Thus the equation of motion takes the familiar form

$$m\frac{d^2x}{dt^2} = |F|. \tag{2.1.1}$$

We shall review some basic definitions and concepts of classical mechanics. Using equation (2.1.1), the chain rule, and the fundamental theorem of calculus, we deduce, for example, the conservation of energy (along a trajectory) and explain why the $\frac{1}{2}$ appears in the definition $T = \frac{1}{2}mv^2$ of the kinetic energy T of a particle with velocity v. Hamiltonian and Lagrangian forms of equations of motion will also be derived.

2.2 Motion in One Dimension

Consider a particle of mass $m > 0$ moving in one dimension with equation of motion $x = \phi(t)$ (where t = time), where ϕ', ϕ'' exist and are continuous on $\mathbb{R}$. The functions $v = \phi'$, $a = \phi''$ evaluated at t give the velocity, acceleration, respectively, of the particle at the instant t. Suppose the particle is under the influence of a continuous force field F: F is a continuous function on $\mathbb{R}$ such that $F(x) =$ force at position x.

By Newton's second law, equation (2.1.1),

$$F(\phi(t)) = ma(t) = mv'(t). \tag{2.2.1}$$

Consider the function $H : \mathbb{R} \to \mathbb{R}$ given by

$$H(t) = \frac{1}{2}mv(t)^2 - \int_{a=\text{any constant}}^{\phi(t)} F(x)\, dx. \tag{2.2.2}$$

If

$$\psi(t) = \int_a^t F(x)\, dx \tag{2.2.3}$$

then (as F is continuous) $\psi'(t) = F(t)$ by the fundamental theorem of calculus, and

$$\int_a^{\phi(t)} F(x)\, dx = \psi(\phi(t)). \tag{2.2.4}$$

That is, by the chain rule (as ψ, ϕ are differentiable on $\mathbb{R}$),

$$\begin{aligned} H'(t) &= \frac{1}{2}m \cdot 2v(t)v'(t) - \psi'(\phi(t))\phi'(t) \\ &= mv(t)v'(t) - F(\phi(t))v(t) = mv(t)v'(t) - mv'(t)v(t) = 0. \end{aligned} \tag{2.2.5}$$

That is, we have the following *conservation law*:

$$\frac{1}{2}mv(t)^2 - \int_{a=\text{any constant}}^{\phi(t)} F(x)\, dx = \text{ a constant} \tag{2.2.6}$$

for all t. The first term $\frac{1}{2}mv(t)^2$ is the *kinetic energy* of the particle at time t. The kinetic energy T of the particle, regarded as a function: $\mathbb{R} \to \mathbb{R}$ of the velocity, is given by $T(v) = \frac{1}{2}mv^2$.

Consider the second term. If we define $V : \mathbb{R} \to \mathbb{R}$ by

$$V(x) = -\int_a^x F(r)\, dr \tag{2.2.7}$$

we obtain a continuously differentiable function V such that $V'(x) = -F(x)$; i.e., $F(x) = -V'(x) \implies F$ is derived from a *potential* V (or *potential energy* function V). The second term in equation (2.2.6) is $V(\phi(t))$. If V_1 is any other continuously differentiable function on $\mathbb{R}$ such that $F(x) = -V_1'(x)$, then of course $V_1 = V+$ constant. We can therefore state equation (2.2.6) as follows:

Let V be any choice of potential energy function for F: V is a continuously differentiable function of $\mathbb{R}$ such that $F(x) = -V'(x)$; example:

$$V(x) = -\int_0^x F(r)\, dr. \tag{2.2.8}$$

Then along the trajectory $t \to \phi(t)$ the *total energy* $E = T + V$ of the particle is conserved:

$$T(v(t)) + V(\phi(t)) = \frac{1}{2}mv(t)^2 + V(\phi(t)) = \text{ a constant for all time } t;$$

$$\frac{1}{2}mv(t)^2 - \int_0^{\phi(t)} F(x)\, dx = \text{ a constant for all time } t. \tag{2.2.9}$$

We see the reason for the factor $1/2$ in the definition of T: it is needed in order to have $H'(t) = 0$, and thus the conservation of energy.

Consider any two instants of time t_1, t_2, say $t_1 < t_2$. Write

$$\begin{aligned}\int_{\phi(t_1)}^{\phi(t_2)} F(x)\, dx &= \int_{\phi(t_1)}^{0} F(x)\, dx + \int_0^{\phi(t_2)} F(x)\, dx \\ &= \int_0^{\phi(t_2)} F(x)\, dx - \int_0^{\phi(t_1)} F(x)\, dx.\end{aligned} \tag{2.2.10}$$

By the conservation of total energy and equation (2.2.10)

$$\begin{aligned}&\frac{1}{2}mv(t_1)^2 - \int_0^{\phi(t_1)} F(x)\, dx = \frac{1}{2}mv(t_2)^2 - \int_0^{\phi(t_2)} F(x)\, dx \\ \Longrightarrow\ &\frac{1}{2}mv(t_2)^2 - \frac{1}{2}mv(t_1)^2 = \int_{\phi(t_1)}^{\phi(t_2)} F(x)\, dx,\end{aligned} \tag{2.2.11}$$

which relates the change in kinetic energy to the *work*

$$\int_{\phi(t_1)}^{\phi(t_2)} F(x)\, dx \tag{2.2.12}$$

done in displacement of the particle from $\phi(t_1)$ to $\phi(t_2)$ in the field F.

From equation (2.2.1)

$$\int_{t_1}^{t_2} F(\phi(t))\, dt = mv(t_2) - mv(t_1) \tag{2.2.13}$$

where $mv(t)$ is the *momentum* of the particle at time t. One can regard the momentum as a function $p : \mathbb{R} \to \mathbb{R}$ of the velocity: $p(v) = mv$. Equation (2.2.13) expresses the fact that the *change in momentum* coincides with the *time-integral* of the force field.

We are given ϕ, ϕ', ϕ'', F continuous on $\mathbb{R}$ with $F(\phi(t)) = m\phi''(t)$. Again choose V continuously differentiable on $\mathbb{R}$ such that $V'(x) = -F(x)$. Define $T, \overline{V}, H : \mathbb{R}^2 \to \mathbb{R}$ by

$$T(q,p) = \frac{p^2}{2m}, \quad \overline{V}(q,p) = V(q), \quad H = T + \overline{V} : H(q,p) = \frac{p^2}{2m} + V(q) \tag{2.2.14}$$

$$\implies \frac{\partial H}{\partial q}(q,p) = V'(q) = -F(q), \quad \frac{\partial H}{\partial p}(q,p) = \frac{p}{m}. \tag{2.2.15}$$

T is C^∞ and $\overline{V}$ is class C^1 (as V is class C^1) $\implies$ H is class C^1. As ϕ, ϕ' are class C^1 the function $t \to H(\phi(t), m\phi'(t))$ is continuously differentiable and

$$\begin{aligned}\frac{d}{dt}H(\phi(t), m\phi'(t)) &= \frac{\partial H}{\partial q}(\phi(t), m\phi'(t))\phi'(t) + \frac{\partial H}{\partial p}(\phi(t), m\phi'(t))m\phi''(t)\\ &= -F(\phi(t))\phi'(t) + \frac{m\phi'(t)}{m}m\phi''(t)\\ &= -m\phi''(t)\phi'(t) + \phi'(t)m\phi''(t) = 0,\end{aligned} \tag{2.2.16}$$

which proves

Theorem 2.1. *$t \to H(\phi(t), m\phi'(t))$ is a constant. Also*

$$\begin{aligned}\frac{\partial H}{\partial q}(\phi(t), m\phi'(t)) &= -\frac{d}{dt}(m\phi'(t)),\\ \frac{\partial H}{\partial p}(\phi(t), m\phi'(t)) &= \frac{d}{dt}(\phi(t)).\end{aligned}$$

The function H is called the *Hamiltonian*, and the equations of the system in Theorem 2.1 are *Hamilton's equations*. For examples involving these equations see Section 2.6 below.

Example: Simple Harmonic Motion

Suppose a point P moves counterclockwise around a circle of radius C with a constant angular speed of ω radians per second. Consider the up and down motion along the x-axis of the projection x of P onto that axis.

To find the equation of motion note that

$$\begin{aligned}&\frac{x}{C} = \sin\theta \implies\\ &\quad x = C\sin\theta = C\cos\left(\theta - \frac{\pi}{2}\right) && \text{where } \theta = \theta(t).\\ &\frac{d\theta}{dt} = \omega \implies\\ &\theta(t) = t\omega + \theta(0) \implies\\ &\quad x = \phi(t) = C\cos(\omega t - \delta) && \text{for } \delta = \frac{\pi}{2} - \theta(0).\end{aligned} \tag{2.2.17}$$

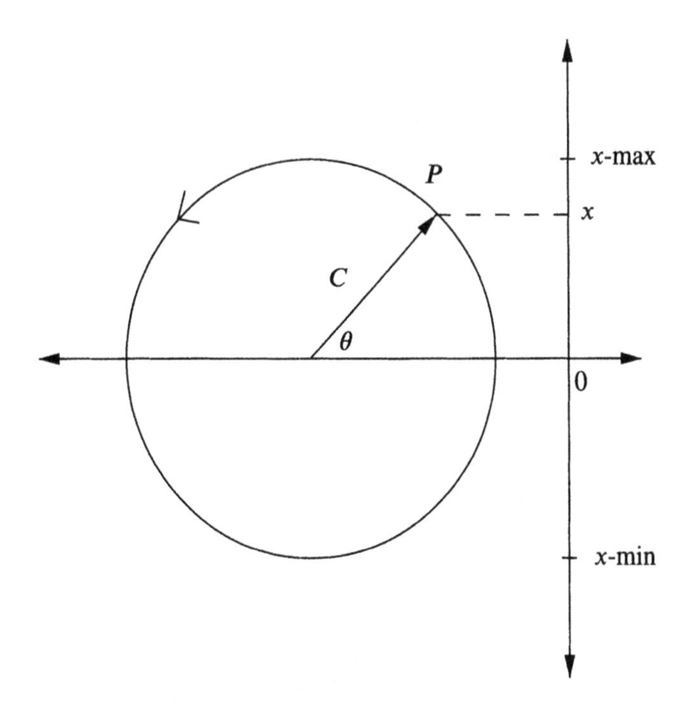

Figure 2.1: Simple harmonic motion (cf. Figure 1.4)

is the desired equation of motion.

The motion along the x-axis is an oscillation between x-max$= C$ and x-min$= -C$. The maximum distance C from 0 is called the *amplitude* of the oscillation. The amount of time T required for one complete oscillation—say for a particle at x to move from x to $x-$max, then down to x-min and back up to x (which is the amount of time for P to complete one revolution)—is called the *period* of the oscillation: Given any time t, $\theta(t+T) = \theta(t) + 2\pi$; i.e., $(t+T)\omega + \theta(0) = t\omega + \theta(0) + 2\pi \implies T\omega = 2\pi \implies T = 2\pi/\omega$, exactly as in Section 1.4 of Chapter 1. As it takes T seconds for one complete revolution of P, the number of revolutions of P per second (possibly a non-integer) is $\nu = 1/T$, called the *frequency* of the oscillation; thus $\nu = \omega/(2\pi)$.

$x(t) = C\cos(\omega t - \delta) \implies v(t) = -\omega C\sin(\omega t - \delta) \implies a(t) = -\omega^2 x(t)$. That is, for $V(x) \stackrel{\text{def}}{=} \frac{\omega^2 x^2 m}{2} = 2\pi^2\nu^2 m x^2$, $m =$ mass, V is the *potential energy* of our particle: $V'(x) = \omega^2 x m \implies -V'(x(t)) = ma(t)$.

Example

It is well known that a freely falling object, relatively near to the surface of the earth, experiences a constant acceleration due to gravity. The magnitude of the acceleration, denoted by g, is approximately 32 ft/sec^2. The object also experiences a force due to air resistance, but we assume here such a force to be negligible. Suppose we wished to determine the depth of a certain well. One way to proceed is as follows. Drop a stone into the well and carefully determine the number of seconds τ it takes to hear the sound of the stone striking the bottom of the well. Suppose the speed of sound is σ ft/sec; one knows that σ is approximately 1100 ft/sec. What is the depth d of the well in terms of τ and σ?

To answer this question note first that choosing the y-coordinate as illustrated in Figure 2.2 we have by the preceding remarks

$$\frac{d^2y}{dt^2} = g \tag{2.2.18}$$

so that the equation of motion of the stone is

$$y(t) = \frac{gt^2}{2} + at + b. \tag{2.2.19}$$

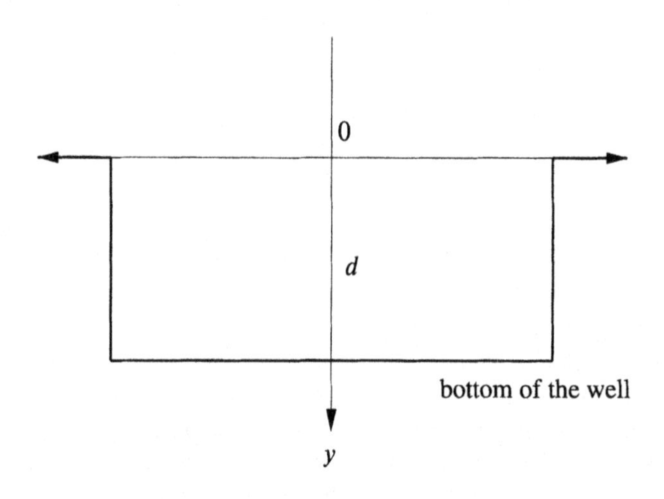

Figure 2.2: A well of depth d

The initial velocity is $y'(0) = a$ which we assume is 0. Also $b = y(0) = 0$: $y(t) = gt^2/2 \implies d = gt^2/2$ provides the amount of time t it takes for the stone to reach the bottom of the well. Since sound travels σ ft/sec, once the stone hits the bottom an additional d/σ seconds will be needed to hear the sound of it striking, as the sound travels upward a distance d feet from the bottom of the well. Thus τ is given by $\tau = d/\sigma + t$ where $gt^2/2 = d$, which means that d is subject to the quadratic equation

$$d = \frac{g}{2}\left[\tau - \frac{d}{\sigma}\right]^2. \tag{2.2.20}$$

The solutions are

$$d = \sigma\tau + \frac{\sigma^2}{g} \pm \frac{\sigma^2}{g}\sqrt{\frac{2\tau g}{\sigma} + 1}. \tag{2.2.21}$$

Now the time t is positive:

$$0 > -t = \frac{d}{\sigma} - \tau = \frac{\sigma}{g} \pm \frac{\sigma}{g}\sqrt{\frac{2\tau g}{\sigma} + 1} \tag{2.2.22}$$

which forces the choice of the minus sign in equation (2.2.21). The solution of the problem is therefore

$$d = \sigma\tau + \frac{\sigma^2}{g} - \frac{\sigma^2}{g}\sqrt{\frac{2\tau g}{\sigma} + 1}. \tag{2.2.23}$$

Note that equation (2.2.23) can also be written

$$d = \left[\sqrt{\sigma\tau + \frac{\sigma^2}{2g}} - \frac{\sigma}{\sqrt{2g}}\right]^2. \tag{2.2.24}$$

2.3 Motion in Three Dimensions

We can extend the discussion of a particle of mass $m > 0$ to its motion in *three* dimensions (or in two dimensions). In this case we have the motion specified by three functions $\phi_j : \mathbb{R} \to \mathbb{R}$ with continuous derivatives ϕ_j', ϕ_j'', $j = 1, 2, 3$. That is, we have a function $\phi : \mathbb{R} \to \mathbb{R}^3$, $\phi = (\phi_1, \phi_2, \phi_3)$ such that at time t, $\phi(t)$ is the position (x, y, z) of the particle. The velocity v and acceleration a of the particle as functions of time are given as maps $v, a : \mathbb{R} \to \mathbb{R}^3$: $v(t) \stackrel{\text{def}}{=} (\phi_1'(t), \phi_2'(t), \phi_3'(t))$, $a(t) \stackrel{\text{def}}{=} (\phi_1''(t), \phi_2''(t), \phi_3''(t))$. One also writes $v = (v_1, v_2, v_3)$, $a = (a_1, a_2, a_3)$.

Assume that we have a continuous force field $F : \mathbb{R}^3 \to \mathbb{R}^3$: for the particle at position (x, y, z) the force acting on it is $F(x, y, z) = (F_1(x, y, z), F_2(x, y, z), F_3(x, y, z))$ where $F_j : \mathbb{R}^3 \to \mathbb{R}$ are the (continuous) components of F. Newton's second law $ma = F$ now takes the form

$$m\phi_j''(t) = F_j(\phi_1(t), \phi_2(t), \phi_3(t)) \text{ at time } t, \text{ for } j = 1, 2, 3. \tag{2.3.1}$$

We shall also assume F is derived from a potential V: $V : \mathbb{R}^3 \to \mathbb{R}$ is a map with continuous first partial derivatives such that $F = -\nabla V$; i.e.,

$$F_1 = -\frac{\partial V}{\partial x}, \qquad F_2 = -\frac{\partial V}{\partial y}, \qquad F_3 = -\frac{\partial V}{\partial z}. \tag{2.3.2}$$

For motion in one dimension, the latter assumption is automatically fulfilled as we have seen. Now $\phi_1, \phi_2, \phi_3 : \mathbb{R} \to \mathbb{R}$ and $V : \mathbb{R}^3 \to \mathbb{R}$ are continuously differentiable. Thus if we define $h : \mathbb{R} \to \mathbb{R}$ by $h(t) = V(\phi_1(t), \phi_2(t), \phi_3(t))$ we obtain (by the chain rule) that h is differentiable, and

$$\begin{aligned} h'(t) &= \frac{\partial V}{\partial x}(\phi(t))\,\phi_1'(t) + \frac{\partial V}{\partial y}(\phi(t))\,\phi_2'(t) + \frac{\partial V}{\partial z}(\phi(t))\,\phi_3'(t) \\ &= -F_1(\phi(t))\,\phi_1'(t) - F_2(\phi(t))\,\phi_2'(t) - F_3(\phi(t))\,\phi_3'(t) \\ &= -m\phi_1''(t)\phi_1'(t) - m\phi_2''(t)\phi_2'(t) - m\phi_3''(t)\phi_3'(t) \qquad \text{by (2.3.1).} \end{aligned} \tag{2.3.3}$$

Therefore define $H : \mathbb{R} \to \mathbb{R}$ by

$$\begin{aligned} H(t) &= \frac{1}{2}m\,\langle v(t), v(t)\rangle + h(t) \\ &= \frac{1}{2}m\left(\phi_1'(t)^2 + \phi_2'(t)^2 + \phi_3'(t)^2\right) + h(t) \end{aligned} \tag{2.3.4}$$

to obtain

$$\begin{aligned} H'(t) &= \frac{1}{2}m\left(2\phi_1'(t)\phi_1''(t) + 2\phi_2'(t)\phi_2''(t) + 2\phi_3'(t)\phi_3''(t)\right) \\ &\quad - m\phi_1''(t)\phi_1'(t) - m\phi_2''(t)\phi_2'(t) - m\phi_3''(t)\phi_3'(t) = 0 \\ &\Longrightarrow H(t) \text{ is a constant.} \end{aligned} \tag{2.3.5}$$

That is, we have the conservation law $\frac{1}{2}m\,\langle v(t), v(t)\rangle + V(\phi(t)) =$ a constant for all time t, for $F = -\nabla V$. The first term here is the *kinetic energy* of the particle at time t. The kinetic energy T as a function $:\mathbb{R}^3 \to \mathbb{R}$ of v is given by

$$T(v) = \frac{1}{2}m\,\langle v, v\rangle = \frac{1}{2}m\left(v_1^2 + v_2^2 + v_3^2\right). \tag{2.3.6}$$

T can also be expressed in terms of the *momentum* $p = mv$ of the particle: $p_j = mv_j \Longrightarrow$

$$\begin{aligned} &\frac{p_j^2}{2m} = \frac{mv_j^2}{2} \\ \Longrightarrow\ &T(v) = \frac{m}{2}(v_1^2 + v_2^2 + v_3^2) = \frac{1}{2m}(p_1^2 + p_2^2 + p_3^2) = \frac{1}{2m}\langle p, p\rangle. \end{aligned} \tag{2.3.7}$$

2.4 Many Particles in Three Dimensions

Consider now a system of n particles moving in three dimensions, with $m_j =$ mass of the jth particle. Let $\phi^{(j)} = (\phi_1^{(j)}, \phi_2^{(j)}, \phi_3^{(j)}) : \mathbb{R} \to \mathbb{R}^3$ specify the position of the jth particle. It is assumed as before that continuous derivatives $(\phi_l^{(j)})', (\phi_l^{(j)})''$, $l = 1, 2, 3$, exist. We simplify the notation by also writing $\phi^{(j)} = (x_j, y_j, z_j)$:

$$\begin{aligned} \dot{x}_j &= (\phi_1^{(j)})' & \ddot{x}_j &= (\phi_1^{(j)})'' \\ \dot{y}_j &= (\phi_2^{(j)})' & \ddot{y}_j &= (\phi_2^{(j)})'' \\ \dot{z}_j &= (\phi_3^{(j)})' & \ddot{z}_j &= (\phi_3^{(j)})'' \end{aligned} \tag{2.4.1}$$

For each j, $1 \le j \le n$, consider the jth particle under the influence of a continuous force field $F^{(j)}$ where now $F^{(j)} = (F_1^{(j)}, F_2^{(j)}, F_3^{(j)}) : \mathbb{R}^{3n} \to \mathbb{R}^3$ (instead of $\mathbb{R}^3 \to \mathbb{R}^3$); also write $F^{(j)} = (X_j, Y_j, Z_j)$; thus $X_j, Y_j, Z_j : \mathbb{R}^{3n} \to \mathbb{R}$ are continuous functions. If $v^{(j)}, a^{(j)} : \mathbb{R} \to \mathbb{R}^3$ denote the velocity, acceleration functions of the

jth particle (thus $v^{(j)} = (\dot{x}_j, \dot{y}_j, \dot{z}_j)$, $a^{(j)} = (\ddot{x}_j, \ddot{y}_j, \ddot{z}_j)$ by the above notation) then Newton's second law takes the form

$$\begin{aligned} m_j a^{(j)}(t) &= F^{(j)}(\phi^{(1)}(t), \ldots, \phi^{(n)}(t)) : \\ m_j \ddot{x}_j(t) &= F_1^{(j)}(\phi^{(1)}(t), \ldots, \phi^{(n)}(t)) \\ &= X_j(x_1(t), y_1(t), z_1(t), \ldots, x_n(t), y_n(t), z_n(t)), \end{aligned} \tag{2.4.2}$$

and similarly

$$\begin{aligned} m_j \ddot{y}_j(t) &= Y_j(x_1(t), y_1(t), z_1(t), \ldots, x_n(t), y_n(t), z_n(t)), \\ m_j \ddot{z}_j(t) &= Z_j(x_1(t), y_1(t), z_1(t), \ldots, x_n(t), y_n(t), z_n(t)), \end{aligned} \tag{2.4.3}$$

for $1 \leq j \leq n$.

As before we shall be interested in the case when a potential energy function V exists for our system: $V : \mathbb{R}^{3n} \to \mathbb{R}$ has continuous first partial derivatives such that

$$X_j = -\frac{\partial V}{\partial x_j}, \qquad Y_j = -\frac{\partial V}{\partial y_j}, \qquad Z_j = -\frac{\partial V}{\partial z_j} \tag{2.4.4}$$

at $(x_1, y_1, z_1, x_2, y_2, z_2, \ldots, x_n, y_n, z_n) \in \mathbb{R}^{3n}$, $1 \leq j \leq n$. We have $\phi_1^{(1)}, \phi_2^{(1)}, \phi_3^{(1)}, \phi_1^{(2)}, \phi_2^{(2)}, \phi_3^{(2)}, \ldots, \phi_1^{(n)}, \phi_2^{(n)}, \phi_3^{(n)} : \mathbb{R} \to \mathbb{R}$, $V : \mathbb{R}^{3n} \to \mathbb{R}$ are continuously differentiable. If we therefore define $h : \mathbb{R} \to \mathbb{R}$ by

$$h(t) = V\left(\phi_1^{(1)}(t), \phi_2^{(1)}(t), \phi_3^{(1)}(t), \ldots, \phi_1^{(n)}(t), \phi_2^{(n)}(t), \phi_3^{(n)}(t)\right) \stackrel{\text{def}}{=} V(\phi(t)), \tag{2.4.5}$$

we obtain by the chain rule

$$\begin{aligned} h'(t) &= \sum_{p=1}^{3n} (D_p V)(\phi(t)) \phi_p'(t) \\ &= \frac{\partial V}{\partial x_1}(\phi(t))(\phi_1^{(1)})'(t) + \frac{\partial V}{\partial y_1}(\phi(t))(\phi_2^{(1)})'(t) + \frac{\partial V}{\partial z_1}(\phi(t))(\phi_3^{(1)})'(t) \\ &+ \frac{\partial V}{\partial x_2}(\phi(t))(\phi_1^{(2)})'(t) + \frac{\partial V}{\partial y_2}(\phi(t))(\phi_2^{(2)})'(t) + \frac{\partial V}{\partial z_2}(\phi(t))(\phi_3^{(2)})'(t) \\ &+ \cdots \\ &+ \frac{\partial V}{\partial x_n}(\phi(t))(\phi_1^{(n)})'(t) + \frac{\partial V}{\partial y_n}(\phi(t))(\phi_2^{(n)})'(t) + \frac{\partial V}{\partial z_n}(\phi(t))(\phi_3^{(n)})'(t) \qquad (2.4.6) \\ &= -X_1(\phi(t))\dot{x}_1(t) - Y_1(\phi(t))\dot{y}_1(t) - Z_1(\phi(t))\dot{z}_1(t) \\ &\quad - X_2(\phi(t))\dot{x}_2(t) - Y_2(\phi(t))\dot{y}_2(t) - Z_2(\phi(t))\dot{z}_2(t) \\ &\quad - \cdots \\ &\quad - X_n(\phi(t))\dot{x}_n(t) - Y_n(\phi(t))\dot{y}_n(t) - Z_n(\phi(t))\dot{z}_n(t) \\ &= -m_1\ddot{x}_1(t)\dot{x}_1(t) - m_1\ddot{y}_1(t)\dot{y}_1(t) - m_1\ddot{z}_1(t)\dot{z}_1(t) \\ &\quad - m_2\ddot{x}_2(t)\dot{x}_2(t) - m_2\ddot{y}_2(t)\dot{y}_2(t) - m_2\ddot{z}_2(t)\dot{z}_2(t) \\ &\quad - -m_n\ddot{x}_n(t)\dot{x}_n(t) - m_n\ddot{y}_n(t)\dot{y}_n(t) - m_n\ddot{z}_n(t)\dot{z}_n(t). \end{aligned}$$

Hence, for $H : \mathbb{R} \to \mathbb{R}$ defined by

$$\begin{aligned} H(t) &= \frac{1}{2}\sum_{j=1}^{n} m_j \langle v^{(j)}(t), v^{(j)}(t)\rangle + h(t) \\ &= \frac{1}{2}\sum_{j=1}^{n} m_j \left((\dot{x}_j(t))^2 + (\dot{y}_j(t))^2 + (\dot{z}_j(t))^2\right) + h(t), \end{aligned} \tag{2.4.7}$$

$H'(t) = 0 \implies$ the conservation law:

$$\frac{1}{2}\sum_{j=1}^{n} m_j \langle v^{(j)}(t), v^{(j)}(t)\rangle + V\left(\phi_1^{(1)}(t), \phi_2^{(1)}(t), \phi_3^{(1)}(t), \ldots, \phi_1^{(n)}(t), \phi_2^{(n)}(t), \phi_3^{(n)}(t)\right) \tag{2.4.8}$$

is a constant for all t. This generalizes (2.3.5).

The *kinetic energy* $T : \mathbb{R}^{3n} \to \mathbb{R}$ of the system as a function of its velocities is given by

$$T(v^{(1)}, \ldots, v^{(n)}) = T(v_1^{(1)}, v_2^{(1)}, v_3^{(1)}, v_1^{(2)}, v_2^{(2)}, v_3^{(2)}, \ldots, v_1^{(n)}, v_2^{(n)}, v_3^{(n)}) \tag{2.4.9}$$

$$= \frac{1}{2}\sum_{j=1}^{n} m_j \langle v^{(j)}, v^{(j)}\rangle \tag{2.4.10}$$

$$= \frac{1}{2}\sum_{j=1}^{n} m_j \left((v_1^{(j)})^2 + (v_2^{(j)})^2 + (v_3^{(j)})^2\right), \tag{2.4.11}$$

(where we write $v^{(j)} = (v_1^{(j)}, v_2^{(j)}, v_3^{(j)})$) or

$$T(v^{(1)}, \ldots, v^{(n)}) = \sum_{j=1}^{n} \left(\frac{(p_1^{(j)})^2}{2m_j} + \frac{(p_2^{(j)})^2}{2m_j} + \frac{(p_3^{(j)})^2}{2m_j} \right) \tag{2.4.12}$$

for $p^{(j)} = m_j v^{(j)}$ the momentum of the jth particle. Definition (2.4.9) is motivated by the first term in (2.4.8).

We have shown in general that the definition of the kinetic energy function is motivated directly by a simple conservation law (which requires the factor $\frac{1}{2}$) based on elementary calculus: Energy is conserved along a trajectory.

2.5 Coulomb and Gravitational Potentials

Consider a system of two particles : $n = 2$. In the present example we will define force fields $F^{(j)}$, $j = 1, 2$, on an open set $U \subset \mathbb{R}^6$ (rather than on $\mathbb{R}^{6=3n}$). Let $U \stackrel{\text{def}}{=} \{(x_1, y_1, z_1, x_2, y_2, z_2) \in \mathbb{R}^6 | (x_1, y_1, z_1) \neq (x_2, y_2, z_2)\}$. If we write $(x_1, y_1, z_1, x_2, y_2, z_2) = (p_1, p_2) \in \mathbb{R}^3 \times \mathbb{R}^3$, and $\pi_1, \pi_2 : \mathbb{R}^6 \to \mathbb{R}^3$ for the projections $(p_1, p_2) \to p_1, p_2$, then as π_1, π_2 are continuous (and as $\mathbb{R}^3$ is Hausdorff), $W = \{x \in \mathbb{R}^6 | \pi_1(x) = \pi_2(x)\}$ is closed in $\mathbb{R}^6$. That is, $U = \mathbb{R}^6 - W$ is open.

For real constants q_1, q_2 fixed define $F^{(1)} : U \to \mathbb{R}^3$ (continuous) by

$$F^{(1)}(x_1, y_1, z_1, x_2, y_2, z_2) = \frac{q_1 q_2 \{(x_1, y_1, z_1) - (x_2, y_2, z_2)\}}{|(x_1, y_1, z_1) - (x_2, y_2, z_2)|^3}, \tag{2.5.1}$$

and define $F^{(2)} : U \to \mathbb{R}^3$ (continuous) by $F^{(2)} = -F^{(1)}$.

The physical meaning here is the following. Think of q_j as the "charge" of the jth particle. For particle one at position (x_1, y_1, z_1) and particle two at position (x_2, y_2, z_2), *Coulomb's inverse square law* (compare equation (0.0.4) in Chapter 0) of electrostatics says that the force on particle one due to particle two is $F^{(1)}(x_1, y_1, z_1, x_2, y_2, z_2)$, and the force on particle two due to particle one is $F^{(2)}(x_1, y_1, z_1, x_2, y_2, z_2) = -F^{(1)}(x_1, y_1, z_1, x_2, y_2, z_2)$. Note that

$$\begin{aligned}
&|F^{(2)}(x_1, y_1, z_1, x_2, y_2, z_2)| = |F^{(1)}(x_1, y_1, z_1, x_2, y_2, z_2)| \\
&= \frac{|q_1 q_2| \left[(x_1 - x_2)^2 + (y_1 - y_2)^2 + (z_1 - z_2)^2\right]^{\frac{1}{2}}}{|(x_1, y_1, z_1) - (x_2, y_2, z_2)|^3 = [(x_1 - x_2)^2 + (y_1 - y_2)^2 + (z_1 - z_2)^2]^{\frac{3}{2}}} \\
&= \frac{|q_1 q_2|}{(x_1 - x_2)^2 + (y_1 - y_2)^2 + (z_1 - z_2)^2} \\
&= \frac{|q_1 q_2|}{[\text{ distance from } (x_1, y_1, z_1) \text{ to } (x_2, y_2, z_2)]^2},
\end{aligned} \tag{2.5.2}$$

which explains the phrase "inverse square." In (2.5.1) we choose the constant κ in (0.0.4) equal to 1.

Writing $F^{(j)} = (X_j, Y_j, Z_j)$, where $X_j, Y_j, Z_j : U \to \mathbb{R}$ are continuous functions, we get $X_2 = -X_1$, $Y_2 = -Y_1$, $Z_2 = -Z_1$, with

$$\begin{aligned}
X_1(x_1, y_1, z_1, x_2, y_2, z_2) &= \frac{q_1 q_2 (x_1 - x_2)}{[(x_1 - x_2)^2 + (y_1 - y_2)^2 + (z_1 - z_2)^2]^{\frac{3}{2}}} \\
Y_1(x_1, y_1, z_1, x_2, y_2, z_2) &= \frac{q_1 q_2 (y_1 - y_2)}{[(x_1 - x_2)^2 + (y_1 - y_2)^2 + (z_1 - z_2)^2]^{\frac{3}{2}}} \\
Z_1(x_1, y_1, z_1, x_2, y_2, z_2) &= \frac{q_1 q_2 (z_1 - z_2)}{[(x_1 - x_2)^2 + (y_1 - y_2)^2 + (z_1 - z_2)^2]^{\frac{3}{2}}}
\end{aligned} \tag{2.5.3}$$

on U. We can construct a potential energy function $V : U \to \mathbb{R}$ for our system by defining

$$V(x_1, y_1, z_1, x_2, y_2, z_2) = \frac{q_1 q_2}{\sqrt{(x_1 - x_2)^2 + (y_1 - y_2)^2 + (z_1 - z_2)^2}}. \tag{2.5.4}$$

Indeed,

$$\begin{aligned}
&\frac{\partial V}{\partial x_1}(x_1, y_1, z_1, x_2, y_2, z_2) \\
&\quad = q_1 q_2 \left(-\frac{1}{2}\right) [(x_1 - x_2)^2 + (y_1 - y_2)^2 + (z_1 - z_2)^2]^{-\frac{3}{2}} 2(x_1 - x_2) \cdot 1 \\
&\Longrightarrow -\frac{\partial V}{\partial x_1} = X_1.
\end{aligned} \tag{2.5.5}$$

Similarly

$$-\frac{\partial V}{\partial y_1} = Y_1, \qquad -\frac{\partial V}{\partial z_1} = Z_1;$$

$$\frac{\partial V}{\partial x_2} = -\frac{\partial V}{\partial x_1}, \qquad \frac{\partial V}{\partial y_2} = -\frac{\partial V}{\partial y_1}, \qquad \frac{\partial V}{\partial z_2} = -\frac{\partial V}{\partial z_1}, \tag{2.5.6}$$

$$\Longrightarrow -\frac{\partial V}{\partial x_2} = -X_1 = X_2, \qquad -\frac{\partial V}{\partial y_2} = -Y_1 = Y_2, \qquad -\frac{\partial V}{\partial z_2} = -Z_1 = Z_2,$$

so that for $j = 1, 2$

$$X_j = -\frac{\partial V}{\partial x_j}, \qquad Y_j = -\frac{\partial V}{\partial y_j}, \qquad Z_j = -\frac{\partial V}{\partial z_j}, \tag{2.5.7}$$

as in (2.4.4).

If q_1, q_2 are *masses*, rather than charges, then V is the *gravitational* potential, up to a constant (the universal gravitational constant), and the inverse square law (2.5.1) is Newton's universal law of gravitational attraction. In (2.5.1) we have omitted an appropriate constant, which some readers might wish to incorporate for later application. This is not a problem since one can replace q_1, for example, by a multiple of itself.

2.6 Hamilton's Equations of Motion

We define the Hamiltonian of our system of n particles moving in $\mathbb{R}^3$ as the total energy function $H : \mathbb{R}^{3n} \times \mathbb{R}^{3n} \to \mathbb{R}$ given by

$$\begin{aligned}
&H(q_1, \ldots, q_{3n}, p^{(1)}, \ldots, p^{(n)}) \\
&\quad = \sum_{j=1}^{n} \left(\frac{(p_1^{(j)})^2}{2m_j} + \frac{(p_2^{(j)})^2}{2m_j} + \frac{(p_3^{(j)})^2}{2m_j} \right) + V(q_1, \ldots, q_{3n})
\end{aligned} \tag{2.6.1}$$

for $p^{(j)} = (p_1^{(j)}, p_2^{(j)}, p_3^{(j)}) \in \mathbb{R}^3$; cf. equations (2.2.14), (2.4.12). Let

$$\phi(t) = (\phi_1^{(1)}(t), \phi_2^{(1)}(t), \phi_3^{(1)}(t), \ldots, \phi_1^{(n)}(t), \phi_2^{(n)}(t), \phi_3^{(n)}(t)) \tag{2.6.2}$$

$$= (x_1(t), y_1(t), z_1(t), \ldots, x_n(t), y_n(t), z_n(t)) \tag{2.6.3}$$

as before, and define

$$\begin{aligned} m\phi'(t) &= (m_1(\phi_1^{(1)})'(t), m_1(\phi_2^{(1)})'(t), m_1(\phi_3^{(1)})'(t), \\ &\quad \ldots, m_n(\phi_1^{(n)})'(t), m_n(\phi_2^{(n)})'(t), m_n(\phi_3^{(n)})'(t)) \qquad (2.6.4) \\ &= (m_1\dot{x}_1(t), m_1\dot{y}_1(t), m_1\dot{z}_1(t), \ldots, m_n\dot{x}_n(t), m_n\dot{y}_n(t), m_n\dot{z}_n(t)) \qquad (2.6.5) \end{aligned}$$

by (2.4.1). Then

$$\begin{aligned} H(\phi(t), m\phi'(t)) &\stackrel{\text{def}}{=} \sum_{j=1}^{n} \left(\frac{m_j^2\dot{x}_j(t)^2}{2m_j} + \frac{m_j^2\dot{y}_j(t)^2}{2m_j} + \frac{m_j^2\dot{z}_j(t)^2}{2m_j} \right) + V(\phi(t)) \\ &= H(t) \end{aligned} \tag{2.6.6}$$

(by (2.6.1) and (2.4.7)), which as we have noted in (2.4.7), (2.4.8) is a constant for all time t.

Moreover for $q^{(j)} = (q_1^{(j)}, q_2^{(j)}, q_3^{(j)}) \in \mathbb{R}^3$ we have

$$\begin{aligned} \frac{\partial H}{\partial q_1^{(j)}}(\phi(t), m\phi'(t)) &= \frac{\partial V}{\partial q_1^{(j)}}(\phi(t)) = -X_j(\phi(t)) \\ &= -m_j\ddot{x}_j(t) = -\frac{d}{dt}[m_j(\phi_1^{(j)})'(t)], \end{aligned} \tag{2.6.7}$$

by (2.6.1), (2.4.4) and (2.4.2). Similarly

$$\frac{\partial H}{\partial q_2^{(j)}}(\phi(t), m\phi'(t)) = -m_j\ddot{y}_j(t) = -\frac{d}{dt}[m_j(\phi_2^{(j)})'(t)], \tag{2.6.8}$$

$$\frac{\partial H}{\partial q_3^{(j)}}(\phi(t), m\phi'(t)) = -m_j\ddot{z}_j(t) = -\frac{d}{dt}[m_j(\phi_3^{(j)})'(t)]. \tag{2.6.9}$$

On the other hand

$$\frac{\partial H}{\partial p_1^{(j)}}(\phi(t), m\phi'(t)) = \frac{m_j(\phi_1^{(j)})'(t)}{m_j} = \frac{d}{dt}[\phi_1^{(j)}(t)]. \tag{2.6.10}$$

Similarly

$$\frac{\partial H}{\partial p_2^{(j)}}(\phi(t), m\phi'(t)) = \frac{d}{dt}[\phi_2^{(j)}(t)], \tag{2.6.11}$$

$$\frac{\partial H}{\partial p_3^{(j)}}(\phi(t), m\phi'(t)) = \frac{d}{dt}[\phi_3^{(j)}(t)]. \tag{2.6.12}$$

Thus we have the following generalization of Theorem 2.1.

Theorem 2.2. *For a system of n particles moving in space, let* $\phi^{(j)} = (\phi_1^{(j)}, \phi_2^{(j)}, \phi_3^{(j)}) : \mathbb{R} \to \mathbb{R}^3$ *specify the position of the jth particle with mass* m_j. *Assume as above the system is governed by a potential* $V : \mathbb{R}^{3n} \to \mathbb{R}$*; see (2.4.4). Let* $H : \mathbb{R}^{3n} \times \mathbb{R}^{3n} \to \mathbb{R}$ *be the Hamiltonian given by (2.6.1), and let* ϕ, $m\phi'$ *be defined on* $\mathbb{R}$ *by (2.6.2) and (2.6.4). Then we have that* $t \to H(\phi(t), m\phi'(t))$ *is a constant function on* $\mathbb{R}$ *(conservation of energy along a trajectory), and Hamilton's equations of motion for the system:*

$$\begin{aligned}\frac{\partial H}{\partial q_k^{(j)}}(\phi(t), m\phi'(t)) &= -\frac{d}{dt}[m_j(\phi_k^{(j)})'(t)],\\ \frac{\partial H}{\partial p_k^{(j)}}(\phi(t), m\phi'(t)) &= \frac{d}{dt}[\phi_k^{(j)}(t)]\end{aligned} \tag{2.6.13}$$

for $k = 1, 2, 3$, $(q^{(1)}, \ldots, q^{(n)}, p^{(1)}, \ldots, p^{(n)}) \in \mathbb{R}^{3n} \times \mathbb{R}^{3n}$ *with* $q^{(j)} = (q_1^{(j)}, q_2^{(j)}, q_3^{(j)})$, $p^{(j)} = (p_1^{(j)}, p_2^{(j)}, p_3^{(j)})$, $j = 1, 2, 3, \ldots, n$.

Examples

As a simple example consider a ball of mass m thrown vertically up from a height z_0 feet above the ground represented by the xy-plane. As we have noted, a freely falling object experiences a constant acceleration, due to gravity, of magnitude g = about 32 ft/sec^2, assuming that the fall is relatively near the earth's surface, and that the force exerted on the object due to air resistance is negligible. This means that the ball will experience a force $F = -mg\kappa$ by Newton's second law where $\kappa = (0, 0, 1)$ is the unit vector along the z-axis. Regarding F as a function $(F_1, F_2, F_3) : \mathbb{R}^3 \to \mathbb{R}^3$, where $F_1(x, y, z) = F_2(x, y, z) = 0$, $F_3(x, y, z) = -mg$, we clearly have that F is derivable from the potential energy function $V : \mathbb{R}^3 \to \mathbb{R}$ given by $V(x, y, z) = mgz : 0 = -\frac{\partial V}{\partial x} = F_1, 0 = -\frac{\partial V}{\partial y} = F_2, -\frac{\partial V}{\partial z} = F_3$.

The corresponding Hamiltonian $H : \mathbb{R}^3 \times \mathbb{R}^3 \to \mathbb{R}$ is given by (2.6.1):

$$H(q_1, q_2, q_3, p_1, p_2, p_3) = \frac{p_1^2 + p_2^2 + p_3^2}{2m} + mgq_3. \tag{2.6.14}$$

Let $\phi(t) = (\phi_1(t), \phi_2(t), \phi_3(t))$ specify the position of the ball at time t. Then Hamilton's equations (2.6.13) read

$$\begin{aligned}\frac{\partial H}{\partial q_k}(\phi(t), m\phi'(t)) &= -\frac{d}{dt}[m\phi_k{}'(t)],\\ \frac{\partial H}{\partial p_k}(\phi(t), m\phi'(t)) &= \frac{d}{dt}[\phi_k(t)]\end{aligned} \tag{2.6.15}$$

for $k = 1, 2, 3$. Now $\frac{\partial H}{\partial q_k} = 0$ for $k = 1, 2$, $\frac{\partial H}{\partial q_3} = mg$, $\frac{\partial H}{\partial p_k}(q_1, q_2, q_3, p_1, p_2, p_3) = \frac{p_k}{m}$ for $k = 1, 2, 3 \implies$

$$m\ddot{\phi}_k(t) = 0 \qquad \text{for } k = 1, 2, \tag{2.6.16}$$

$$m\ddot{\phi}_3(t) = -mg, \tag{2.6.17}$$

$$\dot{\phi}_k(t) = \frac{m\dot{\phi}_k(t)}{m} \qquad \text{for } k = 1, 2, 3. \tag{2.6.18}$$

Integrating equations (2.6.16), (2.6.17) we obtain the equations of motion of the ball

$$\begin{aligned} \phi_1(t) &= a_1 t + b_1, \\ \phi_2(t) &= a_2 t + b_2, \\ \phi_3(t) &= -\frac{gt^2}{2} + at + b. \end{aligned} \tag{2.6.19}$$

Taking initial conditions $\phi(0) = (x_o, y_o, z_o)$, $\phi'(0) = (v_1, v_2, v_3)$ we obtain in the end

$$\begin{aligned} \phi_1(t) &= v_1 t + x_o, \\ \phi_2(t) &= v_2 t + y_o, \\ \phi_3(t) &= -\frac{gt^2}{2} + v_3 t + z_o, \end{aligned} \tag{2.6.20}$$

as is well known. The equations in (2.6.20) are independent of the mass m.

In another example consider the potential $V : \mathbb{R}^3 \to \mathbb{R}$ given by $V(x, y, z) = \frac{\alpha}{2}(x^2+y^2+z^2)$, where $\alpha > 0$ is some fixed constant. The corresponding force field is $F = -\nabla V : F(x, y, z) = -\alpha(x, y, z)$. Suppose a particle moves on the surface of a cylinder of radius r, as pictured below, under the influence of this field. What are its equations of motion?

By (2.6.1) the Hamiltonian $H : \mathbb{R}^3 \times \mathbb{R}^3 \to \mathbb{R}$ is given by

$$H(q_1, q_2, q_3, p_1, p_2, p_3) = \frac{1}{2m}\sum_{j=1}^{3} p_j^2 + \frac{\alpha}{2}(r^2 + q_3^2) \tag{2.6.21}$$

as $x^2 + y^2 = r^2$ on the cylinder. If $\phi(t) = (\phi_1(t), \phi_2(t), \phi_3(t))$ denotes the position of the particle (of mass m) at time t, then Hamilton's equations of motion are as in (2.6.13), where now $\frac{\partial H}{\partial q_k} = 0$ for $k = 1, 2$, $\frac{\partial H}{\partial q_3} = \alpha q_3$, $\frac{\partial H}{\partial p_k} = \frac{p_k}{m}$ for $k = 1, 2, 3$ $\implies \ddot{\phi}_k(t) = 0$ for $k = 1, 2$, $-m\ddot{\phi}_3(t) = \alpha\phi_3(t)$. Integrating, we get

$$\begin{aligned} \phi_k(t) &= a_k t + b_k, & k &= 1, 2 \\ \phi_3(t) &= a\cos(\omega t) + b\sin(\omega t) & \text{for } \omega^2 &= \frac{\alpha}{m}, \text{ since } \alpha > 0. \end{aligned} \tag{2.6.22}$$

Since $\phi(t)$ is a point on the cylinder we must have $(a_1 t + b_1)^2 + (a_2 t + b_2)^2 = r^2\ \forall t : (a_1^2+a_2^2)t^2+2(a_1 b_1+a_2 b_2)t+b_1^2+b_2^2-r^2 = 0 \implies a_1, a_2 = 0, b_1^2+b_2^2 = r^2$. Therefore $\phi_1(t) = b_1$, $\phi_2(t) = b_2$ where $b_1^2 + b_2^2 = r^2$. Also the z-component of the motion is *simple harmonic motion* with frequency $\nu = \omega/2\pi$; cf. Section 2.2. We can write $\phi_3(t) = C\cos(\omega t - \delta)$. Namely $a = C\cos\delta$, $b = C\sin\delta$ for some

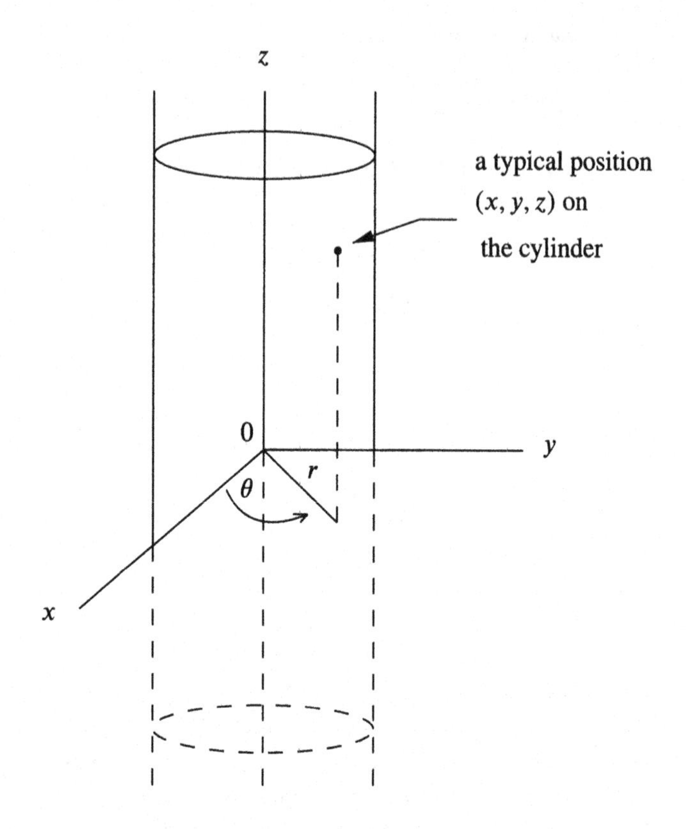

Figure 2.3: Particle on a cylinder

δ, for $C = +\sqrt{a^2 + b^2} : (a/C)^2 + (b/C)^2 = 1$. Then indeed

$$\begin{aligned} C\cos(\omega t - \delta) &= C[\cos(\omega t)\cos\delta + \sin(\omega t)\sin\delta] \\ &= a\cos\omega t + b\sin\omega t = \phi_3(t). \end{aligned} \tag{2.6.23}$$

2.7 Lagrange's Equations of Motion

We have enough notation established to easily derive Lagrange's form of the equations of motion of a system of n particles in $\mathbb{R}^3$ under the influence of a potential $V : \mathbb{R}^{3n} \to \mathbb{R}$. Define the *Lagrangian* $L : \mathbb{R}^{3n} \times \mathbb{R}^{3n} \to \mathbb{R}$ of the system by

$$\begin{aligned} &L(q^{(1)}, \ldots, q^{(n)}, v^{(1)}, \ldots, v^{(n)}) \\ &\quad \stackrel{\text{def}}{=} \frac{1}{2}\sum_{j=1}^{n} m_j \left[(v_1^{(j)})^2 + (v_2^{(j)})^2 + (v_3^{(j)})^2\right] - V(q^{(1)}, \ldots, q^{(n)}) \\ &\quad = T(v^{(1)}, \ldots, v^{(n)}) - V(q^{(1)}, \ldots, q^{(n)}), \end{aligned} \tag{2.7.1}$$

by (2.4.11), where as before $q^{(j)} = (q_1^{(j)}, q_2^{(j)}, q_3^{(j)})$, $v^{(j)} = (v_1^{(j)}, v_2^{(j)}, v_3^{(j)}) \in \mathbb{R}^3$. Then

$$\frac{\partial L}{\partial v_k^{(j)}}(q^{(1)}, \ldots, q^{(n)}, v^{(1)}, \ldots, v^{(n)}) = m_j v_k^{(j)}, \tag{2.7.2}$$

and by (2.4.4)

$$\frac{\partial L}{\partial q_k^{(j)}}(q^{(1)}, \ldots, q^{(n)}, v^{(1)}, \ldots, v^{(n)}) = \begin{bmatrix} X_j(q^{(1)}, \ldots, q^{(n)}) & \text{for } k = 1 \\ Y_j(q^{(1)}, \ldots, q^{(n)}) & \text{for } k = 2 \\ Z_j(q^{(1)}, \ldots, q^{(n)}) & \text{for } k = 3 \end{bmatrix}. \tag{2.7.3}$$

Then for $\phi(t)$ defined by (2.6.2), (2.6.3) and $\dot{\phi}(t)$ defined accordingly

$$\frac{\partial L}{\partial v_k^{(j)}}(\phi(t), \dot{\phi}(t)) = m_j \dot{\phi}_k^{(j)}, \text{ and} \tag{2.7.4}$$

$$\frac{\partial L}{\partial q_k^{(j)}}(\phi(t), \dot{\phi}(t)) = \begin{bmatrix} m_j \ddot{x}_j(t) & \text{for } k = 1 \\ m_j \ddot{y}_j(t) & \text{for } k = 2 \\ m_j \ddot{z}_j(t) & \text{for } k = 3 \end{bmatrix} \tag{2.7.5}$$

by (2.4.2), (2.4.3), and (2.7.3). That is, by (2.4.1)

$$\frac{\partial L}{\partial q_k^{(j)}}(\phi(t), \dot{\phi}(t)) = m_j \ddot{\phi}_k^{(j)}(t). \tag{2.7.6}$$

By (2.7.4), (2.7.6) we therefore obtain

Theorem 2.3 (Lagrange's equations of motion). *As before, let $\phi(t)$ be given by (2.6.2) where $\phi^{(j)}(t) = (x_j(t), y_j(t), z_j(t))$ specifies the position of the jth particle of a system of n particles moving in $\mathbb{R}^3$. Let L be the Lagrangian of the system, given by (2.7.1). Then*

$$\frac{d}{dt}\frac{\partial L}{\partial v_k^{(j)}}(\phi(t), \dot{\phi}(t)) - \frac{\partial L}{\partial q_k^{(j)}}(\phi(t), \dot{\phi}(t)) = 0 \tag{2.7.7}$$

for $k = 1, 2, 3$, $1 \le j \le n$.

Example

Suppose for example a particle of mass m moves in $\mathbb{R}^3$ under the influence of the field $F : \mathbb{R}^3 \to \mathbb{R}^3$ given by $F(x, y, z) = (mx, my, mz)$; here $n = 1$. Then $F = -\nabla V$ for $V(x, y, z) = -\frac{m}{2}(x^2 + y^2 + z^2)$. Equations (2.3.1) become $m\phi_j''(t) = m\phi_j(t)$, $1 \le j \le 3$, so that the equations of motion are

$$\phi_j(t) = a_j \cosh t + b_j \sinh t, \quad 1 \le j \le 3. \tag{2.7.8}$$

The velocity vector is

$$\begin{aligned} v(t) &= (\phi_1'(t), \phi_2'(t), \phi_3'(t)) \\ &= (a_1 \sinh t + b_1 \cosh t, a_2 \sinh t + b_2 \cosh t, a_3 \sinh t + b_3 \cosh t). \end{aligned} \tag{2.7.9}$$

Equations (2.7.8) can easily be derived as well by an application of Theorem 2.3. Namely, by (2.7.1)

$$L(q_1, q_2, q_3, v_1, v_2, v_3) = \frac{m}{2}\sum_{j=1}^{3} v_j^2 + \frac{m}{2}\sum_{j=1}^{3} q_j^2. \tag{2.7.10}$$

Hence for $k = 1, 2, 3$

$$\begin{aligned} \frac{\partial L}{\partial v_k}(\phi(t), \phi'(t)) &= m\phi_k'(t) \\ \frac{\partial L}{\partial q_k}(\phi(t), \phi'(t)) &= m\phi_k(t). \end{aligned} \tag{2.7.11}$$

Differentiating the first equation in (2.7.11) with respect to t we get indeed $m\ddot{\phi}_k(t) - m\phi_k(t) = 0$ by Theorem 2.3, as before.

We can relate the Hamiltonian and the Lagrangian: By (2.7.1), (2.7.2), at a point $(q^{(1)}, \ldots, q^{(n)}, v^{(1)}, \ldots, v^{(n)}) \in \mathbb{R}^{3n} \times \mathbb{R}^{3n}$

$$\begin{aligned} &\sum_{j=1}^{n} \left[\frac{\partial L}{\partial v_1^{(j)}} v_1^{(j)} + \frac{\partial L}{\partial v_2^{(j)}} v_2^{(j)} + \frac{\partial L}{\partial v_3^{(j)}} v_3^{(j)} \right] - L \\ &= \sum_{j=1}^{n} \left[m_j (v_1^{(j)})^2 + m_j (v_2^{(j)})^2 + m_j (v_3^{(j)})^2 \right] \\ &\quad - \frac{1}{2}\sum_{j=1}^{n} m_j \left[(v_1^{(j)})^2 + (v_2^{(j)})^2 + (v_3^{(j)})^2 \right] + V \\ &= \sum_{j=1}^{n} \left[\frac{(m_j v_1^{(j)})^2}{2m_j} + \frac{(m_j v_2^{(j)})^2}{2m_j} + \frac{(m_j v_3^{(j)})^2}{2m_j} \right] + V \\ &= H(q^{(1)}, \ldots, q^{(n)}, m_1 v^{(1)}, \ldots, m_n v^{(n)}), \end{aligned} \tag{2.7.12}$$

where $m_j v^{(j)} = (m_j v_1^{(j)}, m_j v_2^{(j)}, m_j v_3^{(j)})$. In particular since the potential V does not appear in the final form of (2.7.12) which relates H to L, one can use this formula as motivation for a *definition* of H, given L, in situations (more generally) when V might not even exist. We will encounter such a situation, for example, in Chapter 8, where the motion of a charged particle in an electromagnetic field is studied. Also note that by (2.7.2), (2.7.3) given a Lagrangian L in general we can regard the $\frac{\partial L}{\partial v^{(j)}}$ as "generalized momenta," and the $\frac{\partial L}{\partial q^{(j)}}$ as "generalized forces." Thus, for example, we could *define* the momentum $p^{(j)}$ of some jth particle, for some abstract Lagrangian L, by

$$p^{(j)} \overset{\text{def}}{=} \frac{\partial L}{\partial v^{(j)}} = \left(\frac{\partial L}{\partial v_1^{(j)}}, \frac{\partial L}{\partial v_2^{(j)}}, \frac{\partial L}{\partial v_3^{(j)}} \right). \tag{2.7.13}$$

The reader familiar with the calculus of variations recognizes equations (2.7.7) as *Euler–Lagrange* equations—equations satisfied by a local extremum $\phi(t)$ of a suitable functional S. Namely, given two fixed values of time t_1, t_2, say $t_1 < t_2$, let $C^1[t_1, t_2]$ be the space of continuously differentiable functions on $[t_1, t_2]$. Equip the Cartesian product $(C^1[t_1, t_2])^m$ with the norm

$$||\phi|| \stackrel{\text{def}}{=} \sum_{j=1}^{m} \left(||\phi_j||_\infty + ||\phi_j'||_\infty \right) \quad \text{for } \phi = (\phi_1, \ldots, \phi_m) \tag{2.7.14}$$

where $||\cdot||_\infty$ is the sup norm. Given a fixed continuous function $\mathcal{L} : \mathbb{R} \times \mathbb{R}^m \times \mathbb{R}^m \to \mathbb{R}$ of class C^2 in the variables $(x, y) \in \mathbb{R}^m \times \mathbb{R}^m$ on $[t_1, t_2] \times \mathbb{R}^m \times \mathbb{R}^m$, define a functional

$$S : (C^1[t_1, t_2])^m \to \mathbb{R} \tag{2.7.15}$$

by

$$S(\phi) = \int_{t_1}^{t_2} \mathcal{L}(t, \phi_1(t), \ldots, \phi_m(t), \phi_1'(t), \ldots, \phi_m'(t)) \, dt, \tag{2.7.16}$$

$$\text{which we also write as } \int_{t_1}^{t_2} \mathcal{L}(t, \phi(t), \phi'(t)) \, dt.$$

Fix two points $A = (A_1, \ldots, A_m)$, $B = (B_1, \ldots, B_m) \in \mathbb{R}^m$ and let

$$\Omega_{A,B} = \{\phi \in (C^1[t_1, t_2])^m |\ \phi_j(t_1) = A_j, \phi_j(t_2) = B_j \text{ for } 1 \le j \le m\}. \tag{2.7.17}$$

Also let $(t, x_1, \ldots, x_m, y_1, \ldots, y_m)$ denote the typical point in $\mathbb{R} \times \mathbb{R}^m \times \mathbb{R}^m$. Given the above assumptions on $\mathcal{L}$, one can show that

$$t \to \frac{\partial \mathcal{L}}{\partial y_j}(t, \phi(t), \phi'(t))$$

is automatically continuously differentiable on $[t_1, t_2]$, for $\phi \in (C^1[t_1, t_2])^m$, and that if S has a local extremum (maximum or minimum) at a point $\phi \in \Omega_{A,B}$, then ϕ satisfies the Euler–Lagrange equations

$$\frac{d}{dt} \frac{\partial \mathcal{L}}{\partial y_j}(t, \phi(t), \phi'(t)) = \frac{\partial \mathcal{L}}{\partial x_j}(t, \phi(t), \phi'(t)), \quad 1 \le j \le m \tag{2.7.18}$$

on $[t_1, t_2]$.

As an example, for the system of n particles moving in $\mathbb{R}^3$ as above, choose $m = 3n$ and define $\mathcal{L} : \mathbb{R} \times \mathbb{R}^{3n} \times \mathbb{R}^{3n} \to \mathbb{R}$ by

$$\begin{aligned} &\mathcal{L}(t, q^{(1)}, \ldots, q^{(n)}, v^{(1)}, \ldots, v^{(n)}) \\ &= L(q^{(1)}, \ldots, q^{(n)}, v^{(1)}, \ldots, v^{(n)}) \\ &= T(v^{(1)}, \ldots, v^{(n)}) - V(q^{(1)}, \ldots, q^{(n)}), \end{aligned} \tag{2.7.19}$$

as in (2.7.1). The corresponding *action functional* or *action integral* S in (2.7.16) is given by

$$S(\phi) = \int_{t_1}^{t_2} [T(\phi^{(1)'}(t), \ldots, \phi^{(n)'}(t)) - V(\phi^{(1)}(t), \ldots, \phi^{(n)}(t))] \, dt \tag{2.7.20}$$

with $\phi^{(j)}(t) = (\phi_1^{(j)}(t), \phi_2^{(j)}(t), \phi_3^{(j)}(t))$. In this case the Euler–Lagrange equations (2.7.18) coincide with the Lagrange equations (2.7.7), and we have at hand *Hamilton's principle*: Among possible paths $\phi(t) \in (C^1[t_1,t_2])^{3n}$ subject to $\phi_k^{(j)}(t_1) = A_k^{(j)}, \phi_k^{(j)}(t_2) = B_k^{(j)}, 1 \le k \le 3, 1 \le j \le n$, where $A = (A^{(1)}, \dots, A^{(n)})$, $B = (B^{(1)}, \dots, B^{(n)}) \in \mathbb{R}^{3n}$ are fixed, over a fixed interval of time $t_1 \le t \le t_2$ the actual motion of the system is along a path $\phi(t)$ which renders a *stationary value* of the action integral S in (2.7.20); i.e., the motion is described by a path $\phi(t)$ which is a local maximum or minimum of S.

2.8 Center of Mass Coordinates

Consider again $n = 2$ particles in $\mathbb{R}^3$ with masses m_1, m_2. For a point $(x_1, y_1, z_1, x_2, y_2, z_2) \in \mathbb{R}^6$ define (R_1, R_2, R_3) by

$$\begin{aligned} R_1 &= x_1 - \frac{m_2}{m_1+m_2}(x_1 - x_2) \\ R_2 &= y_1 - \frac{m_2}{m_1+m_2}(y_1 - y_2) \\ R_3 &= z_1 - \frac{m_2}{m_1+m_2}(z_1 - z_2). \end{aligned} \tag{2.8.1}$$

That is, for $R = (R_1, R_2, R_3)$ we have

$$\begin{aligned} &(m_1+m_2)R \\ &= (m_1+m_2)\left[(x_1,y_1,z_1) - \frac{m_2}{m_1+m_2}((x_1,y_1,z_1) - (x_2,y_2,z_2))\right] \\ &= m_1(x_1,y_1,z_1) + m_2(x_2,y_2,z_2); \end{aligned} \tag{2.8.2}$$

$$R_1 = \frac{m_1x_1 + m_2x_2}{m_1+m_2}, \quad R_2 = \frac{m_1y_1 + m_2y_2}{m_1+m_2}, \quad R_3 = \frac{m_1z_1 + m_2z_2}{m_1+m_2}. \tag{2.8.3}$$

We obtain a map $\Phi : \mathbb{R}^6 \to \mathbb{R}^6$ by setting $\Phi(x_1, y_1, z_1, x_2, y_2, z_2) = (x_1 - x_2, y_1 - y_2, z_1 - z_2, R_1, R_2, R_3)$. Write $\Phi = (g_1, g_2, g_3, g_4, g_5, g_6)$ where $g_j : \mathbb{R}^6 \to \mathbb{R}$ are C^∞ maps. Thus $g_1(x_1, y_1, z_1, x_2, y_2, z_2) = x_1 - x_2 \implies$

$$\begin{aligned} &\frac{\partial g_1}{\partial x_1} = 1, \quad \frac{\partial g_1}{\partial x_2} = -1, \quad \text{with other first partials of } g_1 = 0, \\ &\frac{\partial g_2}{\partial y_1} = 1, \quad \frac{\partial g_2}{\partial y_2} = -1, \quad \text{with other first partials of } g_2 = 0, \\ &\frac{\partial g_3}{\partial z_1} = 1, \quad \frac{\partial g_3}{\partial z_2} = -1, \quad \text{with other first partials of } g_3 = 0, \end{aligned} \tag{2.8.4}$$

and

$$\frac{\partial g_4}{\partial x_1} = 1 - \frac{m_2}{m_1+m_2} = \frac{m_1}{m_1+m_2}, \tag{2.8.5}$$

$$\frac{\partial g_4}{\partial x_2} = \frac{m_2}{m_1+m_2}, \qquad \text{with other first partials of } g_4 = 0,$$
$$\frac{\partial g_5}{\partial y_1} = 1 - \frac{m_2}{m_1+m_2} = \frac{m_1}{m_1+m_2},$$
$$\frac{\partial g_5}{\partial y_2} = \frac{m_2}{m_1+m_2}, \qquad \text{with other first partials of } g_5 = 0,$$
$$\frac{\partial g_6}{\partial z_1} = 1 - \frac{m_2}{m_1+m_2} = \frac{m_1}{m_1+m_2},$$
$$\frac{\partial g_6}{\partial z_2} = \frac{m_2}{m_1+m_2}, \qquad \text{with other first partials of } g_6 = 0.$$

Let $f : \mathbb{R}^6 \to \mathbb{R}$ have continuous first and second partial derivatives. Define $F : \mathbb{R}^6 \to \mathbb{R}$ by $F(a) = f(g_1(a), \ldots, g_6(a)) = f(\Phi(a))$, $a \in \mathbb{R}^6$. By the chain rule

$$(D_i F)(a) = \sum_{j=1}^{6} (D_j f)(\Phi(a))(D_i g_j)(a) \text{ for } 1 \le i \le 6. \tag{2.8.6}$$

Using the preceding formulas for the partial derivatives $D_i g_j$ we get

$$\begin{aligned}
\frac{\partial F}{\partial x_1}(a) &= \frac{\partial f}{\partial x_1}(\Phi(a)) + \frac{\partial f}{\partial x_2}(\Phi(a))\frac{m_1}{m_1+m_2} \\
\frac{\partial F}{\partial y_1}(a) &= \frac{\partial f}{\partial y_1}(\Phi(a)) + \frac{\partial f}{\partial y_2}(\Phi(a))\frac{m_1}{m_1+m_2} \\
\frac{\partial F}{\partial z_1}(a) &= \frac{\partial f}{\partial z_1}(\Phi(a)) + \frac{\partial f}{\partial z_2}(\Phi(a))\frac{m_1}{m_1+m_2} \\
\frac{\partial F}{\partial x_2}(a) &= -\frac{\partial f}{\partial x_1}(\Phi(a)) + \frac{\partial f}{\partial x_2}(\Phi(a))\frac{m_2}{m_1+m_2} \\
\frac{\partial F}{\partial y_2}(a) &= -\frac{\partial f}{\partial y_1}(\Phi(a)) + \frac{\partial f}{\partial y_2}(\Phi(a))\frac{m_2}{m_1+m_2} \\
\frac{\partial F}{\partial z_2}(a) &= -\frac{\partial f}{\partial z_1}(\Phi(a)) + \frac{\partial f}{\partial z_2}(\Phi(a))\frac{m_2}{m_1+m_2}.
\end{aligned} \tag{2.8.7}$$

Replace f by $\frac{\partial f}{\partial x_1}$, $\frac{\partial f}{\partial x_2}$ (since f has continuous second partial derivatives). That is, let $F_1 = \frac{\partial f}{\partial x_1} \circ \Phi$, $F_4 = \frac{\partial f}{\partial x_2} \circ \Phi$ to get by (2.8.7)

$$\begin{aligned}
\frac{\partial F_1}{\partial x_1}(a) &= \frac{\partial^2 f}{\partial x_1^2}(\Phi(a)) + \frac{\partial^2 f}{\partial x_2 \partial x_1}(\Phi(a))\frac{m_1}{m_1+m_2}, \\
\frac{\partial F_4}{\partial x_1}(a) &= \frac{\partial^2 f}{\partial x_1 \partial x_2}(\Phi(a)) + \frac{\partial^2 f}{\partial x_2^2}(\Phi(a))\frac{m_1}{m_1+m_2}, \\
\Longrightarrow \quad \frac{\partial^2 F}{\partial x_1^2}(a) &= \frac{\partial^2 f}{\partial x_1^2}(\Phi(a)) + \frac{\partial^2 f}{\partial x_2 \partial x_1}(\Phi(a))\frac{m_1}{m_1+m_2} \\
&\quad + \left[\frac{\partial^2 f}{\partial x_1 \partial x_2}(\Phi(a)) + \frac{\partial^2 f}{\partial x_2^2}(\Phi(a))\frac{m_1}{m_1+m_2}\right]\frac{m_1}{m_1+m_2}.
\end{aligned} \tag{2.8.8}$$

Similarly let $F_2 = \frac{\partial f}{\partial y_1} \circ \Phi$, $F_5 = \frac{\partial f}{\partial y_2} \circ \Phi$:

$$\begin{aligned}
\frac{\partial F_2}{\partial y_1}(a) &= \frac{\partial^2 f}{\partial y_1^2}(\Phi(a)) + \frac{\partial^2 f}{\partial y_2 \partial y_1}(\Phi(a))\frac{m_1}{m_1+m_2}, \\
\frac{\partial F_5}{\partial y_1}(a) &= \frac{\partial^2 f}{\partial y_1 \partial y_2}(\Phi(a)) + \frac{\partial^2 f}{\partial y_2^2}(\Phi(a))\frac{m_1}{m_1+m_2}, \\
\Longrightarrow \quad \frac{\partial^2 F}{\partial y_1^2}(a) &= \frac{\partial^2 f}{\partial y_1^2}(\Phi(a)) + \frac{\partial^2 f}{\partial y_2 \partial y_1}(\Phi(a))\frac{m_1}{m_1+m_2} \\
&\quad + \left[\frac{\partial^2 f}{\partial y_1 \partial y_2}(\Phi(a)) + \frac{\partial^2 f}{\partial y_2^2}(\Phi(a))\frac{m_1}{m_1+m_2}\right]\frac{m_1}{m_1+m_2}.
\end{aligned} \tag{2.8.9}$$

Also let $F_3 = \frac{\partial f}{\partial z_1} \circ \Phi$, $F_6 = \frac{\partial f}{\partial z_2} \circ \Phi$:

$$\begin{aligned}
\frac{\partial F_3}{\partial z_1}(a) &= \frac{\partial^2 f}{\partial z_1^2}(\Phi(a)) + \frac{\partial^2 f}{\partial z_2 \partial z_1}(\Phi(a))\frac{m_1}{m_1+m_2}, \\
\frac{\partial F_6}{\partial z_1}(a) &= \frac{\partial^2 f}{\partial z_1 \partial z_2}(\Phi(a)) + \frac{\partial^2 f}{\partial z_2^2}(\Phi(a))\frac{m_1}{m_1+m_2}, \\
\Longrightarrow \quad \frac{\partial^2 F}{\partial z_1^2}(a) &= \frac{\partial^2 f}{\partial z_1^2}(\Phi(a)) + \frac{\partial^2 f}{\partial z_2 \partial z_1}(\Phi(a))\frac{m_1}{m_1+m_2} \\
&\quad + \left[\frac{\partial^2 f}{\partial z_1 \partial z_2}(\Phi(a)) + \frac{\partial^2 f}{\partial z_2^2}(\Phi(a))\frac{m_1}{m_1+m_2}\right]\frac{m_1}{m_1+m_2}.
\end{aligned} \tag{2.8.10}$$

As $\frac{\partial F}{\partial x_2} = -F_1 + F_4 \frac{m_2}{m_1+m_2}$ with

$$\frac{\partial F_1}{\partial x_2} = -\frac{\partial^2 f}{\partial x_1^2} \circ \Phi + \frac{\partial^2 f}{\partial x_2 \partial x_1} \circ \Phi \frac{m_2}{m_1+m_2},$$

we get

$$\frac{\partial^2 F}{\partial x_2^2}(a) = \frac{\partial^2 f}{\partial x_1^2}(\Phi(a)) - \frac{\partial^2 f}{\partial x_2 \partial x_1}(\Phi(a))\frac{m_2}{m_1+m_2} + \frac{\partial F_4}{\partial x_2}(a)\frac{m_2}{m_1+m_2}, \tag{2.8.11}$$

where

$$\frac{\partial F_4}{\partial x_2}(a) = -\frac{\partial^2 f}{\partial x_1 \partial x_2}(\Phi(a)) + \frac{\partial^2 f}{\partial x_2^2}(\Phi(a))\frac{m_2}{m_1+m_2}. \tag{2.8.12}$$

That is,

$$\begin{aligned}
\frac{\partial^2 F}{\partial x_2^2}(a) &= \frac{\partial^2 f}{\partial x_1^2}(\Phi(a)) - \frac{\partial^2 f}{\partial x_2 \partial x_1}(\Phi(a))\frac{m_2}{m_1+m_2} \\
&\quad + \left[-\frac{\partial^2 f}{\partial x_1 \partial x_2}(\Phi(a)) + \frac{\partial^2 f}{\partial x_2^2}(\Phi(a))\frac{m_2}{m_1+m_2}\right]\frac{m_2}{m_1+m_2}.
\end{aligned} \tag{2.8.13}$$

By (2.8.7)

$$\frac{\partial F_2}{\partial y_2}(a) = -\frac{\partial^2 f}{\partial y_1^2}(\Phi(a)) + \frac{\partial^2 f}{\partial y_2 \partial y_1}(\Phi(a))\frac{m_2}{m_1+m_2},$$
$$\frac{\partial^2 F_5}{\partial y_2^2}(a) = -\frac{\partial^2 f}{\partial y_1 \partial y_2}(\Phi(a)) + \frac{\partial^2 f}{\partial y_2^2}(\Phi(a))\frac{m_2}{m_1+m_2}$$
$$\Longrightarrow \frac{\partial^2 F}{\partial y_2^2}(a) = \frac{\partial^2 f}{\partial y_1^2}(\Phi(a)) - \frac{\partial^2 f}{\partial y_2 \partial y_1}(\Phi(a))\frac{m_2}{m_1+m_2}$$
$$+\left[-\frac{\partial^2 f}{\partial y_1 \partial y_2}(\Phi(a)) + \frac{\partial^2 f}{\partial y_2^2}(\Phi(a))\frac{m_2}{m_1+m_2}\right]\frac{m_2}{m_1+m_2}. \tag{2.8.14}$$

Finally,

$$\frac{\partial F_3}{\partial z_2}(a) = -\frac{\partial^2 f}{\partial z_1^2}(\Phi(a)) + \frac{\partial^2 f}{\partial z_2 \partial z_1}(\Phi(a))\frac{m_2}{m_1+m_2},$$
$$\frac{\partial F_6}{\partial z_2}(a) = -\frac{\partial^2 f}{\partial z_1 \partial z_2}(\Phi(a)) + \frac{\partial^2 f}{\partial z_2^2}(\Phi(a))\frac{m_2}{m_1+m_2}$$
$$\Longrightarrow \frac{\partial^2 F}{\partial z_2^2}(a) = \frac{\partial^2 f}{\partial z_1^2}(\Phi(a)) - \frac{\partial^2 f}{\partial z_2 \partial z_1}(\Phi(a))\frac{m_2}{m_1+m_2}$$
$$+\left[-\frac{\partial^2 f}{\partial z_1 \partial z_2}(\Phi(a)) + \frac{\partial^2 f}{\partial z_2^2}(\Phi(a))\frac{m_2}{m_1+m_2}\right]\frac{m_2}{m_1+m_2}. \tag{2.8.15}$$

Therefore,

$$\frac{1}{m_1}\left[\frac{\partial^2 F}{\partial x_1^2}(a) + \frac{\partial^2 F}{\partial y_1^2}(a) + \frac{\partial^2 F}{\partial z_1^2}(a)\right] + \frac{1}{m_2}\left[\frac{\partial^2 F}{\partial x_2^2}(a) + \frac{\partial^2 F}{\partial y_2^2}(a) + \frac{\partial^2 F}{\partial z_2^2}(a)\right]$$
$$= \left(\frac{1}{m_1} + \frac{1}{m_2}\right)\left[\frac{\partial^2 f}{\partial x_1^2}(\Phi(a)) + \frac{\partial^2 f}{\partial y_1^2}(\Phi(a)) + \frac{\partial^2 f}{\partial z_1^2}(\Phi(a))\right]$$
$$+ \frac{1}{m_1+m_2}\left[\frac{\partial^2 f}{\partial x_2^2}(\Phi(a)) + \frac{\partial^2 f}{\partial y_2^2}(\Phi(a)) + \frac{\partial^2 f}{\partial z_2^2}(\Phi(a))\right]. \tag{2.8.16}$$

We have assumed $f : \mathbb{R}^6 \to \mathbb{R}$, but we could have taken the domain of f to be any open subset D of $\mathbb{R}^6$.

Proposition 2.1. *More generally suppose $W, D \subset \mathbb{R}^6$ are open sets and $\Phi : W \to D$, where Φ is defined following (2.8.3). Let $f : D \to \mathbb{R}$ have continuous* *first and second partial differentials. Define F on W by $F(a) = f(\Phi(a))$ for $a \in W$. Then formula* (2.8.16) *holds for $a \in W$:*

$$\frac{1}{m_1}\left[\frac{\partial^2 F}{\partial x_1^2}(a)+\frac{\partial^2 F}{\partial y_1^2}(a)+\frac{\partial^2 F}{\partial z_1^2}(a)\right]+\frac{1}{m_2}\left[\frac{\partial^2 F}{\partial x_2^2}(a)+\frac{\partial^2 F}{\partial y_2^2}(a)+\frac{\partial^2 F}{\partial z_2^2}(a)\right]$$
$$=\left(\frac{1}{m_1}+\frac{1}{m_2}\right)\left[\frac{\partial^2 f}{\partial x_1^2}(\Phi(a))+\frac{\partial^2 f}{\partial y_1^2}(\Phi(a))+\frac{\partial^2 f}{\partial z_1^2}(\Phi(a))\right]$$
$$+\frac{1}{m_1+m_2}\left[\frac{\partial^2 f}{\partial x_2^2}(\Phi(a))+\frac{\partial^2 f}{\partial y_2^2}(\Phi(a))+\frac{\partial^2 f}{\partial z_2^2}(\Phi(a))\right]. \tag{2.8.17}$$

Proposition 2.2. $\Phi : \mathbb{R}^6 \to \mathbb{R}^6$ *is onto and one-to-one.*

Proof. If $a = (a_1, a_2, a_3, a_4, a_5, a_6) \in \mathbb{R}^6$ is given, set

$$\begin{aligned} x_1 &= \frac{m_2}{m_1+m_2}a_1 + a_4, & x_2 &= x_1 - a_1, \\ y_1 &= \frac{m_2}{m_1+m_2}a_2 + a_5, & y_2 &= y_1 - a_2, \\ z_1 &= \frac{m_2}{m_1+m_2}a_3 + a_6, & z_2 &= z_1 - a_3. \end{aligned} \tag{2.8.18}$$

Then $(x_1, y_1, z_1, x_2, y_2, z_2) \in \mathbb{R}^6$ such that $\Phi(x_1, y_1, z_1, x_2, y_2, z_2) = a$. If $\Phi(x) = \Phi(x')$, $x, x' \in \mathbb{R}^6$, then

$$\begin{aligned} x_1 - x_2 &= x_1' - x_2', & y_1 - y_2 &= y_1' - y_2', & z_1 - z_2 &= z_1' - z_2', \\ R_1 &= R_1', & R_2 &= R_2', & \text{and } R_3 &= R_3'; \end{aligned} \tag{2.8.19}$$

i.e., $R_1 = R_1' \implies$

$$x_1 - \frac{m_2}{m_1+m_2}(x_1 - x_2) = x_1' - \frac{m_2}{m_1+m_2}(x_1' - x_2') \tag{2.8.20}$$
$$\implies x_1 = x_1' \quad \text{since } x_1 - x_2 = x_1' - x_2'.$$

Similarly $R_2 = R_2'$ and $y_1 - y_2 = y_1' - y_2' \implies y_1 = y_1'$, $R_3 = R_3'$ and $z_1 - z_2 = z_1' - z_2' \implies z_1 = z_1'$. Then $x_2 = x_2'$, $y_2 = y_2'$, $z_2 = z_2'$ by (2.8.19) $\implies x = x'$.

Corollary 2.1. $\Phi : \mathbb{R}^6 \to \mathbb{R}^6$ *is a diffeomorphism. For $a = (a_1, a_2, a_3, a_4, a_5, a_6) \in \mathbb{R}^6$, $\Phi^{-1}(a) = (x_1, y_1, z_1, x_2, y_2, z_2)$ with*

$$\begin{aligned} x_1 &= \frac{m_2}{m_1+m_2}a_1 + a_4, & x_2 &= x_1 - a_1, \\ y_1 &= \frac{m_2}{m_1+m_2}a_2 + a_5, & y_2 &= y_1 - a_2, \\ z_1 &= \frac{m_2}{m_1+m_2}a_3 + a_6, & z_2 &= z_1 - a_3. \end{aligned} \tag{2.8.21}$$

□

The passage from the coordinates $(x_1, y_1, z_1, x_2, y_2, z_2)$ to the coordinates $\Phi(x_1, y_1, z_1, x_2, y_2, z_2)$, called *Jacobi* or *center of mass* coordinates, is very useful in dealing with Coulomb potentials—as we shall see in the study of the hydrogen atom, for example. Formula (2.8.17) will permit the transformation of Schrödinger's differential operator.

One can formulate classical mechanics, as is well known, in the general framework of a cotangent bundle over a smooth manifold. Some readers may wish to pursue this elegant point of view, which we have not attempted to touch on here as it will not be needed.

3

Quantization and the Schrödinger Equation

The basic equation of quantum mechanics is the Schrödinger equation which expresses the wave function ψ of a quantum system as an eigenfunction of a quantized Hamiltonian operator H: $H\psi = \lambda\psi$ where the (real) eigenvalue λ is the quantum *energy* of the system in the state ψ; see equations (1.3.2), (1.3.4). Embodied already in this equation is the basic quantum mechanical principle that quantum energies cannot take on arbitrary values but are *quantized*: they are given by a discrete set of eigenvalues of a suitable second-order differential operator. This mathematical phenomenon of the discreteness of eigenvalues explains, for example, the observed discreteness of absorption and emission atomic spectral lines; compare remarks in Sections 1.2 and 1.3 of Chapter 1. The Schrödinger theory, and the equivalent theory of Heisenberg, Born and Jordan, represents a distinct advancement of the Bohr theory. Some early basic papers on quantum mechanics are compiled in the book [82], which includes a historic introduction by B. van der Waerden. Also see [6, 8, 9, 24, 75, 76].

As indicated earlier, important physical information (the probability that a particle assumes habitation of a particular region of space, for example) is carried by the norm squared $\|\psi\|^2$ of ψ, rather than by ψ itself. The operator H and the wave function ψ therefore have life, more precisely, on a suitable Hilbert space. We shall elaborate a bit more on the remarks briefly made in Chapter 1 concerning quantization and Schrödinger's equation as we explore various examples.

3.1 The One-Dimensional Schrödinger Equation

For "quantum" or "wave mechanics" considerations of a particle of mass m moving in one dimension we proceed as follows, where for simplicity or for the sake of definiteness we consider a type of idealized situation. Namely, suppose there is assigned to the particle a "wave" function $\psi : \mathbb{R}^2 \to \mathbb{C}$ on $\mathbb{R}^2 = \{(x,t)|x,t \in \mathbb{R}\}$. Assume, for example,

(i) $t \to \psi(x,t)$ is continuously differentiable on $\mathbb{R}$ for $x \in \mathbb{R}$ fixed,

(ii) $x \to \psi(x,t)$ has a continuous second derivative on $\mathbb{R}$ for $t \in \mathbb{R}$ fixed, and

(iii) $x \to \psi(x,t)$ (which is continuous on $\mathbb{R}$, for $t \in \mathbb{R}$ fixed) is in $L^2(\mathbb{R}, dx)$ where dx also denotes Lebesgue measure on $\mathbb{R}$: $\int_{\mathbb{R}} |\psi(x,t)|^2 \, dx < \infty$.

Thus, here, there corresponds to time t a vector ψ_t in a Hilbert space, in contrast to the classical mechanics situation where a real number $\phi(t)$ corresponds to t, $x = \phi(t)$ being the equation of motion: $\psi_t \in L^2(\mathbb{R}, dx)$.

Let V be a potential energy function for the particle, say V for example is a continuous function on $\mathbb{R}$. ψ is required to satisfy *Schrödinger's equation*:

$$i\hbar \frac{\partial \psi}{\partial t} = -\frac{\hbar^2}{2m}\frac{\partial^2 \psi}{\partial x^2} + V\psi \tag{3.1.1}$$

where $\hbar = \frac{h}{2\pi}$ for h =Planck's constant. As shown in Section 1.3 of Chapter 1, one obtains the right-hand side of this equation by a heuristic *quantization* procedure.

Let H denote the *quantized* Hamiltonian operator

$$-\frac{\hbar^2}{2m}\frac{\partial^2}{\partial x^2} + \text{multiplication of } V. \tag{3.1.2}$$

Schrödinger's equation may be written, alternatively, as

$$i\hbar \frac{\partial \psi}{\partial t} = H\psi \quad \text{or} \quad \frac{\hbar}{i}\frac{\partial \psi}{\partial t} + H\psi = 0 \tag{3.1.3}$$

as noted in (1.3.9). We also noted in Chapter 1 that a solution $\psi(x,t)$ of (3.1.3) is obtained immediately if we know a solution $\psi(x)$ of the *time-independent* Schrödinger equation

$$\frac{\hbar^2}{2m}\frac{d^2\psi}{dx^2} + (E - V(x))\psi(x) = 0, \quad \text{or } H\psi = E\psi. \tag{3.1.4}$$

Namely, one simply sets

$$\psi(x,t) \stackrel{\text{def}}{=} \psi(x)\exp(Et/i\hbar). \tag{3.1.5}$$

For $\frac{d^2\psi}{dx^2}$ continuous on $\mathbb{R}$ with $\psi \in L^2(\mathbb{R})$, the latter equation defines a function $\psi : \mathbb{R}^2 \to \mathbb{C}$ which does satisfy conditions (i), (ii), (iii) above.

3.2 The Quantized Harmonic Oscillator

Take the harmonic oscillator for example with equation of motion

$$x = \phi(t) = C\cos(\omega t - \delta), \tag{3.2.1}$$

where C is the amplitude of the oscillation, $\nu = \omega/2\pi$ is its frequency and δ is some phase constant; see (2.2.17). The potential energy V was shown to be given in terms of the constant angular speed ω:

$$V(x) = \frac{\omega^2 m x^2}{2} = 2\pi^2\nu^2 m x^2. \tag{3.2.2}$$

The time independent Schrödinger equation for the simple harmonic oscillator (equation (3.1.4)) therefore is

$$\frac{\hbar^2}{2m}\frac{d^2\psi}{dx^2} + \left(E - \frac{\omega^2 m x^2}{2}\right)\psi = 0. \tag{3.2.3}$$

This is a differential equation of *hypergeometric type*, a type which we will study in Chapter 4. It will turn out that the only possible eigenvalues E (corresponding to a non-zero solution $\psi \in L^2(\mathbb{R})$) are $E = E_n \overset{\text{def}}{=} 2\pi(n + 1/2)\nu\hbar$, $n = 0, 1, 2, 3, \ldots$. The energy levels of the quantum oscillator therefore cannot be arbitrary but are given by the discrete set of real numbers $\{E_n\}_{n=0}^{\infty}$. This compares with initial remarks made above.

Although the potential energy $V : x \to \omega^2 m x^2/2$ is continuously differentiable in this example (it is of course even a C^∞ function), in many practical situations V will have a finite number of discontinuities, or even an infinite number of such. V in the next example will be discontinuous at two points.

3.3 A Finite Potential Well

As an example of a discontinuous potential, fix a, $V_o > 0$ and define

$$V(x) = \begin{cases} -V_o & \text{for } -a \le x \le a \\ 0 & \text{for } |x| > a. \end{cases} \tag{3.3.1}$$

V is continuous except at $x = \pm a$:

The graph of V (Figure 3.1) defines a "well" of length $2a$ of finite depth V_o. We consider the bound states of a particle in this well, i.e., $E < 0$. Roughly we have a one-dimensional model of a neutron in a nucleus or, for example, a model of the potential experienced by an electron in a molecule like acetylene C_2H_2: $H - C \equiv C - H$ with a triple bond between the two carbon atoms.

Though we take $E < 0$ we assume $E > -V_o$. On the open set $x > a$, $V = 0$ and Schrödinger's equation (3.1.4) becomes

$$\frac{d^2\psi}{dx^2} = \alpha^2\psi \quad \text{for } \alpha \overset{\text{def}}{=} +\frac{\sqrt{2m(-E)}}{\hbar}. \tag{3.3.2}$$

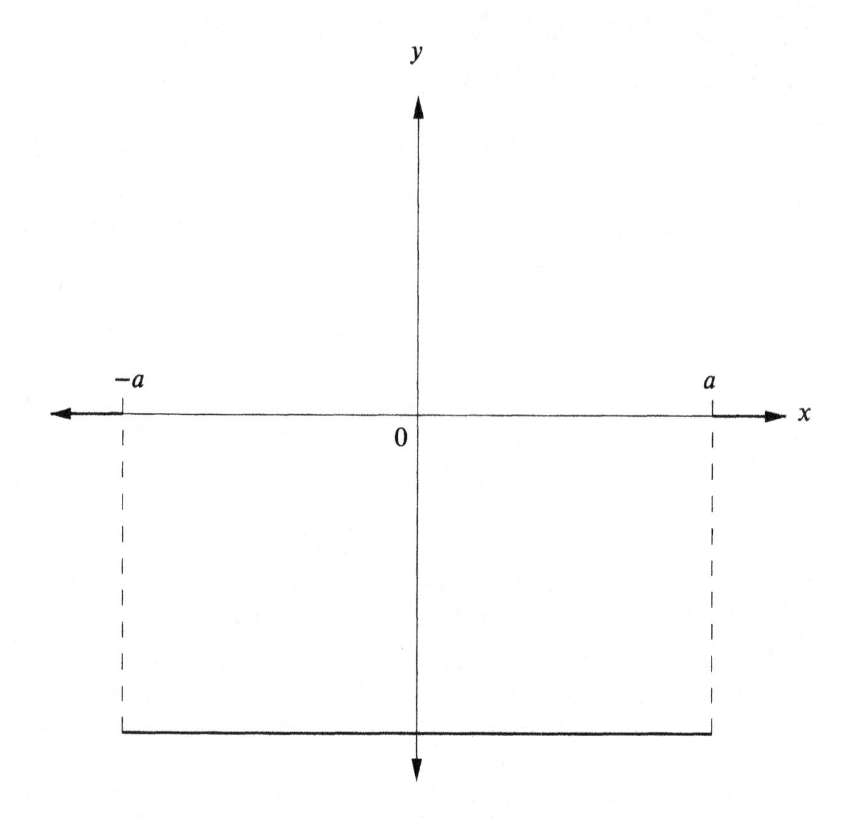

Figure 3.1: Finite potential well

The general solution of equation (3.3.2) is $\psi(x) = A\exp(\alpha x) + B\exp(-\alpha x)$ on $x > a$. Since $\exp(\alpha x) \to \infty$ as $x \to \infty$ (for $\alpha > 0$) we choose $A = 0$; that is, our interest is in square-integrable solutions. Similarly on $x < a$, $V = 0$ and the general solution of the Schrödinger equation is $\psi(x) = C\exp(\alpha x) + D\exp(-\alpha x)$. Since $\exp(-\alpha x) \to \infty$ as $x \to -\infty$ we choose $D = 0$:

$$\psi(x) = \begin{cases} Be^{-\alpha x} & \text{for } x > a \\ Ce^{\alpha x} & \text{for } x < -a. \end{cases} \tag{3.3.3}$$

Let $\beta \stackrel{\text{def}}{=} +\sqrt{2m(E + V_o)}/\hbar$; we have assumed $E + V_o > 0$. On $[-a, a]$, $V = -V_o$, and Schrödinger's equation $\psi''(x) = -\beta^2\psi(x)$ has general solution

$$\psi(x) = \widetilde{E}\cos(\beta x) + F\sin(\beta x) \quad \text{on } [-a, a]. \tag{3.3.4}$$

To make further progress we impose the boundary conditions of continuity of ψ, ψ' at the discontinuities $x = \pm a$ of V:

$$\psi'(x) = \begin{cases} -\alpha B e^{-\alpha x} & \text{on } x > a, \\ \alpha C e^{\alpha x} & \text{on } x < -a, \\ -\beta \widetilde{E} \sin(\beta x) + \beta F \cos(\beta x) & \text{on } (-a, a), \end{cases} \tag{3.3.5}$$

$$\begin{aligned} \implies B e^{-\alpha a} &= \widetilde{E} \cos(\beta a) + F \sin(\beta a), \\ C e^{-\alpha a} &= \widetilde{E} \cos(\beta a) - F \sin(\beta a), \\ -\alpha B e^{-\alpha a} &= -\beta \widetilde{E} \sin(\beta a) + \beta F \cos(\beta a), \\ \alpha C e^{-\alpha a} &= \beta \widetilde{E} \sin(\beta a) + \beta F \cos(\beta a). \end{aligned} \tag{3.3.6}$$

These equations give

$$(B + C) e^{-\alpha a} = 2\widetilde{E} \cos(\beta a), \tag{3.3.7}$$

$$\alpha(C - B) e^{-\alpha a} = 2\beta F \cos(\beta a), \tag{3.3.8}$$

$$(B - C) e^{-\alpha a} = 2F \sin(\beta a), \tag{3.3.9}$$

$$\alpha(B + C) e^{-\alpha a} = 2\beta \widetilde{E} \sin(\beta a). \tag{3.3.10}$$

Note that if both $B + C = 0$ and $B - C = 0$ then of course B and $C = 0$. But also by equations (3.3.7) and (3.3.10), $2\widetilde{E}\cos(\beta a) = 0$ and $2\beta\widetilde{E}\sin(\beta a) = 0$ $\implies 2\widetilde{E}\sin(\beta a) = 0 \implies 4\widetilde{E}^2 = 0 \implies \widetilde{E} = 0$. Similarly $B - C = 0$ $\implies F = 0$. That is, both $B + C,\ B - C = 0 \implies B, C, \widetilde{E}, F = 0 \implies \psi = 0$. Thus assume $\psi \neq 0$. Then either $B + C \neq 0$ or $B - C \neq 0$. Note also that if $\psi \neq 0$, then both $\cos(\beta a) \neq 0$ and $\sin(\beta a) \neq 0$. For if $\cos(\beta a) = 0$, then $B + C = 0$ and $B - C = 0$ by equations (3.3.7) and (3.3.8). Similarly if $\sin(\beta a) = 0$, $B + C = 0$ and $B - C = 0$ by equations (3.3.9) and (3.3.10). Now if $B + C \neq 0$ then $\widetilde{E} \neq 0$ by equation (3.3.7) and we have (dividing the fourth equation by the first) $\alpha = \beta\tan(\beta a)$. Similarly if $B - C \neq 0$, then $F \neq 0$ and $-\alpha = \beta\cot(\beta a)$. Since E is determined by α, β we have proved the following quantization condition.

Proposition 3.1. *For bound states $\psi \neq 0$ given by* (3.3.3), (3.3.4), (3.3.5) *with $-V_o < E < 0$ we have for*

$$\begin{aligned} \alpha &\stackrel{\text{def}}{=} +\frac{\sqrt{2m(-E)}}{\hbar} \in \mathbb{R}, \\ \beta &\stackrel{\text{def}}{=} +\frac{\sqrt{2m(E + V_o)}}{\hbar} \in \mathbb{R} \end{aligned} \tag{3.3.11}$$

that $\sin(\beta a) \neq 0$ and $\cos(\beta a) \neq 0$, and the following condition on the quantum energies E. Either $B + C \neq 0$ or $B - C \neq 0$. If $B + C \neq 0$, then E satisfies $\alpha = \beta\tan(\beta a)$. If $B - C \neq 0$, then E satisfies $-\alpha = \beta\cot(\beta a)$.

Note that in fact we cannot have both cases $B + C \neq 0$ and $B - C \neq 0$ at the same time. For this would give $\alpha = \beta\tan(\beta a)$ and $\alpha = -\beta\cot(\beta a)$. Hence $\alpha^2 = -\beta^2 \implies$ both $\alpha, \beta = 0$ which is a contradiction. Therefore if $B + C \neq 0$, then $B - C = 0$, and hence also $F = 0$ by (3.3.8) or (3.3.9) since $\sin(\beta a) \neq 0$ and $\cos(\beta a) \neq 0$. By equations (3.3.3) and (3.3.4) therefore

$$\psi(x) = \begin{bmatrix} Be^{-\alpha x} & \text{for} & x > a \\ Be^{\alpha x} & \text{for} & x < -a \\ \widetilde{E}\cos(\beta a) & \text{for} & -a \leq x \leq a \end{bmatrix} \text{ for } B + C \neq 0, \tag{3.3.12}$$

and hence ψ is even: $\psi(x) = \psi(-x)\ \forall\ x \in \mathbb{R}$.

Similarly if $B - C \neq 0$, then $B + C = 0$ and $\widetilde{E} = 0 \implies$ (by equations (3.3.3), (3.3.4))

$$\psi(x) = \begin{bmatrix} Be^{-\alpha x} & \text{for} & x > a \\ -Be^{\alpha x} & \text{for} & x < -a \\ F\sin(\beta a) & \text{for} & -a \leq x \leq a \end{bmatrix}; \tag{3.3.13}$$

hence $\psi(-x) = -\psi(x)\ \forall\ x \in \mathbb{R}$ if $B - C \neq 0$. We see that the non-zero wave functions ψ are either *even* or *odd* functions, under the condition $-V_o < E < 0$.

Since $(a\alpha)^2 + (a\beta)^2 = 2mV_o a^2/\hbar^2$ represents a circle about the origin with radius $\gamma = \sqrt{2mV_o}\frac{a}{\hbar}$, we can obtain the energy levels by the intersections of the circle with the curves $y = x\tan(x)$, $y = -x\cot(x)$, illustrated, in part, in Figures 3.2 and 3.3.

For a radius γ with $\gamma < \frac{\pi}{2}$ we see that there are no intersections in the odd case. That is, *odd* bound states cannot exist unless $\gamma^2 \geq \frac{\pi^2}{4}$—i.e., unless $V_o a^2 \geq \frac{\pi^2\hbar^2}{8m}$. However we see that there is always a smallest bound state (called the *ground*

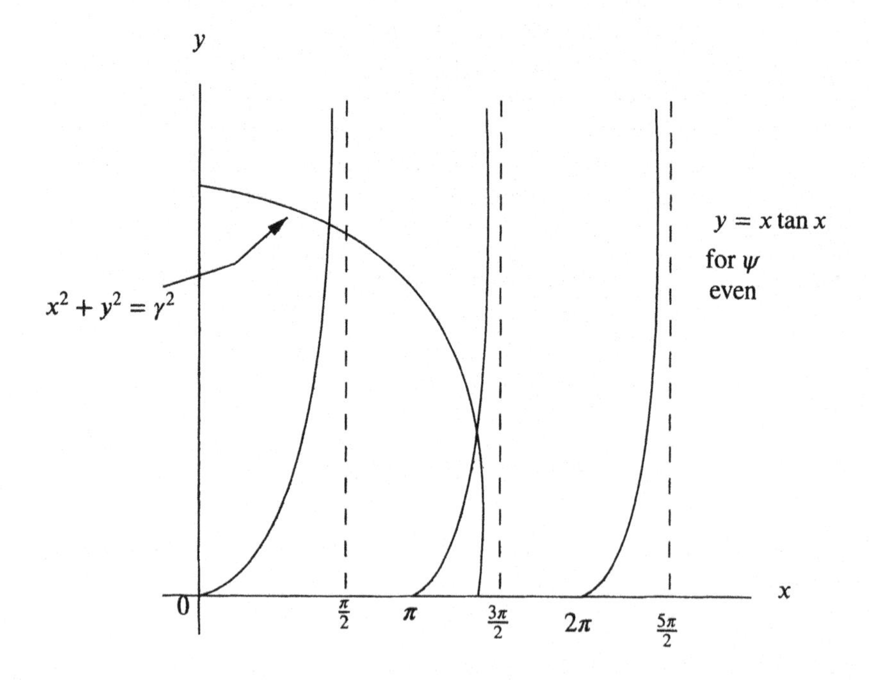

Figure 3.2: Tangent intersections

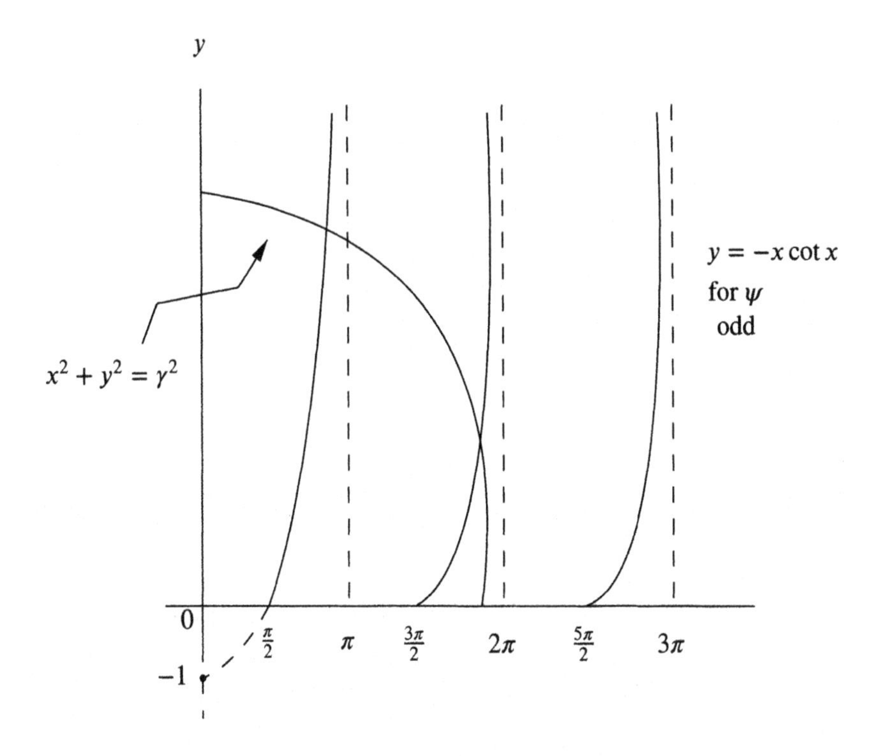

Figure 3.3: Cotangent intersections

energy state) which is *even*. For V_o, m, a fixed (i.e., for a fixed radius γ) there are only *finitely many* bound states (as there are only finitely many intersections). Suppose the radius γ satisfies $\frac{n}{2}\pi < \gamma \leq \frac{(n+1)}{2}\pi$, for n a positive integer. Then one sees that in the even case there are $\frac{n+1}{2}$ intersections for $n+1$ even and $\frac{n+2}{2}$ intersections for $n+1$ odd. Similarly in the odd case there are $\frac{n+1}{2}$ intersections for $n+1$ even and $\frac{n}{2}$ intersections for $n+1$ odd. Since $\gamma = \sqrt{2mV_o}\frac{a}{\hbar}$ we have checked the following.

Proposition 3.2. *For any V_o there are only finitely many bound states. Even bound states always exist, but odd bound states exist only if $V_o a^2 \geq \frac{\pi^2\hbar}{8m}$.*

For $n = 1, 2, 3, \ldots$ fixed suppose that

$$\frac{n\pi}{2} < \sqrt{2mV_o}\frac{a}{\hbar} \leq \left(\frac{n+1}{2}\right)\pi.$$

Then the number of even bound states is $\frac{n+1}{2}$ for $n+1$ even, and $\frac{n+2}{2}$ for $n+1$ odd. The number of odd bound states is $\frac{n+1}{2}$ for $n+1$ even, and $\frac{n}{2}$ for $n+1$ odd. The number of bound states altogether (even and odd) is therefore $n+1$.

Again, in the above analysis the assumption $-V_o < E < 0$, $V_o > 0$ is imposed. Since the radius γ varies directly with V_o (and with m and a) the intersections (and hence the number of energy levels—or bound states) increase in number as V_o is increased, or as m or a is increased.

3.4 Schrödinger's Equation for a System

Consider next a system of n particles moving in three dimensions, with m_j = mass of the jth particle, the motion being influenced by forces derived from a potential energy function $V : \mathbb{R}^{3n} \to \mathbb{R}$ with continuous first partial derivatives.

If $(x_j(t), y_j(t), z_j(t))$ is the position of the jth particle at time t, where the derivatives $\dot{x}_j, \dot{y}_j, \dot{z}_j, \ddot{x}_j, \ddot{y}_j, \ddot{z}_j$ exist and are continuous, then the conservation of total energy $H = T + V$ is expressed as follows:

$$\frac{1}{2}\sum_{j=1}^{n} m_j \left(\dot{x}_j(t)^2 + \dot{y}_j(t)^2 + \dot{z}_j(t)^2\right) \\ + V(x_1(t), y_1(t), z_1(t), \ldots, x_n(t), y_n(t), z_n(t)) \text{ is constant} \quad (3.4.1)$$

for all t, as we have seen in (2.4.8).

For $p^{(j)} = m_j v^{(j)} = m_j(\dot{x}_j, \dot{y}_j, \dot{z}_j)$ the momentum of the jth particle $(= (p_1^{(j)}, p_2^{(j)}, p_3^{(j)}))$, the first term in (3.4.1) (the kinetic energy term) is

$$\sum_{j=1}^{n} \frac{1}{2m_j}\left(p_1^{(j)}(t)^2 + p_2^{(j)}(t)^2 + p_3^{(j)}(t)^2\right) = \sum_{j=1}^{n} \frac{1}{2m_j}\langle p^{(j)}(t), p^{(j)}(t)\rangle. \quad (3.4.2)$$

Turning to quantum mechanical considerations we suppose we have a "wave" function $\psi : \mathbb{R}^{3n} \times \mathbb{R} \to \mathbb{C}$ for our system. For example, in an ideal situation we might have

(i) $t \to \psi(x, t)$ is continuously differentiable on $\mathbb{R}$ for $x \in \mathbb{R}^{3n}$ fixed,

(ii) $x \to \psi(x, t) : \mathbb{R}^{3n} \to \mathbb{C}$ has continuous first and second partial derivatives, for $t \in \mathbb{R}$ fixed, and

(iii) $x \to \psi(x, t)$ (which is continuous on $\mathbb{R}^{3n}$, for $t \in \mathbb{R}$ fixed) is in $L^2(\mathbb{R}^{3n}, dx)$ where dx = Lebesgue measure on $\mathbb{R}^{3n}$:

$$\int_{\mathbb{R}^{3n}} |\psi(x, t)|^2 \, dx < \infty. \quad (3.4.3)$$

It is required that ψ satisfy *Schrödinger's equation*:

$$i\hbar \frac{\partial \psi}{\partial t} = -\frac{\hbar^2}{2} \sum_{j=1}^{n} \frac{1}{m_j}\left(\frac{\partial^2 \psi}{\partial x_j^2} + \frac{\partial^2 \psi}{\partial y_j^2} + \frac{\partial^2 \psi}{\partial z_j^2}\right) + V\psi, \quad (3.4.4)$$

where $(x_1, y_1, z_1, x_2, y_2, z_2, \ldots, x_n, y_n, z_n)$ are the coordinates of a point in $\mathbb{R}^{3n}$. (3.4.4) is an extension of (1.3.10).

Let $\nabla_j^2 = \frac{\partial^2}{\partial x_j^2} + \frac{\partial^2}{\partial y_j^2} + \frac{\partial^2}{\partial z_j^2}$. Then Schrödinger's equation can be written as

$$-\frac{h}{2\pi i}\frac{\partial \psi}{\partial t} = -\frac{h^2}{8\pi^2}\sum_{j=1}^{n}\frac{1}{m_j}\nabla_j^2\psi + V\psi. \tag{3.4.5}$$

One can obtain the right-hand side of the Schrödinger equation by the following heuristic procedure (again as in Chapter 1). To the "classical observables" $p_1^{(j)}, p_2^{(j)}, p_3^{(j)}$ (the momentum coordinates) assign the "operators" $\frac{\hbar}{i}\frac{\partial}{\partial x_j}$, $\frac{\hbar}{i}\frac{\partial}{\partial y_j}$, $\frac{\hbar}{i}\frac{\partial}{\partial z_j}$, and to the observable V assign the operator which is multiplication by V. To the kinetic energy $\sum_{j=1}^{n}\frac{1}{2m_j}\left(p_1^{(j)2} + p_2^{(j)2} + p_3^{(j)2}\right)$ there corresponds therefore the operator $-\sum_{j=1}^{n}\frac{\hbar^2}{2m_j}\left(\frac{\partial^2}{\partial x_j^2} + \frac{\partial^2}{\partial y_j^2} + \frac{\partial^2}{\partial z_j^2}\right)$. And to the total energy $H = T + V$ (or "Hamiltonian") there corresponds the quantized Hamiltonian operator $-\sum_{j=1}^{n}\frac{\hbar^2}{2m_j}\left(\frac{\partial^2}{\partial x_j^2} + \frac{\partial^2}{\partial y_j^2} + \frac{\partial^2}{\partial z_j^2}\right)$ + multiplication by V, which acting on ψ gives the right-hand side of the Schrödinger equation (3.4.4) (or (3.4.5)).

For example, fix $E \in \mathbb{R}$ and let $\psi : \mathbb{R}^{3n} \to \mathbb{C}$ be a function with continuous first and second partial derivatives such that $\psi \in L^2(\mathbb{R}^{3n}, dx)$. Define $\psi : \mathbb{R}^{3n} \times \mathbb{R} \to \mathbb{C}$ by $\psi(x,t) = \psi(x)\exp\left(\frac{E}{i\hbar}t\right)$. Then ψ satisfies (i), (ii), (iii) above, and also satisfies $i\hbar\frac{\partial\psi}{\partial t} = E\psi$.

Also

$$\frac{\partial^2\psi}{\partial x_j^2} + \frac{\partial^2\psi}{\partial y_j^2} + \frac{\partial^2\psi}{\partial z_j^2} = \left(\frac{\partial^2\psi}{\partial x_j^2} + \frac{\partial^2\psi}{\partial y_j^2} + \frac{\partial^2\psi}{\partial z_j^2}\right)\exp\left(\frac{E}{i\hbar}t\right). \tag{3.4.6}$$

We see therefore that ψ satisfies Schrödinger's equation $\iff$

$$E\psi(x)\exp\left(\frac{E}{i\hbar}t\right) = -\frac{\hbar^2}{2}\sum_{j=1}^{n}\frac{1}{m_j}\left(\nabla_j\psi\right)(x)\,\exp\left(\frac{E}{i\hbar}t\right) + V(x)\psi(x)\,\exp\left(\frac{E}{i\hbar}t\right) \tag{3.4.7}$$

$\iff$ ψ satisfies the *time independent* Schrödinger equation:

$$E\psi = -\frac{\hbar^2}{2}\sum_{j=1}^{n}\frac{1}{m_j}\left(\frac{\partial^2\psi}{\partial x_j^2} + \frac{\partial^2\psi}{\partial y_j^2} + \frac{\partial^2\psi}{\partial z_j^2}\right) + V\psi, \tag{3.4.8}$$

or

$$\frac{\hbar^2}{2}\sum_{j=1}^{n}\frac{1}{m_j}\left(\frac{\partial^2\psi}{\partial x_j^2} + \frac{\partial^2\psi}{\partial y_j^2} + \frac{\partial^2\psi}{\partial z_j^2}\right) - (V - E)\psi = 0, \tag{3.4.9}$$

or

$$\sum_{j=1}^{n}\frac{1}{m_j}\left(\frac{\partial^2\psi}{\partial x_j^2} + \frac{\partial^2\psi}{\partial y_j^2} + \frac{\partial^2\psi}{\partial z_j^2}\right) + \frac{8\pi^2}{h^2}(E - V)\psi = 0. \tag{3.4.10}$$

In practice the preceding ideal situation may not prevail. As in previous examples V might be defined on some subset of $\mathbb{R}^{3n}$, or might be a discontinuous function.

3.5 Schrödinger's Equation for an Atom

For a system (= an atom) consisting of n electrons, each with charge $-e$, and a nucleus of charge Ze, (Z a positive integer = number of protons $\stackrel{\text{def}}{=}$ the *atomic number*), there is an easy application of Coulomb's electrostatic law.

Let $x^{(j)} = (x_j, y_j, z_j) \in \mathbb{R}^3$ denote the position of the jth electron and let $x^{(o)} = (x_o, y_o, z_o) \in \mathbb{R}^3$ denote the position of the nucleus. The force on the jth electron due to the nucleus is, by Coulomb's law (2.5.1),

$$(-e)Ze\frac{x^{(j)} - x^{(o)}}{|x^{(j)} - x^{(o)}|^3}. \tag{3.5.1}$$

The other electrons (say $n \geq 2$) contribute to the total force on the jth electron the amount of force

$$\sum_{i, i\neq j}(-e)(-e)\frac{x^{(j)} - x^{(i)}}{|x^{(j)} - x^{(i)}|^3}. \tag{3.5.2}$$

The total force $F^{(j)}$ on the jth electron is therefore

$$F^{(j)} = -Ze^2\frac{x^{(j)} - x^{(o)}}{|x^{(j)} - x^{(o)}|^3} + \sum_{i=1, i\neq j}^{n} e^2\frac{x^{(j)} - x^{(i)}}{|x^{(j)} - x^{(i)}|^3}. \tag{3.5.3}$$

Regard $F^{(j)}$ as a function to $\mathbb{R}^3$ on the set

$$\begin{aligned} U \stackrel{\text{def}}{=} \{x = (x^{(o)}, x^{(1)}, \ldots, x^{(n)}) \in \mathbb{R}^3 \times \mathbb{R}^{3n} = \mathbb{R}^{3(n+1)} | \\ x^{(i)} \neq x^{(o)}, 1 \leq i \leq n, x^{(l)} \neq x^{(i)}, l \neq i, 1 \leq l, i \leq n\}. \end{aligned} \tag{3.5.4}$$

Write $F^{(j)} = (X_j, Y_j, Z_j)$, as in Section 2.4 of Chapter 2: $x^{(j)} - x^{(o)} = (x_j - x_o, y_j - y_o, z_j - z_o)$, $x^{(j)} - x^{(i)} = (x_j - x_i, y_j - y_i, z_j - z_i) \implies$

$$\begin{aligned} X_j(x) &= \frac{-Ze^2(x_j - x_o)}{|x^{(j)} - x^{(o)}|^3} + \sum_{i\neq j} \frac{e^2(x_j - x_i)}{|x^{(j)} - x^{(i)}|^3}, \\ Y_j(x) &= \frac{-Ze^2(y_j - y_o)}{|x^{(j)} - x^{(o)}|^3} + \sum_{i\neq j} \frac{e^2(y_j - y_i)}{|x^{(j)} - x^{(i)}|^3}, \\ Z_j(x) &= \frac{-Ze^2(z_j - z_o)}{|x^{(j)} - x^{(o)}|^3} + \sum_{i\neq j} \frac{e^2(z_j - z_i)}{|x^{(j)} - x^{(i)}|^3}, \end{aligned} \tag{3.5.5}$$

for $x = (x^{(o)}, x^{(1)}, \ldots, x^{(n)}) \in U$.

We show that the $F^{(j)}$ are derived from a potential energy function V on U. Define $V : U \to \mathbb{R}$ by

$$\begin{aligned} V(x) &= V(x^{(o)}, x^{(1)}, \ldots, x^{(n)}) \\ &= \sum_{i=1}^{n} \frac{-Ze^2}{|x^{(i)} - x^{(o)}|} + \sum_{1\leq i<j\leq n} \frac{e^2}{|x^{(i)} - x^{(j)}|}. \end{aligned} \tag{3.5.6}$$

Now

$$\begin{aligned}\frac{\partial}{\partial x_j}\frac{1}{|x^{(i)}-x^{(o)}|} &= \frac{\partial}{\partial x_j}[(x_i-x_o)^2+(y_i-y_o)^2+(z_i-z_o)^2]^{-1/2} \\ &= -\frac{1}{2}[(x_i-x_o)^2+(y_i-y_o)^2+(z_i-z_o)^2]^{-3/2}\,2(x_i-x_o)\delta_{ij}\end{aligned} \tag{3.5.7}$$

and similarly

$$\begin{aligned}&\frac{\partial}{\partial x_j}\frac{1}{|x^{(i)}-x^{(l)}|} \\ &= -\frac{1}{2}[(x_i-x_l)^2+(y_i-y_l)^2+(z_i-z_l)^2]^{-3/2}\,2(x_i-x_l)(\delta_{ij}-\delta_{lj})\end{aligned} \tag{3.5.8}$$

for $1 \le j \le n \implies$

$$\frac{\partial V}{\partial x_j}(x) = \frac{Ze^2(x_j-x_o)}{|x^{(j)}-x^{(o)}|^3} + \sum_{1\le i<l\le n} \frac{-e^2(x_i-x_l)}{|x^{(i)}-x^{(l)}|^3}(\delta_{ij}-\delta_{lj}). \tag{3.5.9}$$

The second term here is

$$\begin{aligned}&-e^2\left[\sum_{i<l}\frac{(x_i-x_l)\delta_{ij}}{|x^{(i)}-x^{(l)}|^3} - \sum_{i<l}\frac{(x_i-x_l)\delta_{lj}}{|x^{(i)}-x^{(l)}|^3}\right] \\ &= -e^2\left[\sum_{j<l}\frac{(x_j-x_l)}{|x^{(j)}-x^{(l)}|^3} - \sum_{i<j}\frac{(x_i-x_j)}{|x^{(i)}-x^{(j)}|^3}\right] \\ &= -e^2\left[\sum_{j<i}\frac{(x_j-x_i)}{|x^{(j)}-x^{(i)}|^3} + \sum_{i<j}\frac{(x_j-x_i)}{|x^{(j)}-x^{(i)}|^3}\right] \\ &= \sum_{i\ne j}\frac{-e^2(x_j-x_i)}{|x^{(j)}-x^{(i)}|^3}\end{aligned} \tag{3.5.10}$$

$$\implies -\frac{\partial V}{\partial x_j}(x) = X_j(x) \quad \text{for } 1 \le j \le n. \tag{3.5.11}$$

Similarly

$$-\frac{\partial V}{\partial y_j}(x) = Y_j(x) \text{ and } -\frac{\partial V}{\partial z_j}(x) = Z_j(x) \text{ for } 1 \le j \le n. \tag{3.5.12}$$

Also

$$\begin{aligned}\frac{\partial}{\partial x_o}\frac{1}{|x^{(i)}-x^{(o)}|} &= \frac{\partial}{\partial x_o}[(x_i-x_o)^2+(y_i-y_o)^2+(z_i-z_o)^2]^{-1/2} \\ &= -\frac{1}{2}[(x_i-x_o)^2+(y_i-y_o)^2+(z_i-z_o)^2]^{-3/2}\,2(x_i-x_o)(-1) \\ &= \frac{(x_i-x_o)}{|x^{(i)}-x^{(o)}|^3} \qquad \text{and}\end{aligned} \tag{3.5.13}$$

$$\frac{\partial}{\partial x_o}\frac{1}{|x^{(i)}-x^{(j)}|} = 0 \implies$$

$$\frac{\partial V}{\partial x_o}(x) = \sum_{i=1}^{n} \frac{-Ze^2(x_i - x_o)}{|x^{(i)} - x^{(o)}|} = \sum_{i=1}^{n} \frac{Ze^2(x_o - x_i)}{|x^{(i)} - x^{(o)}|}. \tag{3.5.14}$$

Similarly

$$\frac{\partial V}{\partial y_o}(x) = \sum_{i=1}^{n} \frac{Ze^2(y_o - y_i)}{|x^{(i)} - x^{(o)}|}, \qquad \frac{\partial V}{\partial z_o}(x) = \sum_{i=1}^{n} \frac{Ze^2(z_o - z_i)}{|x^{(i)} - x^{(o)}|}. \tag{3.5.15}$$

By Coulomb's law the force $F^{(o)}$ on the nucleus (of charge Ze) due to the n electrons is given by

$$F^{(o)}(x) = \sum_{j=1}^{n} \frac{(Ze)(-e)(x^{(o)} - x^{(j)})}{|x^{(o)} - x^{(j)}|^3}. \tag{3.5.16}$$

Thus $F^{(o)} = (X_o, Y_o, Z_o)$ where

$$\begin{aligned} X_o(x) &= \sum_{j=1}^{n} \frac{-Ze^2(x_o - x_j)}{|x^{(o)} - x^{(j)}|^3} = -\frac{\partial V}{\partial x_o}(x), \\ Y_o(x) &= \sum_{j=1}^{n} \frac{-Ze^2(y_o - y_j)}{|x^{(o)} - x^{(j)}|^3} = -\frac{\partial V}{\partial y_o}(x), \quad \text{and} \\ Z_o(x) &= \sum_{j=1}^{n} \frac{-Ze^2(z_o - z_j)}{|x^{(o)} - x^{(j)}|^3} = -\frac{\partial V}{\partial z_o}(x). \end{aligned} \tag{3.5.17}$$

We see that indeed the $F^{(j)}$ are derived from a potential V. Given the potential energy $V : U \subset \mathbb{R}^{3(n+1)} \to \mathbb{R}$ of an atom,

$$\begin{aligned} V(x) &= V(x^{(o)}, x^{(1)}, \ldots, x^{(n)}) \\ &= \sum_{i=1}^{n} \frac{-Ze^2}{|x^{(i)} - x^{(o)}|} + \sum_{1 \le i < j \le n} \frac{e^2}{|x^{(i)} - x^{(j)}|} \end{aligned} \tag{3.5.18}$$

for $x^{(j)} = (x_j, y_j, z_j)$, Schrödinger's time independent equation (3.4.10) for such a system is therefore

$$0 = \frac{1}{m_o}(\nabla_o^2 \psi)(x) + \sum_{j=1}^{n} \frac{1}{m}(\nabla_j^2 \psi)(x) + \frac{8\pi^2}{h^2}(E - V(x))\psi(x), \tag{3.5.19}$$

where m_o = mass of the nucleus and m = electron mass. The time-dependent equation (3.4.4) is

$$i\hbar \frac{\partial \psi}{\partial t}(x,t) = -\frac{\hbar^2}{2} \sum_{j=0}^{n} \frac{1}{m_j}(\nabla_j^2 \psi)(x,t) + V(x)\psi(x,t) \tag{3.5.20}$$

where $m_j = m$ for $1 \le j \le n$.

The helium atom, for example, consists of two electrons, two protons ($Z = 2$), and two neutrons. The symbol for helium is He. The Schrödinger time independent equation for He is therefore

$$\frac{1}{m_o}(\nabla_o^2\psi)(x) + \frac{1}{m}(\nabla_1^2\psi)(x) + \frac{1}{m}(\nabla_2^2\psi)(x) + \frac{8\pi^2}{h^2}(E - V(x))\psi(x) = 0 \quad (3.5.21)$$

for $x = (x^{(o)}, x^{(1)}, x^{(2)}) = (x_o, y_o, z_o, x_1, y_1, z_1, x_2, y_2, z_2)$ where

$$\begin{aligned} V(x) = &\frac{-2e^2}{\sqrt{(x_1 - x_o)^2 + (y_1 - y_o)^2 + (z_1 - z_o)^2}} \\ &+ \frac{-2e^2}{\sqrt{(x_2 - x_o)^2 + (y_2 - y_o)^2 + (z_2 - z_o)^2}} \\ &+ \frac{e^2}{\sqrt{(x_1 - x_2)^2 + (y_1 - y_2)^2 + (z_1 - z_2)^2}}. \end{aligned} \quad (3.5.22)$$

Even for a simple atom like helium, however, the Schrödinger equation has never been solved exactly. We shall therefore study the hydrogen atom (with only one electron and one proton, and for which the Schrödinger equation can be solved exactly) in detail, in Chapter 5.

3.6 Non-Quantization of the Free Particle

For a particle of mass m moving in space under a zero force field ($V : \mathbb{R}^3 \to \mathbb{R}$ is zero) the time independent Schrödinger equation (3.4.10) takes the (simplest) form

$$\frac{\partial^2\psi}{\partial x^2} + \frac{\partial^2\psi}{\partial y^2} + \frac{\partial^2\psi}{\partial z^2} + \frac{8\pi^2}{h^2}mE\psi = 0. \quad (3.6.1)$$

We look for a solution $\psi \neq 0$ with $\psi(x, y, z) = \psi_1(x)\psi_2(y)\psi_3(z)$ where $\psi_j : \mathbb{R} \to \mathbb{C}$ have continuous first and second derivatives. $\psi \overset{\text{def}}{=} \psi_1 \otimes \psi_2 \otimes \psi_3 \implies \frac{\partial^2\psi}{\partial x^2} = \frac{d^2\psi_1}{dx^2} \otimes \psi_2 \otimes \psi_3$, with similar statements for $\frac{\partial^2\psi}{\partial y^2}, \frac{\partial^2\psi}{\partial z^2}$. Equation (3.6.1) becomes

$$\begin{aligned} \psi_1''(x)\psi_2(y)\psi_3(z) + \psi_1(x)\psi_2''(y)\psi_3(z) + \psi_1(x)\psi_2(y)\psi_3''(z)& \\ + \frac{8\pi^2}{h^2}mE\psi_1(x)\psi_2(y)\psi_3(z) = 0& \end{aligned} \quad (3.6.2)$$

for $(x, y, z) \in \mathbb{R}^3$. Choose and fix $(y, z) \in \mathbb{R}^2$ such that $\psi_2(y), \psi_3(z) \neq 0$. Divide by $\psi_2(y)\psi_3(z)$:

$$\psi_1''(x) + \psi_1(x)\frac{\psi_2''(y)}{\psi_2(y)} + \psi_1(x)\frac{\psi_3''(z)}{\psi_3(z)} + C\psi_1(x) = 0 \quad (3.6.3)$$

for $C \overset{\text{def}}{=} \frac{8\pi^2}{h^2}mE$. That is $\psi_1''(x) = \kappa_1\psi_1(x)$ on $\mathbb{R}$ for $\kappa_1 = -\frac{\psi_2''(y)}{\psi_2(y)} - \frac{\psi_3''(z)}{\psi_3(z)} - C =$ a constant (independent of x).

Similarly $\psi_2''(y) = \kappa_2\psi_2(y), \psi_3''(z) = \kappa_3\psi_3(z)$ on $\mathbb{R}$ for suitable constants κ_2, κ_3: $\kappa_2 = -\frac{\psi_1''(x)}{\psi_1(x)} - \frac{\psi_3''(z)}{\psi_3(z)} - C$ for some $(x, z) \in \mathbb{R}^2$ such that $\psi_1(x), \psi_3(z) \neq 0$, $\kappa_3 = -\frac{\psi_1''(x)}{\psi_1(x)} - \frac{\psi_2''(y)}{\psi_2(y)} - C$ with $\psi_1(x), \psi_2(y) \neq 0$.

Note that

$$\begin{aligned}
&\psi_1(x)\psi_2(y)\psi_3(z)(\kappa_1 + \kappa_2 + \kappa_3) \\
&\quad = -\psi_2''(y)\psi_1(x)\psi_3(z) - \psi_3''(z)\psi_1(x)\psi_2(y) - \psi_1''(x)\psi_2(y)\psi_3(z) \\
&\qquad - \psi_3''(z)\psi_1(x)\psi_2(y) - \psi_1''(x)\psi_2(y)\psi_3(z) - \psi_2''(y)\psi_1(x)\psi_3(z) \\
&\qquad - 3\psi_1(x)\psi_2(y)\psi_3(z)C \qquad\qquad (3.6.4) \\
&\quad = -2\psi_1''(x)\psi_2(y)\psi_3(z) - 2\psi_2''(y)\psi_1(x)\psi_3(z) - 2\psi_3''(z)\psi_1(x)\psi_2(y) \\
&\qquad - 2\psi_1(x)\psi_2(y)\psi_3(z)C - \psi_1(x)\psi_2(y)\psi_3(z)C \\
&\quad = -\psi_1(x)\psi_2(y)\psi_3(z)C \qquad \text{by equation (3.6.2)}
\end{aligned}$$

$$\Longrightarrow \kappa_1 + \kappa_2 + \kappa_3 = -C. \qquad (3.6.5)$$

Define E_1, E_2, E_3 by $\kappa_j = -\frac{8\pi^2 m}{h^2}E_j$. Then

$$\begin{aligned}
&\psi_1''(x) + \frac{8\pi^2 m}{h^2}E_1\psi_1(x) = 0 \\
&\psi_2''(y) + \frac{8\pi^2 m}{h^2}E_2\psi_2(y) = 0 \quad \text{on } \mathbb{R} \qquad (3.6.6) \\
&\psi_3''(z) + \frac{8\pi^2 m}{h^2}E_3\psi_3(z) = 0
\end{aligned}$$

are time independent Schrödinger equations. By equation (3.6.5) and the definition of C, $E_1 + E_2 + E_3 = E$.

Conversely if $\psi_j : \mathbb{R} \to \mathbb{C}$ have continuous first and second derivatives, $j = 1, 2, 3$, and satisfy the preceding three Schrödinger equations for constant E_1, E_2, E_3, then $\psi \stackrel{\text{def}}{=} \psi_1 \otimes \psi_2 \otimes \psi_3$ satisfies equation (3.6.1) for $E = E_1 + E_2 + E_3$: Namely, multiply the first Schrödinger equation in (3.6.6) by $\psi_2(y)\psi_3(z)$, the second by $\psi_1(x)\psi_3(z)$ and the third by $\psi_1(x)\psi_2(y)$, and then add:

$$\begin{aligned}
&\psi_1''(x)\psi_2(y)\psi_3(z) + \psi_2''(y)\psi_1(x)\psi_3(z) + \psi_3''(z)\psi_1(x)\psi_2(y) \\
&\qquad + \frac{8\pi^2 m}{h^2}\psi_1(x)\psi_2(y)\psi_3(z)(E_1 + E_2 + E_3) = 0, \quad (3.6.7)
\end{aligned}$$

which is equation (3.6.1) for $E = E_1 + E_2 + E_3$.

Going back to the Schrödinger equations (3.6.6) $\psi_j'' + \frac{8\pi^2 m}{h^2}E_j\psi_j = 0, j = 1, 2, 3$, the general solutions are

$$\begin{aligned}
\psi_j(x) &= A\cos\frac{2\sqrt{2mE_j}\pi x}{h} + B\sin\frac{2\sqrt{2mE_j}\pi x}{h} \\
&= A\cos\frac{p_j x}{\hbar} + B\sin\frac{p_j x}{\hbar},
\end{aligned} \qquad (3.6.8)$$

where $p_j = \sqrt{2mE_j}$. Note: Regard E_j as total energy: $E_j = \frac{1}{2}mv^2 + 0$ since $V = 0$. Then $2mE_j = m^2v^2 \implies p_j = mv$ is the momentum. Since E_j can take on arbitrary values without restriction we see that the *free particle is not quantized.* Clearly, moreover, the ψ_j in (3.6.8) are not square-integrable over $\mathbb{R}$. In contrast we have seen, for example, that in the case of the harmonic oscillator with frequency ν one has the quantization (restriction) of energy given by $E = E_n = (n + \frac{1}{2})\nu h$ for some integer $n \geq 0$.

3.7 Quantization, Large Eigenvalues, and Spectral Zeta Functions

One can relate the quantization of energy of some quantum systems to a general result concerning large eigenvalues λ of the second-order linear differential equation on $\mathbb{R}$

$$\frac{d^2y}{dx^2} + [\lambda - q(x)]y(x) = 0; \tag{3.7.1}$$

$$\text{i.e., } Dy = \lambda y \quad \text{for } D = -\frac{d^2}{dx^2} + q(x). \tag{3.7.2}$$

The result, roughly, is that λ satisfies

$$\int_{a_\lambda}^{b_\lambda} [\lambda - q(x)]^{1/2}\, dx = \left(n + \frac{1}{2}\right)\pi + O\left(\frac{1}{\sqrt{\lambda}}\right) \tag{3.7.3}$$

for some $a_\lambda, b_\lambda \in \mathbb{R}$ and some positive integer n. Disregarding the term $O(\frac{1}{\sqrt{\lambda}})$ for large λ one obtains from (3.7.3) a general kind of (approximate) quantization condition, which when specialized to Schrödinger's equation amounts to the *Bohr–Wilson–Sommerfeld quantization rule*; compare (3.7.10), (3.7.13) below. The result, Theorem 3.1 below, was developed by Kemble, Zwaan, Langer, Titchmarsh, Olver, and others. We will state Olver's version of it, which to some extent follows Titchmarsh. Compare [58, 67, 81] for, example.

The following assumptions are set forth. $\lambda > 0$ and $\lim_{x\to\pm\infty} y(x) = 0$, $\lim_{x\to\pm\infty} q(x) = +\infty$. q is of class C^3 on $\mathbb{R}$ (i.e., q', q'', q''' exist and are continuous on $\mathbb{R}$), and q satisfies

(3.7.4:i) $\exists$ positive constants α, β such that $q(x) \sim \beta x^\alpha$ as $x \to \infty$ (i.e., $\lim_{x\to\infty} \frac{q(x)}{\beta x^\alpha} = 1$), $q'(x) \sim \beta\alpha x^{\alpha-1}$ as $x \to \infty$, $q''(x) = O(x^{\alpha-2})$ as $x \to \infty$ (i.e., $\left|\frac{q''(x)}{x^{\alpha-2}}\right|$ is bounded), $q'''(x) = O(x^{\alpha-3})$ as $x \to \infty$,

(3.7.4:ii) $\exists$ positive constants $\tilde{\alpha}, \tilde{\beta}$ such that $q(x) \sim \tilde{\beta}(-x)^{\tilde{\alpha}}$, $q'(x) \sim -\tilde{\beta}\tilde{\alpha}(-x)^{\tilde{\alpha}-1}$, $q''(x) = O(x^{\tilde{\alpha}-2})$, $q'''(x) = O(x^{\tilde{\alpha}-3})$ as $x \to -\infty$.

These assumptions on q are satisfied if we take $q(x) = Ax^{2n}$ for example, where $A > 0$, $n = 1, 2, 3, \ldots$. Namely, in (i) take $\alpha = 2n, \beta = A$; asymptoticity $\sim$ becomes equality. In (ii), similarly take $\tilde{\alpha} = 2n, \tilde{\beta} = A$.

Using assumptions (i), (ii) one can argue that $\exists\, M > 0$, sufficiently large such that for $\lambda > M$ the equation $q(x) = \lambda$ has exactly two real roots, say $a_\lambda, b_\lambda \in \mathbb{R}$ with $a_\lambda < 0 < b_\lambda$; actually $a_\lambda \sim -\left(\frac{\lambda}{\tilde{\beta}}\right)^{1/\tilde{\alpha}}, b_\lambda \sim \left(\frac{\lambda}{\beta}\right)^{1/\alpha}$; $q(x) - \lambda \leq 0$ on $[a_\lambda, b_\lambda]$. In the example $q(x) = Ax^{2n}$, the equation $q(x) = \lambda$ (for any λ) has exactly one negative root $a_\lambda = -\left(\frac{\lambda}{A}\right)^{1/(2n)}$, and one positive root $b_\lambda = +\left(\frac{\lambda}{A}\right)^{1/(2n)}$ (by Descartes' rule of sign for example). With the above hypotheses and notation in place, the following is proved in [67, Sections 9.2–9.4].

Theorem 3.1. *Let y be a non-zero solution of equation* (3.7.1) *(with* $\lim_{x\to\pm\infty} y(x) = 0$*). Then*

$$\int_{a_\lambda}^{b_\lambda} [\lambda - q(x)]^{1/2}\, dx = \left(n + \frac{1}{2}\right)\pi + O\left(\frac{1}{\sqrt{\lambda}}\right) \tag{3.7.5}$$

for some positive integer n.

Thus for λ sufficiently large

$$\left| \int_{a_\lambda}^{b_\lambda} [\lambda - q(x)]^{1/2}\, dx - \left(n + \frac{1}{2}\right)\pi \right| < \frac{C}{\sqrt{\lambda}} \tag{3.7.6}$$

for some constant $C > 0$ independent of λ.

For the Schrödinger equation (3.1.4)

$$\frac{d^2\psi}{dx^2} + \frac{2m}{\hbar^2}(E - V(x))\psi(x) = 0 \tag{3.7.7}$$

on $\mathbb{R}$ with $\lim_{x\to\pm\infty} \psi(x) = 0$, $\lambda = \frac{2mE}{\hbar^2}$ and $q(x) = \frac{2m}{\hbar^2}V(x)$. Then the equation $q(x) = \lambda$ means that $V(x) = E$, say with real solutions a_E, b_E at least for E sufficiently large; here we take $E > V(x)$. Also $[\lambda - q(x)]^{1/2} = \frac{\sqrt{2m}}{\hbar}[E - V(x)]^{1/2}$, so that equation (3.7.5) assumes the form

$$\frac{\sqrt{2m}}{\hbar}\int_{a_E}^{b_E} [E - V(x)]^{1/2}\, dx = \left(n + \frac{1}{2}\right)\pi + O\left(\frac{\hbar}{\sqrt{2mE}}\right), \tag{3.7.8}$$

or

$$\int_{a_E}^{b_E} \sqrt{2m(E - V(x))}\, dx = \left(n + \frac{1}{2}\right)\pi\hbar + O\left(\frac{1}{\sqrt{E}}\right). \tag{3.7.9}$$

Thus the condition

$$\int_{a_E}^{b_E} \sqrt{2m(E - V(x))}\, dx = \left(n + \frac{1}{2}\right)\pi\hbar, \quad n = 1, 2, \ldots \tag{3.7.10}$$

determines approximately the allowed energy levels.

As an example consider the quantized harmonic oscillator. $V(x) = \frac{\omega^2 m x^2}{2} = 2\pi^2 \nu^2 m x^2$ in (3.2.2) so that $V(x) = E \implies x = \pm\frac{1}{\omega}\sqrt{\frac{2E}{m}}$. That is, $a_E = -\frac{1}{\omega}\sqrt{\frac{2E}{m}}$, $b_E = +\frac{1}{\omega}\sqrt{\frac{2E}{m}} \implies$

$$\begin{aligned} \int_{a_E}^{b_E} \sqrt{2m(E - V(x))}\, dx &= 2\sqrt{2m} \int_0^{b_E} \sqrt{E - \frac{\omega^2 m x^2}{2}}\, dx \\ &= \frac{4}{\omega} \int_0^{\sqrt{E}} \sqrt{E - t^2}\, dt \qquad (3.7.11) \\ &= \frac{4}{\omega}\frac{E\pi}{4} = \frac{\pi E}{\omega} \end{aligned}$$

since

$$\int \sqrt{a^2 - x^2}\, dx = \frac{x}{2}\sqrt{a^2 - x^2} + \frac{a^2}{2} \arcsin\frac{x}{2} \implies \int_0^a \sqrt{a^2 - x^2}\, dx = \frac{a^2 \pi}{4} \qquad (3.7.12)$$

for $a > 0$. Condition (3.7.10) therefore means that $E = (n+\frac{1}{2})\omega\hbar = (n+\frac{1}{2})2\pi\nu\hbar$, so that we obtain precisely the energy levels $\{E_n\}_{n=0}^{\infty}$ mentioned following equation (3.2.3), apart from E_o.

The condition (3.7.10) also amounts to the WKB (Wentzel, Kramers, Brillouin) eigenvalue approximation, for the *classical region* $R_E \stackrel{\text{def}}{=} \{x \in \mathbb{R} | V(x) \leq E\}$, where the solutions a_E, b_E of $V(x) = E$ are called *turning points*; see Figure 3.4.

Classically the particle of momentum $p = mv$ has kinetic energy $\frac{p^2}{2m} = T = E - V$, and is confined to R_E in its motion. That is, $p = \sqrt{2m(E - V)}$ and one also writes (3.7.10) as

$$\int_{a_E}^{b_E} p(x)\, dx = \left(n + \frac{1}{2}\right)\pi\hbar. \qquad (3.7.13)$$

For the special potentials $q(x) = Ax^{2M}$ above, with $A > 0$, $M = 1, 2, 3, \ldots$, the differential operator $D = -\frac{d^2}{dx^2} + q(x)$ in (3.7.2) has the following special properties, as shown for example by Titchmarsh [81]. D is positive and has a compact resolvent. Thus D has a discrete spectrum $\{\lambda_j, n_j\}_{j=0}^{\infty}$ where n_j is the multiplicity of the eigenvalue λ_j; in fact each $n_j = 1$. $0 < \lambda_o < \lambda_1 < \lambda_2 < \ldots$; $\lim_{k\to\infty} \lambda_k = \infty$. The λ_j depend on M of course and we sometimes write $\lambda_j = \lambda_j(M)$. One has (cf. (3.7.5))

$$\int_{-a^{1/(2M)}}^{a^{1/(2M)}} [\lambda_n - Ax^{2M}]^{1/2}\, dx = \left(n + \frac{1}{2}\right)\pi + O\left(\frac{1}{n}\right), \quad \text{where } a = \frac{\lambda_n}{A}. \qquad (3.7.14)$$

We shall compute the integral in (3.7.14) and derive the following spectral asymptotics.

Theorem 3.2 (Titchmarsh).

$$\lambda_n(M) \sim \left[\frac{A^{1/(2M)} M\sqrt{\pi}\,(2n+1)\,\Gamma(\frac{1}{2M} + \frac{3}{2})}{\Gamma(\frac{1}{2M})}\right]^{\frac{2M}{M+1}} \quad \textit{as } n \to \infty. \qquad (3.7.15)$$

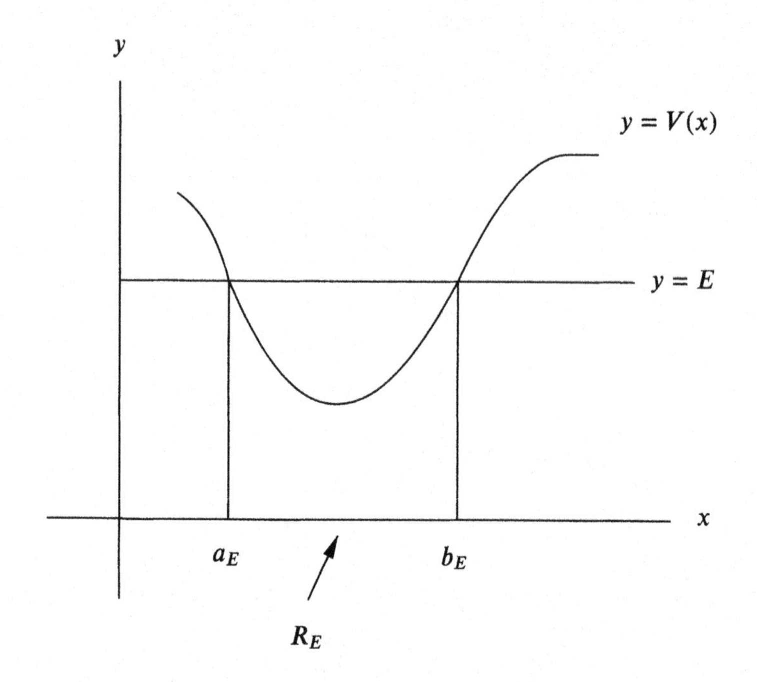

Figure 3.4: Turning points for the classical region R_E

First, for $a > 0$, $n = 1, 2, 3, \ldots$ let

$$I = I_n(a) = \int_{-a^{1/n}}^{a^{1/n}} \sqrt{a^2 - t^{2n}}\, dt, \tag{3.7.16}$$

$$\begin{aligned}
I &= 2\int_0^{a^{1/n}} \sqrt{a^2\left[1 - \left(\frac{x}{a^{1/n}}\right)^{2n}\right]}\, dx \\
&= 2aa^{1/n}\int_0^1 \sqrt{1 - t^{2n}}\, dt \\
&= 2a^{(n+1)/n}\int_0^1 \sqrt{1-x}\, \frac{x^{\frac{1}{2n}-1}}{2n}\, dx \\
&= \frac{2a^{(n+1)/n}}{2n}\int_0^1 x^{p-1}(1-x)^{q-1}\, dx \qquad \text{for } p = \frac{1}{2n}, q = \frac{3}{2}, \\
&= \frac{a^{(n+1)/n}}{n} B(p,q) \stackrel{\text{def}}{=} \frac{a^{(n+1)/n}}{n}\frac{\Gamma(p)\,\Gamma(q)}{\Gamma(p+q)} \\
&= \frac{a^{(n+1)/n}}{n}\frac{\Gamma(\frac{1}{2n})\,\Gamma(\frac{3}{2})}{\Gamma(\frac{1}{2n}+\frac{3}{2})},
\end{aligned} \tag{3.7.17}$$

where B is the Beta function and Γ is the Gamma function:

$$\Gamma(x) \stackrel{\text{def}}{=} \int_0^\infty e^{-t} t^{x-1}\, dt \qquad \text{for } x > 0. \tag{3.7.18}$$

Now $\Gamma(\frac{3}{2} = 1 + \frac{1}{2}) = \frac{1}{2}\,\Gamma(\frac{1}{2}) = \frac{\sqrt{\pi}}{2}$. Hence

Proposition 3.3. *For $a > 0$, $n = 1, 2, 3, \ldots$*

$$\begin{aligned} \int_{-a^{1/n}}^{a^{1/n}} \sqrt{a^2 - x^{2n}}\, dx &= 2\int_0^{a^{1/n}} \sqrt{a^2 - x^{2n}}\, dx \\ &= \frac{a^{(n+1)/n}}{2n}\sqrt{\pi}\frac{\Gamma(\frac{1}{2n})}{\Gamma(\frac{1}{2n} + \frac{3}{2})}. \end{aligned} \tag{3.7.19}$$

For $n = 1$ we obtain the elementary calculus result

$$\int_0^a \sqrt{a^2 - x^2}\, dx = \frac{a^2\pi}{4} \tag{3.7.20}$$

of (3.7.12).

Corollary 3.1. *For $\lambda, A > 0$, $M = 1, 2, 3, \ldots$ let*

$$\widetilde{I}_M(\lambda) = \int_{-\left(\frac{\lambda}{A}\right)^{1/(2M)}}^{\left(\frac{\lambda}{A}\right)^{1/(2M)}} [\lambda - Ax^{2M}]^{(1/2)}\, dx. \tag{3.7.21}$$

Then

$$\widetilde{I}_M(\lambda) = A^{-1/(2M)} \frac{\lambda^{\frac{M+1}{2M}}\sqrt{\pi}}{2M} \frac{\Gamma(\frac{1}{2M})}{\Gamma(\frac{1}{2M} + \frac{3}{2})}. \tag{3.7.22}$$

Proof. By the change of variables $t = A^{1/(2M)}x$, $\widetilde{I}_M(\lambda) = A^{-1/(2M)} \cdot \cdot I_M(\lambda^{1/2})$ for $I_M(\lambda^{1/2})$ given by (3.7.16). Thus Corollary 3.1 follows from Proposition 3.3. □

By Corollary 3.1 and equation (3.7.14)

$$A^{-1/(2M)} \frac{\lambda_n^{\frac{M+1}{2M}}\sqrt{\pi}}{2M} \frac{\Gamma(\frac{1}{2M})}{\Gamma(\frac{1}{2M} + \frac{3}{2})} = \left(n + \frac{1}{2}\right)\pi + O\left(\frac{1}{n}\right). \tag{3.7.23}$$

Let

$$C_M = \frac{A^{-1/(2M)}\sqrt{\pi}\,\Gamma(\frac{1}{2M})}{2M\,\Gamma(\frac{1}{2M} + \frac{3}{2})} > 0, \tag{3.7.24}$$

so that for some natural number n_o and $K > 0$

$$\begin{aligned} &\left| C_M \lambda_n^{\frac{M+1}{2M}} - (n + \frac{1}{2})\pi \right| < K\frac{1}{n} \quad \text{for } n \geq n_o \\ \Longrightarrow &\left| \frac{C_M}{(n + \frac{1}{2})\pi} \lambda_n^{\frac{M+1}{2M}} - 1 \right| < \frac{K}{n(n + \frac{1}{2})\pi} \quad \text{for } n \geq n_o. \end{aligned} \tag{3.7.25}$$

Let $n \to \infty$ and conclude that

$$\lim_{n\to\infty} \frac{C_M}{(n+\frac{1}{2})\pi} \lambda_n^{\frac{M+1}{2M}} = 1 \Longrightarrow \lim_{n\to\infty} \left[\frac{C_M}{(n+\frac{1}{2})\pi}\right]^{\frac{2M}{M+1}} \lambda_n = 1. \tag{3.7.26}$$

That is,

$$\lim_{n\to\infty} \frac{\lambda_n}{\left[\frac{(n+\frac{1}{2})\pi}{C_M}\right]^{\frac{2M}{M+1}}} = 1, \tag{3.7.27}$$

which proves Theorem 3.2 since

$$\frac{(n+\frac{1}{2})\pi}{C_M} \overset{\text{def}}{=} \frac{(2n+1)A^{1/(2M)} M\sqrt{\pi}\,\Gamma(\frac{1}{2M}+\frac{3}{2})}{\Gamma(\frac{1}{2M})}. \tag{3.7.28}$$

Given the discrete spectrum $\{\lambda_j(M), n_j\}$ of $D = -\frac{d^2}{dx^2} + Ax^{2M}$, $A > 0$, with

$$0 < \lambda_o < \lambda_1 < \cdots, \quad \lim_{k\to\infty} \lambda_k = \infty, \lambda_j = \lambda_j(M), \tag{3.7.29}$$

we form the corresponding *spectral zeta function*

$$\zeta_D(s) = \zeta_M(s) \overset{\text{def}}{=} \sum_{j=0}^{\infty} \frac{n_j}{\lambda_j^s} = \sum_{j=0}^{\infty} \frac{1}{\lambda_j^s}, \tag{3.7.30}$$

as we have noted that each $n_j = 1$. Using the spectral asymptotics of Theorem 3.2 we easily deduce

Corollary 3.2. *The zeta function ζ_M (3.7.30) converges absolutely for Re $s > \frac{M+1}{2M}$ and is holomorphic on this half-plane.*

Proof. We have noted that Theorem 3.2 is expressed by equation (3.7.27): as $n \to \infty$, $\alpha = \alpha(M) \overset{\text{def}}{=} 2M/(M+1) \Longrightarrow$

$$\begin{aligned} &\frac{\lambda_n}{\left[\frac{(n+1/2)\pi}{C_M}\right]^\alpha} \to 1 \Longrightarrow \frac{\left[\frac{(2n+1)\pi}{2C_M}\right]^\alpha}{\lambda_n} \to 1 \\ &\Longrightarrow \frac{(2n+1)^\alpha}{\lambda_n} \to \left(\frac{2C_M}{\pi}\right)^\alpha \overset{\text{def}}{=} \widetilde{c}_M. \end{aligned} \tag{3.7.31}$$

Thus for $\epsilon = 1$, choose a natural number N such that $n \geq N \Longrightarrow$

$$\begin{aligned} &\left|\frac{(2n+1)^\alpha}{\lambda_n} - \widetilde{c}_M\right| < 1 \Longrightarrow \frac{(2n+1)^\alpha}{\lambda_n} < 1 + \widetilde{c}_M \\ &\Longrightarrow \frac{1}{\lambda_n} < \frac{1+\widetilde{c}_M}{(2n+1)^\alpha} \Longrightarrow \left|\frac{1}{\lambda_n^s}\right| < \frac{(1+\widetilde{c}_M)^{\text{Re}\, s}}{(2n+1)^{\alpha\, \text{Re}\, s}}, \end{aligned} \tag{3.7.32}$$

where

$$\sum_{n=0}^{\infty} \frac{1}{(2n+1)^{\alpha \operatorname{Re} s}} < \infty \qquad \text{for } \alpha \operatorname{Re} s > 1. \tag{3.7.33}$$

Thus

$$\sum_{n=0}^{\infty} \left| \frac{1}{\lambda_n^s} \right| \tag{3.7.34}$$

also converges for α Re $s > 1$, as asserted in Corollary 3.2, since $\alpha \stackrel{\text{def}}{=} 2M/(M+1)$. To see that ζ_D is holomorphic on the domain $S_M \stackrel{\text{def}}{=} \{s \in \mathbb{C} | \operatorname{Re} s > \frac{M+1}{2M}\}$ it suffices to show, by Weierstrass' theorem, that the series $\sum_{n=0}^{\infty} \frac{1}{\lambda_n^s}$ converges uniformly on compact subsets of S_M. In fact a little bit more is true: If $\delta > 0$ is arbitrary, then the series $\sum_{n=0}^{\infty} \frac{1}{\lambda_n^s}$ converges uniformly (and absolutely) on Re $s > \frac{M+1}{2M} + \delta$. Namely, on this domain we have for n sufficiently large, and $a \stackrel{\text{def}}{=} \frac{M+1}{2M}$, $|\lambda_n^s| \geq \lambda_n^{a+\delta}$, say for $\lambda_n \geq 1$; see (3.7.29). That is,

$$\frac{1}{|\lambda_n^s|} \leq \frac{1}{\lambda_n^{a+\delta}} \tag{3.7.35}$$

where

$$\sum_{n=0}^{\infty} \frac{1}{\lambda_n^{a+\delta}} < \infty, \tag{3.7.36}$$

as $a + \delta > a = \frac{M+1}{2M}$. The M-test therefore applies. □

We remark that for $M = 1$, $A = 1$, i.e.,

$$D = -\frac{d^2}{dx^2} + x^2 \tag{3.7.37}$$

the eigenvalues λ_n are given by $\lambda_n = 2n+1$, $n = 0, 1, 2, \ldots$, as D is essentially the quantized Hamiltonian of (3.1.2) for the harmonic oscillator; cf. (3.2.2), (3.2.3); then clearly (3.7.29) holds. In this case one has in fact *equality* in the asymptotic assertion of Theorem 3.2: The right-hand side there is (for $M, A = 1$)

$$\frac{\sqrt{\pi}(2n+1)\,\Gamma(1/2+3/2)}{\Gamma(1/2) = \sqrt{\pi}} = (2n+1)\,\Gamma(2) = 2n+1. \tag{3.7.38}$$

Also the zeta function $\zeta_D(s)$ of (3.7.30) (which converges absolutely for Re $s > 1$, by Corollary 3.2) is given by

$$\zeta_D(s) = \left(1 - \frac{1}{2^s}\right)\zeta(s) \qquad \text{where } \zeta(s) \stackrel{\text{def}}{=} \sum_{n=1}^{\infty} \frac{1}{n^s} \tag{3.7.39}$$

is the *Riemann zeta function*. Namely

$$\begin{aligned}
\zeta_D(s) &\stackrel{\text{def}}{=} \sum_{n=0}^{\infty} \frac{1}{(2n+1)^s} = 1 + \frac{1}{3^s} + \frac{1}{5^s} + \frac{1}{7^s} + \cdots \\
&= 1 + \frac{1}{2^s} + \frac{1}{3^s} + \frac{1}{4^s} + \frac{1}{5^s} + \frac{1}{6^s} + \frac{1}{7^s} + \cdots \\
&\quad - \frac{1}{2^s} \qquad - \frac{1}{4^s} \qquad - \frac{1}{6^s} \qquad - \frac{1}{8^s} - \cdots \\
&= \left(1 - \frac{1}{2^s}\right)\left[1 + \frac{1}{2^s} + \frac{1}{3^s} + \frac{1}{4^s} + \frac{1}{5^s} + \cdots\right],
\end{aligned} \tag{3.7.40}$$

as desired. In particular we can use (3.7.39), and the theory of Riemann's zeta function to meromorphically continue ζ_D (for $D = -\frac{d^2}{dx^2} + x^2$) to all of $\mathbb{C}$.

For $M \geq 2$ the eigenvalues $\lambda_n(M)$ are *unknown*. However, even for $M \geq 2$ the zeta function ζ_D (for $D = -\frac{d^2}{dx^2} + Ax^{2M}$) does admit a meromorphic continuation to $\mathbb{C}$. We shall return to this zeta function, and others, in later chapters. For example see Chapter 15.

3.8 The Diffeomorphism Φ Again

In Section 2.8 of Chapter 2 we constructed a diffeomorphism Φ of $\mathbb{R}^6$ which provided passage from rectangular coordinates $(x_1, y_1, z_1, x_2, y_2, z_2)$ to Jacobi, or center of mass coordinates, $(x_1 - x_2, y_1 - y_2, z_1 - z_2, R_1, R_2, R_3)$ where R_1, R_2, R_3 are given by (2.8.1) for fixed masses m_1, m_2. The transformation of a "scaled" Laplacian, via Φ, was given by formula (2.8.17). We now apply that formula in connection with the time independent Schrödinger equation.

Let $U \stackrel{\text{def}}{=} \{(x_1, y_1, z_1, x_2, y_2, z_2) \in \mathbb{R}^6 | \ (x_1, y_1, z_1) \neq (x_2, y_2, z_2)\}$ as in Section 2.5 of Chapter 2. For $p = (x_1, y_1, z_1, x_2, y_2, z_2) \in U$, $\Phi(p) \neq (0, 0, 0, x_2', y_2', z_2')$; otherwise $x_1 - x_2 = 0$, $y_1 - y_2 = 0$, $z_1 - z_2 = 0 \implies (x_1, y_1, z_1) = (x_2, y_2, z_2)$. That is, $\Phi(U) \subset (\mathbb{R}^3 - \{0\}) \times \mathbb{R}^3$. Conversely let $a \in \mathbb{R}^6$, $a = (a_1, a_2, a_3, a_4, a_5, a_6)$ with $(a_1, a_2, a_3) \neq (0, 0, 0)$. Write $a = \Phi(x)$, $x = (x_1, y_1, z_1, x_2, y_2, z_2) \in \mathbb{R}^6$. By Corollary 2.1

$$\begin{aligned}
x_1 &= \frac{m_2 a_1}{m_1 + m_2} + a_4, & x_2 &= x_1 - a_1, \\
y_1 &= \frac{m_2 a_2}{m_1 + m_2} + a_5, & y_2 &= y_1 - a_2, \\
z_1 &= \frac{m_2 a_3}{m_1 + m_2} + a_6, & z_2 &= z_1 - a_3.
\end{aligned} \tag{3.8.1}$$

Then $(x_1, y_1, z_1) \neq (x_2, y_2, z_2)$; otherwise $a_1 = a_2 = a_3 = 0$; that is, $x \in U$:

Proposition 3.4. $\Phi(U) = \{(x_1, y_1, z_1, x_2, y_2, z_2) \in \mathbb{R}^6 | (x_1, y_1, z_1) \neq (0, 0, 0)\} = (\mathbb{R}^3 - \{0\}) \times \mathbb{R}^3$.

Recall the potential $V : U \to \mathbb{R}$ given in equation (2.5.4):

$$V(x_1, y_1, z_1, x_2, y_2, z_2) = \frac{q_1 q_2}{\sqrt{(x_1 - x_2)^2 + (y_1 - y_2)^2 + (z_1 - z_2)^2}}. \tag{3.8.2}$$

By (3.8.1) $\Phi^{-1}(x_1, y_1, z_1, x_2, y_2, z_2) =$

$$\left(\frac{m_2x_1}{m_1+m_2}+x_2, \frac{m_2y_1}{m_1+m_2}+y_2, \frac{m_2z_1}{m_1+m_2}+z_2,\right.$$
$$\left.\frac{m_2x_1}{m_1+m_2}+x_2-x_1, \frac{m_2y_1}{m_1+m_2}+y_2-y_1, \frac{m_2z_1}{m_1+m_2}+z_2-z_1\right); \text{ hence} \quad (3.8.3)$$

$$(V\circ\Phi^{-1})(x_1, y_1, z_1, x_2, y_2, z_2) = \frac{q_1q_2}{\sqrt{x_1^2+y_1^2+z_1^2}} \quad \text{on } \Phi(U). \quad (3.8.4)$$

$U, \Phi(U) \subset \mathbb{R}^6$ are open and $\Phi : U \to \Phi(U)$. Thus if $f : \Phi(U) \to \mathbb{R}$ has continuous first and second partial derivatives, and

$$\nabla_j^2 = \frac{\partial^2}{\partial x_j^2}+\frac{\partial^2}{\partial y_j^2}+\frac{\partial^2}{\partial z_j^2}, \qquad j = 1, 2, \quad (3.8.5)$$

then formula (2.8.17) can be written as

$$\frac{1}{m_1}(\nabla_1^2F)(a)+\frac{1}{m_2}(\nabla_2^2F)(a)$$
$$= \left(\frac{1}{m_1}+\frac{1}{m_2}\right)(\nabla_1^2f)(\Phi(a))+\frac{1}{m_1+m_2}(\nabla_2^2f)(\Phi(a)) \quad (3.8.6)$$

for $a \in U$, for $F = f\circ\Phi$ on U.

Suppose, for example, that $\psi : \mathbb{R}^6 \to \mathbb{C}$ is continuous and $\psi|_U$ has continuous first and second partial derivatives. Define $f = \psi\circ\Phi^{-1} : \mathbb{R}^6 \to \mathbb{C}$ so that f is continuous.

As $\Phi^{-1} : \Phi(U) \to U$ is C^∞ and $\psi : U \to \mathbb{C}$ is class C^2, $f = \psi\circ\Phi^{-1} : \Phi(U) \to \mathbb{C}$ is class C^2. Also $\Phi^{-1}(\Phi(a)) = a$ on $U \implies F = \psi$ on U. For $\mu \overset{\text{def}}{=} \frac{m_1m_2}{m_1+m_2}$ (i.e., $\frac{1}{\mu} = \frac{1}{m_1}+\frac{1}{m_2}$) we therefore have

$$\frac{1}{m_1}(\nabla_1^2\psi)(a)+\frac{1}{m_2}(\nabla_2^2\psi)(a)$$
$$= \frac{1}{\mu}(\nabla_1^2(\psi\circ\Phi^{-1}))(\Phi(a))+\frac{1}{m_1+m_2}(\nabla_2^2(\psi\circ\Phi^{-1}))(\Phi(a)) \quad (3.8.7)$$

for $a \in U$. μ is called the *reduced* mass. The time-independent Schrödinger equation (3.4.10) for ψ can therefore be written (where we take $x = \Phi(a)$)

$$\frac{1}{\mu}(\nabla_1^2(\psi\circ\Phi^{-1}))(x)+\frac{1}{m_1+m_2}(\nabla_2^2(\psi\circ\Phi^{-1}))(x)$$
$$+\frac{8\pi^2}{h^2}(E-(V\circ\Phi^{-1})(x))(\psi\circ\Phi^{-1})(x) = 0 \quad (3.8.8)$$

on $\Phi(U) = (\mathbb{R}^3 - \{0\}) \times \mathbb{R}^3$, where (by (3.8.4))

$$(V \circ \Phi^{-1})(x_1, y_1, z_1, x_2, y_2, z_2) = \frac{q_1 q_2}{\sqrt{x_1^2 + y_1^2 + z_1^2}} \quad \text{on } \Phi(U). \tag{3.8.9}$$

We shall write $x = (x_1, y_1, z_1, x_2, y_2, z_2) = (x^{(1)}, x^{(2)}) \in \mathbb{R}^3 \times \mathbb{R}^3$ for $x \in \mathbb{R}^6$. Assume $f \neq 0$ has the form $f = f_1 \otimes f_2$ for $f_1, f_2 : \mathbb{R}^3 \to \mathbb{C}$ continuous, where f_1, f_2 have continuous first and second partial derivatives on $\mathbb{R}^3 - \{0\}, \mathbb{R}^3$:

$$\begin{aligned} f(x) &= f_1(x^{(1)}) f_2(x^{(2)}) \\ \Longrightarrow \quad (\nabla_1^2 f)(x) &= (\nabla_1^2 f_1)(x^{(1)}) f_2(x^{(2)}), \\ (\nabla_2^2 f)(x) &= f_1(x^{(1)})(\nabla_2^2 f_2)(x^{(2)}) \end{aligned} \tag{3.8.10}$$

for $x \in \Phi(U)$—i.e., $x^{(1)} \in \mathbb{R}^3 - \{0\}$.

On $\Phi(U)$ Schrödinger's equation becomes

$$\begin{aligned} \frac{1}{\mu}(\nabla_1^2 f_1)(x^{(1)}) f_2(x^{(2)}) + \frac{8\pi^2}{h^2}\left(E - \frac{q_1 q_2}{|x^{(1)}|}\right) f_1(x^{(1)}) f_2(x^{(2)}) & \\ + \frac{1}{m_1 + m_2}(\nabla_2^2 f_2)(x^{(2)}) f_1(x^{(1)}) = 0. & \end{aligned} \tag{3.8.11}$$

As $f_2 \neq 0$ choose and fix some $x^{(2)} \in \mathbb{R}^3$ such that $f_2(x^{(2)}) \neq 0$. Divide by $f_2(x^{(2)})$ to get $\forall\, x^{(1)} \in \mathbb{R}^3 - \{0\}$,

$$0 \; = \; \frac{1}{\mu}(\nabla_1^2 f_1)(x^{(1)}) \; + \; \frac{8\pi^2}{h^2}\left(E - \frac{q_1 q_2}{|x^{(1)}|}\right) f_1(x^{(1)}) \; + \; c_o f_1(x^{(1)}), \tag{3.8.12}$$

where

$$c_o = \frac{1}{m_1 + m_2} \frac{(\nabla_2^2 f_2)(x^{(2)})}{f_2(x^{(2)})} \tag{3.8.13}$$

is a constant (independent of $x^{(1)}$) :

$$(\nabla_1^2 f_1)(x^{(1)}) + \frac{8\pi^2 \mu}{h^2}\left(E + c - \frac{q_1 q_2}{|x^{(1)}|}\right) f_1(x^{(1)}) = 0 \tag{3.8.14}$$

on $\mathbb{R}^3 - \{0\}$ for $c = \frac{h^2 c_o}{8\pi^2}$.

Similarly $f_1 \neq 0$ on $\mathbb{R}^3 \implies f_1 \neq 0$ on $\mathbb{R}^3 - \{0\}$ by continuity $\implies \exists\, x^{(1)} \in \mathbb{R}^3 - \{0\}$ (which we fix) such that $f_1(x^{(1)}) \neq 0$. Divide by $f_1(x^{(1)})$:

$$\begin{aligned} \frac{\frac{1}{\mu}(\nabla_1^2 f_1)(x^{(1)})}{f_1(x^{(1)})} f_2(x^{(2)}) + \frac{8\pi^2}{h^2}\left(E - \frac{q_1 q_2}{|x^{(1)}|}\right) f_2(x^{(2)}) & \\ + \frac{1}{m_1 + m_2}(\nabla_2^2 f_2)(x^{(2)}) = 0 \text{ on } R^3 : & \end{aligned} \tag{3.8.15}$$

$$(\nabla_2^2 f_2)(x^{(2)}) + df_2(x^{(2)}) + (m_1 + m_2)\frac{8\pi^2}{h^2}\left(E - \frac{q_1 q_2}{|x^{(1)}|}\right) f_2(x^{(2)}) = 0$$

for

$$d = \frac{(m_1 + m_2)}{\mu}\frac{(\nabla_1^2 f_1)(x^{(1)})}{f_1(x^{(1)})} \tag{3.8.16}$$

a constant (independent of $x^{(2)}$); or

$$(\nabla_2^2 f_2)(x^{(2)}) + \frac{8\pi^2}{h^2}(m_1 + m_2)E_2 f_2(x^{(2)}) = 0 \tag{3.8.17}$$

on $\mathbb{R}^3$ for

$$\begin{aligned} E_2 &= E - \frac{q_1 q_2}{|x^{(1)}|} + \frac{dh^2}{(m_1 + m_2)8\pi^2} \\ &= E - \frac{q_1 q_2}{|x^{(1)}|} + \frac{h^2}{\mu 8\pi^2}\frac{(\nabla_1^2 f_1)(x^{(1)})}{f_1(x^{(1)})}. \end{aligned} \tag{3.8.18}$$

Let

$$\begin{aligned} E_1 &= E + c \stackrel{\text{def}}{=} E + \frac{h^2}{8\pi^2}c_o \\ &\stackrel{\text{def}}{=} E + \frac{h^2}{8\pi^2}\frac{1}{m_1 + m_2}\frac{(\nabla_2^2 f_2)(x^{(2)})}{f_2(x^{(2)})}. \end{aligned} \tag{3.8.19}$$

Then we note that

$$\begin{aligned} &\frac{8\pi^2}{h^2} f_1(x^{(1)}) f_2(x^{(2)})(E_1 + E_2) \\ &\quad = \frac{8\pi^2}{h^2} f_1(x^{(1)}) f_2(x^{(2)}) \left[\frac{h^2}{8\pi^2}\frac{1}{m_1 + m_2}\frac{(\nabla_2^2 f_2)(x^{(2)})}{f_2(x^{(2)})} + E \right. \\ &\quad\quad \left. + E - \frac{q_1 q_2}{|x^{(1)}|} + \frac{h^2}{\mu 8\pi^2}\frac{(\nabla_1^2 f_1)(x^{(1)})}{f_1(x^{(1)})}\right] \\ &\quad = \frac{1}{m_1 + m_2}(\nabla_2^2 f_2)(x^{(2)}) f_1(x^{(1)}) + \frac{8\pi^2}{h^2}\left(E - \frac{q_1 q_2}{|x^{(1)}|}\right) f_1(x^{(1)}) f_2(x^{(2)}) \\ &\quad\quad + \frac{1}{\mu}(\nabla_1^2 f_1)(x^{(1)}) f_2(x^{(2)}) + \frac{8\pi^2}{h^2} f_1(x^{(1)}) f_2(x^{(2)}) E \\ &\quad = \frac{8\pi^2}{h^2} f_1(x^{(1)}) f_2(x^{(2)}) E, \end{aligned} \tag{3.8.20}$$

by equation (3.8.11), $\implies E_1 + E_2 = E$:

Theorem 3.3. *Under the separation of variables $f = f_1 \otimes f_2$ with f a solution of the time-independent Schrödinger equation*

$$\frac{1}{\mu}(\nabla_1^2 f)(x) + \frac{8\pi^2}{h^2}\left(E - (V \circ \Phi^{-1})(x)\right) f(x) + \frac{1}{m_1 + m_2}(\nabla_2^2 f)(x) = 0 \quad (3.8.21)$$

on $\Phi(U)$, we can find constants E_1, E_2 such that $E = E_1 + E_2$ and such that

$$(\nabla_1^2 f_1)(x^{(1)}) + \frac{8\pi^2\mu}{h^2}\left(E_1 - \frac{q_1 q_2}{|x^{(1)}|}\right) f_1(x^{(1)}) = 0 \quad (3.8.22)$$

for $x^{(1)} = (x_1, y_1, z_1) \in \mathbb{R}^3 - \{0\}$,

$$(\nabla_2^2 f_2)(x^{(2)}) + \frac{8\pi^2}{h^2}(m_1 + m_2)E_2 f_2(x^{(2)}) = 0 \quad (3.8.23)$$

for $x^{(2)} = (x_2, y_2, z_2) \in \mathbb{R}^3$. Here

$$V(x_1, y_1, z_1, x_2, y_2, z_2) = \frac{q_1 q_2}{\sqrt{(x_1 - x_2)^2 + (y_1 - y_2)^2 + (z_1 - z_2)^2}} \quad (3.8.24)$$

on $U \overset{\text{def}}{=} \{(x_1, y_1, z_1, x_2, y_2, z_2) \in \mathbb{R}^6 | (x_1, y_1, z_1) \neq (x_2, y_2, z_2)\} \implies$

$$(V \circ \Phi^{-1})(x_1, y_1, z_1, x_2, y_2, z_2) = \frac{q_1 q_2}{|x^{(1)}| = \sqrt{x_1^2 + y_1^2 + z_1^2}} \quad (3.8.25)$$

on $\Phi(U) = \{(x_1, y_1, z_1, x_2, y_2, z_2) \in \mathbb{R}^6 | (x_1, y_1, z_1) \neq (0, 0, 0)\}$. $\mu = \frac{m_1 m_2}{m_1 + m_2}$.

Conversely, let $f_1, f_2 : \mathbb{R}^3 \to \mathbb{C}$ (continuous) be given with continuous first and second partial derivatives on $\mathbb{R}^3 - \{0\}, \mathbb{R}^3$. Let E_1, E_2 be constants such that

$$(\nabla_1^2 f_1)(x^{(1)}) + \frac{8\pi^2\mu}{h^2}\left(E_1 - \frac{q_1 q_2}{|x^{(1)}|}\right) f_1(x^{(1)}) = 0 \quad (3.8.26)$$

on $\mathbb{R}^3 - \{0\}$,

$$(\nabla_2^2 f_2)(x^{(2)}) + \frac{8\pi^2}{h^2}(m_1 + m_2)E_2 f_2(x^{(2)}) = 0 \quad (3.8.27)$$

on $\mathbb{R}^3$.

Multiply the first equation by $\frac{1}{\mu} f_2(x^{(2)})$ and the second equation by $\frac{f_1(x^{(1)})}{m_1 + m_2}$:

$$\frac{1}{\mu}(\nabla_1^2 f_1)(x^{(1)}) f_2(x^{(2)}) + \frac{8\pi^2}{h^2}\left(E_1 - \frac{q_1 q_2}{|x^{(1)}|}\right) f_1(x^{(1)}) f_2(x^{(2)}) = 0 \quad (3.8.28)$$

and

$$\frac{1}{m_1 + m_2}(\nabla_2^2 f_2)(x^{(2)}) f_1(x^{(1)}) + \frac{8\pi^2}{h^2} E_2 f_1(x^{(1)}) f_2(x^{(2)}) = 0. \quad (3.8.29)$$

Add the last two equations:

$$\frac{1}{\mu}(\nabla_1^2 f_1)(x^{(1)})f_2(x^{(2)}) + \frac{8\pi^2}{h^2}\left(E_1 + E_2 - \frac{q_1 q_2}{|x^{(1)}|}\right) f_1(x^{(1)})f_2(x^{(2)}) + \frac{1}{m_1+m_2}(\nabla_2^2 f_2)(x^{(2)})f_1(x^{(1)}) = 0. \quad (3.8.30)$$

Thus for $f \stackrel{\text{def}}{=} f_1 \otimes f_2$, $E = E_1 + E_2$, (3.8.21) holds:

$$\frac{1}{\mu}(\nabla_1^2 f)(x) + \frac{8\pi^2}{h^2}\left(E - (V \circ \Phi^{-1})(x)\right) f(x) + \frac{1}{m_1+m_2}(\nabla_2^2 f)(x) = 0 \quad (3.8.31)$$

on $\Phi(U)$. Here $f : \mathbb{R}^6 \to \mathbb{C}$ is continuous with continuous first and second partial derivatives on $\Phi(U)$.

3.9 Some Brief Remarks on Axioms

In the reference [84], J. von Neumann presents a rigorous mathematical foundation of quantum mechanics based on the spectral theory of Hermitian (= self-adjoint) operators on a separable Hilbert space. It is not our purpose to duplicate the rigors of that great treatise. We offer some brief, somewhat leisure comments on some of the basic axioms of the subject. Such comments complement some of the ideas of Chapter 1 and serve as background for some remarks and examples that will arise in Chapters 5, 6, 7, 9, and 10. Other works with sufficient rigor, of particular interest to mathematicians, are those of F. Berezin and M. Shubin [3], G. Mackey [62], E. Prugovečki [71], A. Sudbery [80], B. van der Waerden [83], and H. Weyl [87], for example.

Observables of classical mechanics are the basic dynamical variables such as position, momentum (linear and angular), energy, etc., or functions of such variables. To the observable p = linear momentum, for example, we have assigned the self-adjoint operator $\hat{p} = -\hbar i \frac{d}{dx}$ (in a one-dimensional set-up), though we have not been careful about specifying the (dense) domain of $\hat{p}$. For the case of angular momentum, note the assignment given in (5.4.4). In general one of the basic axioms of quantum mechanics is that to each classical observable O there is assigned a Hermitian operator $\widehat{O}$ on a complex separable Hilbert space H. Here H is the *state space* of some given quantum system, a state of the system being represented by some non-zero vector $\psi \in H$. Compare for example the sentence following condition (iii) in Section 3.1. More precisely, and axiomatically, the states of the quantum system are the one-dimensional subspaces of H; thus ψ and $c\psi$, $c \in \mathbb{C}$, represent the *same* state.

As $\widehat{O}$ is self-adjoint its eigenvalues λ are real. Another basic axiom is the following:

> *The only values that a measurement of O can yield are the eigenvalues λ of $\widehat{O}$.* (3.9.1)

If $\psi \in H - \{0\}$ is an eigenvector corresponding to the eigenvalue λ of $\widehat{O}$, then ψ describes the state of the system in which O assumes the definite value λ. For $\widehat{O}$ equal to the quantized z-component of angular momentum, for example, with eigenvalues $\lambda = \hbar m$, for certain *magnetic quantum numbers m* (see Theorem 5.3 of Chapter 5), we see that by (3.9.1) a measurement of O (which is the z-component of the classical angular momentum of a particle) can yield only the values $\hbar m$. Later we will take $|\hbar m| = \hbar|m|$ to be the *magnitude* of O.

We will need to consider the *expectation value* $\overline{\widehat{O}_\psi}$ of $\widehat{O}$ in the state ψ. One takes as axiomatic the following definition

$$\overline{\widehat{O}_\psi} \stackrel{\text{def}}{=} \frac{\langle \widehat{O}\psi, \psi\rangle}{\langle \psi, \psi\rangle} \tag{3.9.2}$$

where $\langle , \rangle$ denotes the inner product structure on H. See definition (6.2.7) for example, where ψ there is normalized: $\langle \psi, \psi\rangle = 1$. Suppose for example that $\{\psi_n\}_{n=1}^{\infty}$ is a complete orthonormal basis of H consisting of eigenvectors of $\widehat{O}$:

$$\widehat{O}\psi_n = \lambda_n \psi_n. \tag{3.9.3}$$

Then for $\psi \in H$ arbitrary,

$$\psi = \sum_{n=1}^{\infty} a_n \psi_n \qquad \text{for } a_n = \langle \psi, \psi_n\rangle \tag{3.9.4}$$

where convergence is taken in the strong sense:

$$\|\psi - \text{ the } n^{\text{th}} \text{ partial sum}\| \to 0 \quad \text{as } n \to \infty \tag{3.9.5}$$

$$\text{for } \|\psi\|^2 \stackrel{\text{def}}{=} \langle \psi, \psi\rangle = \sum_{n=1}^{\infty} |a_n|^2. \tag{3.9.6}$$

Using (3.9.3) and (3.9.4) we can express the numerator in (3.9.2) as follows:

$$\begin{aligned}\langle \widehat{O}\psi, \psi\rangle &= \sum_{n,m=1}^{\infty} \langle \widehat{O} a_n \psi_n, a_m \psi_m\rangle \\ &= \sum_{n,m=1}^{\infty} a_n \bar{a}_m \lambda_n \delta_{n,m} = \sum_{n=1}^{\infty} \lambda_n |a_n|^2.\end{aligned} \tag{3.9.7}$$

That is, by (3.9.6), (3.9.7)

$$\begin{aligned}&\overline{\widehat{O}_\psi} = \sum_{n=1}^{\infty} \lambda_n p_n \quad \text{for} \\ &p_n \stackrel{\text{def}}{=} \frac{|a_n|^2}{\sum_{n=1}^{\infty} |a_n|^2}; \text{ hence } \sum_{n=1}^{\infty} p_n = 1.\end{aligned} \tag{3.9.8}$$

To understand the meaning of (3.9.8) consider for example a sequence of n discrete values $\{x_1, \ldots, x_n\}$ where x_j occurs with frequency f_j. Then

$$\bar{x} \stackrel{\text{def}}{=} \frac{1}{n} \sum_{j=1}^{n} x_j f_j \tag{3.9.9}$$

is the ordinary mean value of the sequence. Of course we can write

$$\bar{x} = \sum_{j=1}^{n} x_j p_j \quad \text{for } p_j \stackrel{\text{def}}{=} \frac{f_j}{n}; \text{ hence } \sum_{j=1}^{n} p_j = 1. \tag{3.9.10}$$

Thus p_j is the *relative frequency* with which the value x_j occurs in the sequence. Or p_j is the *probability* that the value x_j would be obtained by a random selection from the list $\{x_1, \ldots, x_n\}$. Keeping (3.9.1) in mind, and comparing (3.9.8) with (3.9.10), one is lead to postulate that in (3.9.8), p_n *is the probability that a measurement of the system in the state* ψ *will yield precisely the value* λ_n. Moreover, equation (3.9.8) expresses $\overline{\widehat{O}_\psi}$ as a *weighted average* of a large number of measurements of the system. In particular if the state ψ is normalized we obtain a statistical meaning of the values $|a_n|^2 = p_n$. In Section 7.7 of Chapter 7 we will have, for example, certain time-dependent coefficients $a_n(t)$ where the $|a_n(t)|^2$ will represent certain *transition probabilities*.

It seems fitting at this point to mention also a fundamental result of Kato. The theory and applications of quantum mechanics proceeded for many years with no formal proof of the essential self-adjointness of the basic Hamiltonians $H = -$ Laplacian $+V$ occuring in Schrödinger's equation. This fundamental and critical question for the mathematical foundations of the theory was certainly on J. von Neumann's mind in the late 1920s and afterwards. Finally in 1951, Tosio Kato (1917-1999) established essential self-adjointness for atomic Hamiltonians (defined initially on a dense domain of smooth, compactly supported functions)—i.e., for potentials V given in (3.5.18); see references [22, 56, 57, 71]. Kato and others have extended his result to other types of Hamiltonians to the point that now the issue of self-adjointness is settled for all of the practical cases of interest in quantum mechanics.

4

Hypergeometric Equations and Special Functions

4.1 Initial Remarks

Many of the important partial differential equations of applied mathematics and theoretical physics are second order and are studied by the method of separation of variables which leads to a set of second-order ordinary differential equations. This is the case, for example, for equations of Laplace and Helmholtz and equations of Schrödinger, Dirac, and Klein–Gordon, say when one has a Coulomb type potential. The ordinary differential equations, arising by separation of variables, can be reduced in fact to equations of *hypergeometric type*—a type which we shall describe shortly.

In the standard treatment in books on quantum mechanics these equations, arising from Schrödinger's equation, say, are solved by the method of power series (in the few cases where solutions are possible). One thus obtains the wave functions and the desired energy levels of the quantum system. We shall present an entirely different approach, due to Arnold Nikiforov and Vasilii Uvarov [66] which is more uniform and elegant. For example, as will be seen later, energy levels will have the very general form

$$\lambda_n = -n\frac{d\tau}{dx} - \frac{n(n-1)}{2}\frac{d\sigma^2}{dx^2} \tag{4.1.1}$$

for $n = 0, 1, 2, \ldots$ where σ, τ are polynomial coefficients in a hypergeometric equation (arising from Schrödinger's equation) of the form (4.2.2) below. As $\deg \sigma \leq 2$, $\deg \tau \leq 1$ the λ_n indeed are scalars. Thus we begin with the study of special second-order ordinary differential equations, those of hypergeometric type, a study which provides in fact a uniform approach to the special functions of mathematical physics.

4.2 Differential Equations of Hypergeometric Type

A differential equation of *hypergeometric type* is one of the form

$$\sigma^2 u'' + \sigma\tilde{\tau}u' + \tilde{\sigma}u = 0 \tag{4.2.1}$$

on some domain of real or complex numbers, where σ, $\tilde{\sigma}$, $\tilde{\tau}$ are polynomials with $\deg\sigma$, $\deg\tilde{\sigma} \leq 2$, $\deg\tilde{\tau} \leq 1$. A special case of (4.2.1) is

$$\sigma y'' + \tau y' + \lambda y = 0 \tag{4.2.2}$$

where σ, τ are polynomials, $\deg\sigma \leq 2$, $\deg\tau \leq 1$, and $\lambda \in \mathbb{R}$ or $\mathbb{C}$. Namely write (4.2.2) as $\sigma^2 y'' + \sigma\tau y' + \lambda\sigma y = 0$. Thus take $\tilde{\tau} = \tau$, $\tilde{\sigma} = \lambda\sigma$. We shall call (4.2.2) a *canonical form* of (4.2.1). Following A. Nikiforov and V. Uvarov [66] we show that (4.2.1) can be reduced to (4.2.2). First we consider some classical examples.

Classical Examples

$$(1 - x^2)y''(x) - 2xy'(x) + \lambda y(x) = 0 \text{ on } \mathbb{R} \tag{4.2.3}$$
$$y''(x) - 2xy'(x) + \lambda y(x) = 0 \text{ on } \mathbb{R} \tag{4.2.4}$$
$$xy''(x) + (\alpha + 1 - x)y'(x) + \lambda y(x) = 0 \text{ on } \mathbb{R}^+ \stackrel{\text{def}}{=} \{x \in \mathbb{R} \mid x > 0\} \tag{4.2.5}$$
$$(1 - x^2)y''(x) - xy'(x) + \lambda^2 y(x) = 0 \text{ on } (-1, 1) \tag{4.2.6}$$

$$x^2 y''(x) + xy'(x) + (x^2 - \lambda^2)y(x) = 0 \text{ on } \mathbb{R} - \{0\} \tag{4.2.7}$$
$$x(x - 1)y''(x) + [(\alpha + \beta + 1)x - \gamma]y'(x) + \alpha\beta y(x) = 0 \text{ on } x > 1 \tag{4.2.8}$$
$$\alpha x^2 y''(x) + \beta xy'(x) + \gamma y(x) = 0 \text{ on } \mathbb{R}^+ \tag{4.2.9}$$

are equations of Legendre, Hermite, Laguerre, Chebyshev, Bessel, Gauss and Euler, respectively. Except for equation (4.2.7) all are examples of equation (4.2.2). Equation (4.2.8) of course is the Gauss hypergeometric equation.

We shall also consider Jacobi's differential equation:

$$(1 - x^2)y''(x) + [-(\alpha + \beta + 2)x + \beta - \alpha]\, y'(x) + \mu y(x) = 0 \text{ on } (-1, 1). \tag{4.2.10}$$

Some standard polynomials from advanced calculus are given as follows, in *Rodrigues form*. Let $n, m \geq 0$ be integers. The *Legendre* polynomials P_n of degree n are defined by

$$P_n(x) \stackrel{\text{def}}{=} \frac{1}{2^n n!} \frac{d^n}{dx^n}(x^2-1)^n \tag{4.2.11}$$

for $n \geq 1$; we set $P_o(x) = 1$.

For example $P_1(x) = x$, $P_2(x) = \frac{3}{2}x^2 - \frac{1}{2}$, $P_3(x) = \frac{5}{2}x^3 - \frac{3}{2}x$. Similarly, the *Laguerre* polynomials L_n^m of degree n are defined by

$$L_n^m(x) = e^x \frac{x^{-m}}{n!} \frac{d^n}{dx^n}(e^{-x}x^{n+m}) \tag{4.2.12}$$

for $x \neq 0$; we set $L_n^m(0) = \frac{(n+m)!}{n!m!}$. More generally we will replace m by $\alpha \in \mathbb{R}$.

For example $L_o^m(x) = 1$, $L_1^m(x) = 1+m-x$, $L_2^m(x) = \frac{1}{2}(m+1)(m+2) - (m+2)x + \frac{x^2}{2}$. The *Hermite* polynomials H_n of degree n are defined by

$$H_n(x) = (-1)^n \exp(x^2) \frac{d^n}{dx^n} \exp(-x^2). \tag{4.2.13}$$

$H_o(x) = 1$, $H_1(x) = 2x$, $H_2(x) = 4x^2-2$. A generalization of the Legendre polynomials P_n is given by the three-parameter family of *Jacobi* polynomials $P_n^{(\alpha,\beta)}$ of degree n defined by

$$P_n^{(\alpha,\beta)}(x) = \frac{(-1)^n}{2^n n!}(1-x)^{-\alpha}(1+x)^{-\beta} \frac{d^n}{dx^n}\left[(1-x)^{\alpha+n}(1+x)^{\beta+n}\right] \tag{4.2.14}$$

on $(-1,1)$ for $\alpha, \beta \in \mathbb{R}$; $P_n^{(0,0)}(x) = P_n(x)$ on $(-1,1)$. We shall prove that, for example, $P_n^{(\alpha,\beta)}$ is a solution of Jacobi's differential equation (4.2.10) provided $n(\alpha+\beta+n+1) = \mu$. Similarly H_n, P_n, L_n^m are solutions of equations (4.2.4), (4.2.3), (4.2.5) provided that $\lambda = 2n, n(n+1), n$, respectively, there. Given P_n, the *associated Legendre functions* P_n^m are defined by

$$P_n^m(x) = (1-x^2)^{m/2} \frac{d^m}{dx^m} P_n(x) \tag{4.2.15}$$

on $(-1,1)$.

4.3 Reduction to Canonical Form

Let $f : \mathbb{C} \to \mathbb{C}$ be a polynomial function of degree ≤ 2: $f(z) = az^2 + bz + c$, $a, b, c \in \mathbb{C}$. Let $\Delta = b^2 - 4ac$ be the discriminant of f. If $a \neq 0$ ($\deg f = 2$) then completing the square, one writes $f(z) = a((z+\frac{b}{2a})^2 - (\frac{b^2-4ac}{4a^2})) = a(z+\frac{b}{2a})^2 - \frac{\Delta}{4a}$. Choose a square root a_0 of a: $a_0^2 = a \implies f(z) = (a_0(z+\frac{b}{2a}))^2 - \frac{\Delta}{4a}$. Thus if $\Delta = 0$, $f(z)$ has a square root $a_0 z + \frac{a_0 b}{2a}$ for $a \neq 0$; here $a_0 \neq 0$. If $a = 0$ and $\Delta = 0$, then $b^2 = 4ac \implies b = 0 \implies f(z) = c$. Thus if c_0 is a square root of c, the constant function $z \to c_0$ is a square root of f. This proves

Proposition 4.1. *Let $f : \mathbb{C} \to \mathbb{C}$ be a polynomial function of degree ≤ 2, with discriminant $\Delta : f(z) = az^2 + bz + c$, a, b, $c \in \mathbb{C}$, $\Delta = b^2 - 4ac$. Suppose $\Delta = 0$. Then f has a square root $p : \mathbb{C} \to \mathbb{C}$ which is a polynomial of degree ≤ 1 : If*

$a \neq 0$ and $a_0^2 = a$, then $p(z) = a_0 z + \frac{a_0 b}{2a} = a_0 z + \frac{b}{2a_0} \implies$ degree $p = 1$; $a_0 \neq 0$. If $a = 0$ and $c_0^2 = c$, then $p(z) = c_0 \implies$ degree $p = 0$; here $a = 0$, $\Delta = 0 \implies b = 0 \implies f(z) = c$.

Given polynomial functions $\sigma, \tilde{\sigma}, \tilde{\tau} : \mathbb{C} \to \mathbb{C}$ with degree $\sigma, \tilde{\sigma} \leq 2$, degree $\tilde{\tau} \leq 1$, and given $\kappa \in \mathbb{C}$, define $f_\kappa : \mathbb{C} \to \mathbb{C}$ by $f_\kappa = \frac{(\tilde{\tau}-\sigma')^2}{4} + \kappa\sigma - \tilde{\sigma}$. Thus f_κ is a polynomial function of degree ≤ 2. More explicitly, write $\sigma(z) = a_0 + a_1 z + a_2 z^2$, $\tilde{\sigma}(z) = b_0 + b_1 z + b_2 z^2$, $\tilde{\tau}(z) = c_0 + c_1 z$; $\sigma'(z) = a_1 + 2a_2 z \implies (\tilde{\tau}(z) - \sigma'(z))^2 = (c_0 - a_1 + (c_1 - 2a_2)z)^2 = (c_0 - a_1)^2 + 2(c_0 - a_1)(c_1 - 2a_2)z + (c_1 - 2a_2)^2 z^2 \implies f_\kappa(z) = (\frac{(c_0-a_1)^2}{4} - \kappa a_0 - b_0) + (\frac{(c_0-a_1)(c_1-2a_2)}{2} + \kappa a_1 - b_1)z + (\frac{(c_1-2a_2)^2}{4} + \kappa a_2 - b_2)z^2$.

Therefore if $\Delta(\kappa)$ is the discriminant of f_κ, the equation $\Delta(\kappa) = 0$ is a quadratic equation in κ (which might in some special case reduce to a linear equation in κ). Assume $\Delta(\kappa) = 0$. Then, by Proposition 4.1, we know that we can find a polynomial function p such that $p^2 = f_\kappa$, degree $p \leq 1$: If $f_\kappa(z) = c + bz + az^2$ with $a \neq 0$ we can take $p(z) = a_0 z + \frac{a_0 b}{2a} = a_0 z + \frac{b}{2a_0}$ where $a_0^2 = a$, $a_0 \neq 0$; if $a = 0$, then $\Delta(\kappa) = 0 \implies b = 0 \implies f_\kappa(z) = c$, and we take $p(z) = c_0$, where $c_0^2 = c$.

Let $\pi^\pm = \frac{\sigma'-\tilde{\tau}}{2} \pm p$. Then degree $\pi^\pm \leq 1$ and $(\pi^\pm)^2 + (\tilde{\tau} - \sigma')\pi^\pm = \frac{(\tilde{\tau}-\sigma')^2}{4} \pm (\sigma' - \tilde{\tau})p + p^2 + (\tilde{\tau} - \sigma')(\frac{\sigma'-\tilde{\tau}}{2}) \pm (\tilde{\tau} - \sigma')p = -\frac{(\tilde{\tau}-\sigma')^2}{2} + p^2 = -\frac{(\tilde{\tau}-\sigma')^2}{4} + f_\kappa \overset{\text{def}}{=} \kappa\sigma - \tilde{\sigma}$, which proves

Proposition 4.2. *Let $\kappa \in \mathbb{C}$ such that $\Delta(\kappa) = 0$ where $\Delta(\kappa)$ is the discriminant of $f_\kappa \overset{\text{def}}{=} \frac{(\tilde{\tau}-\sigma')^2}{4} + \kappa\sigma - \tilde{\sigma}$. Thus, by Proposition 4.1, we can choose a polynomial p of degree ≤ 1 such that $p^2 = f_\kappa$. Let $\pi^\pm \overset{\text{def}}{=} \frac{\sigma'-\tilde{\tau}}{2} \pm p$. Then degree $\pi^\pm \leq 1$ and $(\pi^\pm)^2 + (\tilde{\tau} - \sigma')\pi^\pm + \tilde{\sigma} - \kappa\sigma = 0$.*

Again let $\kappa \in \mathbb{C}$ be a solution of $\Delta(\kappa) = 0$, where $\Delta(\kappa)$ is the discriminant of $f_\kappa \overset{\text{def}}{=} \frac{(\tilde{\tau}-\sigma')^2}{4} + \kappa\sigma - \tilde{\sigma}$. Then we can find a solution π of $\pi^2 + (\tilde{\tau} - \sigma')\pi + \tilde{\sigma} - \kappa\sigma \overset{\dagger}{=} 0$, $\pi(z) = \alpha + \beta z$, $\alpha, \beta \in \mathbb{C}$; for example, we have solutions $\pi = \pi^\pm$ by Proposition 4.2. Set $\lambda \overset{\text{def}}{=} \kappa + \pi' \in \mathbb{C}$ and set $\tau \overset{\text{def}}{=} \tilde{\tau} + 2\pi$; then degree $\tau \leq 1$. Also, on a domain D on which σ is non-vanishing, let ϕ be a solution of $\phi' = \phi\pi/\sigma$: $\phi(z) = e^{\int \pi(z)/\sigma(z)\, dz}$ on D. Then

Proposition 4.3.

$$\sigma^2\phi'' + \sigma\tilde{\tau}\phi' + \tilde{\sigma}\phi = \lambda\sigma\phi \quad \text{on } D \tag{4.3.1}$$

and

$$2\sigma^2\phi' + \sigma\tilde{\tau}\phi = \sigma\phi\tau \quad \text{on } D.$$

Proof. We compute on D: $\frac{\sigma\pi'-\pi\sigma'}{\sigma^2} = (\frac{\pi}{\sigma})' \overset{\text{def}}{=} (\frac{\phi'}{\phi})' = \frac{\phi\phi''-\phi'\phi'}{\phi^2} \overset{\text{def}}{=} \frac{\phi''}{\phi} - \frac{\pi^2}{\sigma^2} \implies \sigma\pi' - \pi\sigma' = \sigma^2\frac{\phi''}{\phi} - \pi^2 \implies \sigma^2\frac{\phi''}{\phi} - \sigma\pi' = \pi^2 - \pi\sigma' \overset{\text{by } \dagger}{=} -\tilde{\tau}\pi + \kappa\sigma - \tilde{\sigma} \overset{\text{def}}{=} -\tilde{\tau}\pi + (\lambda - \pi')\sigma - \tilde{\sigma} \implies \sigma^2\frac{\phi''}{\phi} = \lambda\sigma - \tilde{\sigma} - \tilde{\tau}\pi \implies \sigma^2\phi'' = \lambda\sigma\phi - \tilde{\sigma}\phi - \phi\tilde{\tau}\pi \overset{\text{def}}{=} \lambda\sigma\phi - \tilde{\sigma}\phi - \tilde{\tau}\sigma\phi' \implies \sigma^2\phi'' + \sigma\tilde{\tau}\phi' + \tilde{\sigma}\phi = \lambda\sigma\phi$, as claimed. Also $2\sigma^2\phi' + \sigma\tilde{\tau}\phi \overset{\text{def}}{=} 2\sigma^2\phi\frac{\pi}{\sigma} + \sigma\tilde{\tau}\phi = 2\sigma\phi\pi + \sigma\tilde{\tau}\phi = \sigma\phi(2\pi + \tilde{\tau}) \overset{\text{def}}{=} \sigma\phi\tau$. □

Let u, y be functions related by $u = \phi y$ on D. Then $u' = \phi y' + y\phi'$, $u'' = \phi y'' + y'\phi' + y\phi'' + \phi' y' = \phi y'' + 2\phi' y' + y\phi'' \implies \sigma^2 u'' + \sigma\tilde{\tau} u' + \tilde{\sigma} u = \sigma^2 \phi y'' + 2\sigma^2 \phi' y' + \sigma^2 y\phi'' + \sigma\tilde{\tau}\phi y' + \sigma\tilde{\tau}\phi' y + \tilde{\sigma}\phi y = \sigma^2\phi y'' + (2\sigma^2\phi' + \sigma\tilde{\tau}\phi) y' + (\sigma^2\phi'' + \sigma\tilde{\tau}\phi' + \tilde{\sigma}\phi) y = \sigma^2\phi y'' + (\sigma\phi\tau) y' + \lambda\sigma\phi y$ (by Proposition 4.3) $= \sigma\phi(\sigma y'' + \tau y' + \lambda y)$. Since σ never vanishes on D we have thus proved

Theorem 4.1 (Reduction to canonical form). *Let D be a domain on which σ is non-vanishing. Then on D,*

$$\sigma^2 u'' + \sigma\tilde{\tau} u' + \tilde{\sigma} u = 0 \iff \sigma y'' + \tau y' + \lambda y = 0, \tag{4.3.2}$$

for u, y related by $u = \phi y$.

Equivalently

$$u'' + \frac{\tilde{\tau}}{\sigma} u' + \frac{\tilde{\sigma}}{\sigma^2} u = 0 \iff \sigma y'' + \tau y' + \lambda y = 0. \tag{4.3.3}$$

Proposition 4.4. *On the domain D on which σ is non-vanishing, given a solution ϕ of $\phi' = \frac{\phi\pi}{\sigma}$, and a solution h of $h' = \frac{\tilde{\tau}}{\sigma}$, set $\rho = \frac{\phi^2}{\sigma} e^h$. Then ρ is a solution of $(\rho\sigma)' = \rho\tau$ on D. Recall $\tau \stackrel{\text{def}}{=} \tilde{\tau} + 2\pi$.*

Proof. $(\rho\sigma)' = (\phi^2 e^h)' = \phi^2 e^h h' + 2\phi\phi' e^h = \phi^2 e^h \frac{\tilde{\tau}}{\sigma} + 2\phi\left(\frac{\phi\pi}{\sigma}\right) e^h \stackrel{\text{def}}{=} \rho\tilde{\tau} + 2\pi\rho = \rho(\tilde{\tau} + 2\pi) \stackrel{\text{def}}{=} \rho\tau$. □

In applying Theorem 4.1 the reader obviously must be clear about the notation and definitions established. For convenience we summarize the reduction procedure, reiterating how one obtains τ and λ in (4.3.2) (the second equation there being equation (4.2.2) of course). We are given the coefficients $\sigma, \tilde{\sigma}, \tilde{\tau}$ in (4.2.1). Given $\kappa \in \mathbb{C}$ one chooses a square root p of the polynomial

$$f_\kappa \stackrel{\text{def}}{=} \frac{(\tilde{\tau} - \sigma')^2}{4} + \kappa\sigma - \tilde{\sigma}, \tag{4.3.4}$$

assuming that the discriminant $\Delta(\kappa)$ of f_κ vanishes; see Proposition 4.1. Let

$$\pi^{\pm} \stackrel{\text{def}}{=} \frac{\sigma' - \tilde{\tau}}{2} \pm p \tag{4.3.5}$$

and let π be one of the choices π^+ or π^-. Then

$$\tau \stackrel{\text{def}}{=} \tilde{\tau} + 2\pi, \qquad \lambda \stackrel{\text{def}}{=} \kappa + \pi', \tag{4.3.6}$$

with degree $\pi, \tau \leq 1$, $\lambda \in \mathbb{C}$. Also the solutions u, y in (4.3.2) (i.e., in (4.2.1) and (4.2.2)) are related by $u = \phi y$ where ϕ satisfies

$$\phi' = \frac{\phi\pi}{\sigma} \tag{4.3.7}$$

on a domain D on which σ is non-vanishing. For example one can take

$$\phi = \exp\left(\int \frac{\pi(z)}{\sigma(z)}\, dz\right). \tag{4.3.8}$$

The reduction to canonical form involves certain choices of signs. For example, f_κ (with $\Delta(\kappa) = 0$) has two square roots: $\pm p$. There is a more or less standard way of making the choice of signs. In some basic examples from mathematical physics one has in fact that $\tau' < 0$ and τ has a *root* on some distinguished interval a to b (possibly $a = -\infty$, $b = \infty$) on which $\sigma y'' + \tau y + \lambda y = 0$ is defined. Thus we will generally choose our signs so that $\tau' < 0$ and such that τ has a root in (a, b). Further justification for this is given by the following lemma.

Lemma 4.1. *Let $\tau(x)$ be a real-valued polynomial function of x of degree ≤ 1. Let σ, ρ be real-valued continuously differentiable functions on a closed interval $[a, b]$ $(a, b \in \mathbb{R}, a < b)$ such that $(\rho\sigma)'(x) = (\rho\tau)(x)$ with $\rho(x)$, $\sigma(x) > 0$ on (a, b). Suppose also that $\rho(x)\sigma(x)x^k \overset{\ddagger}{=} 0$ at $x = a, b$, for $k = 0, 1$. Then $\tau' < 0$. In particular τ is non-constant; i.e.,* $\deg \tau = 1$. *Also τ has a root in (a, b).*

Proof. On (a, b), $(\rho\tau^2)(x) = ((\rho\tau)\tau)(x) = ((\rho\sigma)'\tau)(x)$. Therefore $(\rho\tau^2)(x) = ((\rho\sigma)'\tau)(x)$ on $[a, b]$, by continuity (i.e., $(\rho\sigma)'(x) = (\rho\tau)(x)$ on $[a, b]$). Here τ, $(\rho\sigma)$ are differentiable on $[a, b]$ with τ', $(\rho\sigma)'$ integrable there. Therefore integrate by parts: $\int_a^b (\rho\tau^2)(x)\, dx = \int_a^b [\tau(\rho\sigma)'](x)\, dx = \tau\rho\sigma(x)]_a^b - \int_a^b \tau'(\rho\sigma)(x)\, dx = -\tau' \int_a^b (\rho\sigma)(x)\, dx$ by ($\ddagger$). We note that $\tau \not\equiv 0$ on (a, b). Otherwise $(\rho\sigma)'(x) = (\tau\rho)(x) = 0$ on $(a, b) \implies (\rho\sigma)(x) = c$ is a constant on (a, b). By continuity $(\rho\sigma)(x) = c$ on $[a, b]$. By ($\ddagger$), $c = (\rho\sigma)(a) = 0 \implies (\rho\sigma)(x) = 0$ on (a, b), where $\rho(x)$, $\sigma(x) > 0$. As this is not possible we must have $\tau \not\equiv 0$ on $(a, b) : \tau^2(x_0) > 0$ for some x_0, $a < x_0 < b$. Then $\int_a^b (\rho\tau^2)(x)\, dx > 0$, as $\rho\tau^2 \geq 0$ on $[a, b]$. Also $\int_a^b (\rho\sigma)(x)\, dx > 0$ [which also follows by the mean-value theorem for integrals]. Therefore

$$\tau' = \frac{-\int_a^b (\rho\tau^2)(x)\, dx}{\int_a^b (\rho\sigma)(x)\, dx} < 0. \tag{4.3.9}$$

We can write (4.3.9) of course as

$$\int_a^b (\rho\tau^2)(x)\, dx = -\tau' \int_a^b (\rho\sigma)(x)\, dx. \tag{4.3.10}$$

Also, as $\rho\sigma$ is continuous in $[a, b]$ and differentiable on (a, b), and as $(\rho\sigma)(a) = (\rho\sigma)(b) (= 0)$ by ($\ddagger$) we can apply Rolle's theorem: $\exists$ a point x_0 in (a, b) for which $(\rho\sigma)'(x_0) = 0$; i.e., $\rho(x_0)\tau(x_0) = 0 \implies \tau(x_0) = 0$. That is, *$\tau$ has a zero in (a, b).* □

Remarks. Suppose in the preceding arguments we start with a polynomial function $f : \mathbb{R} \to \mathbb{R}$ of degree ≤ 2 with *real* coefficients : $f(x) = ax^2 + bx + c$,

$a, b, c \in \mathbb{R}$. If $a, c \geq 0$ the above arguments provide for a *real* square root p of f provided $\Delta = b^2 - 4ac = 0 : p(x) = a_0 x + \frac{b}{2a_0}$ for $a_0 = \sqrt{a} \neq 0$ if $a > 0$; $p(x) = \sqrt{c}$ if $a = 0$. Thus suppose now we are given polynomial functions $\sigma, \tilde{\sigma}$, $\tilde{\tau} : \mathbb{R} \to \mathbb{R}$ with *real* coefficients such that again degree $\sigma, \tilde{\sigma} \leq 2$, degree $\tilde{\tau} \leq 1$. If $\kappa \in \mathbb{R}$, then $f_\kappa \stackrel{\text{def}}{=} \frac{[\tilde{\tau}-\sigma']^2}{4} + \kappa\sigma - \tilde{\sigma}$ has real coefficients; $\deg f_\kappa \leq 2$. Write $f_\kappa(x) = ax^2 + bx + c$, $a, b, c \in \mathbb{R}$, and assume $\Delta(\kappa)(= b^2 - 4ac) = 0$ for $\kappa \in \mathbb{R}$. If $a, c \geq 0$, then f_κ has a real square root $p : p(x) = \sqrt{a}x + \frac{b}{2\sqrt{a}}$ if $a > 0$, $p(x) = \sqrt{c}$ if $a = 0$. Then $\pi^\pm \stackrel{\text{def}}{=} \frac{\sigma'-\tilde{\tau}}{2} \pm p$ has real coefficients. Also $\lambda \stackrel{\text{def}}{=} \kappa + \pi' \in \mathbb{R}$, and $\pi = \pi^\pm$ and $\tau = \tilde{\tau} + 2\pi$ have real coefficients. Choose ϕ = a real solution on D of $\phi' = \phi\pi/\sigma$, where D = a domain on which σ is non-vanishing. For h a real solution of $h' = \tilde{\tau}/\sigma$, $\rho = \frac{\phi^2}{\sigma}e^h$ is real: $(\rho\sigma)' = \rho\tau$ on D, as in Proposition 4.4.

4.4 Solutions of $\sigma y'' + \tau y' + \lambda y = 0$

Given the reduction of the general hypergeometric equation (4.2.1) to a canonical form (4.2.2), provided by Theorem 4.1, we focus a bit on solutions of (4.2.2)—in particular on polynomial solutions. In a unified, uniform manner we obtain the classical orthogonal polynomial solutions to equations of Hermite, Legendre, Laguerre, Jacobi, and others, for examples, as well as recursion properties of these solutions; cf. equations (4.2.3) through (4.2.10). These solutions to (4.2.2) incidentally will be given in a generalized *Rodrigues form*; see (4.4.12) below.

We provide several examples of the reduction technique of Theorem 4.1 to obtain results for later applications, and we compute the wave functions and energy levels of the quantized harmonic oscillator, verifying some assertions made in Chapter 3.

Let $p, q : \mathbb{R} \to \mathbb{C}$ be functions of the form $p(x) = a + bx$, $q(x) = \alpha + \beta x + \gamma x^2$ ($a, b, \alpha, \beta, \gamma \in \mathbb{C}$; possibly $b, \gamma = 0$) and let $f : U^{\text{open}} \subset \mathbb{R} \to \mathbb{C}$ be n times differentiable, $n \geq 2$. Then by Leibniz' rule

$$\frac{d^n}{dx^n}(pf) = \sum_{k=0}^{n} \binom{n}{k} \frac{d^k p}{dx^k} \frac{d^{n-k} f}{dx^{n-k}} = \binom{n}{0} p \frac{d^n f}{dx^n} + \binom{n}{1} p' \frac{d^{n-1} f}{dx^{n-1}}$$

$$= \boxed{p\frac{d^n f}{dx^n} + np'\frac{d^{n-1} f}{dx^{n-1}}} \tag{L1}$$

on U, since $\frac{d^k p}{dx^k} = 0$ for $k \geq 2$. Similarly

$$\frac{d^n}{dx^n}(qf) = \binom{n}{0} q \frac{d^n q}{dx^n} + \binom{n}{1} q' \frac{d^{n-1} f}{dx^{n-1}} + \binom{n}{2} q'' \frac{d^{n-2} f}{dx^{n-2}}$$

$$= \boxed{q\frac{d^n f}{dx^n} + nq'\frac{d^{n-1} f}{dx^{n-1}} + \frac{n(n-1)}{2} q'' \frac{d^{n-2} f}{dx^{n-2}}} \tag{L2}$$

on U. Thus we have equations (L1), (L2) on U for polynomials p, q with $\deg p \leq 1$, $\deg q \leq 2$.

We consider the differential equation

$$\sigma(x)y''(x) + \tau(x)y'(x) + \lambda y(x) = 0 \tag{4.4.1}$$

on some open set $U \subset \mathbb{R}$—an equation of hypergeometric type: $\sigma(x)$ is a polynomial in x of degree ≤ 2 and $\tau(x)$ is a polynomial in x of degree ≤ 1; see equation (4.2.2). We suppose we have a C^∞ function ρ on U which never vanishes and which satisfies the equation

$$(\sigma\rho)' = \tau\rho \text{ on } U. \tag{4.4.2}$$

Our goal is to construct (using ρ) a family of particular (polynomial) solutions y_n, $n = 0, 1, 2, 3, 4, \ldots$, of equation (4.4.1), for $\lambda = \lambda_n \stackrel{\text{def}}{=} -n\tau' - \frac{n(n-1)}{2}\sigma''$. In fact set $f_n \stackrel{\text{def}}{=} \sigma^n\rho$, $y_n = \frac{1}{\rho}\frac{d^n f_n}{dx^n}$; f_n, y_n are C^∞. For $n = 0$, $f_0 = \rho$, $y_0 = \frac{1}{\rho}\rho = 1 \implies y_0$ satisfies equation (4.4.1) with $\lambda = \lambda_0 = 0$. For $n = 1$, $\lambda_1 = -\tau'$, $f_1 = \sigma\rho$, $y_1 = \frac{1}{\rho}(\sigma\rho)' = \frac{\tau\rho}{\rho}$ (by equation (4.4.2)) $= \tau$. Thus $\sigma y_1'' + \tau y_1' + \lambda_1 y_1 = \sigma\tau'' + \tau\tau' + (-\tau')\tau = 0$ as $\tau'' = 0$ (for $\deg\tau \leq 1$) $\implies y_1$ satisfies equation (4.4.1) for $\lambda = \lambda_1$.

Theorem 4.2. *y_n is a solution of equation* (4.4.1) *for $\lambda = \lambda_n$, $n = 0, 1, 2, 3, \ldots$.*

Proof. For the proof we may assume $n \geq 2$, the cases $n = 0, 1$ having already been checked. Define

$$\tau_n = \tau + n\sigma' \tag{4.4.3}$$

so that $\tau_n(x)$ is polynomial in x of degree ≤ 1.

Then

$$\begin{aligned} & f_n \stackrel{\text{def}}{=} \sigma^n\rho = \sigma^{n-1}\sigma\rho \\ \implies & f_n' = \sigma^{n-1}(\sigma\rho)' + (n-1)\sigma^{n-2}\sigma'\sigma\rho \\ & \quad = \sigma^{n-1}\tau\rho + (n-1)\sigma^{n-1}\rho\sigma' && \text{by definition (4.4.2)} \\ & \quad = \tau f_{n-1} + (n-1)\sigma' f_{n-1} = \tau_{n-1}f_{n-1} && \text{by definition (4.4.3),} \end{aligned} \tag{4.4.4}$$

so that

$$\begin{aligned} y_n & \stackrel{\text{def}}{=} \frac{1}{\rho}\frac{d^n f_n}{dx^n} = \frac{1}{\rho}\frac{d^{n-1}}{dx^{n-1}}f_n' = \frac{1}{\rho}\frac{d^{n-1}}{dx^{n-1}}(\tau_{n-1}f_{n-1}) \\ & = \frac{1}{\rho}\left[\tau_{n-1}\frac{d^{n-1}}{dx^{n-1}}f_{n-1} + (n-1)\tau_{n-1}'\frac{d^{n-2}}{dx^{n-2}}f_{n-1}\right] \\ & \qquad \text{by (L1) (since } \deg\tau_{n-1} \leq 1) \\ & = \frac{1}{\rho}\tau_{n-1}\rho y_{n-1} + \frac{n-1}{\rho}\tau_{n-1}'\frac{d^{n-2}}{dx^{n-2}}f_{n-1} \\ & = \tau_{n-1}y_{n-1} + (n-1)\tau_{n-1}'\frac{1}{\rho}\frac{d^{n-2}}{dx^{n-2}}f_{n-1}. \end{aligned} \tag{4.4.5}$$

Similarly write

$$\begin{aligned} y_n &= \frac{1}{\rho}\frac{d^n f_n}{dx^n} = \frac{1}{\rho}\frac{d^n}{dx^n}(\sigma\sigma^{n-1}\rho) = \frac{1}{\rho}\frac{d^n}{dx^n}(\sigma f_{n-1}) \\ &= \frac{1}{\rho}\left[\sigma\frac{d^n f_{n-1}}{dx^n} + n\sigma'\frac{d^{n-1} f_{n-1}}{dx^{n-1}} + \frac{n(n-1)}{2}\sigma''\frac{d^{n-2}}{dx^{n-2}} f_{n-1}\right] \\ &\qquad \text{by (L2) (since } \deg\sigma \le 2), \end{aligned}$$

where

$$\begin{aligned} &\frac{d^n f_{n-1}}{dx^n} = \frac{d}{dx}\frac{d^{n-1} f_{n-1}}{dx^{n-1}} = (\rho y_{n-1})' = \rho y'_{n-1} + \rho' y_{n-1} \\ &\implies y_n = \sigma y'_{n-1} + \frac{\sigma\rho'}{\rho} y_{n-1} + n\sigma' y_{n-1} + \frac{n(n-1)}{2}\sigma''\frac{1}{\rho}\frac{d^{n-2}}{dx^{n-2}} f_{n-1}, \end{aligned} \tag{4.4.6}$$

and where by equation (4.4.2)

$$\sigma\rho' + \sigma'\rho = \tau\rho \implies \frac{\sigma\rho'}{\rho} = \tau - \sigma'. \tag{4.4.7}$$

That is,

$$y_n = \sigma y'_{n-1} + [\tau + (n-1)\sigma'] y_{n-1} + \frac{n(n-1)}{2}\sigma''\frac{1}{\rho}\frac{d^{n-2}}{dx^{n-2}} f_{n-1}. \tag{4.4.8}$$

On the other hand,

$$\begin{aligned} \frac{\lambda_{n-1}}{n-1} + \tau'_{n-1} &\stackrel{\text{def}}{=} \frac{-(n-1)\tau' - \frac{(n-1)(n-2)}{2}\sigma''}{(n-1)} + \tau' + (n-1)\sigma'' \\ &= -\tau' - \frac{(n-2)}{2}\sigma'' + \tau' + (n-1)\sigma'' = \frac{n}{2}\sigma'' \\ \implies \frac{n(n-1)}{2}\sigma'' &= \lambda_{n-1} + (n-1)\tau'_{n-1} \implies \text{(by (4.4.3) and (4.4.5))} \\ y_n &= \sigma y'_{n-1} + [\tau + (n-1)\sigma'] y_{n-1} \\ &\quad + (n-1)\tau'_{n-1}\frac{1}{\rho}\frac{d^{n-2}}{dx^{n-2}} f_{n-1} + \lambda_{n-1}\frac{1}{\rho}\frac{d^{n-2}}{dx^{n-2}} f_{n-1} \\ &= \sigma y'_{n-1} + y_n + \lambda_{n-1}\frac{1}{\rho}\frac{d^{n-2}}{dx^{n-2}} f_{n-1} \end{aligned} \tag{4.4.9}$$

so that

$$\begin{aligned} \sigma\rho y'_{n-1} &= -\lambda_{n-1}\frac{d^{n-2}}{dx^{n-2}} f_{n-1} \qquad &&\text{for } n \ge 2;\ \text{i.e.,} \\ \sigma\rho y'_n &= -\lambda_n\frac{d^{n-1}}{dx^{n-1}} f_n &&\text{for } n \ge 1, \end{aligned} \tag{4.4.10}$$

which we differentiate to obtain (by (4.4.2))

$$\sigma\rho y_n'' + y_n'\tau\rho = (\sigma\rho y_n')' = -\lambda_n \frac{d^n}{dx^n} f_n \stackrel{\text{def}}{=} -\lambda_n \rho y_n; \text{ i.e.,} \qquad (4.4.11)$$
$$\sigma y_n'' + \tau y_n' + \lambda_n y_n = 0,$$

which proves the theorem. □

We reiterate that for ρ such that $(\sigma\rho)' = \tau\rho$, $n = 0, 1, 2, 3, \ldots$

$$y_n \stackrel{\text{def}}{=} \frac{1}{\rho}\frac{d^n}{dx^n}\sigma^n\rho$$
$$\lambda_n \stackrel{\text{def}}{=} -n\tau' - \frac{n(n-1)}{2}\sigma''. \qquad (4.4.12)$$

Note also that by equation (4.4.5), $y_{n+1} = \tau_n y_n + n\tau_n' \frac{1}{\rho}\frac{d^{n-1} f_n}{dx^{n-1}}$ so that by equation (4.4.10), $\lambda_n y_{n+1} = \lambda_n \tau_n y_n + n\tau_n' \frac{1}{\rho}(-\sigma\rho y_n') \implies$

Corollary 4.1.

$$\lambda_n y_{n+1} = \lambda_n \tau_n y_n - n\sigma\tau_n' y_n' \; \textit{for } n \geq 1.$$

Here both sides are zero for $n = 0$. Also (by equation (4.4.10))

$$\sigma\rho y_n' = -\lambda_n \frac{d^{n-1}}{dx^{n-1}}(\sigma^n\rho) \; \textit{for } n \geq 1.$$

Proposition 4.5. *y_n is a polynomial on U of degree $\leq n$.*

Proof. $y_0 = \frac{1}{\rho}\rho = 1$, $y_1 = \frac{1}{\rho}(\sigma\rho)' = \frac{1}{\rho}\tau\rho = \tau \implies$ the proposition is true for $n = 0, 1$. To finish the proof we first observe

Lemma 4.2. *Let $k \geq 1$ be an integer and let r be a differentiable function on U. Let $f_{k,r} = \sigma^k \rho r$. Then $f_{k,r}' = \sigma^{k-1}\rho[(k-1)\sigma' r + \tau r + \sigma r']$. In particular if r is a polynomial on U of degree $\leq l$, then $(k-1)\sigma' r + \tau r + \sigma r'$ is a polynomial on U of degree $\leq l + 1$.*

Assume the lemma for now. Then for $f_n \stackrel{\text{def}}{=} \sigma^n\rho \stackrel{\text{def}}{=} f_{n,1}$ (i.e., take $r = 1$ in the lemma), $f_n' = \sigma^{n-1}\rho r_1$ for $n \geq 1$, where $r_1 = (n-1)\sigma' + \tau$ is a polynomial of degree ≤ 1. For $n \geq 2$ (i.e., $n - 1 \geq 1$) apply the lemma to $f_n' = \sigma^{n-1}\rho r_1 = f_{n-1,r_1} : f_n'' = \sigma^{n-2}\rho r_2$, where r_2 is a polynomial of degree ≤ 2. For $n \geq 3$ (i.e., $n - 2 \geq 1$) apply the lemma to $f_n'' = \sigma^{n-2}\rho r_2 = f_{n-2,r_2} : f_n''' = \sigma^{n-3}\rho r_3$, where r_3 is a polynomial of degree ≤ 3. Similarly for $n \geq p$, $f_n^{(p)} = \sigma^{n-p}\rho r_p$, where r_p is a polynomial of degree $\leq p$. Therefore $f_n^{(n)} = \sigma^{n-n}\rho r_n = \rho r_n$, where r_n is a polynomial of degree $\leq n \implies y_n \stackrel{\text{def}}{=} \frac{1}{\rho} f_n^{(n)} = r_n$ is a polynomial of degree $\leq n$, as desired. □

Now we prove the lemma. If $k = 1$ we simply have

$$\begin{aligned}(\sigma\rho r)' &= (\sigma\rho)'r + \sigma\rho r' = \tau\rho r + \sigma\rho r' \qquad \text{by (4.4.2)} \\ &= \rho(\tau r + \sigma r'),\end{aligned} \tag{4.4.13}$$

as claimed.

Therefore assume $k \geq 2$.

$$\begin{aligned}f'_{k,r} &= (\sigma^{k-1}[\sigma\rho]r)' \\ &= (\sigma^{k-1})'[\sigma\rho]r + \sigma^{k-1}\left\{[\sigma\rho]'r + [\sigma\rho]r'\right\} \\ &= (k-1)\sigma^{k-2}\sigma'\sigma\rho r + \sigma^{k-1}\left\{\tau\rho r + \sigma\rho r'\right\} \qquad \text{by equation (4.4.2)} \\ &= (k-1)\sigma^{k-1}\sigma'\rho r + \sigma^{k-1}\rho\left\{\tau r + \sigma r'\right\} \\ &= \sigma^{k-1}\rho\left\{(k-1)\sigma' r + \tau r + \sigma r'\right\},\end{aligned} \tag{4.4.14}$$

as desired.

Given condition (4.4.2) we note that equation (4.4.1) ($\sigma y'' + \tau y' + \lambda y = 0$) holds $\Longleftrightarrow$

$$\frac{1}{\rho}\frac{d}{dx}[\rho\sigma y'] + \lambda y = 0 \tag{4.4.15}$$

holds. (4.4.15) is called the *adjoint form* of (4.4.1).

4.5 Examples of the Reduction Technique

Consider the time-independent Schrödinger equation for the harmonic oscillator; see Section 3.2, Chapter 3:

$$V(x) = \frac{\beta x^2}{2} \implies \frac{d^2\psi}{dx^2} + \frac{2m}{\hbar^2}\left(E - \frac{\beta x^2}{2}\right)\psi = 0; \quad \beta = 4\pi^2\nu^2 m. \tag{4.5.1}$$

This has the form (4.2.1):

$$u''(x) + \frac{\tilde{\tau}(x)}{\sigma(x)}u'(x) + \frac{\tilde{\sigma}(x)}{\sigma^2(x)}u(x) = 0 \tag{4.5.2}$$

on $\mathbb{R}$ for $u = \psi, \tilde{\tau} = 0, \sigma = 1, \tilde{\sigma}(x) = \frac{2m}{\hbar^2}(E - \frac{\beta x^2}{2})$. We employ the notation (4.3.4)–(4.3.8).

As $\tilde{\tau}, \sigma' = 0$, $f_\kappa \stackrel{\text{def}}{=} \frac{[\tilde{\tau}-\sigma']^2}{4} + \kappa\sigma - \tilde{\sigma} = \kappa - \tilde{\sigma}$ for $\kappa \in \mathbb{R}$: $f_\kappa(x) = \kappa - \theta E + \frac{\theta\beta}{2}x^2$ for $\theta \stackrel{\text{def}}{=} \frac{2m}{\hbar^2}$. Thus $f_\kappa(x) = ax^2 + bx + c$ for $a = \frac{\theta\beta}{2} = \frac{m\beta}{\hbar^2} > 0, b = 0, c = \kappa - \theta E \implies \Delta(\kappa) \stackrel{\text{def}}{=} b^2 - 4ac = -4\frac{\theta\beta}{2}(\kappa - \theta E)$. Thus $\Delta(\kappa) = 0 \implies \kappa - \theta E = 0 \implies c = 0, \kappa = \theta E$.

$p(x) \stackrel{\text{def}}{=} \sqrt{a}x \implies p^2 = ax^2 = \frac{\theta\beta}{2}x^2 = f_\kappa$ (as $\kappa - \theta E = 0$).

$\pi^\pm \stackrel{\text{def}}{=} \frac{\sigma' - \tilde{\tau}}{2} \pm p = \pm p$. If we choose the minus here (we also choose the positive square root of a) then $\tau \stackrel{\text{def}}{=} \tilde{\tau} + 2\pi^- = -2p$ will have a negative derivative: $p(x) =$

$\frac{\sqrt{m\beta}}{\hbar}x \implies \tau(x) = -2\frac{\sqrt{m\beta}}{\hbar}x \implies \tau'(x) = -2\frac{\sqrt{m\beta}}{\hbar}$; compare remarks preceding Lemma 4.1. Take (as in (4.3.8))

$$\begin{aligned}\phi(x) &= e^{\int \pi^-(x)/\sigma(x)\,dx} = e^{-\sqrt{a}x^2/2} = e^{-\frac{\sqrt{m\beta}}{2\hbar}x^2}: \\ \phi'(x) &= \frac{\phi(x)\pi^-(x)}{\sigma(x)} \text{ (which is (4.3.7)).}\end{aligned} \tag{4.5.3}$$

Also $h = 0$ satisfies $h' = \tilde{\tau}/\sigma$ (since $\tilde{\tau} = 0$). Thus we can take $\rho(x) = \frac{\phi(x)^2}{\sigma(x)}e^{h(x)} = e^{-\frac{\sqrt{m\beta}}{\hbar}x^2}$ in Proposition 4.4 : $(\rho\sigma)' = \rho\tau$. For $n \geq 0$ an integer, $\lambda_n \overset{\text{def}}{=} -n\tau' - \frac{1}{2}n(n-1)\sigma'' = 2n\frac{\sqrt{m\beta}}{\hbar}$. Let

$$y_n(x) = \frac{1}{\rho(x)}\frac{d^n}{dx^n}(\sigma(x)^n\rho(x)) = e^{\frac{\sqrt{m\beta}}{\hbar}x^2}\frac{d^n}{dx^n}e^{-\frac{\sqrt{m\beta}}{\hbar}x^2}. \tag{4.5.4}$$

Then, by Theorem 4.2, we know that y_n is a solution of $\sigma y'' + \tau y' + \lambda_n y = 0$ and thus for $\lambda = \lambda_n$ ($\lambda \overset{\text{def}}{=} \kappa + (\pi^-)' = \theta E - \sqrt{a} = \theta E - \frac{\sqrt{m\beta}}{\hbar}$) we obtain a solution $\psi_n = \phi y_n$ of $u'' + \frac{\tilde{\tau}}{\sigma}u' + \frac{\tilde{\sigma}}{\sigma^2}u = 0$, by Theorem 4.1. The condition $\lambda = \lambda_n$ is $\theta E - \frac{\sqrt{m\beta}}{\hbar} = 2n\frac{\sqrt{m\beta}}{\hbar}$, or $E = E_n \overset{\text{def}}{=} \frac{(2n+1)}{\theta}\frac{\sqrt{m\beta}}{\hbar} \overset{\text{def}}{=} \frac{\hbar}{2m}(2n+1)\frac{\sqrt{m\beta}}{\hbar} = (n+\frac{1}{2})\sqrt{\frac{\beta}{m}}\hbar$. For $\beta \overset{\text{def}}{=} 4\pi^2\nu^2 m$ as above and $\hbar \overset{\text{def}}{=} \frac{h}{2\pi}$ we obtain $E = E_n = (n+\frac{1}{2})\nu h$. Thus for $E = E_n$ we have the solution $\psi = \psi_n \overset{\text{def}}{=} \phi y_n$ of equation (4.5.1):

$$\frac{d^2\psi_n}{dx^2} + \frac{2m}{\hbar}\left(E_n - \frac{\beta x^2}{2}\right)\psi_n = 0 \tag{4.5.5}$$

on $\mathbb{R}$. Let H_n be the nth Hermite polynomial of (4.2.13): $H_n(x) \overset{\text{def}}{=} (-1)^n e^{x^2}\frac{d^n}{dx^n}(e^{-x^2})$ on $\mathbb{R}$. If $g(x) = e^{-x^2}$ and $g_b(x) = e^{-bx^2}$ for $b > 0$, then $g_b(x) = g(\sqrt{b}x) \implies g_b'(x) = g'(\sqrt{b}x)\sqrt{b}$, $g_b''(x) = g''(\sqrt{b}x)(\sqrt{b})^2, \ldots, g_b^{(n)}(x) = g^{(n)}(\sqrt{b}x)(\sqrt{b})^n$. Choose $b = \frac{\sqrt{m\beta}}{\hbar}$:

$$y_n(x) \overset{\text{def}}{=} e^{bx^2}g_b^{(n)}(x) = (\sqrt{b})^n e^{bx^2}g^{(n)}(\sqrt{b}x). \tag{4.5.6}$$

But

$$H_n(x) \overset{\text{def}}{=} (-1)^n e^{x^2}g^{(n)}(x) \implies H_n(\sqrt{b}x) = (-1)^n e^{bx^2}g^{(n)}(\sqrt{b}x); \tag{4.5.7}$$

i.e.,

$$\begin{aligned}&y_n(x) = (-1)^n(\sqrt{b})^n H_n(\sqrt{b}x) \\ \implies &\psi_n(x) \overset{\text{def}}{=} \phi(x)y_n(x) = e^{-\frac{\sqrt{m\beta}}{2\hbar}x^2}(-1)^n(\sqrt{b})^n H_n(\sqrt{b}x)\end{aligned} \tag{4.5.8}$$

for $b = \frac{\sqrt{m\beta}}{\hbar} = \frac{2\pi\nu m}{\hbar} = \frac{4\pi^2\nu m}{h}$.

We have thus obtained the energy levels $E = E_n = \left(n + \frac{1}{2}\right) \nu h$ for the quantized harmonic oscillator (as advertised in Section 3.2 of Chapter 3) and the corresponding wave functions ψ_n given by (4.5.8). The ψ_n will be normalized later. We also see that the quantum condition for the energies is precisely the mathematical condition $\lambda = \lambda_n$ of Theorem 4.2.

Example

Consider the differential equation

$$\frac{1}{r^2}\frac{d}{dr}(r^2 F_1')(r) + \frac{a}{r^2}F_1(r) + C\left(E_1 - \frac{c}{r}\right)F_1(r) = 0 \tag{4.5.9}$$

on $\mathbb{R}^+ \stackrel{\text{def}}{=} \{r \in \mathbb{R} | r > 0\}$ for $a, C, c, E_1 \in \mathbb{R}$. We assume $-a = l(l+1)$ for $l \geq 0$ an integer and we assume $C > 0$, c, $E_1 < 0$. Write equation (4.5.9) as

$$F_1''(r) + \frac{2F_1'(r)}{r} + \frac{[a + r^2 C(E_1 - \frac{c}{r})]F_1(r)}{r^2} = 0. \tag{4.5.10}$$

Thus $\sigma(r) = r$, $\tilde{\tau}(r) = 2$, $\tilde{\sigma}(r) = a + r^2 C(E_1 - \frac{c}{r})$. From (4.3.4)

$$\begin{aligned} f_\kappa(r) &\stackrel{\text{def}}{=} \frac{[\tilde{\tau} - \sigma']^2}{4}(r) + \kappa\sigma(r) - \tilde{\sigma}(r) \\ &= \frac{1}{4} + \kappa r - a - r^2 C\left(E_1 - \frac{c}{r}\right) \\ &= C(-E_1)r^2 + (\kappa + Cc)r + \frac{1}{4} - a \\ \Longrightarrow \ \Delta(\kappa) &= (\kappa + Cc)^2 + 4CE_1\left(\frac{1}{4} - a\right), \end{aligned} \tag{4.5.11}$$

which we set equal to 0: $(\kappa + Cc)^2 = 4C(-E_1)(\frac{1}{4} + l(l+1))$ (where the right-hand side is positive by the assumptions $E_1 < 0$, $C > 0$, $l \geq 0$). Let $\nu \stackrel{\text{def}}{=} 4C(-E_1)(\frac{1}{4} + l(l+1))$ so that $\kappa = -Cc \pm \sqrt{\nu}$, with $\frac{1}{4} + l(l+1) = (l + \frac{1}{2})^2$; i.e., $\sqrt{\nu} = \sqrt{\mu}(2l+1)$ for $\mu \stackrel{\text{def}}{=} C(-E_1) \implies \kappa = -Cc \pm \sqrt{\mu}(2l+1)$. We get

$$f_\kappa(r) = \mu r^2 \pm \sqrt{\nu} r + \left(l + \frac{1}{2}\right)^2 = \mu r^2 \pm \sqrt{\mu}(2l+1)r + \left(l + \frac{1}{2}\right)^2, \tag{4.5.12}$$

so that $p(r) = \sqrt{\mu} r \pm (l + \frac{1}{2})$ satisfies $p^2 = f_\kappa$. By (4.3.5), (4.3.6)

$$\begin{aligned} \pi(r) = \pi^\pm(r) &\stackrel{\text{def}}{=} \frac{\sigma' - \tilde{\tau}}{2}(r) \pm p(r) \\ &= -\frac{1}{2} \pm p(r) = -\frac{1}{2} \pm \left[\sqrt{\mu} r \pm \left(l + \frac{1}{2}\right)\right], \\ \tau(r) &\stackrel{\text{def}}{=} \tilde{\tau}(r) + 2\pi(r) = 2 + 2\pi(r). \end{aligned} \tag{4.5.13}$$

If we make the choices $\kappa = -Cc - \sqrt{\nu}$, $\pi = \pi^-$—i.e., $\pi(r) = -\frac{1}{2} - p(r) = -\frac{1}{2} - [\sqrt{\mu}r - (l+\frac{1}{2})] = -\sqrt{\mu}r + l$, then $\tau(r) = -2\sqrt{\mu}r + 2l + 2$ will satisfy $\tau' < 0$, $\tau(r) = 0$ for $r = \frac{l+1}{\sqrt{\mu}} \in (0, \infty)$; compare Lemma 4.1. Then $\lambda \stackrel{\text{def}}{=} \kappa + \pi' = -Cc - \sqrt{\nu} - \sqrt{\mu} = -Cc - \sqrt{\mu}(2l+1) - \sqrt{\mu} = -Cc - 2(l+1)\sqrt{\mu}$, $\lambda_p \stackrel{\text{def}}{=} -p\tau' - \frac{p(p-1)}{2}\sigma'' = 2p\sqrt{\mu}$, $p = 0, 1, 2, 3 \ldots$ (see (4.4.12)). Thus

$$\begin{aligned} & \lambda = \lambda_p \\ \iff \quad & -Cc = 2(l+p+1)\sqrt{\mu} \\ \iff \quad & C^2c^2 = 4(l+p+1)^2\mu \quad \text{(since } |Cc| = -Cc \text{ for } C > 0, c < 0) \\ \iff \quad & E_1 = \frac{-Cc^2}{4(p+l+1)^2}, \qquad \text{as } \mu \stackrel{\text{def}}{=} -CE_1. \end{aligned} \tag{4.5.14}$$

By (4.3.8) and Proposition 4.4

$$\begin{aligned} \phi(r) &= \exp\left(\int \frac{\pi}{\sigma}(r)dr\right) = \exp\left(\int \frac{-\sqrt{\mu}r + l}{r}dr\right) = r^l \exp(-\sqrt{\mu}r), \\ h(r) &= \int \frac{\tilde{\tau}}{\sigma}(r)dr = \int \frac{2}{r}dr = 2\log r \\ \Longrightarrow \rho(r) &= \frac{\phi(r)^2}{\sigma(r)}\exp(h(r)) = \frac{r^{2l}\exp(-2\sqrt{\mu}r)}{r}r^2 \\ &= r^{2l+1}\exp(-2\sqrt{C(-E_1)}r) \quad \text{on } \mathbb{R}^+. \end{aligned} \tag{4.5.15}$$

Then

$$\begin{aligned} y_p(r) &\stackrel{\text{def}}{=} \frac{1}{\rho(r)}\frac{d^p}{dr^p}\left[\sigma(r)^p\rho(r)\right] \\ &= r^{-(2l+1)}\exp(\alpha r)\frac{d^p}{dr^p}\left[r^{p+2l+1}\exp(-\alpha r)\right] \end{aligned} \tag{4.5.16}$$

for $\alpha = 2\sqrt{C(-E_1)} = 2\sqrt{\mu}$. On the other hand, for $q \geq 0$ an integer we have the Laguerre polynomials L_p^q of degree p given in (4.2.12):

$$L_p^q(x) = e^x\frac{x^{-q}}{p!}\frac{d^p}{dx^p}(e^{-x}x^{p+q}) \quad \text{on } \mathbb{R}^+. \tag{4.5.17}$$

Let $f(x) = e^{-x}x^{p+q}$ on $\mathbb{R}^+$, $g(x) = f(\alpha x)$, $\alpha \in \mathbb{R}^+$. Then $g^{(n)}(x) = f^{(n)}(\alpha x)\alpha^n \Longrightarrow f^{(n)}(\alpha x) = \alpha^{-n}g^{(n)}(x)$. That is,

$$\begin{aligned} & L_p^q(x) = e^x\frac{x^{-q}}{p!}f^{(p)}(x) \\ \Longrightarrow \quad & L_p^q(\alpha x) = e^{\alpha x}\frac{\alpha^{-q}x^{-q}}{p!}f^{(p)}(\alpha x) \end{aligned}$$

$$
\begin{aligned}
&= e^{\alpha x}\alpha^{-q}\frac{x^{-q}}{p!}\alpha^{-p}g^{(p)}(x) \qquad (4.5.18)\\
&\stackrel{\text{def}}{=} \frac{e^{\alpha x}x^{-q}}{\alpha^{p+q}\, p!}\frac{d^p}{dx^p}e^{-\alpha x}(\alpha x)^{p+q}\\
&= \frac{e^{\alpha x}}{p!}x^{-q}\frac{d^p}{dx^p}(e^{-\alpha x}x^{p+q}).
\end{aligned}
$$

We can therefore write $y_p(r) = p!\ L_p^{2l+1}(\alpha r)$. By Theorem 4.2, y_p is a solution of the canonical form $\sigma y'' + \tau y' + \lambda y = 0$, equation (4.4.1), provided $\lambda = \lambda_p$— i.e., $E_1 = -\frac{Cc^2}{4(p+l+1)^2}$, in which case we obtain the solution $F_1 = u_p \stackrel{\text{def}}{=} \phi y_p$ of equation (4.5.10), by Theorem 4.1:

$$
\begin{aligned}
u_p(r) &= r^l \exp(-\sqrt{C(-E_1)}r)\ p!\ L_p^{2l+1}(2\sqrt{C(-E_1)}r) \quad \text{for}\\
E_1 &= \frac{-Cc^2}{4(p+l+1)^2}.
\end{aligned} \tag{4.5.19}
$$

Now specialize the parameters; choose $C = \frac{8\pi^2\mu}{h^2} = \frac{2\mu}{\hbar^2}$ where $\mu = \frac{m_1m_2}{m_1+m_2}$ is the reduced mass of two particles, say a nucleus of mass m_1 and an electron of mass m_2; choose $c = q_1q_2$ where $q_1 = Ze$ for $-e$ the electron charge, $q_2 = -e$. Here Z = number of protons in the nucleus (also called the *atomic number*). Thus $c = -Ze^2 < 0$. Given an integer $n \geq l+1$ take $p = n-(l+1) \geq 0$ (an integer). Set

$$
\begin{aligned}
E^{(n)} &= -\frac{Cc^2}{4(p+l+1)^2} \qquad \text{(see equation (4.5.19))}\\
&= -\frac{\mu Z^2e^4}{2\hbar^2n^2}\\
&= -\frac{2\pi^2\mu Z^2e^4}{h^2n^2},
\end{aligned} \tag{4.5.20}
$$

and set

$$
\begin{aligned}
\alpha_n &= 2\sqrt{C(-E^{(n)})} = \frac{2\mu Ze^2}{\hbar^2 n},\\
\rho_n(r) &= \alpha_n r \qquad \text{on } \mathbb{R}^+.
\end{aligned} \tag{4.5.21}
$$

By the above discussion we obtain solutions u_{nl} of equation (4.5.9) (for $C = \frac{8\pi^2\mu}{h^2} = \frac{2\mu}{\hbar^2}$, $a = -l(l+1)$, $c = -Ze^2$) given by

$$
u_{nl}(r) = r^l \exp(-\alpha_n r/2)[n-(l+1)]!\ L_{n-(l+1)}^{2l+1}(\rho_n(r)), \tag{4.5.22}
$$

provided $E_1 = E^{(n)}$. The solutions u_{nl} will play a vital role in the next chapter.

4.6 Examples: Hermite, Legendre, Laguerre, Jacobi

In the Hermite case (see (4.2.4))

$$
\sigma(x) = 1, \qquad \tau(x) = -2x \tag{4.6.1}
$$

$$\Rightarrow \lambda_n = 2n; \qquad \tau_n \stackrel{\text{def}}{=} \tau + n\sigma' = \tau, \qquad \rho(x) = e^{-x^2},$$

and therefore $y_n(x) = e^{x^2} \frac{d^n}{dx^n} e^{-x^2}$. Corollary 4.1 gives $\lambda_n y_{n+1} = \lambda_n \tau_n y_n - n\sigma\tau_n' y_n'$ for $n \geq 0$. That is, $2n y_{n+1}(x) = 2n(-2x) y_n(x) - n(-2) y_n'(x)$, or $y_{n+1}(x) = -2x y_n(x) + y_n'(x)$ for $n \neq 0$, which we multiply by $(-1)^{n+1}$: $H_n = (-1)^n y_n \Rightarrow$

$$H_{n+1}(x) = 2x H_n(x) - H_n'(x) : \tag{4.6.2}$$

Corollary 4.2.

$$H_{n+1}(x) - 2x H_n(x) + H_n'(x) = 0 \, for \quad n \geq 1;$$

this also holds for $n = 0$. *Also* H_n *is a solution of equation* (4.2.4) *for* $\lambda = 2n$ *there.*

In the Legendre case (see (4.2.3))

$$\begin{aligned} &\sigma(x) = 1 - x^2, && \tau(x) = -2x && \qquad (4.6.3)\\ \Rightarrow\ &\lambda_n = n(n+1); && \tau_n(x) = -2(n+1)x, && \rho(x) = 1 \end{aligned}$$

and therefore $y_n = \frac{d^n}{dx^n}(1 - x^2)^n$.

$$\begin{aligned} P_n(x) &\stackrel{\text{def}}{=} \frac{1}{2^n n!} \frac{d^n}{dx^n}(x^2 - 1)^n = \frac{(-1)^n}{2^n n!} \frac{d^n}{dx^n} \sigma(x)^n \\ &= \frac{(-1)^n}{2^n n!} y_n(x) \qquad \text{(as } \rho(x) = 1\text{).} \end{aligned}$$

Corollary 4.1, namely $\lambda_n y_{n+1} = \lambda_n \tau_n y_n - n\sigma\tau_n' y_n'$, becomes (after division by λ_n)

$$y_{n+1}(x) = -2(n+1)x y_n(x) + 2(1 - x^2) y_n'(x) \text{ for } n \neq 0. \tag{4.6.4}$$

For $c_n = \frac{(-1)^n}{2^n n!}$, $c_{n+1} = \frac{(-1)^{n+1}}{2^{n+1}(n+1)!} = \frac{-1}{2(n+1)} c_n$. Thus multiply the formula for $y_{n+1}(x)$ by c_{n+1} to get

$$P_{n+1}(x) = x P_n(x) - \frac{(1 - x^2)}{n+1} P_n'(x) : \tag{4.6.5}$$

Corollary 4.3.

$$P_{n+1}(x) - x P_n(x) + \frac{(1 - x^2)}{n+1} P_n'(x) = 0 \, for \, n \geq 1;$$

this also holds for $n = 0$. *Also* $P_n(x)$ *is a solution of equation* (4.2.3) *for* $\lambda = n(n+1)$ *there.*

In the Laguerre case (see (4.2.5))

$$\begin{aligned} &\sigma(x) = x, \qquad \tau(x) = \alpha + 1 - x, \qquad \rho(x) = x^\alpha e^{-x}, x > 0 \qquad (4.6.6)\\ \Rightarrow\ &\lambda_n = n \end{aligned}$$

so that

$$y_n(x) = x^{-\alpha} e^x \frac{d^n}{dx^n}(x^\alpha e^{-x} x^n) = x^{-\alpha} e^x \frac{d^n}{dx^n}(x^{\alpha+n} e^{-x}) = n! L_n^\alpha(x)$$

is a solution of

$$xy''(x) + (\alpha + 1 - x)y'(x) + ny(x) = 0 \text{ on } \mathbb{R}^+ \text{ by Theorem 4.2.} \tag{4.6.7}$$

Also $\sigma'(x) = 1 \implies \tau_n(x) = \tau(x) + n = \alpha - x + n + 1$.

Corollary 4.1, namely $\lambda_n y_{n+1} = \lambda_n \tau_n y_n - n\sigma\tau_n' y_n'$, gives (after division by n)

$$y_{n+1}(x) = (\alpha - x + n + 1)y_n(x) + xy_n'(x) \text{ for } n \geq 1, \tag{4.6.8}$$

which we divide by $(n+1)! = (n+1)n!$ to get

$$L_{n+1}^\alpha(x) = \frac{(\alpha - x + n + 1)}{n+1} L_n^\alpha(x) + \frac{x}{n+1}(L_n^\alpha)'(x) : \tag{4.6.9}$$

Corollary 4.4. *L_n^α is a solution of equation* (4.2.5) *for $\lambda = n$ there. Also*

$$(n+1)L_{n+1}^\alpha(x) - (\alpha - x + n + 1)L_n^\alpha(x) - x(L_n^\alpha)'(x) = 0 \textit{ for } n \geq 1.$$

We consider next Jacobi polynomials. Define $\sigma(x) = 1 - x^2, \tau(x) = -(\alpha + \beta + 2)x + \beta - \alpha, \rho(x) = (1 - x)^\alpha (1 + x)^\beta$ on $(-1, 1)$ for $\alpha, \beta \in \mathbb{R}$ fixed. Therefore $\rho(x) \neq 0 \forall\ x \in (-1, 1)$. Usually we will take $\alpha, \beta > -1$. By straightforward differentiation one has

Proposition 4.6 (cf. condition (4.4.2)**).**

$$(\rho\sigma)' = \rho\tau \textit{ on } (-1, 1) \textit{ and}$$

$$\lambda_n \overset{\text{def}}{=} -n\tau' - \frac{n(n-1)}{2}\sigma'' = n(\alpha + \beta + n + 1).$$

Let $y_n(x) = \frac{1}{\rho(x)} \frac{d^n}{dx^n}(\sigma^n(x)\rho(x))$. That is, $y_n(x) = (1 - x)^{-\alpha} (1 + x)^{-\beta} \frac{d^n}{dx^n}[(1 - x)^{n+\alpha}(1 + x)^{n+\beta}]$. Then, given Proposition 4.6, we know by Proposition 4.5 and Theorem 4.2 that $y_n(x)$ is a polynomial in x of degree $\leq n$, and that, on $(-1, 1)$, y_n is a solution of

$$(1 - x^2)y''(x) + [-(\alpha + \beta + 2)x + \beta - \alpha]\, y'(x) + \mu y(x) = 0 \tag{4.6.10}$$

(i.e., of the equation $\sigma y'' + \tau y' + \mu y = 0$) *provided* $n(\alpha + \beta + n + 1) = \mu$ (i.e., $\lambda_n = \mu$).

Equation (4.6.10) is *Jacobi's* differential equation; see (4.2.10) . Recall that $\sigma y'' + \tau y' + \mu y = 0 \iff \frac{d}{dx}[\rho\sigma y'] + \mu\rho y = 0$ (= adjoint form). Thus equation (4.6.10) is also written in the adjoint format (see (4.4.15)):

$$\frac{1}{\rho}\frac{d}{dx}\left[\rho\sigma y'\right] + \mu y = 0; \tag{4.6.11}$$

i.e.,

$$(1-x)^{-\alpha}(1+x)^{-\beta}\frac{d}{dx}\left[(1-x)^{\alpha+1}(1+x)^{\beta+1}y'(x)\right]+\mu y(x)=0$$

is the format in some books. From (4.2.14) we have the Jacobi polynomials

$$P_n^{(\alpha,\beta)}(x)\stackrel{\text{def}}{=}\frac{(-1)^n}{2^n n!}(1-x)^{-\alpha}(1+x)^{-\beta}\frac{d^n}{dx^n}\left[(1-x)^{\alpha+n}(1+x)^{\beta+n}\right] \tag{4.6.12}$$

on $(-1,1)$. We see that $P_n^{(\alpha,\beta)}=\frac{(-1)^n}{2^n n!}y_n \implies P_n^{(\alpha,\beta)}$ is a *polynomial* of degree $\leq n$ and $P_n^{(\alpha,\beta)}$ is a solution of Jacobi's differential equation (4.2.10) provided $n(\alpha+\beta+n+1)=\mu$.

One actually has that $\deg P_n^{(\alpha,\beta)}=n$. Note that $P_n^{(0,0)}(x)=\frac{(-1)^n}{2^n n!}\frac{d^n}{dx^n}(1-x^2)^n=\frac{1}{2^n n!}\frac{d^n}{dx^n}(x^2-1)^n\stackrel{\text{def}}{=}P_n(x)$, the nth Legendre polynomial. Similarly for choices of α,β, $P_n^{(\alpha,\beta)}$ reduces to polynomials of Chebyshev, Gegenbauer, and so on.

4.7 Orthogonality

For integers $n,m\geq 0$ define f_{nm} on U in Section 4.4 by

$$\begin{aligned} f_{nm}&=\sigma(x)\rho(x)[y_n(x)y_m'(x)-y_m(x)y_n'(x)]\\ &=\sigma(x)\rho(x)W(y_n,y_m)(x)\\ \text{where } W(y_n,y_m)(x)&=\begin{vmatrix} y_n(x) & y_m(x)\\ y_n'(x) & y_m'(x)\end{vmatrix}; \end{aligned} \tag{4.7.1}$$

i.e., $W(y_n,y_m)$ is the *Wronskian* of (y_n,y_m). Then

$$\begin{aligned} f_{nm}'&=\sigma\rho[y_ny_m''+y_n'y_m'-y_my_n''-y_m'y_n']+(\sigma\rho)'\left[y_ny_m'-y_my_n'\right]\\ &=\rho y_n[\sigma y_m''+\tau y_m']-\rho y_m[\sigma y_n''+\tau y_n'] \qquad \text{by equation (4.4.2)}\\ &=-\lambda_m\rho y_ny_m+\lambda_n\rho y_my_n, \qquad \text{by Theorem 4.2.} \end{aligned} \tag{4.7.2}$$

That is,

$$f_{nm}'=(\lambda_n-\lambda_m)\rho y_ny_m \quad \text{on } U. \tag{4.7.3}$$

4.7.1 Laguerre Orthogonality. Consider the Laguerre case, for example, where we take $U=\mathbb{R}^+=(0,\infty)$: $\sigma(x)=x$, $\tau(x)=\alpha+1-x$, $\rho(x)=x^\alpha e^{-x}$, $\lambda_n=n$, for $x>0$; we take $\alpha>-1$; $y_n=n!L_n^\alpha$, as noted earlier. By Corollary 4.1

$$(\sigma\rho)(x)\, n!(L_n^\alpha)'(x)=-n\frac{d^{n-1}}{dx^{n-1}}(\sigma^n(x)\rho(x))=-n\frac{d^{n-1}}{dx^{n-1}}(x^{n+\alpha}e^{-x}) \tag{4.7.4}$$

for $n\geq 1$. On the other hand, by (4.2.12)

$$L_{n-1}^{\alpha+1}(x)\stackrel{\text{def}}{=}\frac{1}{(n-1)!}x^{-\alpha-1}e^x\frac{d^{n-1}}{dx^{n-1}}(e^{-x}x^{n+\alpha}) \tag{4.7.5}$$

for $n\geq 1$, which gives

Proposition 4.7.

$$(L_n^\alpha)'(x) = -L_{n-1}^{\alpha+1}(x) \quad \text{for } x > 0, n \geq 1;$$

hence

$$(n+1)L_{n+1}^\alpha(x) - (\alpha - x + n + 1)L_n^\alpha(x) + xL_{n-1}^{\alpha+1}(x) = 0. \tag{4.7.6}$$

by Corollary 4.4.

For $\alpha + 1 > 0$ (i.e., $\alpha > -1$) $\lim_{x\to 0^+} x^{\alpha+1} = 0$, and $\int_0^\infty e^{-x}x^\alpha\, dx$ converges (to $\Gamma(\alpha+1)$).

For $b > \epsilon > 0$, equations (4.7.1), (4.7.3) give

$$(n-m)\int_\epsilon^b x^\alpha e^{-x} L_n^\alpha(x)L_m^\alpha(x)dx = \left[x^{\alpha+1}e^{-x}W(L_n^\alpha, L_m^\alpha)(x)\right]_\epsilon^b \tag{4.7.7}$$

where we have cancelled $n!\ m!$ on both sides. We may therefore let $\epsilon \to 0^+$, $b \to \infty$ to obtain

$$(n-m)\int_0^\infty x^\alpha e^{-x} L_n^\alpha(x)L_m^\alpha(x)\, dx = 0, \tag{4.7.8}$$

as $W(L_n^\alpha, L_m^\alpha)(x)$ is polynomial in x. That is,

Proposition 4.8. *For $\alpha > -1$, and integers $n, m \geq 0$*

$$\int_0^\infty x^\alpha e^{-x} L_n^\alpha(x)L_m^\alpha(x)dx \quad \textit{converges,}$$

and is zero for $n \neq m$.

By definition

$$L_n^\alpha(x)x^\alpha e^{-x} = \frac{1}{n!}\frac{d^n}{dx^n}\left(e^{-x}x^{\alpha+n}\right). \tag{4.7.9}$$

Given any polynomial p we have noted in the argument leading to (4.7.8) that $\alpha > -1 \implies [x^{\alpha+1}e^{-x}p(x)]_0^\infty = 0$. Therefore

$$\begin{aligned}\int_0^\infty p(x)L_n^\alpha(x)\, x^\alpha e^{-x}\, dx &= \frac{1}{n!}\int_0^\infty p(x)\frac{d^n}{dx^n}\left(e^{-x}x^{\alpha+n}\right)\, dx \\ &\overset{\text{claim}}{=} \frac{(-1)^n}{n!}\int_0^\infty \left[\frac{d^n}{dx^n}p(x)\right]\left(e^{-x}x^{\alpha+n}\right)\, dx,\end{aligned} \tag{4.7.10}$$

where the claim is true for $n = 1$ (one integrates by parts), and follows by induction on n, using that

$$\begin{aligned}\frac{d^n}{dx^n}\left(e^{-x}x^{\alpha+n+1}\right) = \sum_{k=0}^n \binom{n}{k}&(-1)^k e^{-x}(n+\alpha+1)(n-1+\alpha+1) \\ &\cdot (n-2+\alpha+1)\cdots(k+1+\alpha+1)x^{\alpha+1+k}\end{aligned} \tag{4.7.11}$$

[or more simply that $\frac{d^n}{dx^n}\left(e^{-x}x^{\alpha+n+1}\right) \stackrel{\text{def}}{=} n!e^{-x}x^{\alpha+1}L_n^{\alpha+1}(x)$], which implies

$$\left[p(x)\frac{d^n}{dx^n}\left(e^{-x}x^{\alpha+n+1}\right)\right]_0^\infty = 0. \tag{4.7.12}$$

By Proposition 4.7 $(L_n^\alpha)'(x) = -L_{n-1}^{\alpha+1}(x)$ for $n \geq 1$. It follows that $(L_n^\alpha)^{(k)}(x)$ (the kth derivative) $= (-1)^k L_{n-k}^{\alpha+k}$ for $n \geq k$. In particular

$$(L_n^\alpha)^{(n)}(x) = (-1)^n \text{ and } (L_n^\alpha)^{(n-1)}(x) = (-1)^{n-1}(\alpha+n-x) \tag{4.7.13}$$

as $L_o^\alpha(x) = 1$, $L_1^\alpha(x) = 1+\alpha-x$. For the choice $p = L_n^\alpha$ in (4.7.10) we therefore see that since $\int_0^\infty e^{-x}x^{\alpha+n}\,dx = \Gamma(n+\alpha+1)$ we may conclude in conjunction with Proposition 4.8

Theorem 4.3 (Laguerre Orthogonality).

$$\int_0^\infty x^\alpha e^{-x} L_n^\alpha(x) L_m^\alpha(x)\,dx = \delta_{nm}\frac{\Gamma(n+\alpha+1)}{n!}$$

for $\alpha > -1$, for integers $n, m \geq 0$.

In the study of the hydrogen atom (in Chapter 5) we shall need the following formula, where as usual we assume $\alpha > -1$.

Theorem 4.4. *For $n \geq 0$,*

$$\int_0^\infty x^{\alpha+1}e^{-x}(L_n^\alpha(x))^2\,dx = \frac{(2n+\alpha+1)\,\Gamma(n+\alpha+1)}{n!}.$$

Proof. Choose $p(x) = xL_n^\alpha(x)$ in (4.7.10) and use (4.7.13): For $n \geq 1$

$$\begin{aligned} p^{(n)}(x) &= \binom{n}{0}x(L_n^\alpha)^{(n)}(x) + \binom{n}{1}(L_n^\alpha)^{(n-1)}(x) \\ &= (-1)^n x + n(-1)^{n-1}(\alpha+n-x) \\ &= (-1)^n(n+1)x + (-1)^{n-1}n(\alpha+n) \qquad \text{for } n \geq 1 \implies \end{aligned} \tag{4.7.14}$$

$$\begin{aligned} &\int_0^\infty xL_n^\alpha(x)L_n^\alpha(x)x^\alpha e^{-x}\,dx \\ &\quad = \frac{(-1)^n}{n!}\int_0^\infty [(-1)^n(n+1)x + (-1)^{n-1}n(\alpha+n)]e^{-x}x^{\alpha+n}\,dx \\ &\quad = \frac{1}{n!}[(n+1)\,\Gamma(n+\alpha+2) - n(\alpha+n)\,\Gamma(n+\alpha+1)] \\ &\quad = \frac{1}{n!}\,\Gamma(n+\alpha+1)(2n+\alpha+1), \end{aligned} \tag{4.7.15}$$

since $\Gamma(z+1) = z\Gamma(z)$. We have assumed $n \geq 1$ in appealing to (4.7.10). But for $n = 0$ Theorem 4.4 is simply the assertion

$$\int_0^\infty x^{\alpha+1}e^{-x}\,dx = \Gamma(\alpha+2) \stackrel{\text{i.e.,}}{=} (\alpha+1)\Gamma(\alpha+1). \tag{4.7.16}$$

□

In the preceding proof we chose $p(x) = xL_n^\alpha(x)$. Now choose $p(x) = x^2 L_n^\alpha(x)$:

$$\int_0^\infty x^{\alpha+2} e^{-x} \left(L_n^\alpha(x)\right)^2 \, dx = \frac{(-1)^n}{n!} \int_0^\infty p^{(n)}(x) e^{-x} x^{\alpha+n} \, dx \tag{4.7.17}$$

by (4.7.10). Now

$$p^{(n)}(x) = x^2 (L_n^\alpha)^{(n)}(x) + n2x(L_n^\alpha)^{(n-1)}(x) + \frac{n(n-1)}{2} 2(L_n^\alpha)^{(n-2)}(x)$$

by the Leibniz' rule; cf. (L2) in Section 4.4 of this chapter. That is, using (4.7.13) and more generally $(L_n^\alpha)^{(k)}(x) = (-1)^k L_{n-k}^{\alpha+k}$, $n \geq k$, we get $p^{(n)}(x) = (-1)^n x^2 - (-1)^n 2nx(\alpha+n-x) + n(n-1)(-1)^n L_2^{\alpha+n-2}(x)$, where $L_2^\beta(x) = \frac{1}{2}(\beta+1)(\beta+2) - (\beta+2)x + \frac{x^2}{2}$. After simplification we obtain

$$(-1)^n p^{(n)}(x) = \frac{(n+1)(n+2)}{2} x^2 - n(n+1)(n+\alpha)x + \frac{n(n-1)(\alpha+n-1)(n+\alpha)}{2}, \tag{4.7.18}$$

which we plug into (4.7.17). Using also that

$$\int_0^\infty e^{-x} x^{\alpha+m} \, dx = \Gamma(m+\alpha+1), \quad \alpha > -1, \tag{4.7.19}$$

we see that the right-hand side of (4.7.17) becomes

$$\frac{1}{n!} \left[\frac{(n+1)(n+2)}{2} \Gamma(n+\alpha+3) - n(n+1)(n+\alpha)\, \Gamma(n+\alpha+2) + \frac{n(n-1)(n+\alpha-1)(n+\alpha)}{2} \Gamma(n+\alpha+1) \right], \tag{4.7.20}$$

where $\Gamma(z+1) = z\,\Gamma(z) \implies \Gamma(n+\alpha+3) = (n+\alpha+2)(n+\alpha+1)(n+\alpha)\,\Gamma(n+\alpha)$, $\Gamma(n+\alpha+2) = (n+\alpha+1)(n+\alpha)\,\Gamma(n+\alpha)$, $\Gamma(n+\alpha+1) = (n+\alpha)\,\Gamma(n+\alpha) \implies$ (by (4.7.17))

$$\begin{aligned} n! \int_0^\infty & x^{\alpha+2} e^{-x} (L_n^\alpha(x))^2 \, dx \\ &= (n+\alpha)\,\Gamma(n+\alpha) \left[\frac{(n+1)(n+2)}{2} (n+\alpha+2)(n+\alpha+1) \right. \\ & \left. - n(n+1)(n+\alpha)(n+\alpha+1) + \frac{n(n-1)(n+\alpha-1)(n+\alpha)}{2} \right]. \end{aligned} \tag{4.7.21}$$

The product 2[ ], where [ ] is the bracket on the right-hand side of (4.7.21), simplifies to $12n^2 + 12n\alpha + 12n + 2\alpha^2 + 6\alpha + 4$, for which we obtain

Theorem 4.5. *For* $\alpha > -1$, $n = 0, 1, 2, 3, \ldots$

$$\int_0^\infty x^{\alpha+2} e^{-x} (L_n^\alpha(x))^2 \, dx = \frac{(n+\alpha)\,\Gamma(n+\alpha)}{n!}[6n^2 + 6n\alpha + 6n + \alpha^2 + 3\alpha + 2].$$

Corollary 4.5. *Let* k, m *be integers with* $m \geq k \geq 0$. *Then*

$$\int_0^\infty x^{k+2} e^{-x} (L_{m-k}^k(x))^2 \, dx = \frac{m!}{(m-k)!}[6m^2 - 6mk + k^2 + 6m - 3k + 2]. \quad (4.7.22)$$

Hence for integers n, l *with* $n \geq 1$, $0 \leq l \leq n-1$ *we have*

$$\begin{aligned} \int_0^\infty x^{2l+3} e^{-x} (L_{n-l-1}^{2l+1}(x))^2 \, dx &= \frac{(n+l)!}{(n-l-1)!}[6n^2 - 2l^2 - 2l] \\ &= \frac{4n^2(n+l)!}{(n-l-1)!}\left[1 + \frac{1}{2}\left(1 - \frac{l(l+1)}{n^2}\right)\right]. \end{aligned} \quad (4.7.23)$$

Proof. Theorem 4.5 implies (4.7.22) immediately, and equation (4.7.23) follows from (4.7.22) by the choices $k = 2l + 1$, $m = n + l$. □

Formula (4.7.23) is the key result needed to compute the average distance of the electron from the nucleus, as we shall see later.

4.7.2 Jacobi and Legendre Orthogonality. In the Jacobi case take $U = (-1, 1)$: $\sigma(x) = 1 - x^2$, $\tau(x) = -(\alpha + \beta + 2)x + \beta - \alpha$, $\rho(x) = (1-x)^\alpha(1+x)^\beta$, $\lambda_n = n(\alpha + \beta + n + 1)$; cf. Proposition 4.6; we take $\alpha, \beta > -1$; as before $y_n = (-1)^n \, n! \, 2^n \, P_n^{(\alpha,\beta)}$. By Corollary 4.1,

$$\begin{aligned} &(\sigma\rho)(x)(-1)^n \, n! \, 2^n \, P_n^{(\alpha,\beta)\prime}(x) \\ &\quad = -n(\alpha + \beta + n + 1)\frac{d^{n-1}}{dx^{n-1}}[(1-x^2)^n(1-x)^\alpha(1+x)^\beta]. \end{aligned} \quad (4.7.24)$$

On the other hand by definition (4.2.14)

$$P_{n-1}^{(\alpha+1,\beta+1)}(x) = \frac{(-1)^{n-1}}{2^{n-1}(n-1)!}(1-x)^{-\alpha-1}(1+x)^{-\beta-1}\frac{d^{n-1}}{dx^{n-1}}[(1-x)^{\alpha+n}(1+x)^{\beta+n}] \quad (4.7.25)$$

for $n \geq 1$, which gives

Proposition 4.9.

$$P_n^{(\alpha,\beta)\prime}(x) = \frac{\alpha + \beta + n + 1}{2} P_{n-1}^{(\alpha+1,\beta+1)}(x)$$

on $(-1, 1)$ *for* $n \geq 1$.

For $\alpha, \beta > -1$, $\int_{-1}^1 (1-x)^\alpha(1+x)^\beta \, dx$ converges (to $2^{\alpha+\beta+1} B(\alpha+1, \beta+1)$, for $B =$ the beta function). It follows that

$$\int_{-1}^1 \rho(x) P_n^{(\alpha,\beta)}(x) P_m^{(\alpha,\beta)}(x) \, dx \quad \text{converges.} \quad (4.7.26)$$

By equations (4.7.1), (4.7.3)

$$\begin{aligned}[][n(\alpha+\beta+n+1) - m(\alpha+\beta+m+1)] \int_{-\epsilon}^{\epsilon} P_n^{(\alpha,\beta)}(x)\ P_m^{(\alpha,\beta)}(x)\ \rho(x)\ dx \\ = \left[(1-x)^{\alpha+1}(1+x)^{\beta+1}\ W(P_n^{(\alpha,\beta)}, P_m^{(\alpha,\beta)})(x)\right]_{-\epsilon}^{\epsilon}\end{aligned} \tag{4.7.27}$$

for $0 < \epsilon < 1$. Let $\epsilon \to 1^-$; the right-hand side of (4.7.27) approaches 0 since $\alpha, \beta > -1$. Also if $n \neq m$, then $n(\alpha+\beta+n+1) \neq m(\alpha+\beta+m+1)$; in fact $n(\alpha+\beta+n+1) - m(\alpha+\beta+m+1) = (n-m)[\alpha+\beta+n+m+1]$. That is,

Proposition 4.10. *For $\alpha, \beta > -1$, and integers $n, m \geq 0$*

$$\int_{-1}^{1} (1-x)^{\alpha}(1+x)^{\beta}\ P_n^{(\alpha,\beta)}(x)\ P_m^{(\alpha,\beta)}(x)\ dx \quad \textit{converges,}$$

and is zero for $n \neq m$.

Similar to equation (4.7.10) one has by induction on n (for $\alpha, \beta > -1$)

$$\begin{aligned}&\int_{-1}^{1} p(x) P_n^{(\alpha,\beta)}(x)(1-x)^{\alpha}(1+x)^{\beta}\ dx \\ &\overset{\text{def}}{=} \frac{(-1)^n}{2^n n!}\int_{-1}^{1} p(x)\frac{d^n}{dx^n}[(1-x)^{\alpha+n}(1+x)^{\beta+n}]\ dx \\ &= \frac{(-1)^n}{2^n n!}(-1)^n \int_{-1}^{1} p^{(n)}(x)[(1-x)^{\alpha+n}(1+x)^{\beta+n}]\ dx\end{aligned} \tag{4.7.28}$$

for any polynomial p. Choose $p = P_n^{(\alpha,\beta)}$ so that by Proposition 4.9

$$\begin{aligned}&p^{(n)}(x) \\ &= \frac{(\alpha+\beta+n+1)(\alpha+\beta+n+2)\cdots(\alpha+\beta+n+n)}{2^n} P_o^{(\alpha+n,\beta+n)}(x) \\ &= \frac{(\alpha+\beta+n+1)(\alpha+\beta+n+2)\cdots(\alpha+\beta+n+n)}{2^n},\end{aligned} \tag{4.7.29}$$

which implies by (4.7.28)

$$\begin{aligned}&\int_{-1}^{1} \left[P_n^{(\alpha,\beta)}(x)\right]^2 (1-x)^{\alpha}(1+x)^{\beta}\ dx \\ &= \frac{(\alpha+\beta+n+1)\cdots(\alpha+\beta+n+n)}{2^{2n}n!}\int_{-1}^{1}(1-x)^{\alpha+n}(1+x)^{\beta+n}\ dx \\ &= \frac{(\alpha+\beta+n+1)\cdots(\alpha+\beta+n+n)}{2^{2n}n!} \\ &\quad \cdot 2^{\alpha+\beta+2n+1}\ B(\alpha+n+1, \beta+n+1);\end{aligned} \tag{4.7.30}$$

compare the sentence following Proposition 4.9. Of course

$$B(\alpha+n+1, \beta+n+1) \overset{\text{def}}{=} \frac{\Gamma(\alpha+n+1)\ \Gamma(\beta+n+1)}{\Gamma(\alpha+\beta+2n+2)}. \tag{4.7.31}$$

Now $\Gamma(z+m) = z(z+1)\cdots(z+m-1)\,\Gamma(z)$ for $m = 1, 2, 3, \ldots$, $z \neq 0, -1, -2, -3, \ldots$, again as $\Gamma(z+1) = z\,\Gamma(z)$. Choose $z = \alpha+\beta+n+1$, $m = n+1$: $\Gamma(\alpha+\beta+2n+2) = (\alpha+\beta+n+1)(\alpha+\beta+n+2)\cdots(\alpha+\beta+n+n)(\alpha+\beta+n+n+1)\,\Gamma(\alpha+\beta+n+1)$. Thus given Proposition 4.10 and (4.7.30) the following conclusion is valid:

Theorem 4.6 (Jacobi Orthogonality).

$$\int_{-1}^{1} P_n^{(\alpha,\beta)}(x)P_m^{(\alpha,\beta)}(x)(1-x)^\alpha(1+x)^\beta\,dx = \frac{2^{\alpha+\beta+1}\,\Gamma(\alpha+n+1)\,\Gamma(\beta+n+1)}{n!\,(\alpha+\beta+2n+1)\,\Gamma(\alpha+\beta+n+1)}\delta_{nm}$$

for $\alpha, \beta > -1$, and integers $n, m \geq 0$.

Corollary 4.6 (Legendre Orthogonality).

$$\int_{-1}^{1} P_n(x)P_m(x)\,dx = \frac{2}{2n+1}\delta_{nm}$$

for integers $n, m \geq 0$.

The corollary follows since $P_n = P_n^{(0,0)}$.

4.7.3 Orthogonality for Associated Legendre Functions. By definition (4.2.11)

$$\begin{aligned}\int_{-1}^{1} p(x)P_n^{(m)}(x)\,dx &= \frac{(-1)^n}{2^n\,n!}\int_{-1}^{1} p(x)\frac{d^{n+m}}{dx^{n+m}}(1-x^2)^n\,dx \\ &= \frac{(-1)^n}{2^n\,n!}(-1)^{n+m}\int_{-1}^{1} p^{(n+m)}(x)(1-x^2)^n\,dx\end{aligned} \tag{4.7.32}$$

for any polynomial p where the latter statement of equality follows by integration by parts, as in preceding arguments. In fact, here we can take p to be any C^∞ function since we only need that

$$\left[p(x)\frac{d^k}{dx^k}(1-x^2)^n\right]_{-1}^{1} = 0. \tag{4.7.33}$$

Choose $p(x) = P_n^{(m)}(x)(1-x^2)^m$ for $m \leq n$:

$$p^{(n+m)}(x) = \sum_{j=0}^{n+m}\binom{n+m}{j}P_n^{(m+j)}(x)\frac{d^{n+m-j}}{dx^{n+m-j}}(1-x^2)^m. \tag{4.7.34}$$

Since $P_n^{(m+j)} = 0$ for $m+j > n =$ degree of P_n and similarly

$$\frac{d^{n+m-j}}{dx^{n+m-j}}(1-x^2)^m = 0 \tag{4.7.35}$$

for $n+m-j > 2m$, we obtain a non-zero summand in (4.7.34) only for $m+j \leq n$ and $n+m-j \leq 2m$—i.e., only for $n = m+j$:

$$p^{(n+m)}(x) = \binom{n+m}{n-m} P_n^{(n)}(x) \frac{d^{2m}}{dx^{2m}}(1-x^2)^m. \tag{4.7.36}$$

Now $P_n = P_n^{(0,0)} \implies P_n^{(n)}(x) = P_n^{(0,0)(n)}(x) = (n+1)(n+2)\cdots(n+n)/2^n$ by (4.7.29), and

$$\frac{d^{2m}}{dx^{2m}}(1-x^2)^m = \frac{d^{2m}}{dx^{2m}} \sum_{j=0}^{m} \binom{m}{j} 1^{m-j}(-x^2)^j, \tag{4.7.37}$$

where $\frac{d^{2m}}{dx^{2m}}(x^2)^j = 0$ unless $2m \leq 2j$. That is, only $j = m$ contributes to the sum in (4.7.37):

$$\frac{d^{2m}}{dx^{2m}}(1-x^2)^m = (-1)^m \frac{d^{2m}}{dx^{2m}}(x^2)^m = (-1)^m(2m)! \implies \tag{4.7.38}$$

$$\begin{aligned} p^{(n+m)}(x) &= \binom{n+m}{n-m} \frac{(n+1)(n+2)\cdots(n+n)}{2^n}(-1)^m(2m)! \\ &= \frac{(n+m)!}{(n-m)!} \frac{(n+1)(n+2)\cdots(n+n)(-1)^m}{2^n} \end{aligned} \tag{4.7.39}$$

by (4.7.36). Also by the change of variables $x = \cos\phi$

$$\begin{aligned} \int_{-1}^{1}(1-x^2)^n\,dx &= 2\int_0^1 (1-x^2)^n\,dx = 2\int_0^{\pi/2} \sin^{2n+1}\phi\,d\phi \\ &= B(n+1,\frac{1}{2}) = \frac{2\cdot 2\cdot 4\cdot 6\cdots 2n}{1\cdot 3\cdot 5\cdot 7\cdots(2n+1)} = \frac{2^{n+1}\,n!}{1\cdot 3\cdot 5\cdots(2n+1)}. \end{aligned} \tag{4.7.40}$$

Equations (4.7.32), (4.7.39), (4.7.40) therefore give

$$\begin{aligned} &\int_{-1}^{1} P_n^{(m)}(x)^2(1-x^2)^m\,dx \\ &\quad = \frac{2(n+m)!}{(n-m)!} \frac{(n+1)(n+2)\cdots(n+n)}{2^n\cdot 1\cdot 3\cdot 5\cdots(2n-1)(2n+1)} \\ &\quad = \frac{2(n+m)!}{(n-m)!\,(2n+1)} \end{aligned} \tag{4.7.41}$$

since $(1\cdot 3\cdot 5\cdots(2n-1))\cdot 2\cdot 4\cdot 6\cdots 2n = (2n)! = n!(n+1)(n+2)\cdots(n+n) \implies$

$$\frac{(n+1)(n+2)\cdots(n+n)}{2^n\cdot 1\cdot 3\cdot 5\cdots(2n-1)} = \frac{2\cdot 4\cdot 6\cdots 2n}{2^n\,n!} = 1. \tag{4.7.42}$$

If we now choose $p(x) = P_l^{(m)}(x)(1-x^2)^m$, then for $l < n$, $p^{(n+m)}(x) = 0$, since $n + m > l + m = (l - m) + 2m =$ degree of p. This means that for $l < n$

$$\int_{-1}^{1} P_l^{(m)}(x)\, P_n^{(m)}(x)\, (1-x^2)^m\, dx = 0 \tag{4.7.43}$$

by (4.7.32). Similarly if $l > n$, replace $P_n^{(m)}(x)$ by $P_l^{(m)}(x)$ in (4.7.32) and choose $p(x) = P_n^{(m)}(x)(1-x^2)^m$ to conclude that (4.7.43) holds for $l > n$ also, which with equation (4.7.41) allows the conclusion

Theorem 4.7 (Legendre Orthogonality).

$$\int_{-1}^{1} P_l^{(m)}(x)\, P_n^{(m)}(x)\, (1-x^2)^m\, dx = \delta_{ln} \frac{2}{2n+1} \frac{(n+m)!}{(n-m)!}$$

for $m \le n$. For $m > n$ the integral is zero since then $P_n^{(m)} = 0$. Here $n, l, m = 0, 1, 2, 3, \ldots$. In particular for $m = 0$ we obtain a second proof of Corollary 4.6.

The associated Legendre functions P_n^m defined on $(-1, 1)$ extend to continuous functions on $[-1, 1]$. Namely, for $m > 0$

$$\lim_{x \to \pm 1^{\mp}} (1-x^2)^{m/2} = 0. \tag{4.7.44}$$

Since $P_n^m(x) \stackrel{\text{def}}{=} (1-x^2)^{m/2} P_n^{(m)}(x)$ we set

$$P_n^m(\pm 1) \stackrel{\text{def}}{=} \begin{bmatrix} 0 & \text{for } m = 1, 2, 3, \ldots \\ P_n(\pm 1) & \text{for } m = 0 \end{bmatrix}. \tag{4.7.45}$$

On $(-1, 1)$, $P_n^m(x) P_l^m(x) \stackrel{\text{def}}{=} (1-x^2)^m P_n^{(m)}(x) P_l^{(m)}(x)$, which also holds on $[-1, 1]$ by (4.7.45), which means that Theorem 4.7 can be expressed as follows.

Theorem 4.8 (Orthogonality for associated Legendre functions). *For integers $n, l, m \ge 0$ with $m \le n$*

$$\int_{-1}^{1} P_l^m(x) P_n^m(x)\, dx = \delta_{ln} \frac{2}{2n+1} \frac{(n+m)!}{(n-m)!}. \tag{4.7.46}$$

Here we may extend P_n^m to a continuous function on $[-1, 1]$ by definition (4.7.45). Equation (4.7.46) may be written, for $m \le n$,

$$\int_{0}^{\pi} P_l^m(\cos\phi) P_n^m(\cos\phi) \sin\phi\, d\phi = \delta_{ln} \frac{2}{2n+1} \frac{(n+m)!}{(n-m)!}, \tag{4.7.47}$$

by the change of variables $x = \cos\phi$. For $m > n$, $P_n^m = 0$.

4.7.4 Hermite Orthogonality. As another example consider the Hermite case where we take $U = \mathbb{R}$: $\sigma(x) = 1$, $\tau(x) = -2x$, $\rho(x) = \exp(-x^2)$, $\lambda_n = 2n$, $y_n = (-1)^n H_n$. By Corollary 4.1

$$\begin{aligned}(\sigma\rho)(x)(-1)^n H_n'(x) &= -2n\frac{d^{n-1}}{dx^{n-1}}\exp(-x^2) \\ &\stackrel{\text{def}}{=} (-2n)(-1)^{n-1}\exp(-x^2)H_{n-1}(x)\end{aligned} \tag{4.7.48}$$

for $n \geq 1$; i.e., $H_n'(x) = 2nH_{n-1}(x)$ so that by Corollary 4.2 the following recursion formula holds.

Proposition 4.11. *For $n \geq 1$, $x \in \mathbb{R}$,*

$$H_{n+1}(x) - 2xH_n(x) + 2nH_{n-1}(x) = 0;$$

also $H_n'(x) = 2nH_{n-1}(x)$. Hence $H_n^{(k)}(x) = 2^k n(n-1)(n-2)\cdots(n-k+1)H_{n-k}(x)$ for $1 \leq k \leq n$.

For any $a > 0$

$$2(n-m)\int_{-a}^{a}\exp(-x^2)H_n(x)H_m(x)\,dx = \left[\exp(-x^2)W(H_n, H_m)(x)\right]_{-a}^{a} \tag{4.7.49}$$

because of equations (4.7.1), (4.7.3). As $W(H_n, H_m)(x)$ is a polynomial in x we can let $a \to \infty$ and conclude

Proposition 4.12.

$$\int_{-\infty}^{\infty} H_n(x)H_m(x)\exp(-x^2)\,dx = 0$$

for $n \neq m$.

If p is any polynomial

$$\begin{aligned}\int_{-\infty}^{\infty} p(x)H_n(x)\exp(-x^2)\,dx &\stackrel{\text{def}}{=} (-1)^n\int_{-\infty}^{\infty} p(x)\frac{d^n}{dx^n}\exp(-x^2)\,dx \\ &= \int_{-\infty}^{\infty} p^{(n)}(x)\exp(-x^2)\,dx\end{aligned} \tag{4.7.50}$$

again by induction on n, as in equation (4.7.10), where we integrate by parts and use that

$$\left[p(x)\frac{d^n}{dx^n}\exp(-x^2)\right]_{-\infty}^{\infty} = 0 \tag{4.7.51}$$

since

$$\frac{d^n}{dx^n}\exp(-x^2) = (-1)^n\exp(-x^2)H_n(x). \tag{4.7.52}$$

Now $H_n^{(n)}(x) = 2^n n! H_o(x) = 2^n n!$ by Proposition 4.11. The choice $p = H_n$ therefore provides in conjunction with Proposition 4.12 and the fact that $\int_{-\infty}^{\infty}\exp(-x^2)\,dx = \sqrt{\pi}$

Theorem 4.9 (Hermite Orthogonality).

$$\int_{-\infty}^{\infty} H_n(x)H_m(x)\exp(-x^2)\,dx = 2^n\ n!\ \sqrt{\pi}\ \delta_{nm}.$$

Later we shall also use

Theorem 4.10. *For* $n = 0, 1, 2, 3, \ldots,$

$$\int_{-\infty}^{\infty} x^2 H_n^2(x)e^{-x^2}\,dx = 2^{n-1}n!(2n+1)\sqrt{\pi}. \tag{4.7.53}$$

Proof. For $q(x) \stackrel{\text{def}}{=} x^2$ choose $p = qH_n$ in (4.7.50):

$$\int_{-\infty}^{\infty} x^2 H_n^2(x)e^{-x^2}\,dx = \int_{-\infty}^{\infty} p^{(n)}(x)e^{-x^2}\,dx. \tag{4.7.54}$$

By Proposition 4.11

$$\begin{aligned} H_n^{(n-1)}(x) &= 2^{n-1}n!\ H_1(x) = 2^{n-1}n!\ 2x = 2^n n!\ x, \\ H_n^{(n-2)}(x) &= 2^{n-2}\frac{n!}{2}H_2(x) = 2^{n-2}\frac{n!}{2}(4x^2-2) = 2^{n-1}n!\ x^2 - 2^{n-2}n! \end{aligned} \tag{4.7.55}$$

for $n \geq 3$. These formulas also hold for $n = 2$, by direct calculation, again as $H_2(x) = 4x^2 - 2$. By equation (L2) in Section 4.4,

$$p^{(n)}(x) = x^2 H_n^{(n)}(x) + 2nxH_n^{(n-1)}(x) + \frac{n(n-1)}{2}2H_n^{(n-2)}(x). \tag{4.7.56}$$

Having noted also that $H_n^{(n)}(x) = 2^n n!$ we see that $p^{(n)}(x) = 2^n n!\ x^2 + 2nx2^n n!\ x + n(n-1)(2^{n-1}n!\ x^2 - 2^{n-2}n!) = 2^{n-1}n!(n+1)(n+2)x^2 - n(n-1)2^{n-2}n!$. Since

$$\int_0^{\infty} e^{-x^2}\,dx = \frac{\sqrt{\pi}}{2}, \quad \int_0^{\infty} x^2 e^{-x^2}\,dx = \frac{\sqrt{\pi}}{4} \tag{4.7.57}$$

the right-hand side of equation (4.7.54) becomes (for $n \geq 2$)

$$\begin{aligned} &2 \cdot 2^{n-1}n!(n+1)(n+2)\frac{\sqrt{\pi}}{4} - 2n(n-1)2^{n-2}n!\frac{\sqrt{\pi}}{2} \\ &= 2^{n-1}n!(2n+1)\sqrt{\pi}, \end{aligned} \tag{4.7.58}$$

which proves the theorem for $n \geq 2$. For $n = 1$, we use $H_1^2(x) = 4x^2$, and that

$$\int_0^{\infty} x^4 e^{-x^2}\,dx = \frac{3}{8}\sqrt{\pi} \tag{4.7.59}$$

to check the theorem directly. For $n = 0$, the theorem states that

$$\int_{-\infty}^{\infty} x^2 e^{-x^2}\,dx = \frac{\sqrt{\pi}}{2},$$

which is true by (4.7.57). Note for the record that the generalization of equations (4.7.57), (4.7.59) is

$$\int_0^\infty x^{2n} e^{-ax^2}\, dx = \frac{1\cdot 3\cdot 5\cdots(2n-1)}{2^{n+1}a^n}\sqrt{\frac{\pi}{a}} \tag{4.7.60}$$

for $a > 0$, $n = 1, 2, 3, \ldots$, which follows by properties of the gamma function. □

4.8 Normalized Wave Functions for the Harmonic Oscillator

As an example of the reduction to canonical form technique we obtained earlier the quantization of energy

$$E = E_n \overset{\text{def}}{=} 2\pi\left(n + \frac{1}{2}\right)\nu\hbar, \quad n = 0, 1, 2, 3, \ldots \tag{4.8.1}$$

for the harmonic oscillator with frequency ν, and the corresponding solutions (wave functions) $\psi = \psi_n^o$ of the time-independent Schrödinger equation

$$\frac{d^2\psi}{dx^2} + \frac{2m}{\hbar^2}(E - 2\pi^2\nu^2 m x^2)\psi = 0, \tag{4.8.2}$$

having noted that the quantization condition (4.8.1) is nothing more (nor less) than the general mathematical condition $\lambda = \lambda_n \overset{\text{def}}{=} -n\tau' - \frac{n(n-1)}{2}\sigma''$ of Theorem 4.2. ψ_n^o was given on $\mathbb{R}$ by (4.5.8):

$$\psi_n^o(x) = (-1)^n(\sqrt{b})^n \exp\left(-\frac{b}{2}x^2\right) H_n(\sqrt{b}x), \tag{4.8.3}$$

for $b \overset{\text{def}}{=} \frac{2\pi\nu m}{\hbar}$, where earlier we wrote ψ_n instead of ψ_n^o. We compute:

$$\begin{aligned}\int_{\mathbb{R}} \psi_n^o(x)\psi_m^o(x)dx &= (-1)^{n+m}(\sqrt{b})^{n+m}\int_{\mathbb{R}} e^{-bx^2} H_n(\sqrt{b}x)H_m(\sqrt{b}x)\, dx \\ &= (-1)^{n+m}\frac{(\sqrt{b})^{n+m}}{\sqrt{b}}\int_{\mathbb{R}} e^{-t^2} H_n(t)H_m(t)\, dt \\ &= (-1)^{n+m}\frac{(\sqrt{b})^{n+m}}{\sqrt{b}} 2^n n!\sqrt{\pi}\cdot\delta_{nm}\end{aligned} \tag{4.8.4}$$

by Hermite orthogonality (Theorem 4.9). Therefore we obtain normalized solutions ψ_n of (4.8.2) by setting

$$\begin{aligned}\psi_n(x) &\overset{\text{def}}{=} \psi_n^o(x)\left[\frac{b^n 2^n n!\sqrt{\pi}}{\sqrt{b}}\right]^{-1/2} \\ &= (-1)^n\left[\sqrt{\frac{b}{\pi}}\cdot\frac{1}{2^n n!}\right]^{1/2} \exp\left(-\frac{b}{2}x^2\right) H_n\left(\sqrt{b}x\right):\end{aligned} \tag{4.8.5}$$

$$\int_{\mathbb{R}} \psi_n(x)\psi_m(x)\, dx = \delta_{nm} \tag{4.8.6}$$

for $b \stackrel{\text{def}}{=} \frac{2\pi\nu m}{\hbar}$. Note that by the definition $H_n(x) = (-1)^n e^{x^2} f^{(n)}(x)$ for $f(x) = e^{-x^2}$ we have $H_n(-x) = (-1)^n H_n(x)$ (since $f^{(l)}(-x) = (-1)^l f^{(l)}(x)$ for $l = 0, 1, 2, 3, \ldots$) and hence the wave functions ψ_n satisfy the parity condition

$$\psi_n(-x) = (-1)^n \psi_n(x) \tag{4.8.7}$$

on $\mathbb{R}$.

For $c_n = (-1)^n[\sqrt{\frac{b}{\pi}} \cdot \frac{1}{2^n n!}]^{1/2}$, we apply Theorem 4.10 to compute

$$\begin{aligned} \int_{\mathbb{R}} x^2\psi_n^2(x)\, dx &\stackrel{\text{def}}{=} c_n^2 \int_{\mathbb{R}} x^2 e^{-bx^2} H_n^2(\sqrt{b}x)\, dx \\ &= \frac{c_n^2}{\sqrt{b}} \int_{\mathbb{R}} \frac{t^2}{b} e^{-t^2} H_n^2(t)\, dt \\ &= \frac{c_n^2}{b\sqrt{b}} 2^{n-1} n!(2n+1)\sqrt{\pi}. \end{aligned} \tag{4.8.8}$$

That is,

Theorem 4.11. *The normalized wave functions ψ_n in (4.8.5) satisfy*

$$\int_{\mathbb{R}} x^2\psi_n^2(x)\, dx = \frac{2n+1}{2b} = \frac{E_n}{2\pi\nu\hbar b} = \frac{E_n}{4\pi^2\nu^2 m} = \frac{\hbar(n+\frac{1}{2})}{m\omega},$$

where $E_n = 2\pi(n+\frac{1}{2})\nu\hbar$ is the nth energy level, and $\omega = 2\pi\nu$ is the angular speed of the oscillator.

4.9 Further Examples of the Reduction Technique: The Pöschl–Teller Potential

The following differential equation will arise in the next chapter:

$$u''(x) - \frac{2x}{1-x^2}u'(x) + \frac{-a(1-x^2)-m^2}{(1-x^2)^2}u(x) = 0 \tag{4.9.1}$$

on $(-1, 1)$ where $a \in \mathbb{R}$, $m \in \mathbb{Z}$. Here $\sigma(x) = 1 - x^2, \tilde{\tau}(x) = -2x, \tilde{\sigma}(x) = -a(1-x^2) - m^2$.

In this case, from (4.3.4)

$$\begin{aligned} f_\kappa(x) &\stackrel{\text{def}}{=} \frac{[\tilde{\tau} - \sigma']^2}{4}(x) + \kappa\sigma(x) - \tilde{\sigma}(x) \\ &= \kappa(1-x^2) + a(1-x^2) + m^2 = -(a+\kappa)x^2 + a + \kappa + m^2 \\ &\Longrightarrow \Delta(\kappa) = 4(a+\kappa)(a+\kappa+m^2). \end{aligned} \tag{4.9.2}$$

$$\Delta(\kappa) = 0 \implies a + \kappa = 0 \text{ or } a + \kappa + m^2 = 0$$

$$\implies f_\kappa(x) = \begin{cases} m^2 & \text{if } a + \kappa = 0, \\ m^2 x^2 & \text{if } a + \kappa + m^2 = 0 \end{cases} \tag{4.9.3}$$

$$\implies p(x) = \begin{cases} |m| & \text{if } a + \kappa = 0, \\ |m|x & \text{if } a + \kappa + m^2 = 0; \end{cases}$$

i.e., $p^2 = f_\kappa$.

By our general theory we have the definitions (4.3.5), (4.3.6):

$$\begin{aligned} \pi(x) = \pi^{\pm}(x) &\overset{\text{def}}{=} \frac{\sigma' - \tilde{\tau}}{2}(x) \pm p(x) = \pm p(x), \\ \tau(x) &\overset{\text{def}}{=} \tilde{\tau}(x) + 2\pi(x) = -2x \pm 2p(x). \end{aligned} \tag{4.9.4}$$

We choose τ such that $\tau' < 0$ and $\tau(x) = 0$ for some x in $(-1, 1)$; cf. Lemma 4.1. This can be achieved by taking $a + \kappa + m^2 = 0$ (i.e., $\kappa = -(a + m^2)$) and $\pi = \pi^-$ (that is $\pi(x) = -|m|x$); i.e., $\tau(x) = -2x - 2p(x) = -2x - 2|m|x = -2(|m| + 1)x$ (which vanishes at 0 in $(-1, 1)$). Then

$$\begin{aligned} \lambda &\overset{\text{def}}{=} \kappa + \pi' = -(a + m^2) - |m| = -a - |m|(|m| + 1), \text{ and} \\ \lambda_n &\overset{\text{def}}{=} -n\tau' - \frac{n(n-1)}{2}\sigma'' = n(2|m| + n + 1). \end{aligned} \tag{4.9.5}$$

The differential equation for u is thus reduced to the canonical form

$$(1 - x^2)y''(x) - 2(|m| + 1)xy'(x) + [-a - |m|(|m| + 1)]y(x) = 0 \tag{4.9.6}$$

on $(-1, 1)$, which is Jacobi's differential equation (4.2.10)

$$(1 - x^2)y''(x) + [-(\alpha + \beta + 2)x + \beta - \alpha]\, y'(x) + \mu y(x) = 0$$

for $\mu = -a - |m|(|m| + 1)$, $\alpha = \beta = |m|$, and which thus has $P_n^{(|m|,|m|)}$ as a solution provided $n(\alpha + \beta + n + 1) = \mu$—i.e., provided $\lambda_n = \lambda$ as usual, as shown in Section 4.6. Now

$$\begin{aligned} &\lambda_n = \lambda \iff \\ &-a = |m|(|m| + 1) + n(2|m| + n + 1) = (|m| + n)(|m| + n + 1). \end{aligned}$$

Thus if $\lambda_n = \lambda$, then $-a = l(l + 1)$ where $l \geq 0$ is an integer $(|m| + n) \geq |m|$. Conversely, given an integer $l \geq |m|$, take $n = l - |m| \geq 0$ (an integer), $-a = l(l + 1)$ and thus conclude that $l(l + 1) - |m|(|m| + 1) = n(2|m| + n + 1)$; i.e., $\lambda = \lambda_n \implies P_{l-|m|}^{(|m|,|m|)}$ is a solution of equation (4.9.6), with $-a = l(l + 1)$ there.

A choice of ϕ in Proposition 4.4 such that $\phi' = \phi\pi/\sigma$ on $(-1, 1)$ is given by

$$\phi(x) = \exp\left(\int \frac{\pi}{\sigma}(x)\, dx\right) = \exp\left(\int \frac{-|m|x}{1 - x^2}\, dx\right) = (1 - x^2)^{|m|/2}.$$

Also let $u_l(x) = \phi(x)P_{l-|m|}^{(|m|,|m|)}(x)$ on $(-1,1)$ and conclude by general principles (Theorem 4.1) that u_l is a solution of our original equation (4.9.1), for $-a = l(l+1)$ there. On the other hand, by Proposition 4.9

$$P_l^{(q)} = \left(P_l^{(0,0)}\right)^{(q)} = \frac{(l+q)!}{l!2^q} P_{l-q}^{(q,q)} \tag{4.9.7}$$

for $q \geq 0$ an integer with $l \geq q$. Therefore

$$u_l(x) = (1-x^2)^{|m|/2} \frac{2^{|m|}l!}{(l+|m|)!} P_l^{(|m|)}(x) \overset{\text{def}}{=} \frac{2^{|m|}l!}{(l+|m|)!} P_l^{|m|}(x), \tag{4.9.8}$$

where $P_l^{|m|}$ is the *associated Legendre function* on $(-1,1)$; see definition (4.2.15). In particular

Corollary 4.7. *$P_l^{|m|}$ is a solution of*

$$u''(x) - \frac{2x}{1-x^2}u'(x) + \frac{l(l+1)(1-x^2)-m^2}{(1-x^2)^2}u(x) = 0 \tag{4.9.9}$$

on $(-1,1)$ for integers l, m with $l \geq |m|$.

In the next example we apply the reduction technique to obtain solutions of the Schrödinger equation

$$\psi''(x) + \frac{2m}{\hbar^2}(E - V(x))\psi(x) = 0 \tag{4.9.10}$$

in case V is the *Pöschl–Teller* potential:

$$V(x) \overset{\text{def}}{=} \frac{-V_o}{\cosh^2 ax}, \qquad V_o > 0,\ a \in \mathbb{R} - \{0\}; \tag{4.9.11}$$

thus $V(x) < 0$ on $\mathbb{R}$.

Given the structure of V it is convenient to make a change of variables $x \to v = \tanh ax$. That is, define the function f by

$$\begin{aligned} f(v) &\overset{\text{def}}{=} \frac{1}{a}\tanh^{-1} v = \frac{1}{2a}\log\frac{1+v}{1-v} \quad \text{on } (-1,1); \\ f'(v) &= \frac{1}{a(1-v^2)}, \qquad f''(v) = \frac{2v}{a(1-v^2)^2}. \end{aligned} \tag{4.9.12}$$

Given a function ψ on $\mathbb{R}$ (for example ψ a solution of (4.9.10)) set

$$\Phi(v) = \psi(f(v)) \quad \text{on } (-1,1). \tag{4.9.13}$$

Then

$$\begin{aligned} &\Phi'(v) = \psi'(f(v))f'(v), \quad \Phi''(v) = \psi'(f(v))f''(v) + \psi''(f(v))f'(v)^2 \\ &\Rightarrow \psi'(f(v)) = \Phi'(v)/f'(v) \Rightarrow \psi''(f(v))f'(v)^2 = \Phi''(v) - \frac{\Phi'(v)}{f'(v)}f''(v) \\ &\Rightarrow \psi''(f(v)) = \frac{\Phi''(v)}{f'(v)^2} - \frac{\Phi'(v)}{f'(v)^3}f''(v) \\ &\qquad = a^2(1-v^2)^2\Phi''(v) - 2a^2v(1-v^2)\Phi'(v), \end{aligned} \tag{4.9.14}$$

by (4.9.12). Also

$$\cosh^2(\tanh^{-1} v) \stackrel{\text{def}}{=} \left[\frac{\exp\left(\frac{1}{2}\log\frac{1+v}{1-v}\right) + \exp\left(-\frac{1}{2}\log\frac{1+v}{1-v}\right)}{2}\right]^2 = \frac{1}{1-v^2} \tag{4.9.15}$$

(where one squares the bracket and simplifies) which means that

$$V(f(v)) \stackrel{\text{def}}{=} \frac{-V_o}{\cosh^2\tanh^{-1} v} = -V_o(1-v^2). \tag{4.9.16}$$

Assume that ψ is a solution of (4.9.10); take $x = f(v)$ there to get by (4.9.13), (4.9.14), (4.9.16)

$$a^2(1-v^2)^2\Phi''(v) - 2a^2v(1-v^2)\Phi'(v) + \frac{2m}{\hbar^2}\left[E + V_o(1-v^2)\right]\Phi(v) = 0. \tag{4.9.17}$$

That is, for

$$\sigma(v) \stackrel{\text{def}}{=} 1-v^2, \quad \tilde{\tau}(v) \stackrel{\text{def}}{=} -2v, \quad \tilde{\sigma}(v) \stackrel{\text{def}}{=} \frac{2m}{\hbar^2 a^2}[E + V_o(1-v^2)], \tag{4.9.18}$$

equation (4.9.17) reads

$$\sigma(v)^2\Phi''(v) + \sigma(v)\tilde{\tau}(v)\Phi'(v) + \tilde{\sigma}(v)\Phi(v) = 0 \tag{4.9.19}$$

on $(-1, 1)$, which therefore is an equation of hypergeometric type, by (4.2.1). As in the example of the finite potential well in Section 3.3 of Chapter 3, and in the case of a hydrogenic atom considered in the next chapter, we look for bound states Φ—solutions of (4.9.19) for $E < 0$. Thus $-2mE/a^2\hbar^2 > 0$ and we set

$$\beta \stackrel{\text{def}}{=} +\sqrt{\frac{-2mE}{a^2\hbar^2}}, \qquad \gamma \stackrel{\text{def}}{=} +\sqrt{\frac{2mV_o}{a^2\hbar^2}}. \tag{4.9.20}$$

$\tilde{\sigma}(v)$ in (4.9.18) can be expressed in terms of β, γ:

$$\tilde{\sigma}(v) = -\beta^2 + \gamma^2(1-v^2). \tag{4.9.21}$$

We proceed to reduce equation (4.9.19) to canonical form. For $\kappa \in \mathbb{C}$, $v \in (-1, 1)$, $f_\kappa(v)$ in (4.3.4) is given by

$$f_\kappa(v) = (\gamma^2 - \kappa)v^2 + \beta^2 + \kappa - \gamma^2, \tag{4.9.22}$$

in view of (4.9.18), (4.9.21). The discriminant $\Delta(\kappa)$ is $-4(\beta^2 + \kappa - \gamma^2)(\gamma^2 - \kappa)$, which vanishes for $\kappa = \gamma^2$, $\kappa = \gamma^2 - \beta^2$. If $\kappa \neq \gamma^2$ and $\Delta(\kappa) = 0$ then $\kappa =$

$\gamma^2 - \beta^2 \implies f_\kappa(v) = \beta^2 v^2$ in (4.9.22) $\implies f_\kappa$ has a polynomial square root p given by

$$p(v) \stackrel{\text{def}}{=} \beta v. \tag{4.9.23}$$

In case $\kappa = \gamma^2$ we have $\Delta(\kappa) = 0$, $f_\kappa(v) = \beta^2$ is a constant, and $p(v) \stackrel{\text{def}}{=} \beta \implies p$ is a square root of f_κ. $\pi^\pm(v)$ in (4.3.5) is given by $\pi^\pm(v) = \pm p(v)$, and $\tau(v)$ in (4.3.6) is given by $\tau(v) = \tilde{\tau}(v) + 2\pi(v)$ for one of the choices $\pi = \pi^\pm$. We assume $\kappa \neq \gamma^2$ (so that p is given by (4.9.23)) and we make the choice $\pi = \pi^-$, as this insures that τ satisfies Lemma 4.1:

$$\tau(v) = -2(1+\beta)v \implies \tag{4.9.24}$$

$\tau'(v) = -2(1+\beta) < 0$ and $\tau(v_o) = 0$ for some $v_o \in (-1, 1)$—namely $v_o = 0$. Thus

$$\pi(v) = -\beta v \quad \text{and } \lambda \stackrel{\text{def}}{=} \kappa + \pi' = \gamma^2 - \beta^2 - \beta \tag{4.9.25}$$

in (4.3.6), and we see that a canonical form of (4.9.19) is

$$\sigma y'' + \tau y' + \lambda y = 0 \tag{4.9.26}$$

for σ, τ, λ given by

$$\sigma(v) = 1 - v^2, \quad \tau(v) = -2(1+\beta)v, \quad \lambda = \gamma^2 - \beta^2 - \beta \tag{4.9.27}$$

(by (4.9.18), (4.9.24), (4.9.25)). Also in (4.3.8)

$$\phi(v) = (1 - v^2)^{\beta/2}. \tag{4.9.28}$$

By Theorem 4.1, y solves (4.9.26) $\iff$ Φ solves (4.9.19) provided Φ, y are related by

$$\Phi = \phi y. \tag{4.9.29}$$

A solution h of

$$h'(v) = \frac{\tilde{\tau}(v)}{\sigma(v)} = \frac{-2v}{1 - v^2} \text{ on } (-1, 1) \tag{4.9.30}$$

is given by $h(v) = \log(1 - v^2)$. Therefore by Proposition 4.4 a solution ρ of $(\rho\sigma)' = \rho\tau$ on $(-1, 1)$ is given by

$$\rho(v) = \frac{\phi(v)^2}{\sigma(v)} e^{h(v)} = (1 - v^2)^\beta. \tag{4.9.31}$$

Finally, for $n = 0, 1, 2, 3, \ldots$ let

$$\lambda_n \stackrel{\text{def}}{=} -n\tau' - \frac{n(n-1)}{2}\sigma'', \quad y_n = \frac{1}{\rho}\frac{d^n}{dv^n}(\sigma^n \rho). \tag{4.9.32}$$

That is, by (4.9.18), (4.9.27), (4.9.31), and (4.2.14)

$$\lambda_n = 2n(1+\beta) + n(n-1),$$
$$y_n(v) = (1-v^2)^{-\beta}\frac{d^n}{dv^n}(1-v^2)^{n+\beta} = 2^n\ n!\ (-1)^n\ P_n^{(\beta,\beta)}(v) \tag{4.9.33}$$

where $P_n^{(\alpha,\beta)}$ is the Jacobi polynomial of degree n. Then by Theorem 4.2 we know that y_n is a solution of (4.9.26) provided that λ satisfies the integrality condition $\lambda = \lambda_n$, which provides for the quantization of energy. Namely, by (4.9.27), (4.9.33), $\lambda = \lambda_n$ means that $\gamma^2 - \beta^2 - \beta = 2n(1+\beta) + n(n-1)$, which is a quadratic equation in β with solutions

$$\beta = \beta_n^{\pm} \stackrel{\text{def}}{=} -n - \frac{1}{2} \pm \sqrt{\gamma^2 + \frac{1}{4}}. \tag{4.9.34}$$

As $\beta > 0$ we choose $\beta = \beta_n^+$, which we denote more simply by β_n. That is, by definition (4.9.20) the condition $\lambda = \lambda_n$ gives

$$E = E_n \stackrel{\text{def}}{=} -\frac{a^2\hbar^2\beta_n}{2m}. \tag{4.9.35}$$

Note here that the condition $\beta_n > 0$ forces n to satisfy

$$0 \le n < \sqrt{\gamma^2 + \frac{1}{4}} - \frac{1}{2}, \tag{4.9.36}$$

so that there are only *finitely many* eigenvalues E_n (=negative energy levels) prescribed in (4.9.35). The corresponding wave functions ψ_n (=solutions of (4.9.10)) are given as follows. Going back to (4.9.13), (4.9.29) we take $\psi_n(f(v)) = \Phi_n(v) \stackrel{\text{def}}{=} \phi_n(v)y_n(v)$, where $\phi_n(v) \stackrel{\text{def}}{=} (1-v^2)^{\beta_n/2}$; see (4.9.28). That is, for $x = f(v)$ we have $v = \tanh ax$ (by (4.9.12)) $\Longrightarrow$

$$\begin{aligned}\psi_n(x) &= (1-v^2)^{\beta_n/2}\ y_n(v)\\ &= 2^n\ n!\ (-1)^n(\operatorname{sech} ax)^{\beta_n}\ P_n^{(\beta_n,\beta_n)}(\tanh ax) \quad \text{(by (4.9.33))}\end{aligned} \tag{4.9.37}$$

for

$$\begin{aligned}\beta_n &= -n - \frac{1}{2} + \sqrt{\gamma^2 + \frac{1}{4}} > 0,\\ \gamma^2 &= \frac{2mV_o}{a^2\hbar^2}, \quad E = E_n = \frac{-a^2\hbar^2\beta_n^2}{2m}\end{aligned} \tag{4.9.38}$$

where $\beta_n > 0 \implies n$ is subject to the inequality in (4.9.36).

Appendix 4A Spherical Harmonics

For integers l, m with $l \ge 0$, $-l \le m \le l$ define Z_l^m on $\mathbb{R} \times (0,\pi)$ by

$$Z_l^m(\theta,\phi) = e^{im\theta}P_l^{|m|}(\cos\phi), \tag{4.A.1}$$

where $P_l^{|m|}$ is the associated Legendre function in (4.2.15). Let $g : \mathbb{R}^3 \to \mathbb{R}^3$ be the *spherical coordinates map*:

$$g(\rho, \theta, \phi) = (\rho \cos\theta \sin\phi, \rho \sin\theta \sin\phi, \rho \cos\phi). \tag{4.A.2}$$

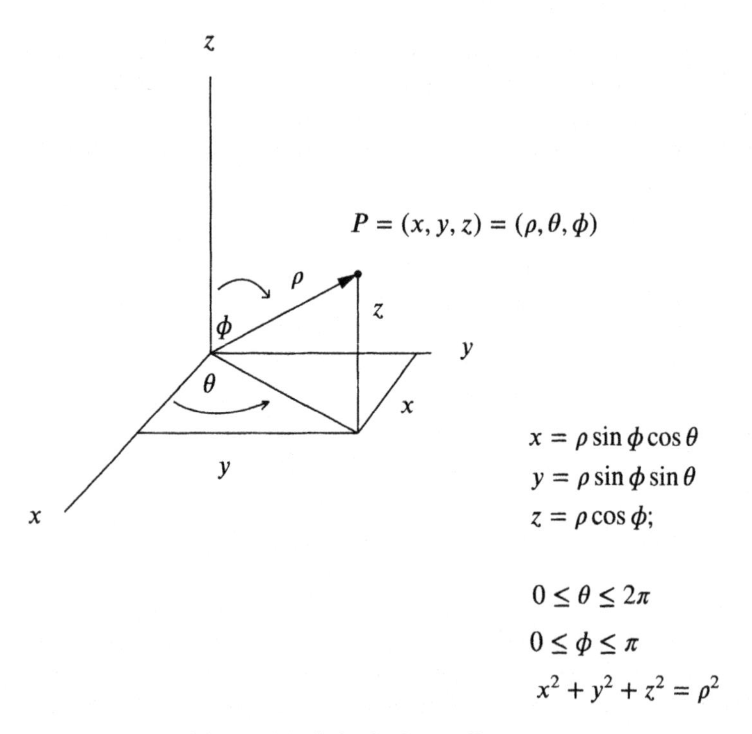

Figure 4.1: Spherical coordinates

We shall need the expression of the Laplacian in spherical coordinates:

Theorem 4.12. *Let* $W = \{(x, y, z) \in \mathbb{R}^3 | (x, y) = (0, 0)\}$, *and let* $T = \{a = (\rho, \theta, \phi) \in \mathbb{R}^3 |$ *both* $\rho, \sin\phi \neq 0\}$, *which is an open set in* $\mathbb{R}^3$ *such that* $g(a) \in \mathbb{R}^3 - W \iff a \in T$ *for* g *in* (4.A.2). *In particular* $g : T \to \mathbb{R}^3 - W$, *so that given a function* $f : \mathbb{R}^3 - W \to \mathbb{R}$ *which has continuous first and second partial derivatives on the open set* $\mathbb{R}^3 - W \subset \mathbb{R}^3 - \{0\}$ *we can define* $F : T \to \mathbb{R}$ *by* $F(a) = f(g(a))$, $a \in T$. *Then for* $a \in T$ *one has the following equation:*

$$\begin{aligned} &\frac{\partial^2 f}{\partial x^2}(g(a)) + \frac{\partial^2 f}{\partial y^2}(g(a)) + \frac{\partial^2 f}{\partial z^2}(g(a)) \\ &= \frac{1}{\rho^2}\left[\frac{\partial}{\partial \rho}\left(\rho^2 \frac{\partial F}{\partial \rho}\right)(a) + \frac{1}{\sin\phi}\frac{\partial}{\partial \phi}\left(\sin\phi \frac{\partial F}{\partial \phi}\right)(a) + \frac{1}{\sin^2\phi}\frac{\partial^2 F}{\partial \theta^2}(a)\right]. \end{aligned} \tag{4.A.3}$$

The proof, as is well known, involves only the chain rule.

Note that for

$$S^+ = \{a = (\rho, \theta, \phi) \in \mathbb{R}^3 | \rho > 0, \theta \in \mathbb{R}, 0 < \phi < \pi\}, \tag{4.A.4}$$

$$V^* = \{(x, y, z) \in \mathbb{R}^3 | (x, y) \neq (0, 0)\} = \mathbb{R}^3 - W$$

one has

Proposition 4.13. *$g : S^+ \to V^*$ is one-to-one and onto.*

In this appendix we prove

Theorem 4.13. *There exists a unique C^∞ function $Y_l^m : \mathbb{R}^3 \to \mathbb{C}$ such that*

$$Y_l^m(g(\rho, \theta, \phi)) = \rho^l Z_l^m(\theta, \phi) \stackrel{\text{def}}{=} \rho^l e^{im\theta} P_l^{|m|}(\cos\phi) \tag{4.A.5}$$

on $\mathbb{R} \times \mathbb{R} \times (0, \pi)$. In fact Y_l^m is a homogeneous polynomial of degree l (i.e., $Y_l^m(tx, ty, tz) = t^l Y_l^m(x, y, z)$ for $(x, y, z) \in \mathbb{R}^3$, $t \in \mathbb{R}$) and is harmonic: $\nabla^2 Y_l^m = 0$ for $\nabla^2 = \frac{\partial^2}{\partial x^2} + \frac{\partial^2}{\partial y^2} + \frac{\partial^2}{\partial z^2}$.

The functions $\{Y_l^m\}_{l,m \in \mathbb{Z}, l \geq 0, -l \leq m \leq l}$ are called *spherical harmonics* of degree l and order m. The uniqueness assertion of Theorem 4.13 is clear since g is surjective.

From (4.2.11) one has

$$P_l^{(m)}(x) = \frac{1}{2^l l!} \frac{d^{l+m}}{dx^{l+m}} (x^2 - 1)^l \tag{4.A.6}$$

for $m \leq l$, where $(x^2 - 1)^l = \sum_{k=0}^{l} \binom{l}{k} (-1)^k x^{2l-2k}$. If $l + m > 2l - 2k$ (i.e., $2k > l - m$) then $\frac{d^{l+m}}{dx^{l+m}} x^{2l-2k} = 0$. Thus

$$\begin{aligned} P_l^{(m)}(x) &= \frac{1}{2^l l!} \sum_{\substack{k \geq 0 \\ 2k \leq l-m}} \binom{l}{k} (-1)^k \frac{d^{l+m}}{dx^{l+m}} x^{2l-2k} \\ &= \frac{1}{2^l l!} \sum_{\substack{0 \leq k \leq l \\ l-m-2k \geq 0}} \binom{l}{k} (-1)^k (2l - 2k)(2l - 2k - 1) \\ &\quad \cdot (2l - 2k - 2) \cdots (2l - 2k - (l + m - 1)) x^{2l-2k-(l+m)} \\ &= \frac{1}{2^l l!} \sum_{\substack{0 \leq k \leq l \\ l-m-2k \geq 0}} \binom{l}{k} (-1)^k (2l - 2k)(2l - 2k - 1) \\ &\quad \cdot (2l - 2k - 2) \cdots (l - m - 2k + 1) x^{l-m-2k}. \end{aligned} \tag{4.A.7}$$

If K is the largest integer $k \in [0, l]$ such that $l - m - 2k \geq 0$ we can therefore write $P_l^{(m)}(x) = \sum_{k=0}^{K} d_k x^{l-m-2k}$ for suitable $d_k = d_{k,l,m} \in \mathbb{R}$; $0 \leq m \leq l$; $K = K_{l,m}$.

For $m \in \mathbb{Z} - \{0\}$, $\theta \in \mathbb{R}$ we can write

$$\begin{aligned}
e^{im\theta} &= \left[e^{i\frac{m}{|m|}\theta}\right]^{|m|} \\
&= \left[\cos\left(\frac{m}{|m|}\theta\right) + i\sin\left(\frac{m}{|m|}\theta\right)\right]^{|m|} \\
&= \left[\cos\theta + i\frac{m}{|m|}\sin\theta\right]^{|m|} \\
&= \sum_{j=0}^{|m|}\binom{|m|}{j}\left(i\frac{m}{|m|}\right)^j \sin^j\theta\,\cos^{|m|-j}\theta \\
&\stackrel{\text{def}}{=} \sum_{j=0}^{|m|} e_j \sin^j\theta\,\cos^{|m|-j}\theta, \qquad e_j = e_{j,m} \in \mathbb{C}.
\end{aligned} \tag{4.A.8}$$

Now for $l \geq 0$, $-l \leq m \leq l$, $l, m \in \mathbb{Z}$,

$$Z_l^m(\theta,\phi) \stackrel{\text{def}}{=} e^{im\theta} P_l^{|m|}(\cos\phi) \tag{4.A.9}$$

on $\mathbb{R} \times (0,\pi)$ with

$$e^{im\theta} = \sum_{j=0}^{|m|} e_j \sin^j\theta\,\cos^{|m|-j}\theta \qquad \text{for } m \neq 0, \tag{4.A.10}$$

$$P_l^{(|m|)}(x) = \sum_{k=0}^{K} d_k x^{l-|m|-2k} \qquad (\text{since } |m| \leq l), \tag{4.A.11}$$

as just seen. For $0 < \phi < \pi$, $\cos\phi \in (-1,1) \implies P_l^{|m|}(\cos\phi) \stackrel{\text{def}}{=} (1-\cos^2\phi)^{|m|/2} P_l^{(|m|)}(\cos\phi) = (\sin^2\phi)^{|m|/2} P_l^{(|m|)}(\cos\phi)$ with $\sin\phi > 0$: $(\sin^2\phi)^{|m|/2} = (\sin\phi)^{|m|}$; i.e.,

$$P_l^{|m|}(\cos\phi) = (\sin\phi)^{|m|} P_l^{(|m|)}(\cos\phi). \tag{4.A.12}$$

For $m = 0$, equation (4.A.10) holds with the right-hand side interpreted as $e_o \stackrel{\text{def}}{=} 1$. That is,

$$\begin{aligned}
Z_l^m(\theta,\phi) &= e^{im\theta} P_l^{|m|}(\cos\phi) \\
&= e^{im\theta}(\sin\phi)^{|m|} P_l^{(|m|)}(\cos\phi) \\
&= \sum_{k=0}^{K}\sum_{j=0}^{|m|} e_j d_k \sin^{|m|}\phi\,\sin^j\theta\,\cos^{|m|-j}\theta\,\cos^{l-|m|-2k}\phi \\
&= \sum_{k=0}^{K}\sum_{j=0}^{|m|} e_j d_k (\cos\theta\sin\phi)^{|m|-j}(\sin\theta\sin\phi)^j(\cos\phi)^{l-|m|-2k}.
\end{aligned} \tag{4.A.13}$$

Therefore if we define $Y_l^m : \mathbb{R}^3 \overset{C^\infty}{\to} \mathbb{C}$ by

$$Y_l^m(x, y, z) = \sum_{k=0}^{K} \sum_{j=0}^{|m|} e_j d_k (x^2 + y^2 + z^2)^k x^{|m|-j} y^j z^{l-|m|-2k}, \tag{4.A.14}$$

we obtain for $0 < \phi < \pi$,

$$\begin{aligned}
&Y_l^m(\rho \sin\phi \cos\theta, \rho \sin\phi \sin\theta, \rho \cos\phi) \\
&= \sum_{k=0}^{K} \sum_{j=0}^{|m|} e_j d_k \rho^{2k} \rho^{|m|-j} (\sin\phi \cos\theta)^{|m|-j} \rho^j (\sin\phi \sin\theta)^j \rho^{l-|m|-2k} (\cos\phi)^{l-|m|-2k} \\
&= \sum_{k=0}^{K} \sum_{j=0}^{|m|} e_j d_k \rho^l (\sin\phi \cos\theta)^{|m|-j} (\sin\phi \sin\theta)^j (\cos\phi)^{l-|m|-2k} \\
&= \rho^l Z_l^m(\theta, \phi) \text{ by (4.A.13)}
\end{aligned} \tag{4.A.15}$$

as desired. By definition of Y_l^m one has $Y_l^m(tx, ty, tz) = t^l Y_l^m(x, y, z)$ for $(x, y, z) \in \mathbb{R}^3, t \in \mathbb{R}$. To conclude the proof we need to show that Y_l^m is harmonic: $\nabla^2 Y_l^m = 0$.

The differential equation in (4.9.9) of Corollary 4.7

$$(1 - x^2) y''(x) - 2x y'(x) + \left[l(l+1) - \frac{m^2}{1 - x^2} \right] y(x) = 0 \tag{4.A.16}$$

on $(-1, 1)$, satisfied by the associated Legendre functions $P_l^m, l, m = 0, 1, 2, 3, \ldots,$ $m \leq l$ can be expressed in an alternate manner: By the change of variables $x = \cos\phi$ for $\phi \in (0, \pi)$ one has

$$\begin{aligned}
\frac{1}{\sin\phi} \frac{d}{d\phi} \left[(\sin\phi) \frac{d}{d\phi} y(\cos\phi) \right] &= (sin^2\phi) y''(\cos\phi) - 2(\cos\phi) y'(\cos\phi) \\
&= (1 - x^2) y''(x) - 2x y'(x),
\end{aligned} \tag{4.A.17}$$

so that (4.A.16) may be expressed as

$$\frac{1}{\sin\phi} \frac{d}{d\phi} \left[(\sin\phi) \frac{d}{d\phi} y(\cos\phi) \right] + \left[l(l+1) - \frac{m^2}{\sin^2\phi} \right] y(\cos\phi) = 0 \tag{4.A.18}$$

on $(0, \pi)$, for $l, m = 0, 1, 2, 3, \ldots$. As $Z_l^m(\theta, \phi) \overset{\text{def}}{=} e^{im\theta} P_l^{|m|}(\cos\phi)$ one obtains immediately from (4.A.18) that Z_l^m satisfies

$$\frac{1}{\sin\phi} \frac{\partial}{\partial\phi} \left[(\sin\phi) \frac{\partial Z_l^m}{\partial\phi} \right] (\theta, \phi) + \frac{1}{\sin^2\phi} \frac{\partial^2 Z_l^m}{\partial\theta^2} (\theta, \phi) = -l(l+1) Z_l^m(\theta, \phi) \tag{4.A.19}$$

on $\mathbb{R} \times (0, \pi)$; $l, m \in \mathbb{Z}, l \geq 0, |m| \leq l$.

By Theorem 4.12 for $a = (r, \theta, \phi) \in \mathbb{R}^3$ with $r \neq 0$, $\sin\phi \neq 0$

$$(\nabla^2 Y_l^m)(g(a)) = \frac{1}{r^2}\left[\frac{\partial}{\partial r}\left(r^2\frac{\partial F}{\partial r}\right)(a) + \frac{1}{\sin\phi}\frac{\partial}{\partial\phi}\left(\sin\phi\frac{\partial F}{\partial\phi}\right)(a) + \frac{1}{\sin^2\phi}\frac{\partial^2 F}{\partial\theta^2}(a)\right] \tag{4.A.20}$$

for

$$F(a) = Y_l^m(g(a)) = r^l Z_l^m(\theta, \phi). \tag{4.A.21}$$

Now

$$\frac{\partial}{\partial r}\left(r^2\frac{\partial F}{\partial r}\right)(a) = l(l+1)r^l Z_l^m(\theta,\phi)(= l(l+1)F(a)), \tag{4.A.22}$$

which by (4.A.19), (4.A.20) gives $(\nabla^2 Y_l^m)(g(a)) = 0$ for $a = (r, \theta, \phi)$, $r \neq 0$, $(\theta, \phi) \in \mathbb{R} \times (0, \pi)$. By Proposition 4.13 we must have $(\nabla^2 Y_l^m)(x,y,z) = 0$ for $(x, y, z) \in \mathbb{R}^3$ such that $(x, y) \neq (0, 0)$. By continuity $(\nabla^2 Y_l^m)(x, y, z) = 0 \quad \forall (x, y, z) \in \mathbb{R}^3$: For example $\lim_{n\to\infty}\left(\frac{1}{n}, \frac{1}{n}, z\right) = (0, 0, z) \implies 0 = \lim_{n\to\infty}\left(\nabla^2 Y_l^m\right)\left(\frac{1}{n}, \frac{1}{n}, z\right) = \left(\nabla^2 Y_l^m\right)(0, 0, z)$. □

By (4.A.2), $g = (g_1, g_2, g_3)$ where

$$g_1(a) = \rho\cos\theta\sin\phi, \quad g_2(a) = \rho\sin\theta\sin\phi, \quad g_3(a) = \rho\cos\phi \tag{4.A.23}$$

for $a = (\rho, \theta, \phi) \in \mathbb{R}^3$. Let $F = Y_l^m \circ g : \mathbb{R}^3 \to \mathbb{C}$. By the chain rule

$$\begin{aligned}\frac{\partial F}{\partial\theta}(a) &= \frac{\partial Y_l^m}{\partial x}(g(a))\frac{\partial g_1}{\partial\theta}(a) + \frac{\partial Y_l^m}{\partial y}(g(a))\frac{\partial g_2}{\partial\theta}(a) + \frac{\partial Y_l^m}{\partial z}(g(a))\frac{\partial g_3}{\partial\theta}(a)\\ &= -\frac{\partial Y_l^m}{\partial x}(g(a))g_2(a) + \frac{\partial Y_l^m}{\partial y}(g(a))g_1(a).\end{aligned} \tag{4.A.24}$$

On the other hand, by definition (4.A.5) $F(a) = \rho^l e^{im\theta}P_l^{|m|}(\cos\phi) \implies$

$$\frac{\partial F}{\partial\theta}(a) = im\rho^l e^{im\theta}P_l^{|m|}(\cos\phi) = imF(a). \tag{4.A.25}$$

Now g maps $\mathbb{R}^3$ onto $\mathbb{R}^3$. Thus for $(x, y, z) \in \mathbb{R}^3$ arbitrary, write $(x, y, z) = g(a)$ for some suitable $a = (\rho, \theta, \phi) \in \mathbb{R}^3$. By (4.A.25) and (4.A.24),

$$imY_l^m(x, y, z) \stackrel{\text{def}}{=} imF(a) = -\frac{\partial Y_l^m}{\partial x}(x, y, z)y + \frac{\partial Y_l^m}{\partial y}(x, y, z)x, \tag{4.A.26}$$

which proves

Proposition 4.14. *For the differential operator* $D \stackrel{\text{def}}{=} x\frac{\partial}{\partial y} - y\frac{\partial}{\partial x}$ *one has*

$$DY_l^m = imY_l^m.$$

Using definition (4.7.45) we can extend Z_l^m to a continuous function on $\mathbb{R} \times [0, \pi]$. Namely

$$Z_l^m(\theta, \phi = 0 \text{ or } \pi) \stackrel{\text{def}}{=} \begin{bmatrix} 0 & \text{for } m \neq 0 \\ P_l(1) \text{ or } P_l(-1) & \text{for } m = 0 \end{bmatrix}. \tag{4.A.27}$$

Since for $m, r \in \mathbb{Z}$

$$\int_0^{2\pi} \exp(im\theta)\exp(-ir\theta)\, d\theta = 2\pi\delta_{mr} \tag{4.A.28}$$

Theorem 4.8 leads to the following.

Theorem 4.14. *Let $l, n, m, r \in \mathbb{Z}$ with $l, n \geq 0$, $-l \leq m \leq l$, $-n \leq r \leq n$. Then*

$$\frac{1}{4\pi}\int_0^{2\pi}\int_0^{\pi} Z_l^m(\theta,\phi)\overline{Z_n^r(\theta,\phi)}\sin\phi\, d\phi\, d\theta = \frac{1}{2l+1}\frac{(l+|m|)!}{(l-|m|)!}\delta_{(l,m),(n,r)}.$$

Appendix 4B The Polynomials L_n

For $n \geq 0$ an integer define

$$L_n = n! L_n^o \tag{4.B.1}$$

where L_n^o is the Laguerre polynomial of degree n and order 0. By (4.2.12)

$$L_n(x) = e^x \frac{d^n}{dx^n}(e^{-x}x^n) \text{ for } x \in \mathbb{R}. \tag{4.B.2}$$

Proposition 4.15. *The jth derivative of L_n is given by*

$$\begin{aligned} L_n^{(j)} &= (-1)^j n! L_{n-j}^j && \text{for } 0 \leq j \leq n, \\ L_n^{(j)} &= 0 && \text{for } j > n. \end{aligned} \tag{4.B.3}$$

Proof. (4.B.3) is true for $j = 0$ by definition (4.B.1). By Proposition 4.7

$$(L_n^\alpha)^j(x) = (-1)^j L_{n-j}^{\alpha+j}(x) \tag{4.B.4}$$

on $(0, \infty)$ for $n \geq j$. There we took $x > 0$ to define $\rho(x) = x^\alpha e^{-x}$, $x^\alpha = e^{\alpha \log x}$, $\alpha \in \mathbb{R}$. In the present situation however $\alpha = 0 \in \mathbb{Z}$, and we may assert that (4.B.4) is valid for all $x \in \mathbb{R}$. Similarly, by Corollary 4.4, for $\alpha \in \mathbb{Z}$

$$x(L_n^\alpha)''(x) + (\alpha + 1 - x)(L_n^\alpha)'(x) + nL_n^\alpha(x) = 0 \tag{4.B.5}$$

for all $x \in \mathbb{R}$. By (4.B.4) therefore, $L_n^{(j)} = n!(L_n^o)^{(j)} = n!(-1)^j L_{n-j}^j$ on $\mathbb{R}$ for $n \geq j$, as desired. Of course $L_n^{(j)} = 0$ for $j > n$, as L_n is a polynomial of degree n. □

By (4.B.4) and (4.B.5) we see moreover the following.

Corollary 4.8. *For integers j, n*

$$x(L_n^{(j)})''(x) + (j + 1 - x)(L_n^{(j)})'(x) + (n - j)L_n^{(j)}(x) = 0$$

on $\mathbb{R}$ for $0 \leq j \leq n$, and hence for all $j, n \geq 0$.

5

Hydrogen-like Atoms

Because of considerable mathematical difficulties the Schrödinger equation can be solved exactly for only a few simple quantum systems. In particular for atoms with at least two electrons no such exact solutions exist. The most important system where an exact solution is indeed possible consists of a single electron with charge $q_1 = -e$ in motion under the influence of a Coulomb potential about a nucleus with a positive charge $q_2 = Ze$, for $Z \geq 1$ an integer. Such a system is called a *hydrogenic* or *hydrogen-like atom*, with *atomic number* Z. One may view the nucleus as consisting of Z protons. For $Z = 1$ we obtain the hydrogen atom. For $Z = 2$, for example, the system is singly ionized helium He^+; this is helium with its two protons as usual, but stripped of one of its two electrons. Since it has only one electron but two protons it is *positively* charged, as indicated by the plus sign +. Similarly the lithium atom Li consists of three electrons and three protons (and four neutrons). If we strip off two electrons we obtain the ion Li^{++}—doubly ionized lithium, which is the case $Z = 3$.

The wave functions of hydrogenic atoms can be used to provide a fairly good qualitative description of properties of multi-electron atoms and molecules—the starting point or model for the study of all other atoms being the hydrogen atom. Hence, the importance of this chapter is clear. If our system had $n \geq 2$ electrons instead, then we could study it by assuming for simplicity that each electron moves independently of the others and is influenced only by the *average* field due to these other electrons, and of course by the field due to the nucleus. A discussion of multi-electron atoms will be taken up in Chapter 10.

In addition to obtaining explicit wave functions for the hydrogenic atom we will obtain the other valuable piece of information concerning the specification of possible energy states the electron can occupy. As in the old theory of Bohr we will discover that the (negative) electron energies are not arbitrary, but are quantized: only certain values $\{E_n\}_{n=1}^{\infty}$ are possible, similarly as we have seen in previous examples. These wave functions will carry three bits of quantum data (n, l, m): the *total quantum number* $n = 1, 2, 3, \ldots$, the *azimuthal* (or *orbital*) *quantum number* $l = 0, 1, 2, \ldots, n-1$, and the *magnetic quantum number* $m = -l, -l+1, \ldots, -1, 0, 1, 2, \ldots, l$. In accord with earlier notation the pair $(l, m) \in \mathbb{Z} \times \mathbb{Z}$, with $l \geq |m|$, indeed will arise as the label for the associated Legendre function $P_l^{|m|}$. The energy levels will depend only on the total quantum number n. Thus we can obtain a rather comprehensive picture of the hydrogenic atom. The above data, however, is not sufficient to account for experimentally observed phenomena such as fine-structure splitting of spectral lines and Zeeman effects. For this one must consider, in addition, the *electron-spin* quantum number $s = \frac{1}{2}$, a point which we will deal with later. Here, as shown by Uhlenbeck and Goudsmit, one assigns to the electron an "angular momentum and magnetic moment," as a charged object spinning about an axis would have. The electron however *does not* literally spin about an axis.

5.1 Solutions of Schrödinger's Equation for Hydrogenic Atoms

Let (x_1, y_1, z_1), (x_2, y_2, z_2), m_1, m_2 denote the positions of the nucleus and the electron, and their masses respectively. Let $U \subset \mathbb{R}^6$ be the open subset $\{(p_1, p_2) | p_1 \neq p_2\}$ considered in Section 2.5 of Chapter 2, where we now write $p_j = (x_j, y_j, z_j)$ for $j = 1, 2$. Define continuous functions $F^{(j)} : U \to \mathbb{R}^3$ by

$$F^{(1)}(p_1, p_2) = \frac{q_1 q_2 (p_1 - p_2)}{|p_1 - p_2|^3} = \frac{-Ze^2 (p_1 - p_2)}{|p_1 - p_2|^3}, \tag{5.1.1}$$

$F^{(2)} = -F^{(1)}$. Then, as discussed in Chapter 2, the electrostatic Coulomb force exerted on the electron by the nucleus is $F^{(2)}(p_1, p_2)$:

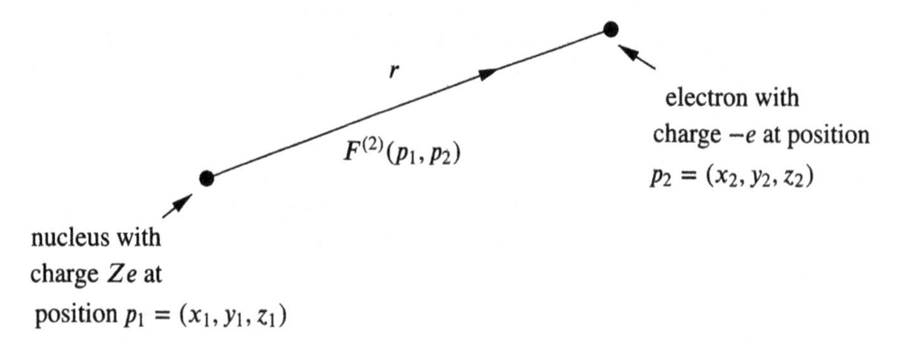

Figure 5.1: Electrostatic coulomb force on an electron

The arrow in Figure 5.1 indicates the direction of $F^{(2)}(p_1, p_2)$, and r is the distance from p_1 to p_2. Of course $F^{(1)}(p_1, p_2)$ (the force exerted on the nucleus by the electron) and $F^{(2)}(p_1, p_2)$ have the same magnitude:

$$|F^{(j)}(p_1, p_2)| = \frac{Ze^2}{r^2} \tag{5.1.2}$$

by equation (5.1.1). If $X_j, Y_j, Z_j : U \to \mathbb{R}$ denote the components of $F^{(j)}$, then

$$X_j = -\frac{\partial V}{\partial x_j}, \qquad Y_j = -\frac{\partial V}{\partial y_j}, \qquad Z_j = -\frac{\partial V}{\partial z_j} \tag{5.1.3}$$

for the potential energy $V : U \to \mathbb{R}$ of our system given by equation (2.5.4) for $q_1 = -e$, $q_2 = Ze$:

$$V(p_1, p_2) = \frac{-Ze^2}{r = [(x_1 - x_2)^2 + (y_1 - y_2)^2 + (z_1 - z_2)^2]^{1/2}}. \tag{5.1.4}$$

By equation (3.4.10), the time-independent Schrödinger equation (TISE) is

$$\frac{1}{m_1}\nabla_1^2\psi + \frac{1}{m_2}\nabla_2^2\psi + \frac{8\pi^2}{h^2}(E - V)\psi = 0 \tag{5.1.5}$$

for

$$\nabla_j^2 = \frac{\partial^2}{\partial x_j^2} + \frac{\partial^2}{\partial y_j^2} + \frac{\partial^2}{\partial z_j^2}. \tag{5.1.6}$$

To study equation (5.1.5) we pass from Cartesian coordinates to center of mass coordinates, via the diffeomorphism Φ of $\mathbb{R}^6$ such that $\Phi(U) = (\mathbb{R}^3 - \{0\}) \times \mathbb{R}^3$, given in Chapters 2, 3; see Section 3.8 of Chapter 3. We look for a continuous function $\psi : \mathbb{R}^6 \to \mathbb{C}$ which solves (5.1.5) such that $\psi|_U$ is class C^2 on U: $\psi|_U$ has continuous first and second partial derivatives on U. Consider $f = \psi \circ \Phi^{-1}$: $\mathbb{R}^6 \to \mathbb{C}$; f is continuous and $f|_{\Phi(U)}$ is class C^2. If $\mu \overset{\text{def}}{=} \frac{m_1 m_2}{m_1 + m_2}$ is the reduced mass of the system, then Φ transforms equation (5.1.5) on U to the equation

$$\frac{1}{\mu}\nabla_1^2 f + \frac{1}{m_1 + m_2}\nabla_2^2 f + \frac{8\pi^2}{h^2}(E - V \circ \Phi^{-1})f = 0 \tag{5.1.7}$$

on $\Phi(U)$, as we have seen, where

$$(V \circ \Phi^{-1})(p_1, p_2) = \frac{-Ze^2}{\sqrt{x_1^2 + y_1^2 + z_1^2}} \tag{5.1.8}$$

on $\Phi(U)$; see equations (3.8.8), (3.8.9). Now apply Theorem 3.3 and the converse argument that followed its statement: Under a separation of variables $f = f_1 \otimes f_2$, f satisfies (5.1.7) if and only if f_1, f_2 satisfy

$$(\nabla_1^2 f_1)(x^{(1)}) + \frac{8\pi^2\mu}{h^2}\left(E_1 + \frac{Ze^2}{\sqrt{x_1^2 + y_1^2 + z_1^2}}\right) f_1(x^{(1)}) = 0, \tag{5.1.9}$$

$$(\nabla_2^2 f_2)(x^{(2)}) + \frac{8\pi^2}{h^2}(m_1 + m_2)E_2 f_2(x^{(2)}) = 0$$

for $x^{(1)} = (x_1, y_1, z_1) \in \mathbb{R}^3 - \{0\}$, $x^{(2)} = (x_2, y_2, z_2) \in \mathbb{R}^3$. Here $E_1 + E_2 = E$, $f_1, f_2 : \mathbb{R}^3 \to \mathbb{C}$ are continuous, and $f_1|_{\mathbb{R}^3-\{0\}}, f_2$ are of class C^2. Of course we have interest only in the case $f \neq 0$, in which case both $f_1, f_2 \neq 0$. Note that the second equation (5.1.9) is the TISE for a free particle of mass $m_1 + m_2$; see Section 3.6 of Chapter 3. The energy E_2 here (called the *translational* energy of the system) is non-quantized, as we know, and generally plays a less important role in quantum mechanics. We will therefore maintain attention on the first equation in (5.1.9) in the discussion which follows. This equation is the TISE on $\mathbb{R}^3 - \{0\}$ of a *single* particle of mass μ influenced by the Coulomb potential

$$V^{(1)}(x_1, y_1, z_1) = \frac{-Ze^2}{\sqrt{x_1^2 + y_1^2 + z_1^2}}. \tag{5.1.10}$$

Again we search for bound states of the electron—i.e., we take $E_1 < 0$.

To make further progress we now pass to spherical coordinates. Let

$$W \stackrel{\text{def}}{=} \{(x, y, z) \in \mathbb{R}^3 | (x, y) = (0, 0)\} \tag{5.1.11}$$

and let

$$\begin{aligned} T &\stackrel{\text{def}}{=} \{a = (r, \theta, \phi) \in \mathbb{R}^3 | \text{ both } r, \sin\phi \neq 0\} \\ &= g^{-1}(\mathbb{R}^3 - W) \end{aligned} \tag{5.1.12}$$

where $g : \mathbb{R}^3 \to \mathbb{R}^3$ denotes the spherical coordinate map:

$$g(r, \theta, \phi) \stackrel{\text{def}}{=} (r\cos\theta\sin\phi, r\sin\theta\sin\phi, r\cos\phi).$$

Given Theorem 4.12 in Appendix 4A we can express the first equation in (5.1.9) as

$$\begin{aligned} \frac{1}{r^2}\left[\frac{\partial}{\partial r}\left(r^2\frac{\partial F}{\partial r}\right)(a) + \frac{1}{\sin\phi}\frac{\partial}{\partial\phi}\left(\sin\phi\frac{\partial F}{\partial\phi}\right)(a) + \frac{1}{\sin^2\phi}\frac{\partial^2 F}{\partial\theta^2}(a)\right] \\ + \frac{8\pi^2\mu}{h^2}\left(E_1 + \frac{Ze^2}{r}\right)F(a) = 0 \end{aligned} \tag{5.1.13}$$

for $F(a) = f_1(g(a))$, where we write $a = (r, \theta, \phi) \in \mathbb{R}^3$, $x^{(1)} = g(a)$. Here we need $x^{(1)} \in \mathbb{R}^3 - W$ (i.e., $(x_1, y_1) \neq (0, 0)$ or equivalently, by (5.1.12), $a \in T$: both r and $\sin\phi \neq 0$). Note that for

$$S^+ = \{a \in \mathbb{R}^3 | r > 0, 0 < \phi < \pi, \theta \in \mathbb{R}\}, \tag{5.1.14}$$

$$V^* = \{(x, y, z) \in \mathbb{R}^3 | (x, y) \neq (0, 0)\} = \mathbb{R}^3 - W \tag{5.1.15}$$

one has (cf. Appendix 4A, Proposition 4.13) the following.

Proposition 5.1. *g maps S^+ onto V^*. In particular* (5.1.13) *holds on S^+ for f_1 satisfying* (5.1.9). *g is one-to-one on S^+.*

For equation (5.1.13) consider again a separation of variables, as follows. Let $F_1 : \mathbb{R}^+ \overset{\text{def}}{=} \{r \in \mathbb{R} | r > 0\} \to \mathbb{C}$, $G : \mathbb{R} \times (0, \pi) \to \mathbb{C}$ be of class C^2 such that

$$\begin{aligned} &\frac{1}{r^2}\frac{d}{dr}(r^2 F_1')(r) + \frac{\lambda F_1(r)}{r^2} + C(E_1 - V_o(r))F_1(r) = 0, \\ &\frac{1}{\sin\phi}\frac{\partial}{\partial\phi}((\sin\phi)\frac{\partial G}{\partial\phi})(\theta,\phi) + \frac{1}{\sin^2\phi}\frac{\partial^2 G}{\partial\theta^2}(\theta,\phi) = \lambda G(\theta,\phi) \end{aligned} \tag{5.1.16}$$

on $\mathbb{R}^+$, $\mathbb{R} \times (0, \pi)$ respectively where $\lambda \in \mathbb{R}$, $C = \frac{8\pi^2\mu}{h^2}$, and V_o is defined on $\mathbb{R}^+$ by $V_o(r) = \frac{c}{r}$ for $c = q_1 q_2 = -Ze^2$.

Put $F = F_1 \otimes G$ on $\mathbb{R}^+ \times \mathbb{R} \times (0, \pi) = S^+$: $F(r, \theta, \phi) \overset{\text{def}}{=} F_1(r)G(\theta, \phi)$. Then F is of class C^2 and satisfies equation (5.1.13) on S^+. Conversely suppose $F = F_1 \otimes G \neq 0$ satisfies equation (5.1.13) on S^+:

$$\begin{aligned} 0 = &\frac{1}{r^2}\frac{d}{dr}(r^2 F_1')(r)G(\theta,\phi) + \frac{F_1(r)}{r^2\sin\phi}(\frac{\partial}{\partial\phi}\sin\phi\frac{\partial G}{\partial\phi})(\theta,\phi) \\ &+ \frac{F_1(r)}{r^2\sin^2\phi}\frac{\partial^2 G}{\partial\theta^2}(\theta,\phi) + C(E_1 - V_o(r))F_1(r)G(\theta,\phi). \end{aligned} \tag{5.1.17}$$

Choose $(\theta_o, \phi_o) \in \mathbb{R} \times (0, \pi)$ such that $G(\theta_o, \phi_o) \neq 0$ and divide (5.1.17) by $G(\theta_o, \phi_o)$. One gets the first equation in (5.1.16) on $\mathbb{R}^+$ for

$$\lambda = \frac{\frac{\partial}{\partial\phi}(\sin\phi\frac{\partial G}{\partial\phi})(\theta_o,\phi_o)}{(\sin\phi_o)G(\theta_o,\phi_o)} + \frac{\frac{\partial^2 G}{\partial\theta^2}(\theta_o,\phi_o)}{(\sin^2\phi_o)G(\theta_o,\phi_o)}. \tag{5.1.18}$$

Similarly choose $r_o \in \mathbb{R}^+$ such that $F_1(r_o) \neq 0$ and divide (5.1.17) by $F_1(r_o)$. Using that

$$\frac{\frac{d}{dr}(r^2 F_1')(r_o)}{r_o^2 F_1(r)} + C(E_1 - V_o(r_o)) = -\frac{\lambda}{r_o^2}, \tag{5.1.19}$$

which is the first equation in (5.1.16), one gets

$$0 = -\frac{\lambda}{r_o^2}G(\theta,\phi) + \frac{\frac{\partial}{\partial\phi}((\sin\phi)\frac{\partial G}{\partial\phi})(\theta,\phi)}{r_o^2\sin\phi} + \frac{\frac{\partial^2 G}{\partial\theta^2}(\theta,\phi)}{r_o^2\sin^2\phi} \tag{5.1.20}$$

on $\mathbb{R} \times (0, \pi)$, which is the second equation in (5.1.16). Therefore consider the first and second equations in (5.1.16). The first equation is exactly equation (4.5.9) which we solved there, where E_1 there was assumed to be negative. Here we also assume $E_1 < 0$. Physically this means that the total energy of our system is insufficient to permit ionization. Let $l, n \geq 0$ be integers with $n \geq l + 1$. For $\lambda = -l(l + 1)$ we obtained the solutions u_{nl} given by (4.5.22):

$$u_{nl}(r) = r^l \exp(-\alpha_n \frac{r}{2})[n - (l + 1)]!\, L^{2l+1}_{n-(l+1)}(\rho_n(r)) \tag{5.1.21}$$

on $\mathbb{R}^+$ for

$$E_1 = E^{(n)} \stackrel{\text{def}}{=} -\frac{2\pi^2\mu Z^2 e^4}{h^2 n^2} = -\frac{\mu Z^2 e^4}{2\hbar^2 n^2} \tag{5.1.22}$$

where

$$\alpha_n = \frac{2\mu Z e^2}{\hbar^2 n}, \qquad \rho_n(r) = \alpha_n r, \qquad r \in \mathbb{R}^+. \tag{5.1.23}$$

$L_{n-(l+1)}^{2l+1}$ is the Laguerre polynomial of degree $n-(l+1)$ and order $2l+1$ defined in (4.2.12).

Let G_1 be of class C^2 which satisfies

$$\frac{1}{\sin\phi}\frac{d}{d\phi}\left((\sin\phi)\frac{dG_1}{d\phi}\right) + \left(-\lambda - \frac{m^2}{\sin^2\phi}\right)G_1(\phi) = 0 \tag{5.1.24}$$

on $(0,\pi)$, where $m \in \mathbb{Z}$. Then we obtain a class C^2 solution G of the second equation in (5.1.16) by setting $G(\theta,\phi) \stackrel{\text{def}}{=} \frac{e^{im\theta}}{\sqrt{2\pi}}G_1(\phi)$ on $\mathbb{R}\times(0,\pi)$. To solve (5.1.24) set $u(x) \stackrel{\text{def}}{=} G_1(\arccos x)$ on $(-1,1)$. Then (5.1.24) is transformed to the following differential equation of hypergeometric type:

$$u''(x) - \frac{2x}{1-x^2}u'(x) + \left[\frac{-\lambda(1-x^2)-m^2}{(1-x^2)^2}\right]u(x) = 0 \tag{5.1.25}$$

on $(-1,1)$. The latter equation was reduced to canonical form (namely to Jacobi's equation) and solved in Chapter 4; see Section 4.9. Given $-\lambda = l(l+1)$ again where $l \geq 0$ is an integer we constructed the solutions $\tilde{u}_{lm}$ of (5.1.25) given by (4.9.8):

$$\tilde{u}_{lm} = \frac{2^{|m|}l!}{(l+|m|)!}P_l^{|m|}(x) \tag{5.1.26}$$

on $(-1,1)$ for $l \geq |m|$, where $P_l^{|m|}$ is the associated Legendre function defined in Section 4.2 of Chapter 4.

The task now is to piece together these various solutions to reach our original goal of obtaining solutions f_1 of the first TISE in (5.1.9), under the assumption that the *internal* energy E_1 is negative. For this we retrace our steps. Working backwards we get solutions $G_1(\phi) \stackrel{\text{def}}{=} u(\cos\phi) = \tilde{u}_{lm}(\cos\phi)$ of (5.1.24) on $(0,\pi)$, $G(\theta,\phi) \stackrel{\text{def}}{=} \frac{e^{im\theta}}{\sqrt{2\pi}}G_1(\phi) = \frac{e^{im\theta}}{\sqrt{2\pi}}\tilde{u}_{lm}(\cos\phi)$ of the second equation in (5.1.16) on $\mathbb{R}\times(0,\pi)$, for $\lambda = -l(l+1)$, and $F(r,\theta,\phi) \stackrel{\text{def}}{=} F_1(r)G(\theta,\phi) = u_{nl}(r)\frac{e^{im\theta}}{\sqrt{2\pi}}\tilde{u}_{lm}(\cos\phi)$ of (5.1.13) on $\mathbb{R}^+\times\mathbb{R}\times(0,\pi)$, for $u_{nl}(r)$ given by (5.1.21), and for $E_1 = E^{(n)}$ in (5.1.22). But for $g : \mathbb{R}^3 \to \mathbb{R}^3$ defined following (5.1.12), $F(a) \stackrel{\text{def}}{=} f_1(g(a))$. We finally obtain solutions (wave functions) $f_1 = \psi_{nlm}^o$ of

$$(\nabla_1^2 f_1)(x,y,z) + \frac{8\pi^2\mu}{h^2}\left(E_1 + \frac{Ze^2}{\sqrt{x^2+y^2+z^2}}\right) f_1(x,y,z) = 0 \quad (5.1.27)$$

(the first equation in (5.1.9)) on $V^* \stackrel{\text{def}}{=} \{(x,y,z) \in \mathbb{R}^3 | (x,y) \neq (0,0)\}$ (noting Proposition 5.1), for $E_1 = E^{(n)}$:

$$\begin{aligned}\psi_{nlm}^o&(r\cos\theta\sin\phi, r\sin\theta\sin\phi, r\cos\phi)\\ &= \frac{[n-(l+1)]!\, 2^{|m|}\, l!}{\sqrt{2\pi}(l+|m|)!} e^{im\theta} e^{-\alpha_n r/2} r^l\, L_{n-(l+1)}^{2l+1}(\rho_n(r))\, P_l^{|m|}(\cos\phi) \quad (5.1.28)\end{aligned}$$

for $(r,\theta,\phi) \in \mathbb{R}^+ \times \mathbb{R} \times (0,\pi)$, where n, l, m are integers with $n \geq 1$, $0 \leq l \leq n-1$, $-l \leq m \leq l$, and where

$$E_1 = E^{(n)} \stackrel{\text{def}}{=} -\frac{2\pi^2\mu Z^2 e^4}{h^2 n^2} = -\frac{\mu Z^2 e^4}{2\hbar^2 n^2}, \quad (5.1.29)$$

$$\alpha_n = \frac{2\mu Z e^2}{\hbar^2 n}, \quad \rho_n(r) = \alpha_n r \quad \text{for } r \in \mathbb{R}^+. \quad (5.1.30)$$

The Laguerre and associated Legendre functions $L_{n-(l+1)}^{2l+1}$, $P_l^{|m|}$ in (5.1.28) are defined in Section 4.2 of Chapter 4. We use the superscript "o" in ψ_{nlm}^o as these functions yet remain to be normalized. It is desirable to express $\psi_{nlm}^o(x,y,z)$ explicitly in terms of Cartesian coordinates x, y, z. Such an expression is rarely (if at all) written down in physics books. Let $Y_l^m : \mathbb{R}^3 \to \mathbb{C}$ be the spherical harmonic of degree l and order m:

$$Y_l^m(g(r,\theta,\phi)) = r^l e^{im\theta} P_l^{|m|}(\cos\phi) \stackrel{\text{def}}{=} r^l Z_l^m(\theta,\phi) \quad (5.1.31)$$

on $\mathbb{R} \times \mathbb{R} \times (0,\pi)$; see Appendix 4A for details. If we write

$$|(x,y,z)| = \sqrt{x^2+y^2+z^2},$$

then by (5.1.28)

$$\begin{aligned}\psi_{nlm}^o(x,y,z) = \frac{[n-(l+1)]!2^{|m|}l!}{\sqrt{2\pi}(l+|m|)!} &\exp(-\alpha_n|(x,y,z)|/2)\\ &\cdot L_{n-(l+1)}^{2l+1}(\alpha_n|(x,y,z)|)\, Y_l^m(x,y,z) \quad (5.1.32)\end{aligned}$$

at least for $(x,y) \neq (0,0)$ (by Proposition 5.1). We use the right-hand side of (5.1.32) in fact to *define* f_1 ($= \psi_{nlm}^o$) on *all* of $\mathbb{R}^3$. Thus $f_1 : \mathbb{R}^3 \to \mathbb{C}$ is continuous, and $f_1 : \mathbb{R}^3 - \{(0,0,0)\} \to \mathbb{C}$ is C^∞, since $\delta : (x,y,z) \stackrel{\text{def}}{\to} |(x,y,z)| = +\sqrt{x^2+y^2+z^2}$ is C^∞ on $\mathbb{R}^3 - \{(0,0,0)\}$. We therefore have that

$$\nabla_1^2 f_1 + \frac{8\pi^2\mu}{h^2}\left(E_1 + \frac{Ze^2}{\delta}\right) f_1 \quad (5.1.33)$$

is a continuous function τ on $\mathbb{R}^3 - \{(0,0,0)\} \supset V^*$ which vanishes on V^*, by (5.1.27). But then τ must vanish on $\mathbb{R}^3 - \{(0,0,0)\}$. Namely given $p = (x,y,z) \in \mathbb{R}^3 - \{(0,0,0)\}$, we already have $\tau(p) = 0$ if $p \in V^*$—i.e., if $(x,y) \neq (0,0)$. If $(x,y) = (0,0)$ then $z \neq 0$ and $p_n \overset{\text{def}}{=} (1/n, 1/n, z)$ $(n = 1,2,3,\dots) \in V^*$ such that $p_n \to p = (0,0,z)$ on $\mathbb{R}^3 - \{(0,0,0)\} \implies \tau(p) = \lim_{n\to\infty} \tau(p_n) = 0$, as each $\tau(p_n) = 0$. Thus we have indeed constructed continuous functions $f_1 = \psi^o_{nlm} : \mathbb{R}^3 \to \mathbb{C}$ which in fact are C^∞ on $\mathbb{R}^3 - \{(0,0,0)\}$ and which on the latter domain solve the first Schrödinger equation in (5.1.9). Because the approach here was based on the general theory of hypergeometric equations, developed in Chapter 4, the ψ^o_{nlm} were obtained without recourse to the traditional method of power series. Note that the quantum energies $E^{(n)}$ in (5.1.22), or in (5.1.29), are exactly those advertised in (1.2.4), where the special case $Z = 1$ was considered.

5.2 Normalized Wave Functions

The solutions u_{nl} in (5.1.21) of the first equation in (5.1.16) satisfy a *square-integrability condition.* First we extend the u_{nl}, defined on $(0,\infty)$ to continuous functions on $[0,\infty)$. Namely, using (5.1.21), we set

$$u_{nl}(0) \overset{\text{def}}{=} \begin{bmatrix} 0 & \text{if } l > 0 \\ (n-1)!L^1_{n-1}(0) = -(n-1)!(L^o_n)'(0) & \text{if } l = 0 \end{bmatrix} \tag{5.2.1}$$

where the formula $(L^\alpha_n)' = -L^{\alpha+1}_{n-1}$, $n \geq 1$ is recalled.

Theorem 5.1.

$$\int_0^\infty u_{nl}(r)^2 r^2\, dr < \infty.$$

The value of this integral in fact is

$$\frac{2n[n-(l+1)]!\,(n+l)!}{\alpha_n^{2l+3}}.$$

Proof. Let

$$\begin{aligned} I &= \int_0^\infty u_{nl}(r)^2 r^2\, dr \\ &\overset{\text{def}}{=} ([n-(l+1)]!)^2 \int_0^\infty r^{2l} e^{-\alpha_n r} \left[L^{2l+1}_{n-(l+1)}(\alpha_n r)\right]^2 r^2\, dr \\ &= \frac{([n-(l+1)]!)^2}{\alpha_n^{2l+3}} \int_0^\infty t^{2l+2} e^{-t} \left[L^{2l+1}_{n-(l+1)}(t)\right]^2 dt. \end{aligned} \tag{5.2.2}$$

By Theorem 4.4, in particular,

$$\int_0^\infty t^{j+1} e^{-t} \left[L^j_m(t)\right]^2 dt = \frac{(2m+j+1)(m+j)!}{m!} \tag{5.2.3}$$

for integers $m, j \geq 0$, which gives

$$I = \frac{([n-(l+1)]!)^2}{\alpha_n^{2l+3}} \frac{(2n)(n+l)!}{[n-(l+1)]!} = \frac{2n[n-(l+1)]!\,(n+l)!}{\alpha_n^{2l+3}}, \tag{5.2.4}$$

as desired. □

Because of Theorem 5.1 we choose the following normalizations of the *radial* solutions u_{nl} in (5.1.21):

$$R_{nl}^{\pm} \stackrel{\text{def}}{=} C_n^{\pm} u_{nl} \qquad \text{for } C_n^{\pm} \stackrel{\text{def}}{=} \pm\sqrt{\frac{\alpha_n}{2n(n+l)!(n-l-1)!}}\,(-\alpha_n^{l+1}). \tag{5.2.5}$$

Thus for $r \geq 0$

$$-R_{nl}^{\pm}(r) = \pm\sqrt{\frac{\alpha_n(n-l-1)!}{2n(n+l)!}}\alpha_n^{l+1} r^l e^{-\alpha_n r/2} L_{n-l-1}^{2l+1}(\alpha_n r), \tag{5.2.6}$$

$$\int_0^{\infty} \left[R_{nl}^{\pm}(r)\right]^2 r^2\, dr = 1.$$

Write

$$\psi_{nlm}^{o}(g(r,\theta,\phi)) = \frac{2^{|m|} l!\, u_{nl}(r)\, Z_l^m(\theta,\phi)}{\sqrt{2\pi}(l+|m|)!} \tag{5.2.7}$$

on $\mathbb{R}^+ \times \mathbb{R} \times (0,\pi)$, by (5.1.21), (5.1.28), and (5.1.31). We apply Theorem 4.14 and Theorem 5.1 to compute

$$\begin{aligned}
&\iint_{\mathbb{R}^3}\!\!\int |\psi_{nlm}^{o}(x,y,z)|^2\, dxdydz \\
&= \int_0^{2\pi}\int_0^{\pi}\int_0^{\infty} \frac{2^{2|m|}(l!)^2}{2\pi[(l+|m|)!]^2} u_{nl}(r)^2\, Z_l^m(\theta,\phi)\overline{Z_l^m(\theta,\phi)}\, r^2 \sin\phi\, drd\phi d\theta \\
&= \frac{2^{2|m|}(l!)^2}{2\pi[(l+|m|)!]^2} \frac{2n(n-l-1)!(n+l)!\,4\pi(l+|m|)!}{\alpha_n^{2l+3}(2l+1)(l-|m|)!} \\
&= \frac{2^{2|m|}(l!)^2 4n(n-l-1)!\,(n+l)!}{\alpha_n^{2l+3}(l+|m|)!\,(l-|m|)!\,(2l+1)}.
\end{aligned} \tag{5.2.8}$$

That is, recalling that we have extended the ψ_{nlm}^{o} continuously to all of $\mathbb{R}^3$, we have

Proposition 5.2. *The wave functions ψ_{nlm}^{o} are square-integrable and continuous over $\mathbb{R}^3$ and thus define elements of the Hilbert space $L^2(\mathbb{R}^3, dxdydz)$, where $dxdydz$ = Lebesgue measure on $\mathbb{R}^3$. Let*

$$\psi_{nlm} \stackrel{\text{def}}{=} C_{nlm}\psi_{nlm}^{o}$$

$$\textit{for } C_{nlm} \stackrel{\text{def}}{=} \left[\frac{\alpha_n^{2l+3}(l+|m|)!(l-|m|)!(2l+1)}{2^{2|m|}(l!)^2 4n(n-l-1)!(n+l)!}\right]^{1/2}. \tag{5.2.9}$$

Then

$$\|\psi_{nlm}\|_2^2 \stackrel{\text{def}}{=} \iiint_{\mathbb{R}^3} |\psi_{nlm}(x,y,z)|^2 \, dxdydz = 1. \tag{5.2.10}$$

The normalized wave functions ψ_{nlm} of Proposition 5.2 can be explicated as follows. For integers l, m with $l \geq 0$, $|m| \leq l$ let

$$c_{lm} \stackrel{\text{def}}{=} \left[\frac{(2l+1)(l-|m|)!}{2(l+|m|)!}\right]^{1/2}. \tag{5.2.11}$$

Then by (5.1.32), (5.2.9), again for $n \geq 1$, $0 \leq l \leq n-1$, $-l \leq m \leq l$, we can write

$$\psi_{nlm}(x,y,z) = c_{lm}\left[\frac{(n-l-1)!\,\alpha_n}{(n+l)!2n}\right]^{1/2} \frac{a_n^{l+1}}{\sqrt{2\pi}} \cdot e^{-\alpha_n|(x,y,z)|/2}\, L_{n-l-1}^{2l+1}(\alpha_n|(x,y,z)|)\, Y_l^m(x,y,z). \tag{5.2.12}$$

Note that for an integer $k \geq 0$ and $x \in \mathbb{R}$

$$\begin{aligned} L_k(x) &\stackrel{\text{def}}{=} e^x \frac{d^k}{dx^k}(e^{-x}x^k) = k!L_k^o(x) \\ \Longrightarrow\ L_k^{(j)}(x) &= (-1)^j k! L_{k-j}^j(x) \qquad \text{for } 0 \leq j \leq k \\ \Longrightarrow\ L_{n+l}^{(2l+1)}(x) &= -(n+l)!\, L_{n-l-1}^{2l+1}(x); \end{aligned} \tag{5.2.13}$$

see Appendix 4B. Therefore for

$$c_{nl}^{\pm} \stackrel{\text{def}}{=} \pm\sqrt{\frac{\alpha_n(n-l-1)!}{2n[(n+l)!]^3}}, \tag{5.2.14}$$

equation (5.2.12) can also be expressed as

$$\psi_{nlm}(x,y,z) = c_{lm}c_{nl}^{-}\frac{\alpha_n^{l+1}}{\sqrt{2\pi}}e^{-\alpha_n|(x,y,z)|/2}L_{n+l}^{(2l+1)}(\alpha_n|(x,y,z)|)Y_l^m(x,y,z) \tag{5.2.15}$$

on $\mathbb{R}^3$.

5.3 The Wave Functions ψ_{nlm} for Small n

In Chapter 0 we introduced the constant

$$a_o = \frac{\hbar^2}{\mu e^2} = \frac{h^2}{4\pi^2\mu e^2} = 0.52946 \times 10^{-8}\text{ cm}; \tag{5.3.1}$$

see (0.0.11), (0.0.12). Historically a_o is *Bohr's radius*, the radius of the smallest possible orbit of an electron about the hydrogen nucleus in the old, or earlier version of quantum theory. By (5.1.30)

$$\alpha_n \stackrel{\text{def}}{=} \frac{2\mu Z e^2}{n\hbar^2} = \frac{2Z}{na_o}, \qquad \rho_n(r) \stackrel{\text{def}}{=} \alpha_n r = \frac{2Zr}{na_o}. \tag{5.3.2}$$

Note that $g(x,0,\pi/2) = (x,0,0)$, or $Y_l^m(x,0,0) = Y_l^m(g(x,0,\pi/2)) \stackrel{\text{def}}{=} x^l Z_l^m(0,\pi/2) \stackrel{\text{def}}{=} x^l e^{im\cdot 0} P_l^{|m|}(\cos(\pi/2)) = x^l P_l^{(|m|)}(0)$; see (5.1.31).

In particular for $n = 1, l = m = 0, Y_o^o(x,0,0) = 1, c_{oo} = 1/\sqrt{2}, c_{10}^- = -\sqrt{\alpha_1/2}$, $\alpha_1 = \frac{2Z}{a_o}$, $L_o^1(x) = 1$. By (5.2.12)

$$\psi_{100}(x,0,0) = \left(\frac{Z}{a_o}\right)^{3/2} \frac{\exp(-\frac{Z}{a_o}|x|)}{\sqrt{\pi}}, \tag{5.3.3}$$

which as a function of x has a *cusp* at the origin: see Figure 5.2, where $c = (Z/a_o)^{3/2}/\sqrt{\pi}$. In particular $\frac{\partial \psi_{100}}{\partial x}(0,0,0)$ *does not* exist.

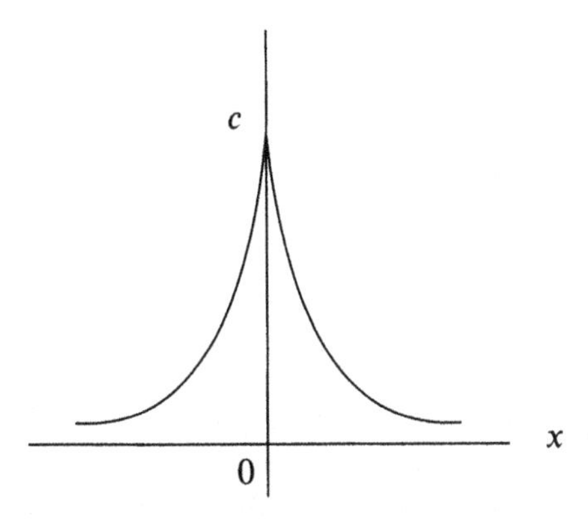

Figure 5.2: Graph of $\psi_{100}(x,0,0)$

Equations (5.2.5), (5.2.6), (5.3.2) imply that the normalized radial wave functions $R_{nl}^-(r)$, $r \geq 0$, are given by

$$R_{nl}^-(r) = \frac{2}{n^2}\left(\frac{Z}{a_o}\right)^{3/2} \sqrt{\frac{(n-l-1)!}{(n+l)!}} \rho_n(r)^l \exp(-\rho_n(r)/2) L_{n-l-1}^{2l+1}(\rho_n(r)); \tag{5.3.4}$$

$$\int_0^\infty R_{nl}^-(r)^2 r^2 \, dr = 1. \tag{5.3.5}$$

For example,

$$\begin{aligned}
R_{10}^-(r) &= 2\left(\frac{Z}{a_o}\right)^{3/2} e^{-\rho/2}, \\
R_{20}^-(r) &= \frac{1}{2\sqrt{2}}\left(\frac{Z}{a_o}\right)^{3/2} e^{-\rho/2}(2-\rho), \\
R_{21}^-(r) &= \frac{1}{2\sqrt{6}}\left(\frac{Z}{a_o}\right)^{3/2} \rho e^{-\rho/2}, \\
R_{30}^-(r) &= \frac{1}{9\sqrt{3}}\left(\frac{Z}{a_o}\right)^{3/2} e^{-\rho/2}(6-6\rho+\rho^2),
\end{aligned} \tag{5.3.6}$$

$$R_{31}^{-}(r) = \frac{1}{9\sqrt{6}}\left(\frac{Z}{a_o}\right)^{3/2} e^{-\rho/2}\rho(4-\rho),$$
$$R_{32}^{-}(r) = \frac{1}{9\sqrt{30}}\left(\frac{Z}{a_o}\right)^{3/2} e^{-\rho/2}\rho^2,$$

where we write

$$\rho = \rho_n(r) = \frac{2Zr}{na_o}. \tag{5.3.7}$$

There are also normalized functions Φ_m, X_{lm} on $\mathbb{R}$, $[0, 2\pi]$ given by

$$\Phi_m(\theta) \stackrel{\text{def}}{=} \frac{e^{im\theta}}{\sqrt{2\pi}}, \qquad X_{lm}(\phi) \stackrel{\text{def}}{=} c_{lm}P_l^{|m|}(\cos\phi) : \tag{5.3.8}$$
$$\int_0^{2\pi} \Phi_m(\theta)\overline{\Phi}_m(\theta)\, d\theta = 1, \qquad \int_0^{\pi} X_{lm}(\phi)^2 \sin\phi\, d\phi = 1,$$

according to Theorem 4.8, given $|m| \leq l$. Again writing $f_1 \otimes f_2$ for the function $(x, y) \to f_1(x)f_2(y)$ we can assert that

$$\psi_{nlm} \circ g = R_{nl}^{-} \otimes \Phi_m \otimes X_{lm} \tag{5.3.9}$$

on $\{(r, \theta, \phi) \in \mathbb{R}^3 | r \geq 0, 0 \leq \phi \leq \pi\}$, by the various definitions involved. The normalized wave function ψ_{100} is computed at the special point $g(r, 0, \pi/2) = (r, 0, 0)$ in (5.3.3). We consider further computations of ψ_{nlm} at points $g(r, \theta, \phi)$ for small values of n, l, m: $1 \leq n \leq 3$; thus $0 \leq l \leq 2$, $-2 \leq m \leq 2$. In what follows we shall write ψ_{nlm} for $\psi_{nlm} \circ g$; i.e., for convenience, ψ_{nlm} *will mean* $R_{nl}^{-} \otimes \Phi_m \otimes X_{lm}$.

We have

$$\begin{aligned}
X_{00}(\phi) &= \frac{\sqrt{2}}{2},\\
X_{10}(\phi) &= \frac{\sqrt{6}}{2}\cos(\phi),\\
X_{1\pm1}(\phi) &= \frac{\sqrt{3}}{2}\sin(\phi),\\
X_{20}(\phi) &= \frac{\sqrt{10}}{4}(3\cos^2(\phi) - 1),\\
X_{2\pm1}(\phi) &= \frac{\sqrt{15}}{2}\sin(\phi)\cos(\phi),\\
X_{2\pm2}(\phi) &= \frac{\sqrt{15}}{4}\sin^2(\phi),\\
X_{30}(\phi) &= \frac{3\sqrt{14}}{4}\left(\frac{5}{3}\cos^3(\phi) - \cos(\phi)\right),\\
X_{3\pm1}(\phi) &= \frac{\sqrt{42}}{8}(\sin(\phi))(5\cos^2(\phi) - 1),
\end{aligned} \tag{5.3.10}$$

for $0 \leq \phi \leq \pi$. Now

$$
\begin{aligned}
\psi_{100} &= R_{10}^{-} \otimes \Phi_0 \otimes X_{00}, \\
\psi_{200} &= R_{20}^{-} \otimes \Phi_0 \otimes X_{00}, \\
\psi_{210} &= R_{21}^{-} \otimes \Phi_0 \otimes X_{10}, \\
\psi_{21\pm1} &= R_{21}^{-} \otimes \Phi_{\pm1} \otimes X_{1\pm1}, \\
\psi_{300} &= R_{30}^{-} \otimes \Phi_0 \otimes X_{00}, \\
\psi_{310} &= R_{31}^{-} \otimes \Phi_0 \otimes X_{10}, \\
\psi_{31\pm1} &= R_{31}^{-} \otimes \Phi_{\pm1} \otimes X_{1\pm1}, \\
\psi_{320} &= R_{32}^{-} \otimes \Phi_0 \otimes X_{20}, \\
\psi_{32\pm1} &= R_{32}^{-} \otimes \Phi_{\pm1} \otimes X_{2\pm1}, \\
\psi_{32\pm2} &= R_{32}^{-} \otimes \Phi_{\pm2} \otimes X_{2\pm2}.
\end{aligned}
\tag{5.3.11}
$$

Therefore,

$$
\begin{aligned}
\psi_{100}(r,\theta,\phi) &= \left(\frac{Z}{a_o}\right)^{3/2} \frac{2e^{-\rho/2}}{\sqrt{2\pi}} \frac{\sqrt{2}}{2} = \left(\frac{Z}{a_o}\right)^{3/2} \frac{1}{\sqrt{\pi}} e^{-\rho/2} \\
&= \left(\frac{Z}{a_o}\right)^{3/2} \frac{1}{\sqrt{\pi}} e^{-(Z/a_o)r}, \\
\psi_{200}(r,\theta,\phi) &= \left(\frac{Z}{a_o}\right)^{3/2} \frac{1}{2\sqrt{2}} e^{-\rho/2}(2-\rho) \frac{1}{\sqrt{2\pi}} \frac{\sqrt{2}}{2} \\
&= \left(\frac{Z}{a_o}\right)^{3/2} \frac{e^{-\rho/2}(2-\rho)}{4\sqrt{2\pi}}, \\
\psi_{210}(r,\theta,\phi) &= \left(\frac{Z}{a_o}\right)^{3/2} \frac{1}{2\sqrt{6}} \rho e^{-\rho/2} \frac{1}{\sqrt{2\pi}} \frac{\sqrt{6}}{2} \cos(\phi) \\
&= \left(\frac{Z}{a_o}\right)^{3/2} \frac{\rho e^{-\rho/2}}{4\sqrt{2\pi}} \cos(\phi), \\
\psi_{21\pm1}(r,\theta,\phi) &= \left(\frac{Z}{a_o}\right)^{3/2} \frac{1}{2\sqrt{6}} \rho e^{-\rho/2} \frac{e^{\pm i\theta}}{\sqrt{2\pi}} \frac{\sqrt{3}}{2} \sin(\phi) \\
&= \left(\frac{Z}{a_o}\right)^{3/2} \frac{1}{8\sqrt{\pi}} \rho e^{-\rho/2} e^{\pm i\theta} \sin(\phi), \\
\psi_{300}(r,\theta,\phi) &= \left(\frac{Z}{a_o}\right)^{3/2} \frac{1}{9\sqrt{3}} e^{-\rho/2}(6-6\rho+\rho^2) \frac{1}{\sqrt{2\pi}} \frac{\sqrt{2}}{2} \\
&= \left(\frac{Z}{a_o}\right)^{3/2} \frac{1}{18\sqrt{3\pi}} e^{-\rho/2}(6-6\rho+\rho^2), \qquad (5.3.12) \\
\psi_{310}(r,\theta,\phi) &= \left(\frac{Z}{a_o}\right)^{3/2} \frac{1}{9\sqrt{6}} \rho e^{-\rho/2}(4-\rho) \frac{1}{\sqrt{2\pi}} \frac{\sqrt{6}}{2} \cos(\phi)
\end{aligned}
$$

$$
\begin{aligned}
&= \left(\frac{Z}{a_o}\right)^{3/2} \frac{1}{18\sqrt{2\pi}} \rho e^{-\rho/2}(4-\rho)\cos(\phi), \\
\psi_{31\pm1}(r,\theta,\phi) &= \left(\frac{Z}{a_o}\right)^{3/2} \frac{1}{9\sqrt{6}} \rho e^{-\rho/2}(4-\rho)\frac{e^{\pm i\theta}}{\sqrt{2\pi}}\frac{\sqrt{3}}{2}\sin(\phi) \\
&= \left(\frac{Z}{a_o}\right)^{3/2} \frac{1}{36\sqrt{\pi}} \rho e^{-\rho/2}(4-\rho)e^{\pm i\theta}\sin(\phi), \\
\psi_{320}(r,\theta,\phi) &= \left(\frac{Z}{a_o}\right)^{3/2} \frac{\rho^2 e^{-\rho/2}}{9\sqrt{30}} \frac{1}{\sqrt{2\pi}} \frac{\sqrt{10}}{4}(3\cos^2(\phi)-1) \\
&= \left(\frac{Z}{a_o}\right)^{3/2} \frac{\rho^2 e^{-\rho/2}}{36\sqrt{6\pi}}(3\cos^2(\phi)-1), \\
\psi_{32\pm1}(r,\theta,\phi) &= \left(\frac{Z}{a_o}\right)^{3/2} \frac{\rho^2 e^{-\rho/2}}{9\sqrt{30}} \frac{e^{\pm i\theta}}{\sqrt{2\pi}} \frac{\sqrt{15}}{2}\sin(\phi)\cos(\phi) \\
&= \left(\frac{Z}{a_o}\right)^{3/2} \frac{\rho^2 e^{-\rho/2}e^{\pm i\theta}}{36\sqrt{\pi}}\sin(\phi)\cos(\phi), \\
\psi_{32\pm2}(r,\theta,\phi) &= \left(\frac{Z}{a_o}\right)^{3/2} \frac{\rho^2 e^{-\rho/2}}{9\sqrt{30}} \frac{e^{\pm 2i\theta}}{\sqrt{2\pi}} \frac{\sqrt{15}}{4}\sin^2(\phi) \\
&= \left(\frac{Z}{a_o}\right)^{3/2} \rho^2 e^{-\rho/2}e^{\pm 2i\theta}\sin^2(\phi)\frac{1}{72\sqrt{\pi}}.
\end{aligned}
\tag{5.3.13}
$$

Note that

$$
\begin{aligned}
&\frac{1}{\sqrt{2}}\left[\psi_{211}(r,\theta,\phi)+\psi_{21(-1)}(r,\theta,\phi)\right] \\
&\quad= \left(\frac{Z}{a_o}\right)^{3/2} \frac{1}{8\sqrt{\pi}} \frac{\rho e^{-\rho/2}}{\sqrt{2}}\sin(\phi)\left[e^{i\theta}+e^{-i\theta}\right] \\
&\quad= \left(\frac{Z}{a_o}\right)^{3/2} \frac{1}{4\sqrt{2\pi}} \rho e^{-\rho/2}\sin(\phi)\cos(\theta),
\end{aligned}
\tag{5.3.14}
$$

$$
\begin{aligned}
&\frac{-i}{\sqrt{2}}\left[\psi_{211}(r,\theta,\phi)-\psi_{21(-1)}(r,\theta,\phi)\right] \\
&\quad= \frac{-i}{\sqrt{2}}\left(\frac{Z}{a_o}\right)^{3/2} \frac{1}{8\sqrt{\pi}} \rho e^{-\rho/2}\sin(\phi)\left[e^{i\theta}-e^{-i\theta}\right] \\
&\quad= \left(\frac{Z}{a_o}\right)^{3/2} \frac{1}{4\sqrt{2\pi}} \rho e^{-\rho/2}\sin(\phi)\sin(\theta).
\end{aligned}
\tag{5.3.15}
$$

Similarly,

$$
\frac{1}{\sqrt{2}}[\psi_{311}(r,\theta,\phi)+\psi_{31(-1)}(r,\theta,\phi)] = \left(\frac{Z}{a_o}\right)^{3/2} \frac{\rho e^{-\rho/2}(4-\rho)\sin(\phi)\cos(\theta)}{18\sqrt{2\pi}},
$$

$$\frac{-i}{\sqrt{2}}\left[\psi_{311}(r,\theta,\phi) - \psi_{31(-1)}(r,\theta,\phi)\right] = \left(\frac{Z}{a_o}\right)^{3/2} \frac{\rho e^{-\rho/2}(4-\rho)\sin(\phi)\sin(\theta)}{18\sqrt{2\pi}},$$

$$\frac{1}{\sqrt{2}}[\psi_{321}(r,\theta,\phi) + \psi_{32(-1)}(r,\theta,\phi)] = \left(\frac{Z}{a_o}\right)^{3/2} \frac{\rho^2 e^{-\rho/2}\sin(\phi)\cos(\phi)\cos(\theta)}{18\sqrt{2\pi}},$$

$$\frac{-i}{\sqrt{2}}\left[\psi_{321}(r,\theta,\phi) - \psi_{32(-1)}(r,\theta,\phi)\right] = \left(\frac{Z}{a_o}\right)^{3/2} \frac{\rho^2 e^{-\rho/2}\sin(\phi)\cos(\phi)\sin(\theta)}{18\sqrt{2\pi}},$$

$$\frac{1}{\sqrt{2}}[\psi_{322}(r,\theta,\phi) + \psi_{32(-2)}(r,\theta,\phi)] = \left(\frac{Z}{a_o}\right)^{3/2} \frac{2\rho^2 e^{-\rho/2}\sin^2(\phi)\cos(2\theta)}{72\sqrt{2\pi}},$$

$$\frac{-i}{\sqrt{2}}\left[\psi_{322}(r,\theta,\phi) - \psi_{32(-2)}(r,\theta,\phi)\right] = \left(\frac{Z}{a_o}\right)^{3/2} \frac{2\rho^2 e^{-\rho/2}\sin^2(\phi)\sin(2\theta)}{72\sqrt{2\pi}}. \quad (5.3.16)$$

In general,

$$\begin{aligned}
&\frac{1}{\sqrt{2}}[\psi_{nlm}(r,\theta,\phi) + \psi_{nl(-m)}(r,\theta,\phi)] \\
&= \frac{1}{\sqrt{2}}\left[R_{nl}^{-}(r)\frac{e^{im\theta}}{\sqrt{2\pi}}X_{lm}(\phi) + R_{nl}^{-}(r)\frac{e^{-im\theta}}{\sqrt{2\pi}}X_{l(-m)}(\phi)\right] \\
&= \frac{1}{\sqrt{2}}R_{nl}^{-}(r)X_{lm}(\phi)\frac{e^{im\theta}+e^{-im\theta}}{\sqrt{2\pi}} \qquad \left(\text{as } X_{l(-m)} = X_{lm}\right) \\
&= \frac{1}{\sqrt{\pi}}R_{nl}^{-}(r)X_{lm}(\phi)\cos(m\theta).
\end{aligned} \quad (5.3.17)$$

Similarly

$$\begin{aligned}
&\frac{-i}{\sqrt{2}}[\psi_{nlm}(r,\theta,\phi) - \psi_{nl(-m)}(r,\theta,\phi)] \\
&= \frac{-i}{\sqrt{2}}R_{nl}^{-}(r)X_{lm}(\phi)\frac{e^{im\theta}-e^{-im\theta}}{\sqrt{2\pi}} \\
&= \frac{1}{\sqrt{\pi}}R_{nl}^{-}(r)X_{lm}(\phi)\sin(m\theta).
\end{aligned} \quad (5.3.18)$$

In particular

$$\begin{aligned}
&\int_0^\infty\int_0^{2\pi}\int_0^\pi \left(\frac{1}{\sqrt{2}}[\psi_{nlm}(r,\theta,\phi) + \psi_{nl(-m)}(r,\theta,\phi)]\right)^2 r^2\sin(\phi)\,d\phi\,d\theta\,dr \\
&= \int_0^\infty\int_0^{2\pi}\int_0^\pi \frac{1}{\pi}R_{nl}^{-}(r)^2X_{lm}(\phi)^2\cos^2(m\theta)r^2\sin(\phi)\,d\phi\,d\theta\,dr \\
&= \frac{1}{\pi}\int_0^{2\pi}\cos^2(m\theta)\,d\theta = 1, \text{ by (5.3.5), (5.3.8).}
\end{aligned} \quad (5.3.19)$$

Similarly

$$\frac{-i}{\sqrt{2}}[\psi_{nlm}(r,\theta,\phi) - \psi_{nl(-m)}(r,\theta,\phi)]$$

is normalized, as

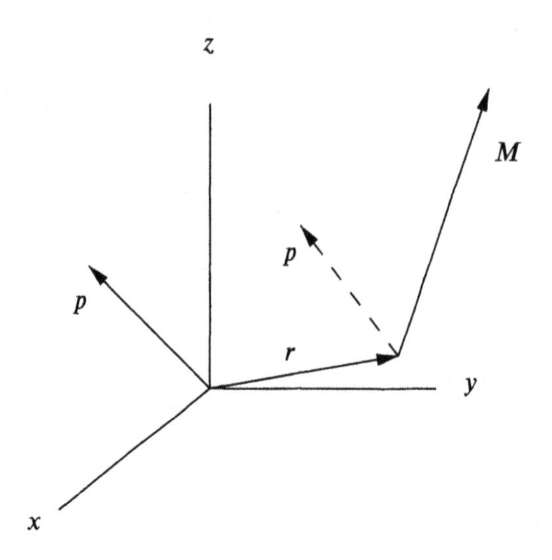

Figure 5.3: Angular momentum

$$\frac{1}{\pi}\int_0^{2\pi} \sin^2(m\theta)\, d\theta = 1. \tag{5.3.20}$$

5.4 Quantization of Angular Momentum

For $n = 1, 2, 3, \ldots, l = 0, 1, 2, \ldots, n-1, m = -l, -l+1, \ldots, -1, 0, 1, 2, \ldots, l$ there corresponds the normalized wave function ψ_{nlm} of (5.2.12), or (5.2.15). As in the introductory remarks of this chapter, for the quantum data (n, l, m), n is called the *total quantum number*, l is called the *azimuthal* or *orbital quantum number*, and m is called the *magnetic quantum number*. Note that the energy $E^{(n)} = -\frac{\mu Z^2 e^4}{2\hbar^2 n^2}$ in (5.1.22) of the quantum system described by ψ_{nlm} depends only on the total quantum number n. The azimuthal quantum number l determines the magnitude $\hbar\sqrt{l(l+1)}$ of the quantized angular momentum M of the electron with wave function ψ_{nlm}, a statement which we now proceed to attach a clear meaning. The magnetic quantum number m will prescribe, for example, the magnitude $\hbar|m|$ (as it will turn out) of the z-component of M and the *spatial quantization* of M given by $\cos\phi = \frac{m}{\sqrt{l(l+1)}}$ where ϕ is the "angle between M and the z-axis."

The classical angular momentum M of a particle of mass m is defined by $M = r \times p$ where $r = (x, y, z)$ is its position vector and $p = (p_x, p_y, p_z) = m\vec{v}$ is its linear momentum vector (cf. Figure 5.3):

$$M = \begin{vmatrix} i & j & k \\ x & y & z \\ p_x & p_y & p_z \end{vmatrix} = (M_x, M_y, M_z), \tag{5.4.1}$$

where

$$M_x = yp_z - zp_y, \qquad M_y = zp_x - xp_z, \qquad M_z = xp_y - yp_x. \tag{5.4.2}$$

The quantization rules

$$\begin{aligned} x, y, z &\to \text{ multiplication by } x, y, z, \\ p_x, p_y, p_z &\to \frac{\hbar}{i}\frac{\partial}{\partial x}, \frac{\hbar}{i}\frac{\partial}{\partial y}, \frac{\hbar}{i}\frac{\partial}{\partial z} \end{aligned} \tag{5.4.3}$$

of Chapter 3 therefore lead to the quantization rules

$$\begin{aligned} M_x &\to \frac{\hbar}{i}\left(y\frac{\partial}{\partial z} - z\frac{\partial}{\partial y}\right), \\ M_y &\to \frac{\hbar}{i}\left(z\frac{\partial}{\partial x} - x\frac{\partial}{\partial z}\right), \\ M_z &\to \frac{\hbar}{i}\left(x\frac{\partial}{\partial y} - y\frac{\partial}{\partial x}\right). \end{aligned} \tag{5.4.4}$$

Let $M^2 = M_x^2 + M_y^2 + M_z^2$, which is the differential operator which represents the length$^2 = |r \times p|^2$. We will show that on $\mathbb{R}^3 - \{0\}$,

$$M^2\psi_{nlm} = \hbar^2 l(l+1)\psi_{nlm}, M_z\psi_{nlm} = \hbar m\psi_{nlm}. \tag{5.4.5}$$

Given (5.4.5), we therefore take $\hbar\sqrt{l(l+1)}$ as the *magnitude of the (quantum) angular momentum* M, and we take $\hbar|m|$ as the magnitude of the z-component of M; see (3.9.1). Graphically one has the following (Figure 5.4):

$$\cos\phi = \frac{M_z}{M} = \frac{\hbar m}{\hbar\sqrt{l(l+1)}} = \frac{m}{\sqrt{l(l+1)}}. \tag{5.4.6}$$

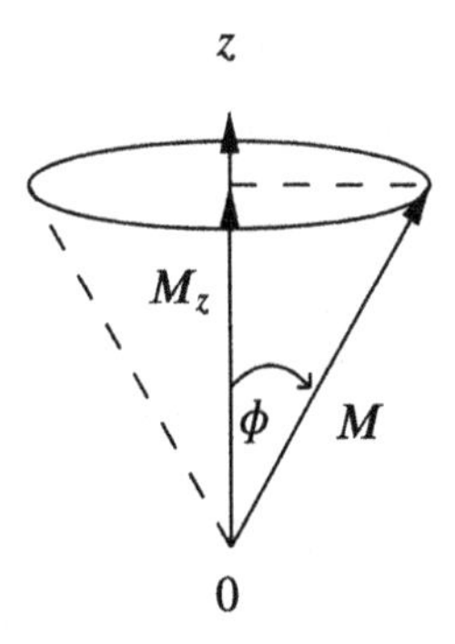

Figure 5.4: Quantum angular momentum

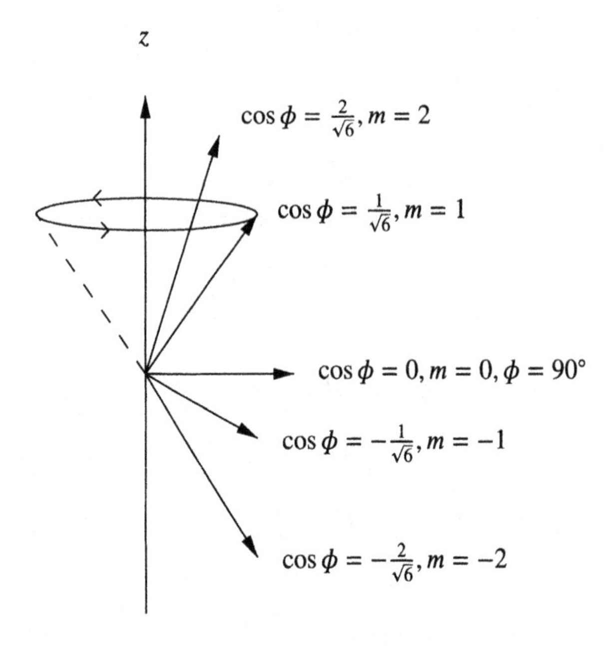

Figure 5.5: Spatial quantization of angular momentum

Since the inequality $|m| \leq l$ restricts m to taking on only the $2l + 1$ values $-l, -l+1, \ldots, 0, 1, \ldots, l$, the "vector" M can assume $2l + 1$ distinct orientations only, with the angle of orientation ϕ given by (5.4.6) as illustrated in Figure 5.4. For example, take $l = 2 : m = -2, -1, 0, 1, 2$, so that $l(l+1) = 6 \implies \cos\phi = \frac{-2}{\sqrt{6}}, \frac{-1}{\sqrt{6}}, 0, \frac{1}{\sqrt{6}}, \frac{2}{\sqrt{6}}$. The following graphic (Figure 5.5 for $l = 2$) therefore illustrates the *spatial quantization* of angular momentum.

The five possible orientations for M are shown. $\cos\phi = \frac{2}{\sqrt{6}}, \frac{1}{\sqrt{6}} \implies \phi \simeq 34^o, 66^o$ respectively. The corresponding magnitudes for M_z are (for $m = -2, -1, 0, 1, 2$) $2\hbar, \hbar, 0$. For $M^2 = M_x^2 + M_y^2 + M_z^2$ with M_x, M_y, M_z given by (5.4.4) we now establish that indeed (5.4.5) holds.

Theorem 5.2. *$M^2\psi_{nlm} = \hbar^2 l(l+1)\psi_{nlm}$ on $\mathbb{R}^3 - \{0\}$.*

It is certainly not obvious that M^2 should act as a scalar on the wave function ψ_{nlm}.

Let $\psi : U^{open} \subset \mathbb{R}^3 \to \mathbb{C}$ be a function with continuous first and second partial derivatives on U. In particular,

$$\frac{\partial^2\psi}{\partial y \partial z} = \frac{\partial^2\psi}{\partial z \partial y}, \tag{5.4.7}$$

$$\frac{\partial^2\psi}{\partial z \partial x} = \frac{\partial^2\psi}{\partial x \partial z}, \tag{5.4.8}$$

$$\frac{\partial^2\psi}{\partial x \partial y} = \frac{\partial^2\psi}{\partial y \partial x}. \tag{5.4.9}$$

On U,

$$M_x^2\psi = -\hbar^2\left(y\frac{\partial}{\partial z} - z\frac{\partial}{\partial y}\right)\left(y\frac{\partial\psi}{\partial z} - z\frac{\partial\psi}{\partial y}\right) \tag{5.4.10}$$

$$\Rightarrow \frac{M_x^2\psi}{-\hbar^2} = y\frac{\partial}{\partial z}\left(y\frac{\partial\psi}{\partial z} - z\frac{\partial\psi}{\partial y}\right) - z\frac{\partial}{\partial y}\left(y\frac{\partial\psi}{\partial z} - z\frac{\partial\psi}{\partial y}\right) \tag{5.4.11}$$

$$= y\left[y\frac{\partial^2\psi}{\partial z^2} - z\frac{\partial^2\psi}{\partial z\partial y} - \frac{\partial\psi}{\partial y}\right] - z\left[y\frac{\partial^2\psi}{\partial y\partial z} + \frac{\partial\psi}{\partial z} - z\frac{\partial^2\psi}{\partial y^2}\right]$$

$$= y^2\frac{\partial^2\psi}{\partial z^2} - 2yz\frac{\partial^2\psi}{\partial y\partial z} - y\frac{\partial\psi}{\partial y} - z\frac{\partial\psi}{\partial z} + z^2\frac{\partial^2\psi}{\partial y^2}$$

by equation (5.4.7). Similarly, permuting x, y, z cyclically, one has

$$\frac{M_y^2\psi}{-\hbar^2} = z^2\frac{\partial^2\psi}{\partial x^2} - 2zx\frac{\partial^2\psi}{\partial z\partial x} - z\frac{\partial\psi}{\partial z} - x\frac{\partial\psi}{\partial x} + x^2\frac{\partial^2\psi}{\partial z^2} \tag{5.4.12}$$

$$\frac{M_z^2\psi}{-\hbar^2} = x^2\frac{\partial^2\psi}{\partial y^2} - 2xy\frac{\partial^2\psi}{\partial x\partial y} - x\frac{\partial\psi}{\partial x} - y\frac{\partial\psi}{\partial y} + y^2\frac{\partial^2\psi}{\partial x^2} \tag{5.4.13}$$

as equations (5.4.8) (5.4.9) apply.

Adding, one gets

$$\begin{aligned}\frac{M^2\psi}{-\hbar^2} &= \left(z^2+y^2\right)\frac{\partial^2\psi}{\partial x^2} + \left(x^2+z^2\right)\frac{\partial^2\psi}{\partial y^2} + \left(y^2+x^2\right)\frac{\partial^2\psi}{\partial z^2}\\ &\quad - 2yz\frac{\partial^2\psi}{\partial y\partial z} - 2zx\frac{\partial^2\psi}{\partial z\partial x} - 2xy\frac{\partial^2\psi}{\partial x\partial y}\\ &\quad - 2x\frac{\partial\psi}{\partial x} - 2y\frac{\partial\psi}{\partial y} - 2z\frac{\partial\psi}{\partial z}\\ &= \left(r^2-x^2\right)\frac{\partial^2\psi}{\partial x^2} + \left(r^2-y^2\right)\frac{\partial^2\psi}{\partial y^2} + \left(r^2-z^2\right)\frac{\partial^2\psi}{\partial z^2}\\ &\quad - 2yz\frac{\partial^2\psi}{\partial y\partial z} - 2zx\frac{\partial^2\psi}{\partial z\partial x} - 2xy\frac{\partial^2\psi}{\partial x\partial y}\\ &\quad - 2\left[x\frac{\partial}{\partial x} + y\frac{\partial}{\partial y} + z\frac{\partial}{\partial z}\right]\psi\end{aligned} \tag{5.4.14}$$

for $r^2 = x^2 + y^2 + z^2$.

For $\nabla^2 = \frac{\partial^2}{\partial x^2} + \frac{\partial^2}{\partial y^2} + \frac{\partial^2}{\partial z^2}$ and $E = x\frac{\partial}{\partial x} + y\frac{\partial}{\partial y} + z\frac{\partial}{\partial z}$ (= Euler's differential operator) we can write (5.4.14) as

$$\begin{aligned}\frac{M^2\psi}{-\hbar^2} &= r^2\nabla^2\psi - x^2\frac{\partial^2\psi}{\partial x^2} - y^2\frac{\partial^2\psi}{\partial y^2} - z^2\frac{\partial^2\psi}{\partial z^2}\\ &\quad - 2yz\frac{\partial^2\psi}{\partial y\partial z} - 2zx\frac{\partial^2\psi}{\partial z\partial x} - 2xy\frac{\partial^2\psi}{\partial x\partial y} - 2E\psi.\end{aligned} \tag{5.4.15}$$

On the other hand note that

$$
\begin{aligned}
E^2\psi &= \left(x\frac{\partial}{\partial x} + y\frac{\partial}{\partial y} + z\frac{\partial}{\partial z}\right)\left(x\frac{\partial\psi}{\partial x} + y\frac{\partial\psi}{\partial y} + z\frac{\partial\psi}{\partial z}\right) \\
&= x\frac{\partial}{\partial x}\left(x\frac{\partial\psi}{\partial x} + y\frac{\partial\psi}{\partial y} + z\frac{\partial\psi}{\partial z}\right) + y\frac{\partial}{\partial y}\left(x\frac{\partial\psi}{\partial x} + y\frac{\partial\psi}{\partial y} + z\frac{\partial\psi}{\partial z}\right) \\
&\quad + z\frac{\partial}{\partial z}\left(x\frac{\partial\psi}{\partial x} + y\frac{\partial\psi}{\partial y} + z\frac{\partial\psi}{\partial z}\right) \\
&= x\left[x\frac{\partial^2\psi}{\partial x^2} + \frac{\partial\psi}{\partial x} + y\frac{\partial^2\psi}{\partial x\partial y} + z\frac{\partial^2\psi}{\partial x\partial z}\right] \\
&\quad + y\left[x\frac{\partial^2\psi}{\partial y\partial x} + y\frac{\partial^2\psi}{\partial y^2} + \frac{\partial\psi}{\partial y} + z\frac{\partial^2\psi}{\partial y\partial z}\right] \\
&\quad + z\left[x\frac{\partial^2\psi}{\partial z\partial x} + y\frac{\partial^2\psi}{\partial z\partial y} + z\frac{\partial^2\psi}{\partial z^2} + \frac{\partial\psi}{\partial z}\right] \\
&= x^2\frac{\partial^2\psi}{\partial x^2} + y^2\frac{\partial^2\psi}{\partial y^2} + z^2\frac{\partial^2\psi}{\partial z^2} + \left(x\frac{\partial}{\partial x} + y\frac{\partial}{\partial y} + z\frac{\partial}{\partial z}\right)\psi \\
&\quad + 2xy\frac{\partial^2\psi}{\partial x\partial y} + 2xz\frac{\partial^2\psi}{\partial x\partial z} + 2yz\frac{\partial^2\psi}{\partial y\partial z}
\end{aligned}
\tag{5.4.16}
$$

by equations (5.4.7), (5.4.8), (5.4.9) again. By equation (5.4.15), then,

$$
\begin{aligned}
&\frac{M^2\psi}{-\hbar^2} - r^2\nabla^2\psi + E\psi \\
&\quad = -E\psi - x^2\frac{\partial^2\psi}{\partial x^2} - y^2\frac{\partial^2\psi}{\partial y^2} - z^2\frac{\partial^2\psi}{\partial z^2} - 2yz\frac{\partial^2\psi}{\partial y\partial z} - 2zx\frac{\partial^2\psi}{\partial z\partial x} - 2xy\frac{\partial^2\psi}{\partial x\partial y} \\
&\quad = -E^2\psi \implies
\end{aligned}
\tag{5.4.17}
$$

Proposition 5.3.

$$\frac{M^2\psi}{\hbar^2} = -r^2\nabla^2\psi + E^2\psi + E\psi$$

for

$$\nabla^2 = \tfrac{\partial^2}{\partial x^2} + \tfrac{\partial^2}{\partial y^2} + \tfrac{\partial^2}{\partial z^2}, \qquad E = x\tfrac{\partial}{\partial x} + y\tfrac{\partial}{\partial y} + z\tfrac{\partial}{\partial z},$$
$$\psi \in C^2(U), \qquad U^{open} \subset \mathbb{R}^3, \qquad r^2 = x^2 + y^2 + z^2.$$

Now, $\psi_{nlm} : \mathbb{R}^3 \to \mathbb{C}$ is given by equation (5.2.15) in Section 5.2:

$$\psi_{nlm}(x, y, z) = \alpha^l \exp\left(-\frac{\alpha}{2}r\right) L_{n+l}^{(2l+1)}(\alpha r)\, \alpha c^{-} \frac{c_{lm}}{\sqrt{2\pi}}\, Y_l^m(x, y, z) \tag{5.4.18}$$

for $c = c_{nl}$, $\alpha = \alpha_n$. $\phi \overset{\text{def}}{=} \psi_{nlm}$ is a C^∞ function on $\mathbb{R}^3 - \{0\}$.

Let

$$C = \alpha^l \alpha c^- \frac{c_{lm}}{\sqrt{2\pi}},$$
$$H_1(x, y, z) = \exp\left(-\frac{\alpha}{2}r\right), \tag{5.4.19}$$
$$H_2(x, y, z) = L_{n+l}^{(2l+1)}(\alpha r)\, Y_l^m(x, y, z)$$

so that $\phi = C H_1 H_2$.

On $\mathbb{R}^3 - \{0\}$,

$$\begin{aligned} x\frac{\partial H_1}{\partial x} &= x \exp\left(-\frac{\alpha}{2}r\right)\left(-\frac{\alpha}{2}\right)\frac{1}{2}\frac{2x}{\sqrt{x^2+y^2+z^2}} \\ &= -\frac{\alpha}{2}\exp\left(-\frac{\alpha}{2}r\right)\frac{x^2}{\sqrt{x^2+y^2+z^2}}. \end{aligned} \tag{5.4.20}$$

Similarly

$$y\frac{\partial H_1}{\partial y} = -\frac{\alpha}{2}\exp\left(-\frac{\alpha}{2}r\right)\frac{y^2}{r}, \qquad z\frac{\partial H_1}{\partial z} = -\frac{\alpha}{2}\exp\left(-\frac{\alpha}{2}r\right)\frac{z^2}{r}, \tag{5.4.21}$$
$$\Longrightarrow\ EH_1 = -\frac{\alpha}{2}\exp\left(-\frac{\alpha}{2}r\right) r = -\frac{\alpha}{2} r H_1. \tag{5.4.22}$$

If f_1, f_2 are nice functions on an open set in $\mathbb{R}^3$ we clearly have $E(f_1 f_2) = (Ef_1)f_2 + f_1 E f_2$ since the product rule holds for $\frac{\partial}{\partial x}, \frac{\partial}{\partial y}, \frac{\partial}{\partial z}$. In particular, on $\mathbb{R}^3 - \{0\}$,

$$\begin{aligned} E^2 H_1 &= E(EH_1) = E\left(-\frac{\alpha}{2} r H_1\right) \qquad \text{(by (5.4.22))} \\ &= -\frac{\alpha}{2}\left[(Er)H_1 + rEH_1\right] \end{aligned} \tag{5.4.23}$$

where

$$x\frac{\partial r}{\partial x} = \frac{x^2}{r} \qquad y\frac{\partial r}{\partial y} = \frac{y^2}{r}, \qquad z\frac{\partial r}{\partial z} = \frac{z^2}{r} \Longrightarrow \tag{5.4.24}$$

$$Er = \frac{x^2+y^2+z^2}{r} = r. \tag{5.4.25}$$

Therefore

$$\begin{aligned} E^2 H_1 &= -\frac{\alpha}{2}\left[rH_1 + rEH_1\right] \\ &= -\frac{\alpha}{2}\left[rH_1 + r\left(-\frac{\alpha}{2}\right) rH_1\right] \\ &= \left[-\frac{\alpha}{2}r + \frac{\alpha^2}{4}r^2\right] H_1. \end{aligned} \tag{5.4.26}$$

Next we compute EH_2, E^2H_2.

If $H_3(x,y,z) = L_{n+l}^{(2l+1)}(\alpha r)$, $H_4 = Y_l^m$, then $H_2 = H_3H_4$ with H_4 homogeneous of degree l on $\mathbb{R}^3$:

$$\begin{aligned} H_4(tx,ty,tz) = t^l H_4(x,y,z) \implies& \qquad \text{(by Euler's theorem)} \\ EH_4 = lH_4 \implies& EH_2 = (EH_3)H_4 + H_3EH_4 \\ &= (EH_3)H_4 + lH_3H_4. \end{aligned} \tag{5.4.27}$$

Now

$$x\frac{\partial}{\partial x}H_3 = \left(L_{n+l}^{(2l+1)}\right)'(\alpha r) \quad \alpha x\frac{\partial r}{\partial x} \tag{5.4.28}$$

and similarly

$$y\frac{\partial}{\partial y}H_3 = \left(L_{n+l}^{(2l+1)}\right)'(\alpha r) \quad \alpha y\frac{\partial r}{\partial y} \tag{5.4.29}$$

$$\begin{aligned} z\frac{\partial}{\partial z}H_3 &= \left(L_{n+l}^{(2l+1)}\right)'(\alpha r) \quad \alpha z\frac{\partial r}{\partial z} \implies \\ EH_3 &= \left(L_{n+l}^{(2l+1)}\right)'(\alpha r) \quad \alpha Er \\ &= \left(L_{n+l}^{(2l+1)}\right)'(\alpha r) \quad \alpha r \qquad \text{(by equation (5.4.25)).} \end{aligned} \tag{5.4.30}$$

By the same argument (replace $L_{n+l}^{(2l+1)}(\alpha r)$ by $\left(L_{n+l}^{(2l+1)}\right)'(\alpha r)$)

$$E\left(L_{n+l}^{(2l+1)}\right)'(\alpha r) = \left(L_{n+l}^{(2l+1)}\right)''(\alpha r)\,\alpha r \qquad \implies \tag{5.4.31}$$

$$\begin{aligned} E^2H_3 &= E\left(EH_3 = \left(L_{n+l}^{(2l+1)}\right)'(\alpha r)\,\alpha r\right) \\ &= \left(L_{n+l}^{(2l+1)}\right)''(\alpha r)\,\alpha r\alpha r + \left(L_{n+l}^{(2l+1)}\right)'(\alpha r)E\alpha r \\ &= \alpha^2r^2\left(L_{n+l}^{(2l+1)}\right)''(\alpha r) + \alpha r\left(L_{n+l}^{(2l+1)}\right)'(\alpha r). \end{aligned} \tag{5.4.32}$$

By equation (5.4.27),

$$\begin{aligned} EH_2 &= (EH_3)H_4 + lH_3H_4 = \left(L_{n+l}^{(2l+1)}\right)'(\alpha r)\,\alpha rH_4 + lH_3H_4 \\ &= \left(L_{n+l}^{(2l+1)}\right)'(\alpha r)\,\alpha rH_4 + lH_2, \end{aligned} \tag{5.4.33}$$

and

$$\begin{aligned} E^2H_2 &= (E^2H_3)H_4 + (EH_3)EH_4 + lEH_2 \\ &= \left[\alpha^2r^2\left(L_{n+l}^{(2l+1)}\right)''(\alpha r) + \alpha r\left(L_{n+l}^{(2l+1)}\right)'(\alpha r)\right]H_4 \\ &\quad + \left(\left(L_{n+l}^{(2l+1)}\right)'(\alpha r)\,\alpha r\right)lH_4 \\ &\quad + l\left[\left(L_{n+l}^{(2l+1)}\right)'(\alpha r)\,\alpha rH_4 + lH_2\right]. \end{aligned} \tag{5.4.34}$$

By the formulas for EH_1, E^2H_1, EH_2 and $E^2H_2, \phi/C = H_1H_2$ on $\mathbb{R}^3-\{0\}$ $\Longrightarrow$

$$\begin{aligned} E\frac{\phi}{C} &= (EH_1)H_2 + H_1EH_2 \\ &= \left(-\frac{\alpha}{2}rH_1\right)H_2 + H_1\left[\left(L_{n+l}^{(2l+1)}\right)'(\alpha r)\,\alpha rH_4 + lH_2\right] \\ &= -\frac{\alpha}{2}r\frac{\phi}{C} + l\frac{\phi}{C} + \alpha r\left(L_{n+l}^{(2l+1)}\right)'(\alpha r)H_1H_4, \end{aligned} \tag{5.4.35}$$

$$\begin{aligned} E^2\frac{\phi}{C} &= (E^2H_1)H_2 + (EH_1)EH_2 + (EH_1)EH_2 + H_1E^2H_2 \\ &= \left[-\frac{\alpha}{2}r + \frac{\alpha^2}{4}r^2\right]H_1H_2 \\ &\quad + 2\left(-\frac{\alpha}{2}rH_1\right)\left(\left(L_{n+l}^{(2l+1)}\right)'(\alpha r)\,\alpha rH_4 + lH_2\right) \\ &\quad + H_1\left\{\left[\alpha^2r^2\left(L_{n+l}^{(2l+1)}\right)''(\alpha r) + \alpha r\left(L_{n+l}^{(2l+1)}\right)'(\alpha r)\right]H_4\right. \\ &\quad \left. + \left(L_{n+l}^{(2l+1)}\right)'(\alpha r)\,\alpha rlH_4 + l\left[\left(L_{n+l}^{(2l+1)}\right)'(\alpha r)\,\alpha rH_4 + lH_2\right]\right\} \\ &= \left[-\frac{\alpha}{2}r + \frac{\alpha^2}{4}r^2 - \alpha rl + l^2\right]\frac{\phi}{C} \\ &\quad + \left[-(\alpha r)^2 + \alpha r + \alpha rl2\right]\left(L_{n+l}^{(2l+1)}\right)'(\alpha r)H_1H_4 \\ &\quad + \alpha^2r^2\left(L_{n+l}^{(2l+1)}\right)''(\alpha r)H_1H_4 \Longrightarrow \end{aligned} \tag{5.4.36}$$

$$\begin{aligned} (E+E^2)\frac{\phi}{C} &= \left[-\frac{\alpha}{2}r + l - \frac{\alpha}{2}r + \frac{\alpha^2}{4}r^2 - \alpha rl + l^2\right]\frac{\phi}{C} \\ &\quad + \left[\alpha r - \alpha^2r^2 + \alpha r + 2\alpha rl\right]\left(L_{n+l}^{(2l+1)}\right)'(\alpha r)H_1H_4 \\ &\quad + \alpha^2r^2\left(L_{n+l}^{(2l+1)}\right)''(\alpha r)H_1H_4 \\ &= \left[-\alpha r + l + \frac{\alpha^2r^2}{4} - \alpha rl + l^2\right]\frac{\phi}{C} \\ &\quad + \alpha rH_1H_4\left[\alpha r\left(L_{n+l}^{(2l+1)}\right)''(\alpha r) + (2+2l-\alpha r)\left(L_{n+l}^{(2l+1)}\right)'(\alpha r)\right]. \end{aligned} \tag{5.4.37}$$

By the differential equation for $L_{n+l}^{(2l+1)}$ (see Appendix 4B)

$$x\left(L_{n+l}^{(2l+1)}\right)''(x) + (2l+2-x)\left(L_{n+l}^{(2l+1)}\right)'(x) + (n-l-1)L_{n+l}^{(2l+1)}(x) = 0 \tag{5.4.38}$$

we obtain for $x = \alpha r$,

$$(E + E^2)\frac{\phi}{C} = \left[-\alpha r + l + \frac{\alpha^2 r^2}{4} - \alpha r l + l^2\right]\frac{\phi}{C} + \alpha r H_1 H_4 (l + 1 - n) L_{n+l}^{(2l+1)}(\alpha r), \quad (5.4.39)$$

where $H_4 L_{n+l}^{(2l+1)}(\alpha r) = H_4 H_3 = H_2$. That is,

$$(E + E^2)\frac{\phi}{C} = \left[-\alpha r + l + \frac{\alpha^2 r^2}{4} - \alpha r l + l^2 + \alpha r(l + 1 - n)\right]\frac{\phi}{C} \quad (5.4.40)$$

$$\Longrightarrow \quad (E + E^2)\psi_{nlm} = \left[l + \frac{\alpha^2 r^2}{4} + l^2 - \alpha r n\right]\psi_{nlm} \quad (5.4.41)$$

on $\mathbb{R}^3 - \{0\}$!

On the other hand, on $\mathbb{R}^3 - \{0\}$, ϕ satisfies the Schrödinger equation (5.1.27):

$$\nabla^2\phi + \frac{8\pi^2\mu}{h^2}\left(E_1 - \frac{q_1 q_2}{r}\right)\phi = 0, \qquad \text{or} \quad (5.4.42)$$

$$r^2\nabla^2\phi = \left[-\frac{8\pi^2\mu}{h^2} r^2 E_1 + \frac{8\pi^2\mu}{h^2} q_1 q_2 r\right]\phi,$$

for $q_1 = -e$, $q_2 = Ze$. By Proposition 5.3

$$\begin{aligned}\frac{M^2\phi}{\hbar^2} &= -r^2\nabla^2\phi + (E + E^2)\phi \quad \left(\text{on } \mathbb{R}^3 - \{0\} \quad \text{where } \phi \text{ is } C^\infty\right) \\ &= \left[\frac{8\pi^2\mu}{h^2} r^2 E_1 - \frac{8\pi^2\mu}{h^2} q_1 q_2 r + l + \frac{\alpha^2 r^2}{4} + l^2 - \alpha r n\right]\phi\end{aligned} \quad (5.4.43)$$

by equation (5.4.41). Here, by (5.1.22), (5.1.23),

$$E_1 = E^{(n)} = -\frac{q_1^2 q_2^2 \mu}{2n^2\hbar^2}, \qquad \text{and } \alpha = \alpha_n = -\frac{2q_1 q_2 \mu}{n\hbar^2} \quad (5.4.44)$$

$$\Longrightarrow \left(\frac{8\pi^2\mu}{h^2} E_1 + \frac{\alpha^2}{4}\right) r^2 = 0 \cdot r^2 = 0. \quad (5.4.45)$$

Similarly

$$\left(\frac{-8\pi^2\mu}{h^2} q_1 q_2 - \alpha n\right) r = 0 \cdot r = 0, \quad (5.4.46)$$

so that we finally arrive at the equation

$$\frac{M^2}{\hbar^2}\psi_{nlm} = (l + l^2)\psi_{nlm}, \quad (5.4.47)$$

$$\text{or} \quad M^2\psi_{nlm} = (l + l^2)\hbar^2\psi_{nlm} \qquad \text{on } \mathbb{R}^3 - \{0\}, \quad (5.4.48)$$

as desired. □

Maintaining the above notation, we have $H_4 = Y_l^m$. Recall the formula

$$\left(x\frac{\partial}{\partial y} - y\frac{\partial}{\partial x}\right) Y_l^m = imY_l^m \tag{5.4.49}$$

of Appendix 4A. Hence by (5.4.4)

Proposition 5.4. $M_z Y_l^m = \hbar m Y_l^m$.

As a corollary we now prove the second equation in (5.4.5):

Theorem 5.3. $M_z \psi_{nlm} = \hbar m \psi_{nlm}$ *on* $\mathbb{R}^3 - \{0\}$.

Proof. In the above notation,

$$M_z \frac{\phi}{C} = M_z(H_1 H_2) = (M_z H_1) H_2 + H_1 M_z H_2. \tag{5.4.50}$$

We claim $M_z H_1 = 0$: By the above calculations (or by direct recalculation) on $\mathbb{R}^3 - \{0\}$,

$$\frac{\partial H_1}{\partial x} = -\frac{\alpha}{2}\exp\left(-\frac{\alpha}{2}r\right)\frac{x}{r}, \qquad \frac{\partial H_1}{\partial y} = -\frac{\alpha}{2}\exp\left(-\frac{\alpha}{2}r\right)\frac{y}{r}, \tag{5.4.51}$$

so that clearly

$$\begin{aligned} x\frac{\partial H_1}{\partial y} - y\frac{\partial H_1}{\partial x} &= -\frac{\alpha}{2}\exp\left(-\frac{\alpha}{2}r\right)\frac{xy}{r} + \frac{\alpha}{2}\exp\left(-\frac{\alpha}{2}r\right)\frac{yx}{r} = 0 \\ \Longrightarrow\ M_z H_1 = 0. \end{aligned} \tag{5.4.52}$$

Hence by (5.4.50)

$$M_z \frac{\phi}{C} = H_1 M_z H_2 = H_1 M_z (H_3 H_4) = H_1[(M_z H_3) H_4 + H_3 M_z H_4], \tag{5.4.53}$$

where $M_z H_4 = \hbar m H_4$ by Proposition 5.4. We claim that $M_z H_3 = 0$, which thus would give

$$\begin{aligned} M_z \frac{\phi}{C} &= H_1 H_3 \hbar m H_4 = \hbar m H_1 H_2 \qquad \text{or} \\ M_z \phi &= \hbar m C H_1 H_2 = \hbar m \phi, \end{aligned} \tag{5.4.54}$$

as desired.

By the above calculations (see (5.4.28), (5.4.29))

$$\frac{\partial H_3}{\partial x} = \left(L_{n+l}^{(2l+1)}\right)'(\alpha r)\,\alpha\frac{\partial r}{\partial x}, \qquad \frac{\partial H_3}{\partial y} = \left(L_{n+l}^{(2l+1)}\right)'(\alpha r)\,\alpha\frac{\partial r}{\partial y}, \tag{5.4.55}$$

where

$$\frac{\partial r}{\partial x} = \frac{x}{r}, \qquad \frac{\partial r}{\partial y} = \frac{y}{r} \tag{5.4.56}$$

on $\mathbb{R}^3 - \{0\}$. Hence

$$x\frac{\partial H_3}{\partial y} - y\frac{\partial H_3}{\partial x} = \left(L_{n+l}^{(2l+1)}\right)'(\alpha r)\,\alpha\frac{xy}{r} - \left(L_{n+l}^{(2l+1)}\right)'(\alpha r)\,\alpha\frac{yx}{r} = 0$$
$$\Longrightarrow M_z H_3 = 0, \tag{5.4.57}$$

as claimed. □

The proof of Theorem 5.2 was a bit long and non-conceptual. It is possible to give a group-theoretic proof based on some results of Chapter 7 and ideas in Appendix 9C.

5.5 The Average Distance of the Electron from the Nucleus

Consider for example the normalized wave function $\psi = \psi_{100}$ with total quantum number 1 and azimuthal and magnetic quantum numbers equal 0. ψ governs the properties of the hydrogenic atom in its normal or *ground* state—i.e., the state of lowest energy; compare remarks in Section 5.8. By Max Born's basic postulate, $|\psi|^2$ is the probability distribution for the electron about the nucleus:

$$\iiint_{\mathfrak{R}} |\psi|^2\, dxdydz = \begin{array}{l}\text{the probability of finding the electron}\\ \text{in a region of space } \mathfrak{R}.\end{array} \tag{5.5.1}$$

For example we compute the probability P of finding the electron within a sphere about the origin of radius a_o = the Bohr radius. By (5.5.1), (5.3.6), (5.3.8), and (5.3.9)

$$\begin{aligned} P &= \int_0^{\pi}\int_0^{2\pi}\int_0^{a_o} R_{10}^{-}(r)^2|\Phi_o(\theta)|^2 X_{oo}(\phi)^2 r^2 \sin\theta\; dr d\theta d\phi \\ &= 4\left(\frac{Z}{a_o}\right)^3 \int_0^{a_o} \exp\left(-\frac{2Zr}{a_o}\right) r^2\, dr. \end{aligned} \tag{5.5.2}$$

Now

$$\int e^{-ar} r^2\, dr = -\left(\frac{r^2 e^{-ar}}{a} + \frac{2}{a^2}e^{-ar} + \frac{2}{a^3}e^{-ar}\right) \tag{5.5.3}$$

for $a \neq 0$ by two integrations by parts. Choose $a = 2Z/a_o$ to get

$$P = 4\left(\frac{Z}{a_o}\right)^3 \left[-\frac{a_o^3 e^{-2Z}}{2Z} - \frac{2a_o^3 e^{-2Z}}{4Z^2} - \frac{2a_o^3 e^{-2Z}}{8Z^3} + \frac{2a_o^3}{8Z^3}\right]; \text{ i.e.,} \tag{5.5.4}$$

$$P = -2Z^2 e^{-2Z} - 2Ze^{-2Z} - e^{-2Z} + 1. \tag{5.5.5}$$

For the hydrogen atom, for example, $Z = 1 \Longrightarrow P = -5e^{-2} + 1 = -5 \cdot (0.1353) + 1 = 0.3235$.

The expectation value of the position of the electron, i.e., the average distance $\bar{r}$ of the electron from the nucleus is given by

$$\bar{r} \stackrel{\text{def}}{=} \iiint_{\mathbb{R}^3} \sqrt{x^2 + y^2 + z^2}\, |\psi(x, y, z)|^2\, dxdydz. \tag{5.5.6}$$

For the ground state we obtain, as in (5.5.2)

$$\bar{r} = 4\left(\frac{Z}{a_o}\right)^3 \int_0^\infty \exp\left(-\frac{2Zr}{a_o}\right) r^3\, dr. \tag{5.5.7}$$

Now

$$\int_0^\infty e^{-ar} r^n\, dr = \frac{n!}{a^{n+1}} \tag{5.5.8}$$

for $a > 0, n = 0, 1, 2, \ldots$. Thus choose $a = 2Z/a_o$ to obtain $\bar{r} = \frac{3}{2}\frac{a_o}{Z}$; thus $\bar{r} = \frac{3}{2}a_o$ for hydrogen. The electron does *not* move in a circular orbit of radius a_o about the nucleus of hydrogen but maintains an average distance $\frac{3}{2}a_o > a_o$ from the nucleus, as pointed out in Chapter 0. We have shown, moreover, that (for hydrogen) the probability of finding the electron within a sphere about the origin of radius a_o is 0.3235, in the ground state.

We can compute the average distance $\bar{r}_{nlm}$ of the electron from the nucleus for an arbitrary state ψ_{nlm} of the hydrogenic atom. As in (5.5.6)

$$\bar{r}_{nlm} \stackrel{\text{def}}{=} \iiint_{\mathbb{R}^3} \sqrt{x^2 + y^2 + z^2}\, |\psi_{nlm}(x, y, z)|^2\, dxdydz. \tag{5.5.9}$$

By (5.3.4), (5.3.8), and (5.3.9), as $\rho_n(r) \stackrel{\text{def}}{=} \frac{2Zr}{na_o}$

$$\begin{aligned}
\bar{r}_{nlm} &= \int_0^\infty R_{nl}^-(r)^2 r^3\, dr \\
&= \frac{4}{n^4}\left(\frac{Z}{a_o}\right)^3 \frac{(n-l-1)!}{(n+l)!} \\
&\quad \cdot \int_0^\infty \left(\frac{2Zr}{na_o}\right)^{2l} \exp\left(-\frac{2Zr}{a_o}\right) L_{n-l-1}^{2l+1}\left(\frac{2Zr}{na_o}\right)^2 r^3\, dr \qquad (5.5.10)\\
&= \frac{4}{n^4}\left(\frac{Z}{a_o}\right)^3 \frac{(n-l-1)!}{(n+l)!}\left(\frac{2Z}{na_o}\right)^{-4} \int_0^\infty e^{-x} L_{n-l-1}^{2l+1}(x)^2 x^{2l+3}\, dx \\
&= \frac{a_o n^2}{Z}\left[1 + \frac{1}{2}\left(1 - \frac{l(l+1)}{n^2}\right)\right]
\end{aligned}$$

by formula (4.7.23) of Corollary 4.5:

Theorem 5.4.

$$\bar{r}_{nlm} = \frac{a_o n^2}{Z}\left[1 + \frac{1}{2}\left(1 - \frac{l(l+1)}{n^2}\right)\right]$$

is the average distance of the electron from the nucleus of a hydrogenic atom with quantum state ψ_{nlm}.

In particular for the ground state ψ_{100} we obtain $\bar{r}_{100} = \frac{a_0}{Z}[1 + \frac{1}{2}] = \frac{3a_0}{2Z}$, as before.

5.6 The Ritz–Rydberg Formula: Further Examples

In the discussion in Section 1.2 of Chapter 1 of the Bohr frequency rule and emission spectra we were led to Balmer's formula and, more generally, the Ritz–Rydberg formula. We saw that the proof of those formulas (given the Bohr frequency rule) depended on the energy levels formula (1.2.4) stated there without proof. Now that we have established this energy levels formula (in a slightly more general form in equation (5.1.29)) the discussion of Chapter 1 attains a further sense of completeness. By equation (5.1.29) we obtain in fact the following slightly more general Ritz–Rydberg formula for a hydrogenic atom:

$$\frac{1}{\lambda} = Z^2 R\left(\frac{1}{n_1^2} - \frac{1}{n_2^2}\right) = Z^2(109678)\left(\frac{1}{n_1^2} - \frac{1}{n_2^2}\right)\ \text{cm}^{-1} \tag{5.6.1}$$
$$\text{for the transition } E^{(n_2)} \to E^{(n_1)}, \qquad n_2 > n_1,$$

from a higher to a lower energy state, in which case radiation of wavelength λ is emitted. If $n_2 < n_1$ then, as we have seen, radiation is absorbed in the transition $E^{(n_2)} \to E^{(n_1)}$ (from a lower to a higher energy state), the wavelength λ being given by

$$\frac{1}{\lambda} = Z^2(109678)\left(\frac{1}{n_2^2} - \frac{1}{n_1^2}\right)\ \text{cm}^{-1}. \tag{5.6.2}$$

As we consider further computational examples of the Ritz–Rydberg formula the reader may wish to review some of the content and notation of Chapter 1.

For the Lyman series of hydrogen (where $Z = 1$) choose $n_1 = 1$. For the first Lyman line $L_2^{(1)}$ choose $n_2 = 2$ to obtain the wavelength of the emitted radiation, by (5.6.1):

$$\frac{1}{\lambda} = 109678\left(\frac{1}{1^2} - \frac{1}{2^2}\right)\ \text{cm}^{-1} = 82258.5\ \text{cm}^{-1}$$
$$\Longrightarrow \lambda = 1216\ \text{Å}. \tag{5.6.3}$$

Note that as n_2 increases, $\frac{1}{\lambda}$ increases; i.e., λ decreases. The value 1216 Å in (5.6.3) is therefore the longest wavelength in the Lyman series; the shortest wavelength λ_∞ is obtained by letting $n_2 \to \infty$: $\lambda_\infty^{-1} = 109678\ \text{cm}^{-1} \Longrightarrow \lambda_\infty = 911.7$ Å. The frequency ν of the first Lyman line of hydrogen with wavelength $\lambda = 1216$ Å is given by

$$\nu = \frac{c}{\lambda} = \frac{3 \times 10^{10}\ \text{cm/sec}}{1216\ \text{Å}} = 2.467 \times 10^{15}\ \text{Hz}. \tag{5.6.4}$$

During the transition $E^{(3)} \to E^{(1)}$ for hydrogen a photon is emitted. We calculate the energy and the momentum of this photon. By equation (1.1.1) the energy

E is given by $E = h\nu$; here $\nu = \frac{c}{\lambda}$ is given by (5.6.1):

$$\begin{aligned}
\nu &= c109678\left(\frac{1}{1^2} - \frac{1}{3^2}\right)\ \mathrm{cm}^{-1} = 3 \times 10^{10} \times 109678 \times \frac{8}{9}\ \mathrm{sec}^{-1} \\
&= 292474.67 \times 10^{10}\ \mathrm{sec}^{-1} \qquad (5.6.5) \\
\Longrightarrow E &= 6.62608 \times 10^{-27}\ \text{erg-sec} \times 292474.67 \times 10^{10}\ \mathrm{sec}^{-1} \\
&= 1937960.5 \times 10^{-17}\ \text{ergs} \simeq 12.1\ \text{eV; see Appendix 8A.}
\end{aligned}$$

The momentum p of the photon is given by de Broglie's wavelength formula $\lambda = hp^{-1}$. That is, as $E = h\nu = hc\lambda^{-1} = cp$, $p = Ec^{-1} = 1937960.5 \times 10^{-17}$ ergs $\times(3 \times 10^{10}\ \mathrm{cm\ sec}^{-1})^{-1} = 645986.83 \times 10^{-27}\ \mathrm{g\ cm\ sec}^{-1}$.

In the next example we compute the wavelength λ of the second line $L_5^{(3)}$ in the Paschen series for hydrogen. Thus in (5.6.1) take $n_1 = 3$, $n_2 = 5$ to obtain

$$\begin{aligned}
\frac{1}{\lambda} &= 109678\left(\frac{1}{9} - \frac{1}{25}\right) = 7799.3244\ \mathrm{cm}^{-1};\ \text{i.e.,} \\
\lambda &= .0001282\ \mathrm{cm} = 12,820\ \text{Å}.
\end{aligned} \qquad (5.6.6)$$

A *μ-meson* (or *muon*) has the same charge $-e$ as an electron but its mass is 207 times the electron mass m_e. Consider a μ-mesonic lead atom with $Z = 82$ protons. To obtain the nth energy level $E^{(n)}$ of this atom replace m_e by $207m_e$ in (5.1.29). In fact replace the relative mass $m_e m_p/(m_e + m_p) = .99m_e$ (by (0.0.10)) by m_e to obtain

$$E^{(n)} = -\frac{(207)(82)^2 m_e e^4}{2\hbar^2 n^2} = 1391868\ E^{(n)}_{\mathrm{hyd}} \qquad (5.6.7)$$

where $E^{(n)}_{\mathrm{hyd}}$ is the nth energy level of hydrogen. The transition $E^{(2)} \to E^{(1)}$ (again from a higher to a lower energy level, where radiation is emitted) gives rise to the first Lyman line (of μ-mesonic lead) with frequency

$$\begin{aligned}
\nu &= \frac{E^{(2)} - E^{(1)}}{h} = 1391868\left(\frac{E^{(2)}_{\mathrm{hyd}} - E^{(1)}_{\mathrm{hyd}}}{h}\right) \\
&= 1391868 \times \text{(the frequency of the first Lyman line for hydrogen)} \qquad (5.6.8) \\
&= 1391868 \times 2.467 \times 10^{15}\ \mathrm{cm}^{-1} \qquad \text{(by (5.6.4))} \\
&= 3433738.4 \times 10^{15}\ \mathrm{cm}^{-1}.
\end{aligned}$$

In 1930, Ericson and Edlen (in Zeits. f. Phys. 59, 679) reported that the first Lyman line of doubly ionized lithium Li^{++} (i.e., lithium with two electrons removed, with $Z = 3$, as we have seen) and the first Lyman line of triply ionized beryllium Be^{+++} (beryllium with three electrons removed, with $Z = 4$) have wavelengths $\lambda_e = 135.02$ Å, 75.94 Å, respectively, in the high ultraviolet range; cf. Table 1.1. We can compare these experimental values with theoretical values λ_t predicted by quantum mechanics. Namely, for $Z = 3$, we obtain by (5.6.1), (5.6.3)

$$\frac{1}{\lambda_t} = Z^2(82258.5)\ \mathrm{cm}^{-1}$$
$$\Longrightarrow\ \lambda_t = \frac{1}{9}\ 1216\ \text{Å} = 135.11\ \text{Å}, \tag{5.6.9}$$

which is in good agreement with the value $\lambda_e = 135.02$ Å for Li^{++}. Similarly for $Z = 4$, equations (5.6.1), (5.6.3) provide the value

$$\lambda_t = \frac{1}{16}\ 1216\ \text{Å} = 76\ \text{Å} \tag{5.6.10}$$

for Be^{+++}, which is in good agreement with the experimental value $\lambda_e = 75.94$ Å.

As a final example, the following question is posed: Does there exist an $n > 1$ such that a hydrogenic atom can absorb radiation in the x-ray range (say $\lambda = 2$ Å) via the transition $E^{(1)} \to E^{(n)}$?; recall Table 1.1. The answer is easy to come by. Namely, equation (5.6.2) requires that the wavelength λ of the absorbed radiation be given by

$$\frac{1}{\lambda} = Z^2(109678)\left(1 - \frac{1}{n^2}\right). \tag{5.6.11}$$

For $\lambda = 2 \times 10^{-8}$ cm this means that

$$\frac{1}{n^2} = 1 - \frac{1}{2 \times 10^{-8} Z^2(109678)}$$
$$= 1 - \frac{460}{Z^2} = \frac{Z^2 - 460}{Z^2}. \tag{5.6.12}$$

Since $\frac{1}{n^2} > 0$ we must have $Z^2 > 460$, or $Z \geq 22$. The answer is therefore "no" if $1 \leq Z \leq 21$. On the other hand, for $Z^2 > 460$

$$\frac{Z^2}{Z^2 - 460} = n^2 \geq 4 \ \Longrightarrow\ Z^2 \geq 4(Z^2 - 460)$$
$$\Longrightarrow\ 3Z^2 \leq 1840 \ \Longrightarrow\ Z \leq 24.766 \tag{5.6.13}$$
$$\Longrightarrow\ Z = 22, 23,\ \text{or}\ 24.$$

However, the three values $Z = 22, 23, 24$ force $n^2 = 484/24,\ 529/69,\ 576/116$, respectively, which is not possible for n integral. Thus in general the answer to the posed question is "*no.*"

5.7 Wave Function Orthogonality

Equation (4.8.6) expresses orthogonality of the normalized wave functions $\{\psi_n\}_{n=0}^{\infty}$ of the quantum harmonic oscillator. Similarly, one can establish orthogonality for normalized hydrogenic wave functions. From equation (5.3.9), $\psi_{nlm} \circ g = R_{nl}^{-} \otimes \Phi_m \otimes X_{lm}$ on $\{(r, \theta, \phi) \in \mathbb{R}^3 \mid r \geq 0, 0 \leq \phi \leq \pi\}$. As a consequence one indeed has

Proposition 5.5 (Wave Function Orthogonality).

$$\iiint_{\mathbb{R}^3} \psi_{nlm}\overline{\psi}_{nl'm'}\, dxdydz = \delta_{(l,m),(l',m')}. \tag{5.7.1}$$

Proof. The left-hand side of (5.7.1) is

$$\begin{aligned} I &\stackrel{\text{def}}{=} \langle \psi_{nlm}, \psi_{nl'm'} \rangle \\ &= \int_0^{2\pi}\int_0^{\pi}\int_0^{\infty} \psi_{nlm}(g(r,\theta,\phi))\overline{\psi}_{nl'm'}(g(r,\theta,\phi))\, r^2 \sin\phi\, d\phi d\theta dr \\ &= \int_0^{2\pi}\int_0^{\pi}\int_0^{\infty} R^-_{nl}(r)\Phi_m(\theta)X_{lm}(\phi)R^-_{nl'}(r)\overline{\Phi}_{m'}(\theta)X_{l'm'}(\phi)\, r^2 \sin\phi\, d\phi d\theta dr. \end{aligned} \tag{5.7.2}$$

Now since

$$\int_0^{2\pi} \Phi_m(\theta)\overline{\Phi_{m'}}(\theta)\, d\theta \stackrel{\text{def. (5.3.8)}}{=} \frac{1}{2\pi}\int_0^{2\pi} e^{i(m-m')\theta}\, d\theta = \delta_{mm'} \tag{5.7.3}$$

I vanishes unless $m = m'$, in which case

$$I = \int_0^{\infty} R^-_{nl}(r)R^-_{nl'}(r)\, r^2 dr \int_0^{\pi} X_{lm}(\phi)X_{l'm}(\phi)\sin\phi\, d\phi, \tag{5.7.4}$$

where

$$\begin{aligned} &\int_0^{\pi} X_{lm}(\phi)X_{l'm}(\phi)\sin\phi\, d\phi \\ \stackrel{\text{def. (5.3.8)}}{=}\; & c_{lm}c_{l'm}\int_0^{\pi} P_l^{|m|}(\cos\phi)P_{l'}^{|m|}(\cos\phi)\ \sin\phi\, d\phi \\ =\; & c_{lm}c_{l'm}\delta_{ll'}\frac{2}{2l'+1}\frac{(l'+|m|)!}{(l'-|m|)!} = \delta_{ll'} \end{aligned} \tag{5.7.5}$$

by Theorem 4.8 of Chapter 4, and by definition (5.2.11). That is, I vanishes unless $m = m'$ and $l = l'$, in which case

$$I = \int_0^{\infty} R^-_{nl}(r)R^-_{nl'}(r)\, r^2 dr = 1 \tag{5.7.6}$$

by (5.2.6). □

Since orthogonality implies linear independence we have

Corollary 5.1. *The normalized wave functions* $\{\psi_{nlm}\}_{m=-l}^{l}$ *are linearly independent over* $\mathbb{C}$.

5.8 Ground State Energy of Hydrogen-like Atoms

The normal state of a hydrogen atom is determined by the wave function ψ_{100}, where $n = 1$:

$$\psi_{100}(r, \theta, \phi) = \left(\frac{1}{a_o}\right)^{3/2} \frac{e^{-r/a_o}}{\sqrt{\pi}} \tag{5.8.1}$$

by (5.3.12). Its lowest, or ground state, energy $-E_H$ is $E^{(1)}$, given by (5.1.29):

$$\begin{aligned} -E_H \overset{\text{def}}{=} E^{(1)} &= -\frac{\mu e^4}{2\hbar^2} = -\frac{4845.9354 \times 10^{-68} \text{ dyne}^2 \text{ cm}^4 \text{ gr}}{2.2242358 \times 10^{-54} \text{ erg}^2 \sec^2} \\ &= -2.1786968 \times 10^{-11} \text{ ergs} \\ &= -13.59 \text{ eV} \end{aligned} \tag{5.8.2}$$

by (0.0.2), (0.0.3), (0.0.10), where 1 erg $= 6.242 \times 10^{11}$ electron volts (eV). See Appendix 8A.

By (5.1.29) again, the ground state energy of a hydrogenic atom with atomic number Z is given by

$$E = -Z^2 E_H. \tag{5.8.3}$$

Appendix 5A Commutation Relations for Angular Momentum Operators

The angular momentum operators M_x, M_y, M_z of definition (5.4.4) are subject to the commutation relations

$$\begin{aligned} [M_x, M_y] &= i\hbar M_z, \\ [M_y, M_z] &= i\hbar M_x, \\ [M_z, M_x] &= i\hbar M_y, \end{aligned} \tag{5.A.1}$$

which are easy to check. For example, by the product rule,

$$y\frac{\partial}{\partial z} z\frac{\partial}{\partial x} = y\left[z\frac{\partial^2}{\partial z \partial x} + \frac{\partial}{\partial x}\right] \tag{5.A.2}$$

$$\begin{aligned} \Longrightarrow M_x M_y &\overset{\text{def}}{=} -\hbar^2 \left(y\frac{\partial}{\partial z} - z\frac{\partial}{\partial y}\right)\left(z\frac{\partial}{\partial x} - x\frac{\partial}{\partial z}\right) \\ &= -\hbar^2 \left[y\frac{\partial}{\partial z} z\frac{\partial}{\partial x} - z\frac{\partial}{\partial y} z\frac{\partial}{\partial x} - y\frac{\partial}{\partial z} x\frac{\partial}{\partial z} + z\frac{\partial}{\partial y} x\frac{\partial}{\partial z}\right] \\ &= -\hbar^2 \left\{y\left[z\frac{\partial^2}{\partial z \partial x} + \frac{\partial}{\partial x}\right] - z^2\frac{\partial^2}{\partial y \partial x} - yx\frac{\partial^2}{\partial z^2} + zx\frac{\partial^2}{\partial y \partial z}\right\}. \end{aligned} \tag{5.A.3}$$

Similarly

$$
\begin{aligned}
M_y M_x &\stackrel{\text{def}}{=} -\hbar^2 \left(z\frac{\partial}{\partial x} - x\frac{\partial}{\partial z} \right) \left(y\frac{\partial}{\partial z} - z\frac{\partial}{\partial y} \right) \\
&= -\hbar^2 \left[z\frac{\partial}{\partial x} y\frac{\partial}{\partial z} - x\frac{\partial}{\partial z} y\frac{\partial}{\partial z} - z\frac{\partial}{\partial x} z\frac{\partial}{\partial y} + x\frac{\partial}{\partial z} z\frac{\partial}{\partial y} \right] \\
&= -\hbar^2 \left\{ zy\frac{\partial^2}{\partial x\partial z} - xy\frac{\partial^2}{\partial z^2} - z^2\frac{\partial^2}{\partial x\partial y} + x\left[z\frac{\partial^2}{\partial z\partial y} + \frac{\partial}{\partial y} \right] \right\}.
\end{aligned}
\tag{5.A.4}
$$

Using the equality of mixed partial derivatives one obtains from (5.A.3) and (5.A.4)

$$
\begin{aligned}
[M_x, M_y] &\stackrel{\text{def}}{=} M_x M_y - M_y M_x \\
&= -\hbar^2 \left\{ y\frac{\partial}{\partial x} - x\frac{\partial}{\partial y} \right\} = \hbar i M_z.
\end{aligned}
\tag{5.A.5}
$$

The other two commutation relations in (5.A.1) are similarly verified.

In addition to the relations in (5.A.1) one has the commutation relations

$$
[M^2, M_x] = [M^2, M_y] = [M^2, M_z] = 0. \tag{5.A.6}
$$

For example

$$
\begin{aligned}
[M_x^2, M_z] &\stackrel{\text{def}}{=} M_x M_x M_z - M_z M_x M_x \\
&= M_x(M_z M_x - i\hbar M_y) - (M_x M_z + i\hbar M_y)M_x \\
&= -i\hbar(M_x M_y + M_y M_x) \qquad \text{(by (5.A.1))}.
\end{aligned}
\tag{5.A.7}
$$

Similarly by (5.A.1)

$$
\begin{aligned}
[M_y^2, M_z] &\stackrel{\text{def}}{=} M_y M_y M_z - M_z M_y M_y \\
&= i\hbar(M_y M_x + M_x M_y).
\end{aligned}
\tag{5.A.8}
$$

As $[M_z^2, M_z] = 0$ we see that

$$
[M^2, M_z] \stackrel{\text{def}}{=} [M_x^2 + M_y^2 + M_z^2, M_z] = 0
$$

by (5.A.7), (5.A.8). Similarly $[M^2, M_y] = 0$ and $[M^2, M_x] = 0$.

The commutation relations of (5.A.1) will serve as the starting point in Chapter 9 for the construction of certain *spin* operators—a construction that will be based on the representation theory of the complex Lie algebra $sl(2, \mathbb{C})$.

6

Heisenberg's Uncertainty Principle

6.1 Heisenberg's Inequality

We establish in this section an integral inequality

$$\left[\int_{\mathbb{R}} (x-b)^2 |\psi(x)|^2\, dx\right] \left[\int_{\mathbb{R}} |\psi'(x)|^2\, dx - 2ia \int_{\mathbb{R}} \psi'(x)\overline{\psi}(x)\, dx \right.$$
$$\left. + a^2 \int_{\mathbb{R}} |\psi(x)|^2\, dx\right] \geq \frac{1}{4} \left[\int_{\mathbb{R}} |\psi(x)|^2\, dx\right]^2, \quad (6.1.1)$$

referred to as *Heisenberg's inequality*, for a sufficiently nice function $\psi : \mathbb{R} \to \mathbb{C}$ where $a, b \in \mathbb{R}$. Even though ψ is allowed to be complex-valued the second factor on the left-hand side of (6.1.1) will be non-negative. This inequality, which is a purely mathematical result of course, will be shown to form the basis for a precise formulation of Heisenberg's uncertainty principle. The discussion of this principle in Chapter 1 was brief and non-rigorous. For now, we proceed strictly along mathematical lines.

Before getting into a proof of (6.1.1) consider first an example. Fix $\alpha > 0$ and define $\psi : \mathbb{R} \to \mathbb{R}$ by

$$\begin{aligned} \psi(x) &= \sin \frac{\pi}{\alpha} x \quad \text{on } [0, \alpha], \\ \psi(x) &= 0 \quad \text{off } [0, \alpha]. \end{aligned} \quad (6.1.2)$$

For $b \in \mathbb{R}$ consider

$$I_b \stackrel{\text{def}}{=} \int_{\mathbb{R}} (x-b)^2 |\psi(x)|^2 \, dx = \int_0^\alpha (x^2 - 2xb + b^2) \sin^2 \frac{\pi}{\alpha} x \, dx. \tag{6.1.3}$$

Now for $\beta \in \mathbb{R} - \{0\}$

$$\int x^2 \sin^2 \beta x \, dx = \frac{x^3}{6} - \frac{x}{4\beta^2} \cos 2\beta x - \frac{1}{4\beta^3}\left(\beta^2 x^2 - \frac{1}{2}\right) \sin 2\beta x + c. \tag{6.1.4}$$

Choose $\beta = \pi/\alpha$ to obtain

$$\int_0^\alpha x^2 \sin^2 \frac{\pi}{\alpha} x \, dx = \frac{\alpha^3}{2}\left[\frac{1}{3} - \frac{1}{2\pi^2}\right]. \tag{6.1.5}$$

Also

$$\int x \sin^2 \beta x \, dx = \frac{x^2}{4} - \frac{x \sin 2\beta x}{4\beta} - \frac{\cos 2\beta x}{8\beta^2} + c, \tag{6.1.6}$$

$$\int \sin^2 \beta x \, dx = -\frac{\sin 2\beta x}{4\beta} + \frac{x}{2} + c, \tag{6.1.7}$$

so that

$$\int_0^\alpha x \sin^2 \frac{\pi}{\alpha} x \, dx = \frac{\alpha^2}{4}, \tag{6.1.8}$$

$$\int_0^\alpha \sin^2 \frac{\pi}{\alpha} x \, dx = \frac{\alpha}{2}, \tag{6.1.9}$$

together with (6.1.5) implies that

$$I_b = \frac{\alpha^3}{2}\left[\frac{1}{3} - \frac{1}{2\pi^2}\right] - \frac{b\alpha^2}{2} + \frac{b^2\alpha}{2}. \tag{6.1.10}$$

We shall choose, for example, $b = \frac{\alpha}{2}$ which is the *mean value*

$$\frac{\int_0^\alpha x|\psi(x)|^2 \, dx}{\int_0^\alpha |\psi(x)|^2 \, dx} = \frac{\frac{\alpha^2}{4}}{\frac{\alpha}{2}} = \frac{\alpha}{2} \qquad \text{(by (6.1.8), (6.1.9))}. \tag{6.1.11}$$

Then (by (6.1.10))

$$-\frac{b\alpha^2}{2} + \frac{b^2\alpha}{2} = -\frac{\alpha^3}{8} \implies I_{\frac{\alpha}{2}} = \frac{\alpha^3}{4}\left[\frac{1}{6} - \frac{1}{\pi^2}\right]. \tag{6.1.12}$$

Next

$$\int \cos^2 \beta x \, dx = \frac{x}{2} + \frac{\sin 2\beta x}{4\beta} + c \quad \text{and} \tag{6.1.13}$$

$$\int \cos \beta x \sin \beta x \, dx = \frac{\sin^2 \beta x}{2\beta} + c \tag{6.1.14}$$

imply that

$$\int_{\mathbb{R}} |\psi'(x)|^2 \, dx = \frac{\pi^2}{\alpha^2} \int_0^\alpha \cos^2 \frac{\pi}{\alpha} x \, dx = \frac{\pi^2}{2\alpha}, \tag{6.1.15}$$

$$\int_{\mathbb{R}} \psi'(x)\overline{\psi}(x) \, dx = \frac{\pi}{\alpha} \int_0^\alpha \cos \frac{\pi}{\alpha} x \sin \frac{\pi}{\alpha} x \, dx = 0. \tag{6.1.16}$$

For the choice $b = \alpha/2$ we therefore have by (6.1.9), (6.1.12), (6.1.15), (6.1.16)

$$\begin{aligned} 4\left[\int_{\mathbb{R}} (x-b)^2 |\psi(x)|^2 \, dx\right] &\left[\int_{\mathbb{R}} |\psi'(x)|^2 \, dx - 2ia \int_{\mathbb{R}} \psi'(x)\overline{\psi}(x) \, dx \right. \\ &\left. + a^2 \int_{\mathbb{R}} |\psi(x)|^2 \, dx\right] \\ &\stackrel{\text{def}}{=} 4I_{\frac{\alpha}{2}} \left[\frac{\pi^2}{2\alpha} + a^2 \frac{\alpha}{2}\right] \\ &= \alpha^3 \left[\frac{1}{6} - \frac{1}{\pi^2}\right] \left[\frac{\pi^2}{2\alpha} + a^2 \frac{\alpha}{2}\right] \\ &\geq \alpha^3 \left[\frac{1}{6} - \frac{1}{\pi^2}\right] \frac{\pi^2}{2\alpha} = \frac{\alpha^2}{2} \left[\frac{\pi^2}{6} - 1\right] \\ &\geq \frac{\alpha^2}{2} \left[\frac{9}{6} - 1\right] = \frac{\alpha^2}{4} \\ &= \left[\int_{\mathbb{R}} |\psi(x)|^2 \, dx\right]^2, \end{aligned} \tag{6.1.17}$$

which is a verification of Heisenberg's inequality for the example at hand.

Consider now a proof of (6.1.1). Although we could get by for less we shall assume for convenience (as Weyl does in [87]) that $\psi : \mathbb{R} \to \mathbb{C}$ is continuously differentiable with $\psi, \psi', x\psi \in L^2(\mathbb{R}, dx)$; dx = Lebesgue measure on $\mathbb{R}$. By Hölder's inequality $x\psi\overline{\psi'} \in L^1(\mathbb{R}, dx)$; hence $x\overline{\psi}\psi' \in L^1(\mathbb{R}, dx)$. Similarly by Hölder's inequality $x^2\psi\overline{\psi} \in L^1(\mathbb{R}, dx)$, and we can define three real numbers

$$\begin{aligned} A &= \int_{\mathbb{R}} x^2 |\psi(x)|^2 \, dx, \\ B &= \int_{\mathbb{R}} [x\psi(x)\overline{\psi'(x)} + x\overline{\psi(x)}\psi'(x)] \, dx, \\ C &= \int_{\mathbb{R}} |\psi'(x)|^2 \, dx. \end{aligned} \tag{6.1.18}$$

For $t, x \in \mathbb{R}$ arbitrary

$$\begin{aligned} |tx\psi(x) + \psi'(x)|^2 &= [tx\psi(x) + \psi'(x)]\overline{[tx\psi(x) + \psi'(x)]} \\ &= t^2x^2|\psi(x)|^2 + tx\psi'(x)\overline{\psi}(x) + tx\psi(x)\overline{\psi'}(x) + |\psi'(x)|^2 \end{aligned} \tag{6.1.19}$$

which implies that

$$\int_{\mathbb{R}} |tx\psi(x) + \psi'(x)|^2 \, dx \stackrel{\text{def}}{=} At^2 + Bt + C. \tag{6.1.20}$$

That is, $At^2 + Bt + C \geq 0 \quad \forall\, t \in \mathbb{R} \implies B^2 - 4AC \leq 0$. Note that the integrand for B is a continuous element in $L^2(\mathbb{R}, dx)$ by our hypotheses on ψ. In particular

$$\begin{aligned}
B &= \lim_{\alpha\to\infty} \int_{-\alpha}^{\alpha} [x\psi(x)\overline{\psi'}(x) + x\overline{\psi}(x)\psi'(x)] \, dx \\
&= \lim_{\alpha\to\infty} \int_{-\alpha}^{\alpha} x\frac{d}{dx}[\psi(x)\overline{\psi}(x)] \, dx \\
&= \lim_{\alpha\to\infty} \left\{ [x\psi(x)\overline{\psi}(x)]_{-\alpha}^{\alpha} - \int_{-\alpha}^{\alpha} \psi(x)\overline{\psi}(x) \, dx \right\} \quad \text{(integration by parts)} \\
&= -\int_{\mathbb{R}} |\psi(x)|^2 \, dx
\end{aligned} \tag{6.1.21}$$

as $x|\psi(x)|^2 \to 0$ when $x \to \pm\infty$, by the properties of ψ, a short proof of which can be found in Appendix 1 of [87]. In fact, in general, for $\psi \in L^2(\mathbb{R}, dx)$, $\lim_{\alpha\to\infty} x[|\psi(x)|^2 + |\psi(-x)|^2] = 0$.

The statement $B^2 - 4AC \leq 0$ becomes

$$\left[\int_{\mathbb{R}} |\psi(x)|^2 \, dx\right]^2 \leq 4 \left[\int_{\mathbb{R}} x^2|\psi(x)|^2 \, dx\right] \left[\int_{\mathbb{R}} |\psi'(x)|^2 \, dx\right] \tag{6.1.22}$$

which is (6.1.1) for $a, b = 0$, and which we can generalize a bit as follows. For $a, b \in \mathbb{R}$ arbitrary let $f(x) = e^{iax}\psi(x+b)$. Then $f'(x) = e^{iax}\psi'(x+b) + iae^{iax}\psi(x+b) \implies f, f' \in L^2(\mathbb{R}, dx)$ with f' continuous. Also $xf \in L^2(\mathbb{R}, dx)$:

$$\begin{aligned}
\int_{\mathbb{R}} x^2|f(x)|^2 \, dx &= \int_{\mathbb{R}} x^2|\psi(x+b)|^2 \, dx \\
&= \int_{\mathbb{R}} (x-b)^2|\psi(x)|^2 \, dx \\
&= \int_{\mathbb{R}} (x^2 - 2xb + b^2)|\psi(x)|^2 \, dx < \infty
\end{aligned} \tag{6.1.23}$$

by Hölder's inequality, as above, as $\psi, x\psi \in L^2(\mathbb{R}, dx)$. That is, f satisfies the conditions assumed for ψ. As

$$\int_{\mathbb{R}} |f(x)|^2 \, dx = \int_{\mathbb{R}} |\psi(x+b)|^2 \, dx = \int_{\mathbb{R}} |\psi(x)|^2 \, dx \tag{6.1.24}$$

we conclude by (6.1.22) applied to f

$$\left[\int_{\mathbb{R}} |\psi(x)|^2 \, dx\right]^2 \leq 4 \left[\int_{\mathbb{R}} x^2|\psi(x+b)|^2 \, dx\right] \left[\int_{\mathbb{R}} |f'(x)|^2 \, dx\right] \tag{6.1.25}$$

where

$$\begin{aligned} |f'(x)|^2 &= e^{iax}[\psi'(x+b) + ia\psi(x+b)]e^{-iax}[\overline{\psi'}(x+b) - ia\overline{\psi}(x+b)] \\ &= |\psi'(x+b)|^2 + ia\psi(x+b)\overline{\psi'}(x+b) - ia\psi'(x+b)\overline{\psi}(x+b) \\ &\quad + a^2|\psi(x+b)|^2. \end{aligned} \tag{6.1.26}$$

Applying the integration by parts argument again we get

$$\begin{aligned} \int_{\mathbb{R}} \psi(x)\overline{\psi'}(x)\, dx &= \lim_{\alpha\to\infty} \int_{-\alpha}^{\alpha} \psi(x)\overline{\psi'}(x)\, dx \\ &= \lim_{\alpha\to\infty} \left\{ [\psi(x)\overline{\psi}(x)]_{-\alpha}^{\alpha} - \int_{-\alpha}^{\alpha} \psi'(x)\overline{\psi}(x)\, dx \right\} \\ &= -\int_{\mathbb{R}} \psi'(x)\overline{\psi}(x)\, dx, \end{aligned} \tag{6.1.27}$$

and using again the translation invariance of dx we see by (6.1.26) that

$$\int_{\mathbb{R}} |f'(x)|^2 dx = \int_{\mathbb{R}} |\psi'(x)|^2 dx - 2ia \int_{\mathbb{R}} \psi'(x)\overline{\psi}(x) dx + a^2 \int_{\mathbb{R}} |\psi(x)|^2 dx. \tag{6.1.28}$$

In particular the right-hand side of (6.1.28) is non-negative. By the inequality in (6.1.25) we obtain indeed the Heisenberg inequality in (6.1.1). □

The non-negative factor

$$\begin{aligned} F_\psi &\stackrel{\text{def}}{=} \int_{\mathbb{R}} |\psi'(x)|^2\, dx - 2ia \int_{\mathbb{R}} \psi'(x)\overline{\psi}(x)\, dx + a^2 \int_{\mathbb{R}} |\psi(x)|^2\, dx \\ &= \int_{\mathbb{R}} |f'(x)|^2\, dx \end{aligned} \tag{6.1.29}$$

in (6.1.28) (for $f(x) \stackrel{\text{def}}{=} e^{iax}\psi(x+b)$) can be expressed in an alternate way if we assume a bit more about ψ. Assume, for example, that ψ' satisfies the conditions imposed on ψ: ψ' is continuously differentiable with $\psi', \psi'', x\psi' \in L^2(\mathbb{R}, dx)$. In particular $\psi''\overline{\psi} \in L^1(\mathbb{R}, dx)$ by Hölder's inequality, and as above $|\psi(x)|^2$, $|\psi(-x)|^2$, $|\psi'(x)|^2$, $|\psi'(-x)|^2 \to 0$ as $x \to \infty$. It follows also that $\psi'(x)\overline{\psi}(x)$, $\psi'(-x)\overline{\psi}(-x) \to 0$ as $x \to \infty$. Integration by parts therefore gives

$$\int_{\mathbb{R}} \psi'(x)\overline{\psi'}(x)\, dx = -\int_{\mathbb{R}} \psi''(x)\overline{\psi}(x)\, dx \tag{6.1.30}$$

which permits us to write

$$\begin{aligned} F_\psi &= \int_{\mathbb{R}} [-\psi''(x) - 2ia\psi'(x) + a^2\psi(x)]\overline{\psi}(x)\, dx \\ &= \int_{\mathbb{R}} \left(\left[i\frac{d}{dx} - a1 \right]^2 \psi \right)(x)\overline{\psi}(x)\, dx \end{aligned} \tag{6.1.31}$$

to obtain the following version of Heisenberg's inequality (6.1.1):

Theorem 6.1. *For a sufficiently nice $\psi \in L^2(\mathbb{R}, dx)$ (ψ real or complex-valued) let*

$$F_\psi(a) = \int_{\mathbb{R}} (D_a^2\psi)(x)\overline{\psi}(x)\, dx \tag{6.1.32}$$

for $a \in \mathbb{R}$, where $D_a = i\frac{d}{dx} - a1$; $(D_a^2\psi)(x) = -\psi''(x) - 2ia\psi'(x) + a^2\psi(x)$, $x \in \mathbb{R}$. Then $F_\psi(a) \geq 0$ and for $b \in \mathbb{R}$

$$F_\psi(a)\int_{\mathbb{R}} (x-b)^2|\psi(x)|^2\, dx \geq \frac{1}{4}\left[\int_{\mathbb{R}} |\psi(x)|^2\, dx\right]. \tag{6.1.33}$$

Equation (6.1.33) holds if, for example, $\psi, \psi', \psi'', x\psi, x\psi' \in L^2(\mathbb{R}, dx)$ with ψ'' continuous.

6.2 Heisenberg's Uncertainty Principle

As indicated in Section 1.5 of Chapter 1 the uncertainties Δx, Δp in the position and momentum x, p of an electron are subject to Heisenberg's uncertainty principle

$$\Delta x \Delta p \geq \frac{\hbar}{2}. \tag{6.2.1}$$

Thus as the uncertainty in x decreases, the uncertainty in p increases, and similarly Δx increases as Δp decreases. However the argument which led to (6.2.1) was heuristic only. We did not formally define Δx, Δp in fact. The purpose of this section is to provide a meaning for Δx, Δp and thus to establish (6.2.1) on more precise grounds. We shall provide two formulations of (6.2.1), and two proofs.

To this end suppose in general that ϕ is some distribution function on $\mathbb{R}$:

$$\phi \in L^1(\mathbb{R}, dx) \text{ such that } \int_{\mathbb{R}} \phi(x)\, dx = 1, \quad \phi \geq 0. \tag{6.2.2}$$

If f is a Lebesgue measurable function on $\mathbb{R}$ such that $f\phi \in L^1(\mathbb{R}, dx)$, then we can define the *mean* or *expected* value $\bar{f}$ of f with respect to ϕ by

$$\bar{f} = \int_{\mathbb{R}} f(x)\phi(x)\, dx. \tag{6.2.3}$$

Of course $\bar{f}$ also denotes the complex conjugate of f, which could cause confusion at times. In particular for $\phi = |\psi|^2$ where ψ is some normalized wave function, and for $f(x) = x$ we have, as considered before, the mean value

$$\bar{x} = \int_{\mathbb{R}} x|\psi(x)|^2\, dx \in \mathbb{R}. \tag{6.2.4}$$

For example, compare Section 5.5 of Chapter 5.

There are quantities in classical physics like momentum p that can (at times) be expressed in terms of the coordinate x, or the coordinates x, y, z. This is *not* possible in quantum mechanics however as one would have a violation of Heisenberg's uncertainty principle. The equation analogous to (6.2.4)

$$\bar{p} = \int_{\mathbb{R}} p(x)|\psi(x)|^2 \, dx \tag{6.2.5}$$

thus has no meaning at the quantum level. The following question arises therefore: how should the mean value $\bar{p}$ of p (with respect to $|\psi|^2$) be defined? To answer this question we first quantize the classical observables $x, p : x \to M_x =$ multiplication by x, $p \to \frac{\hbar}{i}\frac{d}{dx}$, as before. Next think of $\bar{x}$ in (6.2.4) as an inner product:

$$\bar{x} = \int_{\mathbb{R}} (M_x\psi)(x)\overline{\psi}(x) \, dx = \langle M_x\psi, \psi\rangle. \tag{6.2.6}$$

Then the quantum analogy of (6.2.6) for p would be

$$\bar{p} \stackrel{\text{def}}{=} \langle \frac{\hbar}{i}\frac{d}{dx}\psi, \psi\rangle = \int_{\mathbb{R}} \frac{\hbar}{i}\left[\frac{d\psi}{dx}(x)\right]\overline{\psi}(x) \, dx. \tag{6.2.7}$$

For the example

$$\psi(x) = \begin{bmatrix} \sqrt{\frac{2}{\alpha}}\sin\frac{\pi}{\alpha}x & x \in [0, \alpha] \\ 0 & x \in \mathbb{R} - [0, \alpha] \end{bmatrix}, \tag{6.2.8}$$

where $\alpha > 0$ (compare definition (6.1.2)),

$$\bar{p} = \frac{\hbar}{i}\frac{2\pi}{\alpha^2}\int_0^\alpha \cos\frac{\pi}{\alpha}x \sin\frac{\pi}{\alpha}x \, dx = 0; \tag{6.2.9}$$

cf. equation (6.1.16).

Note that by (6.1.27)

$$\int_{\mathbb{R}} \psi'(x)\overline{\psi}(x) \, dx = -\int_{\mathbb{R}} \psi(x)\overline{\psi'}(x) \, dx \tag{6.2.10}$$

(for ψ sufficiently nice). Hence in (6.2.7)

$$\begin{aligned} \bar{p} &= -\int_{\mathbb{R}} \frac{\hbar}{i}\psi(x)\overline{\psi'}(x) \, dx = \langle \psi, \frac{\hbar}{i}\frac{d}{dx}\psi\rangle \\ &= \text{the complex conjugate of } \bar{p}. \end{aligned} \tag{6.2.11}$$

That is, $\bar{p} \in \mathbb{R}$; $\frac{\hbar}{i}\frac{d}{dx}$ is "self-adjoint."

Given the notion of the mean value $\bar{f}$ of f with respect to some distribution ϕ as in (6.2.3), one automatically has a corresponding notion of the *variance* $\sigma^2(f)$ of f and the *standard deviation* $\sigma(f)$ of f. Namely

$$\sigma^2(f) \stackrel{\text{def}}{=} \overline{(f - \bar{f})^2}, \qquad \sigma(f) \stackrel{\text{def}}{=} +\sqrt{\sigma^2(f)}. \tag{6.2.12}$$

A formal meaning of the uncertainties Δx, Δp can now be given. For a sufficiently nice wave function ψ, Δx, Δp are the standard deviations

$$\Delta x \stackrel{\text{def}}{=} \sigma(x), \qquad \Delta p \stackrel{\text{def}}{=} \sigma(p). \tag{6.2.13}$$

Thus

$$\begin{aligned} 0 &\leq (\Delta x)^2 \stackrel{\text{def}}{=} \overline{(x-\bar{x})^2} \stackrel{\text{def}}{=} \int_{\mathbb{R}} (x-\bar{x})^2 |\psi(x)|^2 \, dx \\ &= \int_{\mathbb{R}} x^2 |\psi(x)|^2 \, dx - 2\bar{x} \int_{\mathbb{R}} x|\psi(x)|^2 \, dx + (\bar{x})^2 \int_{\mathbb{R}} |\psi(x)|^2 \, dx \\ &= \bar{x^2} - 2\bar{x}\bar{x} + (\bar{x})^2 = \bar{x^2} - (\bar{x})^2 \end{aligned} \tag{6.2.14}$$

and

$$\begin{aligned} (\Delta p)^2 \stackrel{\text{def}}{=} \overline{(p-\bar{p})^2} &\stackrel{\text{def}}{=} \left\langle \left(\frac{\hbar}{i} \frac{d}{dx} - \bar{p}1 \right)^2 \psi, \psi \right\rangle \\ &= \int_{\mathbb{R}} \left[\left(\frac{\hbar}{i} \frac{d}{dx} - \bar{p}1 \right)^2 \psi \right] (x) \overline{\psi}(x) \, dx, \end{aligned} \tag{6.2.15}$$

since the quantization rule $p \to \frac{\hbar}{i}\frac{d}{dx}$ leads to $(p-\bar{p})^2 \to \left(\frac{\hbar}{i}\frac{d}{dx} - \bar{p}1\right)^2$; cf. (3.9.2). Choose $a = -\bar{p}/\hbar \in \mathbb{R}$ and note that

$$a\left(\frac{\hbar}{i} \frac{d}{dx} - \bar{p}1 \right) = \bar{p} \left(i \frac{d}{dx} - a1 \right) \tag{6.2.16}$$

$$\Longrightarrow a^2 \left(\frac{\hbar}{i} \frac{d}{dx} - \bar{p}1 \right)^2 = \bar{p}^2 \left(i \frac{d}{dx} - a1 \right)^2, \qquad \text{or}$$

$$\left(\frac{\hbar}{i} \frac{d}{dx} - \bar{p}1 \right)^2 = \hbar^2 \left(i \frac{d}{dx} - a1 \right)^2 \tag{6.2.17}$$

$$\Longrightarrow (\Delta p)^2 = \hbar^2 \int_{\mathbb{R}} \left[\left(i \frac{d}{dx} - a1 \right)^2 \psi \right] (x) \overline{\psi}(x) \, dx. \tag{6.2.18}$$

By Theorem 6.1 we see that $(\Delta p)^2 \geq 0$, so that $\Delta p \stackrel{\text{def}}{=} +\sqrt{(\Delta p)^2}$ is indeed well defined, and that for the choice $b = \bar{x} \in \mathbb{R}$ we have (using (6.1.33)) $(\Delta x)^2 (\Delta p)^2 / \hbar^2 \geq 1/4$. Hence we have proved the following.

Theorem 6.2 (Heisenberg's Uncertainty Principle). *Let $\psi \in L^2(\mathbb{R}, dx)$ satisfy the conditions of Theorem 6.1 and satisfy*

$$\int_{\mathbb{R}} |\psi(x)|^2 \, dx = 1.$$

Let $\bar{x}$, $\bar{p} \in \mathbb{R}$ be the mean values with respect to ψ (or expectation values) defined in (6.2.6) *and* (6.2.7), *interpreted physically as the quantum mean values of the* *position and the momentum x, p of some particle. Define the* uncertainties Δx, Δp *(with respect to* ψ*) by equations* (6.2.13)*:* $\Delta x \stackrel{\text{def}}{=} +\sqrt{\overline{(\Delta x)^2}}$, $\Delta p = +\sqrt{\overline{(\Delta p)^2}}$*; indeed* $\overline{(\Delta x)^2}, \overline{(\Delta p)^2} \geq 0$*. Then* $\Delta x, \Delta p$ *satisfy the inequality*

$$(\Delta x)(\Delta p) \geq \frac{\hbar}{2}. \tag{6.2.19}$$

We shall see that the continuity of ψ' in Theorems 6.1, 6.2 need not be assumed. In fact see (6.2.38) for a more general version of Theorem 6.2. Also see (6.2.33).

The Heisenberg uncertainty principle (6.2.19) is, by far, one of the most fundamental principles in all of quantum theory, especially as this theory is founded on laws of measurement. In some texts the uncertainty principle is stated (more or less precisely) as follows: Two observables O_1, O_2 can be simultaneously (accurately) measured only if their commutator $[O_1, O_2]$ vanishes; i.e., only if O_1 and O_2 commute. If $[O_1, O_2] \neq 0$, then these measurements can be made only with an uncertainty $(\Delta O_1)(\Delta O_2)$, the magnitude of which is dependent on the commutator $[O_1, O_2]$. For $O_1 = M_x$ and $O_2 = \frac{\hbar}{i}\frac{d}{dx}$ as above, for example, $-\frac{i}{2}[O_1, O_2] = \frac{\hbar}{2}1$, as we have seen in Section 1.3 of Chapter 1. Thus in (6.2.19) we indeed see the dependence of $(\Delta x)(\Delta p)$ on the value of the commutator $[M_x, \frac{\hbar}{i}\frac{d}{dx}]$ and that in particular since $[M_x, \frac{\hbar}{i}\frac{d}{dx}] \neq 0$ the position and momentum of a particle, at the quantum microscopic level, cannot be measured simultaneously with complete accuracy—a statement which is in stark contrast to classical mechanics situations where several dynamic quantities can be simultaneously measured with full accuracy.

Among the many philosophical implications of quantum mechanics one of the most striking is the nullification of *classical determinism*, wherein a system's evolution is determined by initial data and acting forces. Newton's deterministic laws predict future events from past knowledge. Given the positions and momenta of a system of mass particles at some initial time $t = t_o$ and the forces of interaction of these particles, for example, the state of the system at any future time t can be determined exactly. Quantum mechanics, on the other hand, points to indeterminacy as a feature of our universe. There are measurements of physical systems which are not determined by prior states of the system. In a scattering experiment, for example, it is not possible to know precisely where a particle will land after it meets the scattering object. We noted earlier that one cannot determine in advance at what point an electron passing through a small slit will strike a screen beyond that slit, although one can determine an interval $(-x_1, x_1)$ on the screen where the probability of the electron striking therein is maximal; recall the discussion in Section 1.5 of Chapter 1. The uncertainty principle highlights the importance of M. Born's basic postulate, which introduces indeterminacy into quantum mechanics at the outset.

It is common to employ the *Fourier transform* in the formulation of Heisenberg's inequality, and thus of his uncertainty principle. We indicate how this is done, assuming some of the basic elements of Fourier analysis stated here without proof. In fact we give a second proof of the basic inequality (6.1.1).

There are various definitions (normalizations) of the Fourier transform, each chosen for a particular convenience. For $\psi \in L^1(\mathbb{R}, dx)$ we will set

$$\widehat{\psi}(x) \stackrel{\text{def}}{=} \int_{\mathbb{R}} e^{ixt}\psi(t)\, dt. \tag{6.2.20}$$

With this normalization ψ is recovered from its Fourier transform $\widehat{\psi}$ via the *inversion formula*

$$\psi(x) = \frac{1}{2\pi}\int_{\mathbb{R}} e^{-ixt}\widehat{\psi}(t)\, dt \tag{6.2.21}$$

for almost all $x \in \mathbb{R}$, assuming that $\widehat{\psi} \in L^1$. Equation (6.2.21) holds at a point x of continuity of ψ (for $\widehat{\psi} \in L^1$). A key result is *Plancherel's* formula.

Theorem 6.3. *Suppose $\psi \in L^1(\mathbb{R}, dx) \cap L^2(\mathbb{R}, dx)$. Then $\widehat{\psi} \in L^2(\mathbb{R}, dx)$ and*

$$\|\widehat{\psi}\|_2^2 \stackrel{\text{def}}{=} \int_{\mathbb{R}} |\widehat{\psi}(x)|^2\, dx = 2\pi \int_{\mathbb{R}} |\psi(x)|^2\, dx \stackrel{\text{def}}{=} 2\pi\|\psi\|_2^2.$$

The *Riemann–Lebesgue Lemma* states that $\widehat{\psi}$ is a *continuous* function on $\mathbb{R}$ such that

$$\lim_{x\to\pm\infty} \widehat{\psi}(x) = 0. \tag{6.2.22}$$

Theorem 6.4.

(i) *Let $\psi \in L^1 \stackrel{\text{def}}{=} L^1(\mathbb{R}, dx)$ such that $x\psi \in L^1$. Then $\widehat{\psi}$ is differentiable on $\mathbb{R}$ and $\widehat{\psi}\,'(x) = i\widehat{x\psi}(x)$.*

(ii) *Let $\psi \in L^1$ be differentiable on $\mathbb{R}$ such that $\psi' \in L^1$. Then* $\lim_{x\to\pm\infty} \psi(x) =$ 0, *and $\widehat{\psi}\,'(x) = -ix\widehat{\psi}(x)$.*

(iii) *For $\psi \in L^1$, $a, b \in \mathbb{R}$ define $f_\psi(x) = e^{iax}\psi(x+b)$ on $\mathbb{R}$. Then $f_\psi \in L^1$ and $\widehat{f_\psi}(x) = e^{-ib(x+a)}\widehat{\psi}(x+a)$ [by the translation invariance of dx].*

Assume now that $\psi : \mathbb{R} \to \mathbb{C}$ is differentiable on $\mathbb{R}$ such that $\psi, \psi', x\psi \in L^2$. By Hölder's inequality $\psi'(x\overline{\psi}) \in L^1$. We integrate by parts:

$$\begin{aligned}\int_{\mathbb{R}} \psi'(x)x\overline{\psi}(x)\, dx &= \lim_{\alpha\to\infty} \int_{-\alpha}^{\alpha} \psi'(x)x\overline{\psi}(x)\, dx \\ &= \lim_{\alpha\to\infty} \left\{ \left[\psi(x)x\overline{\psi}(x)\right]_{-\alpha}^{\alpha} - \int_{-\alpha}^{\alpha} [\psi(x)(x\overline{\psi}'(x) + \overline{\psi}(x))]\, dx \right\}.\end{aligned} \tag{6.2.23}$$

Again by Hölder's inequality $\psi(x\overline{\psi}) \in L^1$; also $[\psi(x\overline{\psi})]' = \psi'(x\overline{\psi}) + \psi(x\overline{\psi}') + \psi\overline{\psi} \in L^1$ (since $\psi(x\overline{\psi}') = (x\psi)\overline{\psi}'$). Therefore, by **(ii)** of Theorem 6.4, $\lim_{x\to\pm\infty} \psi(x)x\overline{\psi}(x) = 0$; hence by (6.2.23)

$$\int_{\mathbb{R}} \psi'(x)x\overline{\psi}(x)\, dx = -\int_{\mathbb{R}} \psi(x)x\overline{\psi'}(x)\, dx - \int_{\mathbb{R}} |\psi(x)|^2\, dx. \tag{6.2.24}$$

That is,

$$\int_{\mathbb{R}} |\psi(x)|^2\, dx = -2 \operatorname{Re} \int_{\mathbb{R}} \psi'(x)\overline{x\psi(x)}\, dx \tag{6.2.25}$$

which with the Cauchy–Schwarz inequality gives

$$\begin{aligned}\left[\int_{\mathbb{R}} |\psi(x)|^2\, dx\right]^2 &= 4(\operatorname{Re}\langle\psi', x\psi\rangle)^2 \le 4|\langle\psi', x\psi\rangle|^2 \\ &\le 4\|\psi'\|_2^2\|x\psi\|_2^2 = 4\int_{\mathbb{R}} |\psi'(x)|^2\, dx \cdot \int_{\mathbb{R}} x^2|\psi(x)|^2\, dx.\end{aligned} \tag{6.2.26}$$

By considering $f_\psi(x) \stackrel{\text{def}}{=} e^{iax}\psi(x+b)$ for $a, b \in \mathbb{R}$ we have already noted that the inequality (6.2.26) implies more generally that

$$\begin{aligned}\left[\int_{\mathbb{R}} |\psi(x)|^2\, dx\right]^2 \le 4 &\left[\int_{\mathbb{R}} (x-b)^2|\psi(x)|^2\, dx\right] \\ &\cdot \int_{\mathbb{R}} [|\psi'(x)|^2 - 2ia\psi'(x)\overline{\psi}(x) + a^2|\psi(x)|^2]\, dx,\end{aligned} \tag{6.2.27}$$

which is Heisenberg's inequality (6.1.1) (without a continuity assumption on ψ'). Here again the second integral on the right-hand side of (6.2.27) is

$$\int_{\mathbb{R}} |f'_\psi(x)|^2\, dx \ge 0. \tag{6.2.28}$$

Assume in addition that $\psi, \psi' \in L^1$: $\psi, \psi' \in L^1 \cap L^2$. Then $f'_\psi(x) = e^{iax}\psi'(x+b) + \psi(x+b)e^{iax}ia \implies f_\psi, f'_\psi \in L^1 \cap L^2$. By Plancherel's theorem, Theorem 6.3, $\widehat{f'_\psi}, \widehat{\psi} \in L^2$ and

$$\|\widehat{f'_\psi}\|^2 = 2\pi\|f'_\psi\|_2^2, \qquad \|\widehat{\psi}\|_2^2 = 2\pi\|\psi\|_2^2. \tag{6.2.29}$$

By **(ii)**, **(iii)** of Theorem 6.4 $\widehat{f'_\psi}(x) = -ix\widehat{f_\psi}(x) = -ixe^{-ib(x+a)}\widehat{\psi}(x+a) \implies$

$$\|\widehat{f'_\psi}\|_2^2 = \int_{\mathbb{R}} x^2|\widehat{\psi}(x+a)|^2\, dx = \int_{\mathbb{R}} (x-a)^2|\widehat{\psi}(x)|^2\, dx, \tag{6.2.30}$$

so that

$$\begin{aligned}&\int_{\mathbb{R}} [|\psi'(x)|^2 - 2ia\psi'(x)\overline{\psi}(x) + a^2|\psi(x)|^2]\, dx \\ &\quad= \int_{\mathbb{R}} |f'_\psi(x)|^2\, dx = \frac{1}{2\pi}\|\widehat{f'_\psi}\|_2^2 = \frac{1}{2\pi}\int_{\mathbb{R}} (x-a)^2|\widehat{\psi}(x)|^2\, dx,\end{aligned} \tag{6.2.31}$$

which allows us to express the Heisenberg inequality (6.1.1) as

$$\left[\int_{\mathbb{R}} |\psi(x)|^2\, dx\right]^2 \leq \frac{2}{\pi}\left[\int_{\mathbb{R}} (x-b)^2|\psi(x)|^2\, dx\right]\left[\int_{\mathbb{R}} (x-a)^2|\widehat{\psi}(x)|^2\, dx\right], \tag{6.2.32}$$

or more symmetrically as

$$\begin{aligned}\left[\int_{\mathbb{R}} |\psi(x)|^2\, dx\right]&\left[\int_{\mathbb{R}} |\widehat{\psi}(x)|^2\, dx\right]\\ &\leq 4\left[\int_{\mathbb{R}} (x-b)^2|\psi(x)|^2\, dx\right]\left[\int_{\mathbb{R}} (x-a)^2|\widehat{\psi}(x)|^2\, dx\right]\end{aligned} \tag{6.2.33}$$

by (6.2.29). The validity of (6.2.33) has been established for $\psi : \mathbb{R} \to \mathbb{C}$ differentiable which satisfies the conditions $\psi, \psi' \in L^1(\mathbb{R}, dx) \cap L^2(\mathbb{R}, dx)$, $x\psi \in L^2(\mathbb{R}, dx)$, though it could be established for more general ψ.

We consider also the dilation of the Fourier transform

$$\widehat{\psi_\hbar}(p) \stackrel{\text{def}}{=} \frac{1}{\sqrt{2\pi\hbar}}\widehat{\psi}\left(\frac{p}{\hbar}\right) \stackrel{\text{def}}{=} \frac{1}{\sqrt{2\pi\hbar}}\int_{\mathbb{R}} e^{\frac{ip}{\hbar}x}\psi(x)\, dx \tag{6.2.34}$$

for $\psi \in L^1$, $p \in \mathbb{R}$. If $\psi \in L^1 \cap L^2$, then by Theorem 6.3, $\widehat{\psi_\hbar} \in L^2$ and

$$\begin{aligned}||\widehat{\psi_\hbar}||_2^2 &\stackrel{\text{def}}{=} \frac{1}{2\pi\hbar}\int_{\mathbb{R}} |\widehat{\psi}\left(\frac{p}{\hbar}\right)|^2\, dp\\ &= \frac{1}{2\pi}\int_{\mathbb{R}} |\widehat{\psi}(p)|^2\, dp = ||\psi||_2^2.\end{aligned} \tag{6.2.35}$$

Similarly

$$\begin{aligned}\int_{\mathbb{R}} (p-a)^2|\widehat{\psi_\hbar}(p)|^2\, dp &= \frac{1}{2\pi\hbar}\int_{\mathbb{R}} (p-a)^2\left|\widehat{\psi}\left(\frac{p}{\hbar}\right)\right|^2 dp\\ &= \frac{1}{2\pi}\int_{\mathbb{R}} (p\hbar-a)^2|\widehat{\psi}(p)|^2\, dp = \frac{\hbar^2}{2\pi}\int_{\mathbb{R}}\left(p-\frac{a}{\hbar}\right)^2|\widehat{\psi}(p)|^2\, dp.\end{aligned} \tag{6.2.36}$$

By equations (6.2.35), (6.2.36) and the inequality (6.2.33)

$$\begin{aligned}4&\left[\int_{\mathbb{R}} (x-b)^2|\psi(x)|^2\, dx\right]\left[\int_{\mathbb{R}} (p-a)^2|\widehat{\psi_\hbar}(p)|^2\, dp\right]\\ &= 4\left[\int_{\mathbb{R}} (x-b)^2|\psi(x)|^2\, dx\right]\frac{\hbar^2}{2\pi}\left[\int_{\mathbb{R}}\left(p-\frac{a}{\hbar}\right)^2|\widehat{\psi}(p)|^2\, dp\right]\\ &\geq \hbar^2\left[\int_{\mathbb{R}} |\psi(x)|^2\, dx\right]\frac{1}{2\pi}\left[\int_{\mathbb{R}} |\widehat{\psi}(p)|^2\, dp\right]\\ &= \hbar^2\left[\int_{\mathbb{R}} |\psi(x)|^2\, dx\right]\left[\int_{\mathbb{R}} |\widehat{\psi_\hbar}(p)|^2\, dp\right],\end{aligned} \tag{6.2.37}$$

or

$$\frac{\left[\int_{\mathbb{R}}(x-b)^2|\psi(x)|^2\,dx\right]}{\left[\int_{\mathbb{R}}|\psi(x)|^2\,dx\right]}\frac{\left[\int_{\mathbb{R}}(p-a)^2|\widehat{\psi_\hbar}(p)|^2\,dp\right]}{\left[\int_{\mathbb{R}}|\widehat{\psi_\hbar}(p)|^2\,dp\right]}\geq\frac{\hbar^2}{4}, \tag{6.2.38}$$

which is a slightly more general version of the Heisenberg uncertainty principle. Equation (6.2.38) is established for $\psi\in L^1(\mathbb{R},dx)\cap L^2(\mathbb{R},dx)$ differentiable such that $\psi'\in L^1(\mathbb{R},dx)\cap L^2(\mathbb{R},dx)$, $x\psi\in L^2(\mathbb{R},dx)$.

6.3 Example

Let ψ_n be the normalized wave function of Chapter 4 for the harmonic oscillator of frequency ν:

$$\begin{aligned}\psi_n(x)&=c_n\exp\left(-\frac{b}{2}x^2\right)H_n\left(\sqrt{b}x\right),\\ \text{for } c_n&=(-1)^n\left[\sqrt{\frac{b}{\pi}}\frac{1}{2^n\,n!}\right]^{1/2},b=\frac{2\pi\nu m}{\hbar},\end{aligned} \tag{6.3.1}$$

$n=0,1,2,3,\ldots$; see definition (4.8.5) there. By (4.8.7), $\psi_n(-x)=(-1)^n\psi_n(x)$ $\implies |\psi_n(-x)|^2=|\psi_n(x)|^2\implies$

$$\bar{x}=\int_{\mathbb{R}}x|\psi_n(x)|^2\,dx=0 \tag{6.3.2}$$

as the integrand is an odd function. To compute $\bar{p}$, consider

$$\begin{aligned}\psi_n'(x)&=c_n\left[\exp\left(-\frac{b}{2}x^2\right)\sqrt{b}\,H_n'\left(\sqrt{b}x\right)-bx\exp\left(-\frac{b}{2}x^2\right)H_n\left(\sqrt{b}x\right)\right]\\&=c_n\exp\left(-\frac{b}{2}x^2\right)\left[-\sqrt{b}\,H_{n+1}\left(\sqrt{b}x\right)+2bx\,H_n\left(\sqrt{b}x\right)-bx\,H_n\left(\sqrt{b}x\right)\right]\end{aligned}$$

(by Corollary 4.2)

$$\begin{aligned}&=c_n\exp\left(-\frac{b}{2}x^2\right)\left[-\sqrt{b}\,H_{n+1}\left(\sqrt{b}x\right)+bx\,H_n\left(\sqrt{b}x\right)\right]\\&=-\frac{c_n\sqrt{b}}{c_{n+1}}\psi_{n+1}(x)+bx\psi_n(x).\end{aligned} \tag{6.3.3}$$

By wave function orthogonality (4.8.6) and equation (6.3.3)

$$\begin{aligned}\bar{p}&\overset{\text{def}}{=}-\hbar i\int_{\mathbb{R}}\psi_n'(x)\psi_n(x)\,dx\\&=-\hbar i\int_{\mathbb{R}}bx\psi_n(x)\psi_n(x)\,dx\\&=-\hbar ib\bar{x}=0 \qquad\qquad \text{by (6.3.2).}\end{aligned} \tag{6.3.4}$$

Then by definition (6.2.15)

$$(\Delta p)^2 = \int_{\mathbb{R}} \left[\left(\frac{\hbar}{i}\frac{d}{dx} - \bar{p} \right)^2 \psi_n \right](x)\, \psi_n(x)\, dx \\ = -\hbar^2 \int_{\mathbb{R}} \psi_n''(x)\psi_n(x)\, dx. \tag{6.3.5}$$

But ψ_n is a solution of Schrödinger 's equation

$$\frac{d^2\psi}{dx^2} + \frac{2m}{\hbar^2}(E - 2\pi^2\nu^2 m x^2)\psi = 0 \tag{6.3.6}$$

for $E = E_n = 2\pi(n+\frac{1}{2})\nu\hbar$, as we know. Thus, since ψ_n is normalized, one has by Theorem 4.11

$$\begin{aligned}(\Delta p)^2 &= 2m \int_{\mathbb{R}} (E_n - 2\pi^2\nu^2 m x^2)\psi_n(x)\psi_n(x)\, dx \\ &= 2mE_n - \hbar^2 b^2 \int_{\mathbb{R}} x^2\psi_n^2(x)\, dx \\ &= 2mE_n - \frac{\hbar^2 b^2 E_n}{4\pi^2\nu^2 m} = 2mE_n - mE_n \\ &= mE_n.\end{aligned} \tag{6.3.7}$$

Again since $\bar{x} = 0$

$$(\Delta x)^2 \stackrel{\text{def}}{=} \int_{\mathbb{R}} (x - \bar{x})^2 |\psi_n(x)|^2\, dx \tag{6.3.8}$$

$$= \int_{\mathbb{R}} x^2\psi_n^2(x)\, dx = \frac{E_n}{4\pi^2\nu^2 m} \tag{6.3.9}$$

as in (6.3.7). We can conclude that

$$(\Delta x)(\Delta p) = \frac{\sqrt{E_n}}{2\pi\nu\sqrt{m}}\sqrt{mE_n} = \frac{E_n}{2\pi\nu} = \frac{\hbar}{2}(2n+1), \tag{6.3.10}$$

which verifies Heisenberg's uncertainty principle, $(\Delta x)(\Delta p) \geq \frac{\hbar}{2}$, for every quantum state ψ_n of the harmonic oscillator. Moreover we see that equality $(\Delta x)(\Delta p) = \frac{\hbar}{2}$ is attained $\iff$ $n = 0$; i.e., only for the ground state $\psi_o(x) = \left(\frac{b}{\pi}\right)^{1/4} \exp(-\frac{b}{2}x^2)$ is equality attained.

While we are at it, we can also compute the mean value $\bar{V}$ of the potential energy V for any quantum state ψ_n. According to definition (6.2.3)

$$\bar{V} = \int_{\mathbb{R}} V(x)|\psi_n(x)|^2\, dx \tag{6.3.11}$$

$$= 2\pi^2\nu^2 m \int_{\mathbb{R}} x^2\psi_n^2(x)\, dx = \frac{E_n}{2}, \tag{6.3.12}$$

by (6.3.9). That is, the mean value of the potential energy of a state ψ_n is half the quantum energy of that state.

7

Group Representations and Selection Rules

7.1 Group Representations

One of the main efforts of Chapter 5 was to determine explicitly the normalized wave functions ψ_{nlm} of a hydrogenic atom; see equation (5.2.12). Here $n \in \{1, 2, 3, \ldots\}, l \in \{0, 1, 2, \ldots, n-1\}, m \in \{-l, -l+1, \ldots, -1, 0, 1, 2, \ldots, l\}$ are the total, azimuthal, and magnetic quantum numbers, respectively. The triples $(n, l, m) = (5, 2, -2), (5, 4, -2)$ define, for example, two quantum states with the *same* energy $E^{(5)}$ (by (5.1.29)) of the hydrogenic atom. One could ask the following question: Is it possible for the atom to proceed directly from the state $(5, 2, -2)$ to the state $(5, 4, -2)$: is the transition $(5, 2, -2) \to (5, 4, -2)$ always physically possible? The answer is *no*, as we shall see in this chapter. Moreover the answer comes by way of some beautiful applications of group theory. This example, of whether a certain transition is allowable or forbidden, illustrates one of many *selection rules* in quantum mechanics. The answer *no* just asserted turns out to depend on whether or not a certain integral (depending on the quantum numbers $(5, 2, -2), (5, 4, -2)$) vanishes.

In this chapter therefore we shall consider very broadly, and abstractly, a selection rule to be some rule or means by which one can ascertain in advance that certain integrals vanish. From this viewpoint the familiar statement

$$\int_{-a}^{a} f(x)\, dx = 0 \text{ for } f \text{ continuous on } [-a, a] \text{ with } f(-x) = -f(x) \qquad (7.1.1)$$

would be an example of a selection rule. Our goal is to formulate, purely in terms of group theory, a general abstract selection rule—which indeed will have concrete applications. This we do in the main theorem, Theorem 7.2. Also see in fact Theorem 7.6. In Section 7.7 we specialize our discussion and consider, more concretely, *spectroscopic* selection rules. To prepare for Theorem 7.2 we review some elementary definitions and facts about group representation theory—material available in many standard texts, collected here for the reader's convenience.

Let G denote a fixed abstract group. A *representation* of G on a finite-dimensional vector space V over some field $\mathbb{F}$ is a group homomorphism $\pi : G \to \mathrm{Gl}(V)$ from G to the group of non-singular linear operators $\mathrm{Gl}(V)$ on V. Thus $\pi(x_1x_2) = \pi(x_1) \circ \pi(x_2)$ for $x_1, x_2 \in G$. In practice G will be a topological group (even a Lie group) and V will be defined over $\mathbb{F} = \mathbb{R}$ or $\mathbb{C}$, in which case π will be required to be *continuous*: if V^* denotes the dual space of V then for each pair $(v, f) \in V \times V^*$ the map $\pi_{v,f} : G \to \mathbb{F}$ given by

$$\pi_{v,f}(x) = f(\pi(x)v) \text{ for } x \in G \tag{7.1.2}$$

is continuous. The *degree* or *dimension* of π is the dimension of V over $\mathbb{F}$. Some examples of continuous representations are as follows.

1. Take $G = \mathbb{R}, V = \mathbb{R}^2$ and define $\pi : \mathbb{R} \to \mathrm{Gl}(\mathbb{R}^2)$ by $\pi(x)(a, b) = (a + bx, b)$ for $x \in G$, $(a, b) \in \mathbb{R}^2$. Clearly $\pi(x) : \mathbb{R}^2 \to \mathbb{R}^2$ is a non-singular linear map such that $\pi(x_1 + x_2) = \pi(x_1) \circ \pi(x_2)$ for $x_1, x_2 \in \mathbb{R}$. If $f \in (\mathbb{R}^2)^*$, then for $e_1 = (1, 0)$, $e_2 = (0, 1)$, $(a_1, b_1) \in \mathbb{R}^2$ one has $f((a_1, b_1)) = a_1 f(e_1) + b_1 f(e_2)$. Therefore $\pi_{(a,b),f} : \mathbb{R} \to \mathbb{R}$ in equation (7.1.2) is given by $\pi_{(a,b),f}(x) \stackrel{\text{def}}{=} f(a+bx, b) = (a+bx)f(e_1)+bf(e_2)$, which is continuous in x. Thus π is a continuous 2-dimensional representation of $\mathbb{R}$ on $\mathbb{R}^2$.

2. Take $G = \mathbb{R}/2\pi\mathbb{Z}$ (= the circle group S^1) and $V = \mathbb{C}$; $\mathbb{Z}$ = the ring of integers. For each fixed $n \in \mathbb{Z}$ define $\pi = \pi^{(n)} : \mathbb{R}/2\pi\mathbb{Z} \to \mathrm{Gl}(\mathbb{C})$ by $\pi^{(n)}(\overline{x})z = e^{inx}z$ for $z \in \mathbb{C}$, $\overline{x}$ = the class $x + 2\pi\mathbb{Z}$ of $x \in \mathbb{R}$. The non-singular linear operator $\pi^{(n)}(\overline{x}) : \mathbb{C} \to \mathbb{C}$ depends only on the class $\overline{x}$ and not the representative x of the class, since $e^{2\pi im} = 1 \ \forall\ m \in \mathbb{Z}$. For $(z, f) \in \mathbb{C} \times \mathbb{C}^*$ the map $\pi^{(n)}_{z,f} : \mathbb{R}/2\pi\mathbb{Z} \to \mathbb{C}$ in equation (7.1.2) is given by $\pi^{(n)}_{z,f}(\overline{x}) \stackrel{\text{def}}{=} f(e^{inx}z) = f(z)e^{inx}$, which is continuous, since the composition $\mathbb{R} \to \mathbb{R}/2\pi\mathbb{Z} \xrightarrow{\pi^{(n)}_{z,f}} \mathbb{C}$ $(x \to f(z)e^{inx})$ is continuous. Each $n \in \mathbb{Z}$ therefore gives rise to a continuous 1-dimensional representation $\pi^{(n)}$ of $G = \mathbb{R}/2\pi\mathbb{Z}$ on $\mathbb{C}$. With respect to the standard inner product $\langle\ ,\ \rangle$ on $\mathbb{C}$ (i.e., $\langle z_1, z_2\rangle \stackrel{\text{def}}{=} z_1\overline{z_2}$) each $\pi^{(n)}$ is *unitary*: $\langle \pi^{(n)}(\overline{x})z_1, \pi^{(n)}(\overline{x})z_2\rangle = \langle z_1, z_2\rangle$ for $\overline{x} \in G$, $z_1, z_2 \in \mathbb{C}$. It is not hard to show that the family $\{\pi^{(n)}\}_{n\in\mathbb{Z}}$ gives *all* continuous, 1-dimensional, unitary representations of G, up to unitary equivalence. Namely if $\pi : G \to \mathrm{Gl}(V)$ is any continuous, 1-dimensional unitary representation of G, then for some $n \in \mathbb{Z}$ there is a unitary transformation $U : V \to \mathbb{C}$ such that the diagram

$$\begin{array}{ccc} V & \xrightarrow{\pi(\overline{x})} & V \\ {\scriptstyle U}\downarrow & & \downarrow{\scriptstyle U} \\ \mathbb{C} & \xrightarrow[\pi^{(n)}(\overline{x})]{} & \mathbb{C} \end{array}$$

commutes $\forall\, \overline{x} \in G$.

3. Let $G = \mathrm{O}(n)$ be the topological group of real $n \times n$ orthogonal matrices: $AA^t = 1$ for $1 =$ the $n \times n$ identity matrix, where A^t is the transpose of $A \in G$. For $V = \mathbb{R}^n$ or $\mathbb{C}^n$ let $\pi : G \to \mathrm{Gl}(V)$ be the canonical map given by $\pi(A)(x_1, \ldots, x_n) = (x_1, \ldots, x_n)A^t$ for $x = (x_1, \ldots, x_n) \in V$. Then π is a representation of G on V. Let $e_i = (0, \ldots, 1, 0, \ldots, 0) \in V$ where 1 occurs in the ith slot. Similar to example 1, one has $f(x) = \sum_{i=1}^n x_i f(e_i)$ for any $f \in V^*$ so that for $v = (v_1, \ldots, v_n) \in V$ the map $\pi_{v,f} : \mathrm{O}(n) \to \mathbb{R}$ or $\mathbb{C}$ in equation (7.1.2) is given by $\pi_{v,f}(A) = \sum_{i=1}^n x_i f(e_i)$, where $x = vA^t$. That is, $x_i = \sum_{j=1}^n v_j a_{ij}$ for $A = [a_{ij}] \implies \pi_{v,f}(A) = \sum_{i,j=1}^n v_j f(e_i) a_{ij}$ is a continuous function of $A \implies \pi$ is a continuous n-dimensional representation. In particular if $\mathrm{SO}(n)$ is the subgroup of $\mathrm{O}(n)$ of matrices A with $\det A = 1$ ($\mathrm{SO}(n)$ is called the *special* orthogonal group), then by restriction we obtain the continuous representation $\pi : \mathrm{SO}(n) \to \mathrm{Gl}(V)$ of $\mathrm{SO}(n)$ on V.

4. Let π be the continuous representation of $G = \mathrm{O}(3)$ on $\mathbb{C}^3$ given in example 3. Given any function $\phi : \mathbb{R}^3 \to \mathbb{R}$ or $\mathbb{C}$ define $A \cdot \phi : \mathbb{R}^3 \to \mathbb{R}$ or $\mathbb{C}$ for $A \in G$ by

$$(A \cdot \phi)(x, y, z) \stackrel{\text{def}}{=} \phi(\pi(A^{-1})(x, y, z)) \stackrel{\text{def}}{=} \phi((x, y, z)A), \tag{7.1.3}$$

since $A^{-1} = A^t$. In particular for the projections $p_1, p_2, p_3 : \mathbb{R}^3 \to \mathbb{R}$ onto the first, second, third coordinates respectively, $A = [a_{ij}] \implies (x, y, z)A = (a_{11}x + a_{21}y + a_{31}z, a_{12}x + a_{22}y + a_{32}z, a_{13}x + a_{23}y + a_{33}z) \implies (A \cdot p_1)(x, y, z) = a_{11}x + a_{21}y + a_{31}z = a_{11}p_1(x, y, z) + a_{21}p_2(x, y, z) + a_{31}p_3(x, y, z) \implies$

$$A \cdot p_1 = a_{11}p_1 + a_{21}p_2 + a_{31}p_3. \tag{7.1.4}$$

Similarly,

$$A \cdot p_2 = a_{12}p_1 + a_{22}p_2 + a_{32}p_3 \quad \text{and} \tag{7.1.5}$$

$$A \cdot p_3 = a_{13}p_1 + a_{23}p_2 + a_{33}p_3. \tag{7.1.6}$$

Let $\mathrm{O}(3) \cdot p_j$, $\langle p_1, p_2, p_3 \rangle$ denote the complex subspaces of $C(\mathbb{R}^3, \mathbb{C})$ (the space of $\mathbb{C}$-valued continuous functions on $\mathbb{R}^3$) spanned by the vectors $\{A \cdot p_j | A \in \mathrm{O}(3)\}$, $\{p_1, p_2, p_3\}$ respectively; $1 \le j \le 3$. By equations (7.1.4), (7.1.5), (7.1.6) we have

$$\mathrm{O}(3) \cdot p_j \subset \langle p_1, p_2, p_3 \rangle. \tag{7.1.7}$$

Consider

$$A_1 \stackrel{\text{def}}{=} \begin{bmatrix} 0 & -1 & 0 \\ 1 & 0 & 0 \\ 0 & 0 & 1 \end{bmatrix}, A_2 \stackrel{\text{def}}{=} \begin{bmatrix} 0 & 0 & -1 \\ 0 & 1 & 0 \\ 1 & 0 & 0 \end{bmatrix}, A_3 \stackrel{\text{def}}{=} \begin{bmatrix} 1 & 0 & 0 \\ 0 & 0 & -1 \\ 0 & 1 & 0 \end{bmatrix} \in \mathrm{O}(3). \tag{7.1.8}$$

Note in fact that $A_1, A_2, A_3 \in \mathrm{SO}(3)$. By (7.1.4), $A_1 \cdot p_1 = p_2$, $A_2 \cdot p_1 = p_3$, $1 \cdot p_1 = p_1 \implies p_1, p_2, p_3 \in \mathrm{SO}(3) \cdot p_1 \implies \langle p_1, p_2, p_3\rangle \subset \mathrm{SO}(3) \cdot p_1 \implies \langle p_1, p_2, p_3\rangle = \mathrm{SO}(3) \cdot p_1 = \mathrm{O}(3) \cdot p_1$ by (7.1.7). Similarly $A_3 \cdot p_2 = p_3$, $A_1 \cdot p_2 = -p_1$, $1 \cdot p_2 = p_2 \implies p_1, p_2, p_3 \in \mathrm{SO}(3) \cdot p_2 \implies \langle p_1, p_2, p_3\rangle = \mathrm{SO}(3) \cdot p_2 = \mathrm{O}(3) \cdot p_2$, and $A_3 \cdot p_3 = -p_2$, $A_2 \cdot p_3 = -p_1$, $1 \cdot p_3 = p_3 \implies p_1, p_2, p_3 \in \mathrm{SO}(3) \cdot p_3 \implies \langle p_1, p_2, p_3\rangle = \mathrm{SO}(3) \cdot p_3 = \mathrm{O}(3) \cdot p_3$. That is, $\langle p_1, p_2, p_3\rangle = \mathrm{SO}(3)\cdot p_1 = \mathrm{SO}(3)\cdot p_2 = \mathrm{SO}(3)\cdot p_3 = \mathrm{O}(3)\cdot p_1 = \mathrm{O}(3)\cdot p_2 = \mathrm{O}(3)\cdot p_3$, with $\dim_{\mathbb{C}}\langle p_1, p_2, p_3\rangle = 3$ as p_1, p_2, p_3 are $\mathbb{C}$-linearly independent. For $A \in \mathrm{O}(3)$ and $\phi : \mathbb{R}^3 \to \mathbb{C}$ we have defined $A \cdot \phi : \mathbb{R}^3 \to \mathbb{C}$ by (7.1.3). If $\phi \in \langle p_1, p_2, p_3\rangle$, then $A \cdot \phi \in \langle p_1, p_2, p_3\rangle$ by (7.1.4), (7.1.5), (7.1.6). Thus we have a linear operator $\pi^{(3)}(A) : \langle p_1, p_2, p_3\rangle \to \langle p_1, p_2, p_3\rangle$ given by $\pi^{(3)}(A)\phi = A \cdot \phi$. By (7.1.4), (7.1.5), (7.1.6) the matrix of $\pi^{(3)}(A)$ relative to the ordered basis $\{p_1, p_2, p_3\}$ is A. Hence $\pi^{(3)}(A)$ is non-singular. Indeed the map $\pi^{(3)} : \mathrm{O}(3) \to \mathrm{Gl}(\langle p_1, p_2, p_3\rangle)$ is a representation of O(3) on $\langle p_1, p_2, p_3\rangle$. We note however that the two representations $\pi, \pi^{(3)}$ of O(3) are *equivalent*: $\exists$ a linear isomorphism T of $\langle p_1, p_2, p_3\rangle$ onto $\mathbb{C}^3$ such that the diagram

$$\begin{array}{ccc} \langle p_1, p_2, p_3\rangle & \xrightarrow{\pi^{(3)}(A)} & \langle p_1, p_2, p_3\rangle \\ T\downarrow & & \downarrow T \\ \mathbb{C}^3 & \xrightarrow[\pi(A)]{} & \mathbb{C}^3 \end{array}$$

commutes $\forall\ A \in G = \mathrm{O}(3)$; compare the similar situation in example 2. In particular π is also continuous (although continuity is easy to check directly). T is given by $Tp_j = e_j$ where $e_1 = (1,0,0)$, $e_2 = (0,1,0)$, $e_3 = (0,0,1)$. Indeed for $1 \le j \le 3$, $T\pi^{(3)}(A)p_j \stackrel{\text{def}}{=} TA \cdot p_j = T\sum_{i=1}^3 a_{ij}p_i$ (by (7.1.4), (7.1.5), (7.1.6)) $\stackrel{\text{def}}{=} \sum_{i=1}^3 a_{ij}e_i = (a_{1j}, a_{2j}, a_{3j})$. On the other hand $\pi(A)Tp_j \stackrel{\text{def}}{=} \pi(A)e_j \stackrel{\text{def}}{=} e_jA^t = (a_{1j}, a_{2j}, a_{3j})$, as desired.

5. Let $p_1, p_2, p_3 : \mathbb{R}^3 \to \mathbb{R}$ be the projections of example 4; each $p_j \in C^\infty(\mathbb{R}^3)$. Given an integer $l \ge 0$ let $P^{(l)}$ be the space of polynomial functions p in x, y, z that are homogeneous of degree l: $p = \sum_I c_I p_I$ where $c_I \in \mathbb{C}$, $I = (a, b, c)$ for integers $a, b, c \ge 0$, with $p_I \stackrel{\text{def}}{=} p_1^a p_2^b p_3^c$ and where, by the homogeneity condition $p(tx, ty, tz) = t^l p(x, y, z)$ for $t \in \mathbb{R}$, one has $a + b + c = l$—by Euler's theorem for example which says that $x\frac{\partial p}{\partial x} + y\frac{\partial p}{\partial y} + z\frac{\partial p}{\partial z} = lp$. Conversely, if $a + b + c = l$ for each I, then p is clearly homogeneous

of degree l. Let

$$H^{(l)} \stackrel{\text{def}}{=} \{p \in P^{(l)} | \nabla^2 p = 0\} \tag{7.1.9}$$

where ∇^2 is the Laplacian $\frac{\partial^2}{\partial x^2} + \frac{\partial^2}{\partial y^2} + \frac{\partial^2}{\partial z^2}$ on $\mathbb{R}^3$. By Theorem 4.13 in Appendix 4A the spherical harmonics $\{Y_l^m\}_{m=-l}^{l}$ are elements of $H^{(l)}$, for example. More generally if $\nabla^2 = \sum_{j=1}^n \frac{\partial^2}{\partial x_j^2}$ is the Laplacian on $\mathbb{R}^n$, then for $f : \mathbb{R}^n \to \mathbb{R}$ or $\mathbb{C}$ a function with continuous first and second partial derivatives one has (essentially by the chain rule)

$$\nabla^2(A \cdot f) = A \cdot \nabla^2 f \tag{7.1.10}$$

for $A \in \mathrm{O}(n)$ where $(A \cdot f)(x) \stackrel{\text{def}}{=} f(xA)$ for $x \in \mathbb{R}^n$ as in (7.1.3). A proof of (7.1.10) is presented in Appendix 7C. In particular if f is harmonic (i.e., $\nabla^2 f = 0$), then so is $A \cdot f$. Also by equations (7.1.4), (7.1.5), (7.1.6) in example 4 one has $A \cdot p \in P^{(l)}$ for $p \in P^{(l)}, A \in \mathrm{O}(3)$. Namely for $A = [a_{ij}]$

$$p = \sum_I c_I p_I \implies A \cdot p = \sum_I c_I A \cdot p_I, \tag{7.1.11}$$

where for $I = (a, b, c)$ with $a + b + c = l$,

$$\begin{aligned} A \cdot p_I &= (A \cdot p_1)^a (A \cdot p_2)^b (A \cdot p_3)^c \\ &= (a_{11}p_1 + a_{21}p_2 + a_{31}p_3)^a \cdot (a_{12}p_1 + a_{22}p_2 + a_{32}p_3)^b \\ &\quad \cdot (a_{13}p_1 + a_{23}p_2 + a_{33}p_3)^c. \end{aligned} \tag{7.1.12}$$

By the binomial theorem

$$(a_{11}p_1 + a_{21}p_2 + a_{31}p_3)^a = \sum_{\substack{0 \le r \le a \\ 0 \le s \le r}} \binom{a}{r} \binom{r}{s} a_{11}^s a_{21}^{r-s} a_{31}^{a-r} p_1^s p_2^{r-s} p_3^{a-r}. \tag{7.1.13}$$

Similarly $(a_{12}p_1 + a_{22}p_2 + a_{32}p_3)^b$, $(a_{13}p_1 + a_{23}p_2 + a_{33}p_3)^c$ are linear combinations of terms $p_1^u p_2^{t-u} p_3^{b-t}$, $p_1^w p_2^{v-w} p_3^{c-v}$, respectively, where $\substack{0 \le t \le b \\ 0 \le u \le t}$, $\substack{0 \le v \le c \\ 0 \le w \le v}$, and where the coefficients are polynomial functions of the a_{ij}. Since

$$\begin{aligned} &p_1^s p_2^{r-s} p_3^{a-r} p_1^u p_2^{t-u} p_3^{b-t} p_1^w p_2^{v-w} p_3^{c-v} \\ &\quad = p_1^{s+u+w} p_2^{r-s+t-u+v-w} p_3^{a-r+b-t+c-v} \in P^{(l)} \end{aligned}$$

(i.e., $s+u+w+r-s+t-u+v-w+a-r+b-t+c-v = a+b+c = l$) we see that by (7.1.12), (7.1.13) $A \cdot p \in P^{(l)}$ (which follows by simpler arguments), but we also see that if $\omega \in P^{(l)*}$, then the function $A \to \omega(A \cdot p)$ is continuous on $\mathrm{O}(3)$, and is even C^∞ of course, and is even polynomial in the entries a_{ij} of A. That is, the equation

$$\pi^{(l)}(A)p \stackrel{\text{def}}{=} A \cdot p, \qquad (A, p) \in \mathrm{O}(3) \times P^{(l)} \tag{7.1.14}$$

defines a *continuous* representation $\pi^{(l)}$ of O(3) on $P^{(l)}$ which by (7.1.10) has $H^{(l)}$ as an invariant subspace: $\pi^{(l)}(A)H^{(l)} \subset H^{(l)}\ \forall\ A \in O(3)$. We can therefore regard $\pi^{(l)}$ also as a continuous representation of O(3) on $H^{(l)}$ and by restriction to SO(3), $\pi^{(l)}$ is a continuous representation of SO(3) on $H^{(l)}$.

We state without proof the following standard facts: $\dim P^{(l)} = (l+2)(l+1)/2$ and $\dim H^{(l)} = 2l+1$; in fact $\{Y_l^m\}_{m=-l}^{l}$ is a $\mathbb{C}$ basis of $H^{(l)}$. The representation $\pi^{(l)}$ of SO(3) on $H^{(l)}$ is *irreducible*: if $V \subset H^{(l)}$ is any subspace such that $\pi^{(l)}(A)V \subset V\ \forall\ A \in SO(3)$ then $V =$ either $\{0\}$ or $H^{(l)}$. In particular $\pi^{(l)}$ is an irreducible representation of O(3) on $H^{(l)}$. If π is *any* continuous, irreducible representation of SO(3) on some complex vector space V then π is equivalent to $\pi^{(l)}$ for some integer $l \geq 0$: As in examples 2 and 4 the diagram

$$\begin{array}{ccc} H^{(l)} & \xrightarrow{\pi^{(l)}(A)} & H^{(l)} \\ T\big\downarrow & & \big\downarrow T \\ V & \xrightarrow[\pi(A)]{} & V \end{array}$$

is commutative for some suitable linear isomorphism T. We remark that the representation π of SO(3) on $\mathbb{C}^3$, in example 3, is known to be irreducible. Since $\dim \pi = 3 = \dim \pi^{(l)} = 2l+1 \implies l = 1$ we must therefore have that π is equivalent to $\pi^{(1)}$!

6. By Corollary 5.1 the wave functions $\{\psi_{nlm}\}_{m=-l}^{l}$ are linearly independent over $\mathbb{C}$. The $\mathbb{C}$-span S_{nl} in $L^2(\mathbb{R}^3, dxdydz)$ of the $\{\psi_{nlm}\}_{m=-l}^{l}$ is therefore a $(2l+1)$-dimensional complex vector space, which we claim moreover is O(3)-invariant. For $A \in O(3)$, $(x,y,z) \in \mathbb{R}^3$, $|(x,y,z)|^2 \stackrel{\text{def}}{=} x^2+y^2+z^2 = |(x,y,z)A|^2$, so that by (5.2.15) and (7.1.3)

$$\begin{aligned}(A\cdot\psi_{nlm})(x,y,z) &\stackrel{\text{def}}{=} \psi_{nlm}((x,y,z)A) \\ &= c_{lm}c_{nl}^{-}\frac{\alpha_n^{l+1}}{\sqrt{2\pi}}\exp\left(-\frac{\alpha_n}{2}|(x,y,z)|\right) \\ &\quad \cdot L_{n+l}^{(2l+1)}(\alpha_n|(x,y,z)|)\, Y_l^m((x,y,z)A).\end{aligned} \tag{7.1.15}$$

By example 5 we can write

$$\begin{aligned}Y_l^m((x,y,z)A) &= (\pi^{(l)}(A)Y_l^m)(x,y,z) \\ &= \sum_{j=-l}^{l} c_{jm}(A)\, Y_l^j(x,y,z)\end{aligned} \tag{7.1.16}$$

for suitable $c_{jm}(A) \in \mathbb{C}$. Then

$$(A\cdot\psi_{nlm})(x,y,z) = \sum_{j=-l}^{l} c_{jm}(A)c_{lm}c_{nl}^{-}\frac{\alpha_n^{l+1}}{\sqrt{2\pi}}\frac{c_{lj}}{c_{lj}}\exp\left(-\frac{\alpha_n}{2}|(x,y,z)|\right)$$

$$\cdot L_{n+l}^{(2l+1)}(\alpha_n|(x,y,z)|)\, Y_l^j(x,y,z)$$

$$= c_{lm} \sum_{j=-l}^{l} \frac{c_{jm}(A)}{c_{lj}} \psi_{nlj}(x,y,z) \tag{7.1.17}$$

$$\Longrightarrow A \cdot \psi_{nlm} = c_{lm} \sum_{j=-l}^{l} \frac{c_{jm}(A)}{c_{lj}} \psi_{nlj} \in S_{nl}, \tag{7.1.18}$$

as claimed. The equation

$$\pi^{(n.l)}(A)f \stackrel{\text{def}}{=} A \cdot f, \qquad A \in \mathrm{O}(3), f \in S_{nl}, \tag{7.1.19}$$

therefore defines a representation $\pi^{(n.l)}$ of O(3) on S_{nl}. If $\{\omega_j^{(l)}\}_{j=-l}^{l}$ is the basis of $H^{(l)}$ dual to $\{Y_l^j\}_{j=-l}^{l}$ then the $c_{jm}(A)$ in (7.1.16) are given by $c_{jm}(A) = \omega_j^{(l)}(\pi^{(l)}(A)Y_l^m)$ and are therefore continuous functions of A, since the representation $\pi^{(l)}$ is continuous. By (7.1.18) it follows that $\forall$ m, and $\tau \in S_{nl}{}^*$ the function

$$A \to \tau(\pi^{(n,l)}(A)\psi_{nlm}) = c_{lm} \sum_{j=-l}^{l} \frac{c_{jm}(A)}{c_{lj}} \tau(\psi_{nlj}) \tag{7.1.20}$$

is continuous on O(3). Hence $\pi^{(n.l)}$ is a continuous representation.

7. Define $\pi : G \to \mathrm{Gl}(V)$ by $\pi(x)v = v$ for $x \in G$, $v \in V$. π is called the *trivial* representation. Of some interest is the trivial one-dimensional representation $\pi : G \to \mathrm{Gl}(\mathbb{C})$.

   The *character* of a finite-dimensional group representation $\pi : G \to \mathrm{Gl}(V)$ is the function $\chi_\pi : G \to \mathbb{F}$ = the base field of V given by $\chi_\pi(g) \stackrel{\text{def}}{=} \operatorname{trace} \pi(g)$ for $g \in G$. If G is a compact topological group it is known that two finite-dimensional, continuous representations of G with the same character are equivalent; clearly, equivalent representations have the same character. For example, by (7.1.18)

$$\begin{aligned} \pi^{(n,l)}(A)\psi_{nlm} &= \sum_{j=-l}^{l} \pi^{(n,l)}(A)_{jm}\, \psi_{nlj}, \qquad \text{where} \\ \pi^{(n,l)}(A)_{jm} &= c_{lm} \frac{c_{jm}(A)}{c_{lj}}. \end{aligned} \tag{7.1.21}$$

   That is, for every $A \in \mathrm{O}(3)$

$$\begin{aligned} \chi_{\pi^{(n,l)}}(A) &\stackrel{\text{def}}{=} \sum_{j=-l}^{l} \pi^{(n,l)}(A)_{jj} = \sum_{j=-l}^{l} c_{jj}(A) \\ &= \operatorname{trace} \pi^{(l)}(A) \stackrel{\text{def}}{=} \chi_{\pi^{(l)}}(A) \end{aligned}$$

(by (7.1.16)) which shows that the representations $\pi^{(n,l)}$ and $\pi^{(l)}$ are equivalent. In particular, $\pi^{(n,l)}$ is irreducible. In fact, by example 5, the restriction of $\pi^{(n,l)}$ to SO(3) defines an *irreducible* representation of SO(3).

Note, for example, that the representation $\pi : \mathbb{R} \to \mathrm{Gl}(\mathbb{R}^2)$ in example 1 is *not* irreducible. Namely for $x, a \in \mathbb{R}$, $\pi(x)(a,0) \stackrel{\text{def}}{=} (a,0)$. That is, $R \stackrel{\text{def}}{=} \{(a,0)|a \in \mathbb{R}\} \subset \mathbb{R}^2$ is a subspace which is $\pi(x)$-invariant $\forall\, x \in \mathbb{R}$, and which is neither $\{0\}$ nor $\mathbb{R}^2$.

It is clear that we could also define a finite-dimensional representation of the group G as a homomorphism $\pi : G \to \mathrm{Gl}(n, \mathbb{F})$ = the group of non-singular $n \times n$ matrices over the field $\mathbb{F}$: $\pi(x_1x_2) = \pi(x_1)\pi(x_2)$ for $x_1, x_2 \in G$. If G has a topology and $\mathbb{F} = \mathbb{R}$ or $\mathbb{C}$ (or if $\mathbb{F}$ is a topological field), then continuity of π could be taken to mean that for $1 \le i, j \le n$ the functions $g \to \pi(g)_{ij} : G \to \mathbb{F}$ are continuous where $\pi(g) = \left[\pi(g)_{ij}\right]$. n is the degree or dimension of π. In example 1, for example, we obtain the "matrix version" of the representation $\pi : \mathbb{R} \to \mathrm{Gl}(\mathbb{R}^2)$ via the homomorphism $\pi : \mathbb{R} \to \mathrm{Gl}(2, \mathbb{R})$ given by $\pi(x) = \left[\begin{smallmatrix} 1 & x \\ 0 & 1 \end{smallmatrix}\right]$, $x \in \mathbb{R}$, which is simply the matrix of $\pi(x) \in \mathrm{Gl}(\mathbb{R}^2)$ relative to the standard ordered basis $\{(1,0), (0,1)\}$ of $\mathbb{R}^2$.

We will also have occasions to consider *infinite-dimensional* representations, say, for example, G has a locally compact Hausdorff topology and $\pi : G \to \mathrm{U}(H)$ is a homomorphism to the group of unitary operators $\mathrm{U}(H)$ on a Hilbert space H. Since every continuous (i.e., bounded) linear functional $f : H \to \mathbb{C}$ on H has the form $f(\alpha) = \langle \alpha, \beta \rangle$ for some unique $\beta \in H$, $\langle\ ,\ \rangle$ being the inner product structure on H, we can (with definition (7.1.2) in mind) take continuity of π in this case to mean that the functions $x \to \langle \pi(x)v, \beta \rangle$ on G are continuous $\forall\, v, \beta \in H$. Here *irreducibility* of π would mean that if $H_1 \subset H$ is any *closed* subspace such that $\pi(x)H_1 \subset H_1$ $\forall\, x \in G$, then H_1 = either $\{0\}$ or H. π is called a *unitary* representation since each operator $\pi(x) : H \to H$, $x \in G$, is unitary: $\langle \pi(x)v_1, \pi(x)v_2 \rangle = \langle v_1, v_2 \rangle$ for $v_1, v_2 \in H$.

7.2 Contragredient and Tensor Product Representations

Given representations π, π_1, π_2 of a group G we can form other representations $\pi_1 \oplus \pi_2$, $\pi_1 \otimes \pi_2$, π^*, etc. of G. We review these constructions as they are needed for later applications. Assume G is a topological group and that we are dealing with finite-dimensional, continuous representations on complex vector spaces, for specificity. If $\pi_j : G \to \mathrm{Gl}(V_j)$, $j = 1, 2$, are two such representations we can form the *external direct sum* $\pi_1 \oplus \pi_2$ which is a representation of G on $V_1 \times V_2$:

$$(\pi_1 \oplus \pi_2)(x)(v_1, v_2) \stackrel{\text{def}}{=} (\pi_1(x)v_1, \pi_2(x)v_2) \tag{7.2.1}$$

for $x \in G$, $(v_1, v_2) \in V_1 \times V_2$. Let $i_1 : V_1 \to V_1 \times V_2$, $i_2 : V_2 \to V_1 \times V_2$ be the linear injections $v_1 \to (v_1, 0)$, $v_2 \to (0, v_2)$. If $F \in (V_1 \times V_2)^*$, then $F \circ i_j \in V_j^* \implies x \to F \circ i_j(\pi_j(x)v_j) : G \to \mathbb{C}$ is continuous, since π_j is continuous

$\Longrightarrow x \to F((\pi_1 \oplus \pi_2)(x)(v_1, v_2)) = (F \circ i_1)(\pi_1(x)v_1) + (F \circ i_2)(\pi_2(x)v_2)$ is continuous. That is, $\pi_1 \oplus \pi_2$ is a continuous representation.

Given the representation $\pi : G \to \mathrm{Gl}(V)$ there is a naturally associated representation $\pi^* : G \to \mathrm{Gl}(V^*)$ of G on V^* defined by

$$(\pi^*(x)f)(v) = f(\pi(x^{-1})v) \qquad \text{for } (v, f) \in V \times V^*, x \in G. \tag{7.2.2}$$

Thus $\pi^*(x)$ is the *transpose* of $\pi(x^{-1})$. π^* is called the representation *dual* or *contragredient* to π. To see that π^* is continuous we use the duality $(V^*)^* \simeq V$; i.e., given $\psi \in (V^*)^*$ there is a unique $v \in V$ such that $\psi(f) = f(v)\ \forall\ f \in V^*$. Then the function $\pi^*_{f,\psi} : G \to \mathbb{C}$ in (7.1.2) is given by $x \to \psi(\pi^*(x)f) = (\pi^*(x)f)(v) \overset{\text{def}}{=} f(\pi(x^{-1})v)$ which is continuous since π is continuous and the map $x \to x^{-1}$ of G is continuous, G being a topological group.

Next we construct the *tensor product* $\pi_1 \otimes \pi_2$ of π_1 and π_2. Let (T, b) be the tensor product over $\mathbb{C}$ of V_1 with V_2. Here we take V_1, V_2 to be arbitrary (even infinite-dimensional) vector spaces over $\mathbb{C}$. That is, T is a $\mathbb{C}$-vector space and $b : V_1 \times V_2 \to T$ is a bilinear map, where the pair (T, b) satisfies the following universal condition: Given *any* bilinear map $b_1 : V_1 \times V_2 \to W =$ any $\mathbb{C}$-vector space, $\exists$ a *unique* linear map $\tilde{b}_1 : T \to W$ such that the diagram

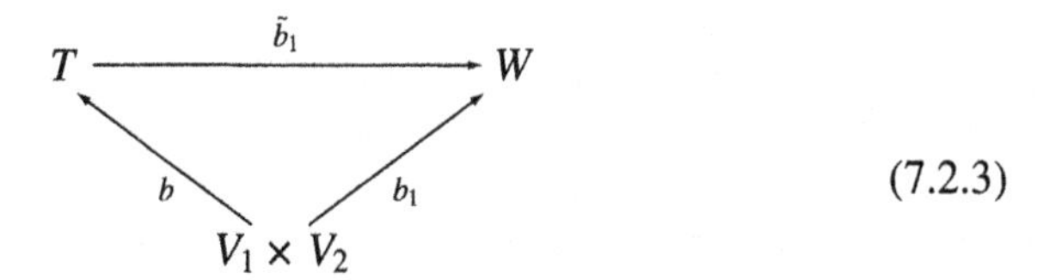

(7.2.3)

is commutative. (T, b) exists and is unique up to a natural equivalence. One writes $T = V_1 \otimes_{\mathbb{C}} V_2 = V_1 \otimes V_2$, $b(v_1, v_2) = v_1 \otimes_{\mathbb{C}} v_2 = v_1 \otimes v_2$, for $(v_1, v_2) \in V_1 \times V_2$, as usual. If V_3, V_4 are also $\mathbb{C}$-vector spaces and $T : V_1 \to V_3$, $S : V_2 \to V_4$ are linear maps, then the map $b_{T,S} : V_1 \times V_2 \to V_3 \otimes V_4$ given by $b_{T,S}(v_1, v_2) \overset{\text{def}}{=} Tv_1 \otimes Sv_2$ is bilinear, since b is bilinear. Hence $\exists$ a unique linear map $\widetilde{b_{T,S}} : V_1 \otimes V_2 \to V_3 \otimes V_4$ such that

$$\widetilde{b_{T,S}}(v_1 \otimes v_2) = b_{T,S}(v_1, v_2) \tag{7.2.4}$$

for $(v_1, v_2) \in V_1 \times V_2$ by (7.2.3). Define $T \otimes S \overset{\text{def}}{=} \widetilde{b_{T,S}}$. Thus $T \otimes S : V_1 \otimes V_2 \to V_3 \otimes V_4$ is the unique $\mathbb{C}$-linear map such that $(T \otimes S)(v_1 \otimes v_2) = Tv_1 \otimes Sv_2$ for $(v_1, v_2) \in V_1 \times V_2$, by (7.2.4).

Now for the finite-dimensional representations $\pi_j : G \to \mathrm{Gl}(V_j)$, $j = 1, 2$, for each $x \in G$, $\pi_1(x) : V_1 \to V_1$, $\pi_2(x) : V_2 \to V_2$ are linear maps; in fact they are automorphisms. By the preceding discussion, there is a unique linear map $\pi_1(x) \otimes \pi_2(x) : V_1 \otimes V_2 \to V_1 \otimes V_2$ such that

$$(\pi_1(x) \otimes \pi_2(x))(v_1 \otimes v_2) = \pi_1(x)v_1 \otimes \pi_2(x)v_2 \tag{7.2.5}$$

for $(v_1, v_2) \in V_1 \times V_2$. We note that $\pi_1(x) \otimes \pi_2(x)$ is surjective [and thus is automatically one-to-one by dimension]. Namely any $t \in V_1 \otimes V_2$ is a finite sum of terms $v_1 \otimes v_2$ with $(v_1, v_2) \in V_1 \times V_2$. Since $\pi_1(x)$, $\pi_2(x)$ are surjective we can write $v_j = \pi_j(x)v_j'$ for suitable $v_j' \in V_j$. Then $v_1 \otimes v_2 = \pi_1(x)v_1' \otimes \pi_2(x)v_2' = (\pi_1(x) \otimes \pi_2(x))(v_1' \otimes v_2')$, by (7.2.5), $\Longrightarrow$ $\pi_1(x) \otimes \pi_2(x)$ is indeed surjective. That is, we have a map $\pi_1 \otimes \pi_2 : G \to \mathrm{Gl}(V_1 \otimes V_2)$ defined by

$$(\pi_1 \otimes \pi_2)(x) = \pi_1(x) \otimes \pi_2(x) \quad \text{for } x \in G, \tag{7.2.6}$$

which indeed is a representation:

$$\begin{aligned}
&[(\pi_1 \otimes \pi_2)(x_1x_2)](v_1 \otimes v_2) \\
&\quad \stackrel{\text{def}}{=} [\pi_1(x_1x_2) \otimes \pi_2(x_1x_2)](v_1 \otimes v_2) \\
&\quad \stackrel{\text{def}}{=} \pi_1(x_1x_2)v_1 \otimes \pi_2(x_1x_2)v_2 = \pi_1(x_1)\pi_1(x_2)v_1 \otimes \pi_2(x_1)\pi_2(x_2)v_2 \\
&\quad \stackrel{\text{def}}{=} (\pi_1(x_1) \otimes \pi_2(x_1))[\pi_1(x_2)v_1 \otimes \pi_2(x_2)v_2] \\
&\quad = [\pi_1(x_1) \otimes \pi_2(x_1)] \circ [\pi_1(x_2) \otimes \pi_2(x_2)](v_1 \otimes v_2) \\
&\quad \stackrel{\text{def}}{=} [(\pi_1 \otimes \pi_2)(x_1)] \circ [(\pi_1 \otimes \pi_2)(x_2)](v_1 \otimes v_2) \\
&\Longrightarrow [(\pi_1 \otimes \pi_2)(x_1x_2)]t \\
&\quad = [(\pi_1 \otimes \pi_2)(x_1)] \circ [(\pi_1 \otimes \pi_2)(x_2)]t \quad \forall t \in V_1 \otimes V_2.
\end{aligned} \tag{7.2.7}$$

Finally, we check that $\pi_1 \otimes \pi_2$ is continuous. Choose dual bases $\{e_i^{(j)}\}_i$, $\{e_i^{(j)*}\}_i$ of $V_j, V_j^*, j = 1, 2$. For $\omega \in (V_1 \otimes V_2)^*$, $t \in V_1 \otimes V_2$ we use again that t is a finite sum of terms $v_1 \otimes v_2$ with $(v_1, v_2) \in V_1 \times V_2$. Thus to see that $x \to \omega((\pi_1 \otimes \pi_2)(x)t)$ is continuous on G it suffices to see that $x \to \omega((\pi_1 \otimes \pi_2)(x)(v_1 \otimes v_2))$ is continuous on G. Write, for $x \in G$,

$$v_j = \sum_i e_i^{(j)*}(v_j)\, e_i^{(j)}, \qquad \pi_j(x)e_i^{(j)} = \sum_p e_p^{(j)*}(\pi_j(x)e_i^{(j)})\, e_p^{(j)}. \tag{7.2.8}$$

Then

$$\begin{aligned}
&\omega((\pi_1 \otimes \pi_2)(x)(v_1 \otimes v_2)) \\
&\quad \stackrel{\text{def}}{=} \omega(\pi_1(x)v_1 \otimes \pi_2(x)v_2) \\
&\quad = \omega\left(\sum_i e_i^{(1)*}(v_1)\, \pi_1(x)e_i^{(1)} \otimes \sum_l e_l^{(2)*}(v_2)\, \pi_2(x)e_l^{(2)}\right) \\
&\quad = \sum_{i,l} \sum_{p,q} e_i^{(1)*}(v_1)\, e_l^{(2)*}(v_2)\, e_p^{(1)*}(\pi_1(x)e_i^{(1)})\, e_q^{(2)*}(\pi_2(x)e_l^{(2)})\, \omega(e_p^{(1)} \otimes e_q^{(2)}),
\end{aligned} \tag{7.2.9}$$

which is a continuous function of x as π_1, π_2 are continuous.

Since we have certain notation in place and equation (7.2.9) we establish a simple linear algebraic formula for later use. Let $\epsilon : \mathbb{C} \otimes \mathbb{C} \to \mathbb{C}$ be the canonical vector space isomorphism given by $\epsilon(z_1 \otimes z_2) = z_1 z_2$ for $z_1, z_2 \in \mathbb{C}$. Then $\omega_{ij} \stackrel{\text{def}}{=}$

$\epsilon \circ (e_i^{(1)^*} \otimes e_j^{(2)^*}) \in (V_1 \otimes V_2)^*$ satisfies $\omega_{ij}(e_r^{(1)} \otimes e_s^{(2)}) = \epsilon(\delta_{ir} \otimes \delta_{js}) = \delta_{ir}\delta_{js} = \delta_{(i,j),(r,s)}$. That is, $\{\omega_{ij}\}$ is the basis of $(V_1 \otimes V_2)^*$ dual to the basis $\{e_i^{(1)} \otimes e_j^{(2)}\}_{i,j}$ of $V_1 \otimes V_2$. By (7.2.9), for any $x \in G$

$$\begin{aligned} &\omega_{ij}((\pi_1 \otimes \pi_2)(x)(e_r^{(1)} \otimes e_s^{(2)})) \\ &= \sum_{m,l,p,q} \delta_{mr}\delta_{ls} e_p^{(1)^*}(\pi_1(x)e_m^{(1)})\, e_q^{(2)^*}(\pi_2(x)e_l^{(2)})\, \delta_{(i,j),(p,q)} \\ &= \sum_{p,q} e_p^{(1)^*}(\pi_1(x)e_r^{(1)})\, e_q^{(2)^*}(\pi_2(x)e_s^{(2)})\, \delta_{(i,j),(p,q)} \\ &= e_i^{(1)^*}(\pi_1(x)e_r^{(1)})\, e_j^{(2)^*}(\pi_2(x)e_s^{(2)}), \end{aligned} \tag{7.2.10}$$

which is the formula we desire.

7.3 Subrepresentations

If $\pi : G \to \mathrm{Gl}(V)$ is a representation of G and $V_1 \subset V$ is a π-invariant subspace (i.e., $\pi(x)V_1 \subset V_1 \ \forall\ x \in G$) then we can construct a representation $\pi_1 : G \to \mathrm{Gl}(V_1)$ of G on V_1 by setting

$$\pi_1(x) \stackrel{\text{def}}{=} \pi(x)|_{V_1} \tag{7.3.1}$$

for $x \in G$. π_1 is called a *subrepresentation* of π. In example 5, for example, the representation $\pi^{(l)}$ of O(3) on $H^{(l)}$ is a subrepresentation of the representation $\pi^{(l)}$ of O(3) on $P^{(l)}$. Note that π_1 is continuous, assuming that π is continuous, since any $f_1 \in V_1^*$ is the restriction to V_1 of some $f \in V^*$. If V were an infinite-dimensional Hilbert space, for example, we would take V_1 to be a *closed* subspace and $f_1 : V_1 \to \mathbb{C}$ to be a *bounded* linear function. Then we could choose a bounded linear functional f on V such that $f|_{V_1} = f_1$ by the Hahn–Banach theorem.

More generally, a representation $\pi_1 : G \to \mathrm{Gl}(V_1)$ is said to be *contained in* $\pi : G \to \mathrm{Gl}(V)$ if π_1 is equivalent to some subrepresentation of π: For some π-invariant subspace $V_o \subset V$ the diagram

$$\begin{array}{ccc} V_1 & \xrightarrow{\pi_1(x)} & V_1 \\ {\scriptstyle T}\downarrow & & \downarrow{\scriptstyle T} \\ V_o & \xrightarrow[\pi(x)|_{V_o}]{} & V_o \end{array} \tag{7.3.2}$$

is commutative, where T is a suitable linear isomorphism. Note that the trivial one-dimensional representation $\pi_1 : G \to \mathrm{Gl}(\mathbb{C})$ of example 7 is contained in $\pi \iff \exists$ a one-dimensional subspace $V_o \subset V$ such that

$$\pi(x)v_o = v_o \quad \forall\ v_o \in V_o. \tag{7.3.3}$$

By example 6, the representation $\pi^{(n,l)}$ of O(3) on S_{nl} is contained in the representation $\pi^{(l)}$ of O(3) on $P^{(l)}$. As we have seen π is *irreducible* if $\{0\}$ and V are the only π-invariant subspaces of V. Here we assume $V \neq \{0\}$. Thus irreducibility of π means that π has no non-trivial subrepresentations.

Suppose we have $V = V_1 \oplus \cdots \oplus V_m$ as a direct sum of m subspaces V_i with each V_i π-invariant. Then we have m subrepresentations $\pi_i = \pi|_{V_i}$ of π and we say that π is the *internal direct sum* of the π_i. Using the same notation for the external direct sum (see Section 7.2) we write $\pi = \pi_1 \oplus \cdots \oplus \pi_m$. By example 6 of Section 7.1 for each integer $l \geq 0$ there is exactly one continuous, irreducible representation $\pi^{(l)}$ of SO(3) of dimension $2l + 1$, up to equivalence. A natural question arises: What is the structure of the tensor product $\pi^{(l_1)} \otimes \pi^{(l_2)}$ for two integers $l_1, l_2 \geq 0$? The answer is given by

Theorem 7.1 (The Clebsch–Gordan Formula). *Up to equivalence*

$$\pi^{(l_1)} \otimes \pi^{(l_2)} = \pi^{(l_1+l_2)} \oplus \pi^{(l_1+l_2-1)} \oplus \pi^{(l_1+l_2-2)} \oplus \cdots \oplus \pi^{(|l_1-l_2|)}. \tag{7.3.4}$$

Formula (7.3.4) exhibits all the possible irreducible representations of SO(3) contained in $\pi^{(l_1)} \otimes \pi^{(l_2)}$. For example $\pi^{(3)} \otimes \pi^{(5)} = \pi^{(8)} \oplus \pi^{(7)} \oplus \pi^{(6)} \oplus \pi^{(5)} \oplus \pi^{(4)} \oplus \pi^{(3)} \oplus \pi^{(2)}$.

7.4 Group Actions

G is again assumed to be a topological group. G is said to *act* on a topological space X on the *left* if $\forall\ g \in G$ there is a homeomorphism $h_g : X \to X$ of X such that the following conditions hold. Set $g \cdot x \stackrel{\text{def}}{=} h_g(x)$ for $x \in X$, $g \in G$. We require that $(g_1g_2) \cdot x = g_1 \cdot (g_2 \cdot x)$ [or $(g_1g_2) \cdot x = g_2 \cdot (g_1 \cdot x)$ for a *right* action], $1 \cdot x = x$ for 1 the identity element of G, $g_1, g_2 \in G$, $x \in X$. We also require that the map $G \times X \to X$ given by $(g, x) \to h_g(x) = g \cdot x$ be continuous. Given a left action, note that the map

$$(g, x) \to g^{-1} \cdot x$$

defines a *right* action.

Some examples of left group actions are the following.

1. Let G act on $X = G$ by left multiplication: $h_g(g_1) \stackrel{\text{def}}{=} gg_1$ for $g, g_1 \in G$. By definition of a topological group, the above map $G \times X \to X$ (i.e., $(g, g_1) \to h_g(g_1) = gg_1$) is continuous.

2. More generally, for a closed subgroup H of G let X be the coset space G/H with the quotient topology. For $g \in G$, define $h_g : X \to X$ by $h_g(g_1H) = (gg_1)H$ for $g, g_1 \in G$.

3. Let $\pi : G \to \mathrm{Gl}(V)$ be a continuous representation of G on V where V is finite-dimensional over $\mathbb{F} = \mathbb{R}$ or $\mathbb{C}$. Let τ be the weak topology on V generated by the $f \in V^*$. This is the same as the topology on V obtained by a metric on V induced by an isomorphism $V \simeq \mathbb{F}^{\dim V}$ for some choice

of basis of V. Let $X = (V, \tau)$ and define $h_g : V \to V$ by $h_g(v) = \pi(g)v$ for $g \in G$, $v \in V$. Such actions, coming by way of a representation of G are called *linear* actions. From example 3 in Section 7.1 we obtain, for example, the linear action of $O(n)$ on $\mathbb{R}^n$ or on $\mathbb{C}^n$ by setting

$$g \cdot x \overset{\text{i.e.,}}{=} h_g(x) = xg^t$$

for $g \in O(n)$, $x \in \mathbb{R}^n, \mathbb{C}^n$ a row vector.

4. Define $h_g : G \to G$ by $h_g(x) = gxg^{-1}$ for $g, x \in G$. Thus h_g is the inner automorphism of G induced by g, and one has indeed that $(g_1g_2) \cdot x = g_1 \cdot (g_2 \cdot x)$ for $g_1, g_2, x \in G$.

7.5 An Abstract Selection Rule

As indicated in introductory remarks to this chapter, our goal is to formulate in broad, abstract terms a "selection rule"—a group theoretic result by which we may determine that certain integrals vanish (of course without actually computing these integrals). Such a result seems to be in use in physics and chemistry implicitly where often enough no clear general statement, or careful set of hypotheses, is set forth. Toward achieving our goal we consider the following context, and set of hypotheses.

As before G is a fixed topological group which acts on a topological space X on the left. We assume now that G and X are locally compact (with Hausdorff topologies). Eventually we will take G to be *compact*. We also assume that on X there is some G-invariant measure dx: For $C_c(X)$ = the space of continuous, compactly supported complex-valued functions on X

$$\int_X (g \cdot f)(x)\, dx = \int_X f(x)\, dx, \quad \text{where} \tag{7.5.1}$$

$$(g \cdot f)(x) \overset{\text{def}}{=} f(g^{-1} \cdot x) \tag{7.5.2}$$

for $(g, f) \in G \times C_c(X)$. By "measure" we mean that dx is a Radon measure; one could also take dx to be a regular Baire, or Borel measure. The G-invariance of dx in (7.5.1) means that the equation

$$(\pi_L(g)f)(x) \overset{\text{def}}{=} f(g^{-1} \cdot x) \overset{\text{def}}{=} (g \cdot f)(x) \tag{7.5.3}$$

for $(g, x, f) \in G \times X \times L^2(X, dx)$ defines a unitary representation $\pi = \pi_L$ of G on the Hilbert space $L^2(X, dx)$ of square-integrable complex-valued functions (with respect to dx) on X:

$$||\pi(g)f||_2^2 \overset{\text{def}}{=} \int_X |(g \cdot f)(x)|^2\, dx = \int_X |f(x)|^2\, dx = ||f||_2^2. \tag{7.5.4}$$

Proposition 7.1. *The representation π_L of G on $L^2(X, dx)$ is continuous.*

First we prove a standard lemma.

Lemma 7.1. *Let $f \in C_c(X)$, $g_o \in G$. Then for $\epsilon > 0$ $\exists$ a neighborhood U_o of g_o in G (i.e., U_o is an open set in G containing g_o) such that*

$$g \in U_o \implies \left| f(g^{-1} \cdot x) - f(g_o^{-1} \cdot x) \right| < \epsilon \quad \forall\, x \in X.$$

Proof. As f has compact support let $f(X - K) = 0$ for $K \subset X$ compact. Let U be a neighborhood of g_o in G such that its closure $\overline{U}$ is compact (since G is locally compact). Then $\overline{U} \cdot K \subset X$ is compact. Take $x \in \overline{U} \cdot K$. As $y \to f(g_o^{-1} \cdot y)$ is continuous on X, $\exists$ a neighborhood A_x of x in X such that $y \in A_x \implies |f(g_o^{-1} \cdot y) - f(g_o^{-1} \cdot x)| < \frac{\epsilon}{2}$, given $\epsilon > 0$. $(g, y) \to f(g^{-1} \cdot y)$ is continuous on $G \times X$ so that, similarly, $\exists$ neighborhoods U_x, V_x of g_o, x in G, X such that $(g, y) \in U_x \times V_x \implies |f(g^{-1} \cdot y) - f(g_o^{-1} \cdot x)| < \frac{\epsilon}{2}$. Then $\{A_x \cap V_x\}_{x \in \overline{U} \cdot K}$ is an open covering of $\overline{U} \cdot K$ in X. The compactness of $\overline{U} \cdot K$ guarantees the existence of some finite open subcovering: $\overline{U} \cdot K \subset \bigcup_{j=1}^{n} A_{x_j} \cap V_{x_j}$ for suitable $x_1, x_2, \ldots, x_n \in X$. Define $U_o = U \cap \left(\bigcap_{j=1}^{n} U_{x_j}\right)$, which is open in G and which contains g_o. Fix $g \in U_o$ and let $x \in X$ be arbitrary. Consider two cases: (i) $x \in \overline{U} \cdot K$, (ii) $x \notin \overline{U} \cdot K$.

In case (i), $x \in A_{x_j} \cap V_{x_j}$ for some x_j. We have $g \in U_{x_j}, x \in V_{x_j} \implies |f(g^{-1} \cdot x) - f(g_o^{-1} \cdot x_j)| < \frac{\epsilon}{2}$, and $x \in A_{x_j} \implies |f(g_o^{-1} \cdot x) - f(g_o^{-1} \cdot x_j)| < \frac{\epsilon}{2} \implies |f(g^{-1} \cdot x) - f(g_o^{-1} \cdot x)| \le |f(g^{-1} \cdot x) - f(g_o^{-1} \cdot x_j)| + |f(g_o^{-1} \cdot x_j) - f(g_o^{-1} \cdot x)| < \frac{\epsilon}{2} + \frac{\epsilon}{2} = \epsilon$.

On the other hand in case (ii) we have $g^{-1} \cdot x \notin K$; otherwise $x \in g \cdot K \subset U \cdot K \subset \overline{U} \cdot K$ is a contradiction. Similarly $g_o \in U \implies g_o^{-1} \cdot x \notin K$ so that $f(X - K) = 0 \implies |f(g^{-1} \cdot x) - f(g_o^{-1} \cdot x)| = 0 < \epsilon$, which proves the lemma.

To prove Proposition 7.1 we first show that $g \to \langle \pi(g) f_1, f_2 \rangle$ is continuous on G for $f_1 \in C_c(X)$, $f_2 \in L^2(X, dx)$ fixed: For $g_o \in G$

$$\begin{aligned}
&|\langle \pi(g) f_1, f_2 \rangle - \langle \pi(g_o) f_1, f_2 \rangle|^2 \\
&\quad = |\langle \pi(g) f_1 - \pi(g_o) f_1, f_2 \rangle|^2 \\
&\quad \le \|f_2\|_2^2 \|\pi(g) f_1 - \pi(g_o) f_1\|_2^2 \\
&\qquad \text{(which is the Cauchy–Bunyakovskiĭ–Schwarz inequality)} \qquad (7.5.5) \\
&\quad \overset{\text{def}}{=} \|f_2\|_2^2 \int_X \left| f_1(g^{-1} \cdot x) - f_1(g_o^{-1} \cdot x) \right|^2 dx \\
&\quad = \|f_2\|_2^2 \int_{\overline{U} \cdot K} \left| f_1(g^{-1} \cdot x) - f_1(g_o^{-1} \cdot x) \right|^2 dx,
\end{aligned}$$

where $f_1(X - K) = 0$ for $K \subset X$ compact and where U = a neighborhood of g_o in G such that $\overline{U}$ is compact, as in the proof of Lemma 7.1, where we saw in case (ii) there that $x \notin \overline{U} \cdot K \implies f_1(g^{-1} \cdot x), f_1(g_o^{-1} \cdot x) = 0$ for $g \in U$. Thus given $\epsilon > 0$ choose U_o containing g_o as in Lemma 7.1, $U_o \subset U$. In particular $\forall\, x \in \overline{U} \cdot K$, $g \in U_o$

$$\int_{\overline{U} \cdot K} \left| f_1(g^{-1} \cdot x) - f_1(g_o^{-1} \cdot x) \right|^2 dx < \epsilon^2 \cdot \text{ measure of } \overline{U} \cdot K,$$

where the measure of $\overline{U} \cdot K < \infty$ as $\overline{U} \cdot K$ is compact and dx is regular. Hence $g \to \langle \pi(g)f_1, f_2 \rangle$ is continuous for $f_1 \in C_c(X)$, $f_2 \in L^2(X, dx)$. Since $C_c(X)$ is dense in $L^2(X, dx)$ (again for dx regular) we have continuity of $g \to \langle \pi(g)f_1, f_2 \rangle$ for $f_1, f_2 \in L^2(X, dx)$, as desired. □

π_L is called the *left-regular representation* of G on $L^2(X, dx)$ since it coincides with the familiar left regular representation of G on $L^2(G, dg)$ in case $X = G$, with the left action being left multiplication, where $dx = dg$ = a left Haar measure on G. Since the Haar measure of any non-empty open subset of G is *positive* we impose similarly the condition that

$$\text{the } dx\text{-measure of any non-empty open subset of } X \text{ is positive.} \tag{7.5.6}$$

From the condition (7.5.6) it follows in particular that if $C_2(X)$ is the space of continuous, complex-valued, square-integrable functions on X, then the equation

$$\langle f_1, f_2 \rangle = \int_X f_1(x)\overline{f_2}(x)\, dx \tag{7.5.7}$$

defines a complex inner product on $C_2(X)$. Namely if $f \in C_2(X)$ with $\langle f, f \rangle = 0$, then $f(x) = 0$ dx-almost everywhere: $N \stackrel{\text{def}}{=} \{x \in X | f(x) \neq 0\} \subset X$ is a subset of dx-measure zero. But since f is continuous, N is an open subset of X. By (7.5.6) we therefore must have $N = \emptyset \implies f = 0$. If $f \in C_2(X)$, then $g \cdot f \in C_2(X)$ $\forall\, g \in G$ by the G-invariance of dx; $(g \cdot f)(x) \stackrel{\text{def}}{=} f(g^{-1} \cdot x)$ for $x \in X$, as in (7.5.2). Thus as in (7.1.14) the equation

$$\pi(g)f \stackrel{\text{def}}{=} g \cdot f \quad \text{for } (g, f) \in G \times C_2(X) \tag{7.5.8}$$

defines a unitary representation π of G on $C_2(X)$ (with respect to the inner product defined in (7.5.7)) with the following continuity property: $g \to \langle \pi(g)f_1, f_2 \rangle_{C_2(X)} = \langle \pi_L(g)f_1, f_2 \rangle_{L^2(X,dx)}$ is continuous $\forall\, f_1, f_2 \in C_2(X)$, by Proposition 7.1.

Now let $V_1, V_2 \subset C_2(X)$ be finite-dimensional, G-invariant complex subspaces: $f \in V_j, g \in G \implies \pi(g)f = g \cdot f \in V_j$, $j = 1, 2$. Then by restriction of π to V_j we obtain finite-dimensional, continuous, unitary subrepresentations $\pi^{(j)}$ of π, of G on V_j. The abstract selection rule will involve the representations $\pi^{(1)}$, $\pi^{(2)}$, and a third representation $\pi^{(3)}$, constructed as follows. Let ϕ be a continuous, real or complex-valued function on X such that $V_3 \stackrel{\text{def}}{=}$ the $\mathbb{C}$-span of the vectors $\{g \cdot \phi | g \in G\}$ is finite-dimensional. In practice ϕ might *not* be square-integrable with respect to dx. We assume however that ϕf_1 is square-integrable $\forall\, f_1 \in V_1$. Note then that for $g \in G$, $f_1 \in V_1$ we have $g^{-1} \cdot f_1 \in V_1$ (as V_1 is G-invariant) $\implies \phi g^{-1} \cdot f_1$ is square-integrable $\implies g \cdot [\phi g^{-1} \cdot f_1]$ is square-integrable (as dx is G-invariant). That is, $(g \cdot \phi) f_1 = g \cdot [\phi(g^{-1} \cdot f_1)]$ is square-integrable $\forall\, g \in G \implies \psi f_1$ is square-integrable for $(\psi, f_1) \in V_3 \times V_1$:

$$V_3 V_1 \subset C_2(X). \tag{7.5.9}$$

The space V_3 is clearly G-invariant so that we obtain a finite-dimensional representation $\pi^{(3)}$ of G on V_3 given by

$$\pi^{(3)}(g)f = g \cdot f \quad \text{for } (g, f) \in G \times V_3. \tag{7.5.10}$$

It will be clear in concrete examples that $\pi^{(3)}$ is continuous. Rather than formulating a general statement about continuity we simply assume, for convenience, that $\pi^{(3)}$ is indeed continuous.

Since $V_2 \subset C_2(X)$ we see by (7.5.9) that $\langle \psi f_1, f_2 \rangle$ is well defined (i.e., is finite) for $(\psi, f_1, f_2) \in V_3 \times V_1 \times V_2$. In particular if $\{e_j^{(1)}\}_{j=1}^p$, $\{e_j^{(2)}\}_{j=1}^r$, $\{\phi_j\}_{j=1}^q$ are $\mathbb{C}$-bases of V_1, V_2, V_3 respectively, the complex numbers

$$\begin{aligned} I_{sij} &\stackrel{\text{def}}{=} \langle \phi_s e_i^{(1)}, e_j^{(2)} \rangle \qquad \begin{smallmatrix} 1 \le s \le q \\ 1 \le i \le p \\ 1 \le j \le r \end{smallmatrix} \\ &= \int_X \phi_s(x) e_i^{(1)}(x) \overline{e_j^{(2)}(x)}\, dx \end{aligned} \tag{7.5.11}$$

are well defined. Assume, finally, that G is actually *compact*. Thus, to review our assumptions, G is a compact topological group which acts on a locally compact topological space X on the left (where the topologies of G and X are Hausdorff), and dx is a regular Baire, Borel, or Radon measure on X which is G-invariant and subject to condition (7.5.6). For $C_2(X)$ the space of continuous, complex-valued, square-integrable functions on X (with respect to the measure dx) let $V_1, V_2 \subset C_2(X)$ be finite-dimensional G-invariant subspaces (where $(g \cdot f)(x) \stackrel{\text{def}}{=} f(g^{-1} \cdot x)$ for $g \in G$, for a function f on X, and for $x \in X$) and let $\pi^{(1)}, \pi^{(2)}$ be the corresponding finite-dimensional continuous, unitary representations (with respect to the inner product on $C_2(X)$ defined in (7.5.7)) of G on V_1, V_2 given by $\pi^{(j)}(g)f \stackrel{\text{def}}{=} g \cdot f$ for $g \in G$, $f \in V_j$, $j = 1, 2$, as described above. Let ϕ be a continuous function on X such that

1. ϕ is G-finite—i.e., $V_3 \stackrel{\text{def}}{=}$ the $\mathbb{C}$-span of the vectors $\{g \cdot \phi | g \in G\}$ is finite-dimensional,

2. ϕf_1 is square-integrable with respect to dx $\forall$ $f_1 \in V_1$,

3. the corresponding representation $\pi^{(3)} : G \to \mathrm{Gl}(V_3)$ of G on V_3 given by (7.5.10) is continuous.

As we have noted assumption 2 implies that $V_3 V_1 \subset C_2(X)$ so that the inner product $\langle \psi f_1, f_2 \rangle$ of $\psi f_1, f_2 \in C_2(X)$ is well defined for $(\psi, f_1, f_2) \in V_3 \times V_1 \times V_2$, and in particular the integrals I_{sij} in (7.5.11) are well defined. With these assumptions (just reviewed) in place one has the following main theorem, where we recall that π^* denotes the contragredient representation of a representation π of G. Also see Theorem 7.6 for an alternate version of the main theorem.

Theorem 7.2 (An Abstract Selection Rule). *Suppose the representation $(\pi^{(1)} \otimes \pi^{(3)}) \otimes \pi^{(2)*}$ does not contain the trivial one-dimensional representation of G; see* *example 7 above. Then the integrals I_{sij} in (7.5.11) vanish for all s, i, j. If $\pi^{(2)}$ is actually irreducible and is not contained in $\pi^{(1)} \otimes \pi^{(3)}$ we may also conclude (more simply) that all of the integrals I_{sij} vanish.*

A proof of Theorem 7.2, based on the classical *orthogonality relations* for representations of compact groups, is presented in Appendix 7A. Before proceeding to some concrete applications of Theorem 7.2 we first present an alternative expression for the integrals I_{sij}.

Let $\{e_j^{(1)*}\}_{j=1}^p$, $\{e_j^{(2)*}\}_{j=1}^r$, $\{\phi_j^*\}_{j=1}^q$ be the $\mathbb{C}$-bases of V_1^*, V_2^*, V_3^* dual to $\{e_j^{(1)}\}_{j=1}^p$, $\{e_j^{(2)}\}_{j=1}^r$, $\{\phi_j\}_{j=1}^q$ respectively. Then for $g \in G$

$$\pi^{(1)}(g)e_j^{(1)} = \sum_{m=1}^{p} e_m^{(1)*}(\pi^{(1)}(g)e_j^{(1)})e_m^{(1)} \tag{7.5.12}$$

$$\pi^{(2)}(g)e_l^{(2)} = \sum_{s=1}^{r} e_s^{(2)*}(\pi^{(2)}(g)e_l^{(2)})e_s^{(2)} \tag{7.5.13}$$

$$\pi^{(3)}(g)\phi_t = \sum_{n=1}^{q} \phi_n^*(\pi^{(3)}(g)\phi_t)\phi_n. \tag{7.5.14}$$

As $\langle\ ,\ \rangle$ is G-invariant (i.e., dx is G-invariant) one has $\forall\ g \in G$

$$\begin{aligned} I_{tjl} &\stackrel{\text{def}}{=} \langle \phi_t e_j^{(1)}, e_l^{(2)} \rangle = \langle g \cdot (\phi_t e_j^{(1)}), g \cdot e_l^{(2)} \rangle \\ &= \langle (g \cdot \phi_t)(g \cdot e_j^{(1)}), g \cdot e_l^{(2)} \rangle \\ &\stackrel{\text{def}}{=} \langle (\pi^{(3)}(g)\phi_t)(\pi^{(1)}(g)e_j^{(1)}), \pi^{(2)}(g)e_l^{(2)} \rangle \\ &= \sum_{n,m,s} \phi_n^*(\pi^{(3)}(g)\phi_t)\, e_m^{(1)*}(\pi^{(1)}(g)e_j^{(1)})\, \overline{e_s^{(2)*}(\pi^{(2)}(g)e_l^{(2)})}\, I_{nms} \end{aligned} \tag{7.5.15}$$

since $\langle \phi_n e_m^{(1)}, e_s^{(2)} \rangle = I_{nms}$ by definition (7.5.11). Let $\{\omega_{ij}\}_{\substack{1\le i\le p \\ 1\le j\le q}}$ be the basis of $(V_1 \otimes V_3)^*$ dual to the basis $\{e_i^{(1)} \otimes \phi_j\}_{ij}$ of $V_1 \otimes V_3$. Note that for $\rho = \pi^{(1)} \otimes \pi^{(3)}$

$$\omega_{mn}(\rho(g)(e_j^{(1)} \otimes \phi_t)) = e_m^{(1)*}(\pi^{(1)}(g)e_j^{(1)})\phi_n^*(\pi^{(3)}(g)\phi_t) \tag{7.5.16}$$

by (7.2.10). If dg denotes normalized Haar measure on G, then by (7.5.15) and (7.5.16)

$$\begin{aligned} I_{tjl} &= \int_G I_{tjl}\, dg \\ &= \sum_{n,m,s} I_{nms} \int_G \omega_{mn}((\pi^{(1)} \otimes \pi^{(3)})(g)(e_j^{(1)} \otimes \phi_t))\, \overline{e_s^{(2)*}(\pi^{(2)}(g)e_l^{(2)})}\, dg, \end{aligned} \tag{7.5.17}$$

where the integrands are continuous as functions on G since the representations $\pi^{(j)}$, $j = 1, 2, 3$, are continuous. Equation (7.5.17) provides for the desired alternative expression of the integral I_{tjl} where the integration over G, versus the integration over X in the definition (7.5.11), has the advantage that the group theory of G readily applies.

7.6 Some Preliminary Applications of Theorem 7.2

Fix integers n, l, n', l' with $n, n' \geq 1$, $0 \leq l \leq n-1$, $0 \leq l' \leq n'-1$. We take $G = SO(3)$ with its left action on $X = \mathbb{R}^3$ given in example 3 of Section 7.4: $g \cdot x \overset{\text{def}}{=} xg^t$ for $x \in \mathbb{R}^3$ a row vector, $g \in SO(3)$. In (7.5.1) we choose $dx = dxdydz$ = Lebesgue measure on $\mathbb{R}^3$. If A is any real 3×3 non-singular matrix one has by the transformation of integrals formula

$$\int_{\mathbb{R}^3} f(xA)dx = \frac{1}{|\det A|} \int_{\mathbb{R}^3} f(x)dx \tag{7.6.1}$$

for $f \in L^1(\mathbb{R}^3, dx)$. An analogous result holds for $\mathbb{R}^n$ of course. In particular dx is indeed G-invariant; compare definition (7.5.1). Also dx satisfies condition (7.5.6); in fact dx is a Haar measure on $\mathbb{R}^3$. Let $\pi^{(n,l)} \simeq \pi^{(l)}$, $\pi^{(n',l')} \simeq \pi^{(l')}$ be the continuous, irreducible representations of G on the spaces $V_1 = S_{nl}$, $V_2 = S_{n'l'}$ generated by the hydrogenic wave functions $\{\psi_{nlm}\}_{m=-l}^{l}$, $\{\psi_{n'l'm'}\}_{m'=-l'}^{l'}$, as described in example 6 of Section 7.1. $V_1, V_2 \subset C_2(X)$, the space of continuous, square-integrable functions on $\mathbb{R}^3$, by Proposition 5.2. For $p_j : \mathbb{R}^3 \to \mathbb{R}$ the projection of (x, y, z) to x, y, z for $j = 1, 2, 3$, respectively, we have seen in example 4 of Section 7.1 that $V_3^{(j)} \overset{\text{def}}{=}$ the $\mathbb{C}$-span of the vectors $\{g \cdot p_j | g \in G\}$ (also denoted $G \cdot p_j$) = the $\mathbb{C}$-span of the vectors $\{p_1, p_2, p_3\}$ in the space of continuous functions on $\mathbb{R}^3$. Thus $V_3^{(1)} = V_3^{(2)} = V_3^{(3)}$. Moreover the natural representation π of G on $V_3^{(j)}$ is continuous and is equivalent to the natural representation π of G on $\mathbb{C}^3$ given in example 3 of Section 7.1. We have seen in fact in example 5 that π is in fact irreducible and is equivalent to $\pi^{(1)}$. Each $p_j\psi_{nlm}$ is square-integrable. Thus all the hypotheses leading up to the statement of Theorem 7.2 are fulfilled. Since $\pi^{(n',l')} \simeq \pi^{(l')}$ is actually irreducible we may conclude by Theorem 7.2 that if $\pi^{(l')}$ is *not* contained in $\pi^{(l)} \otimes \pi^{(1)}$, then all of the corresponding integrals I_{sij} in (7.5.11) must vanish. In the present case these integrals are the integrals

$$I_{jmm'}^{nn'll'} \overset{\text{def}}{=} \int_{\mathbb{R}^3} p_j(x, y, z)\, \psi_{nlm}(x, y, z)\, \overline{\psi}_{n'l'm'}(x, y, z)\, dxdydz, \tag{7.6.2}$$

where $1 \leq j \leq 3$, $-l \leq m \leq l$, $-l' \leq m' \leq l'$. On the other hand by the Clebsch–Gordan formula (Theorem 7.1)

$$\pi^{(l)} \otimes \pi^{(1)} = \begin{bmatrix} \pi^{(l+1)} \oplus \pi^{(l)} \oplus \pi^{(l-1)} & \text{if } l \geq 1 \\ \pi^{(1)} & \text{if } l = 0 \end{bmatrix}. \tag{7.6.3}$$

Therefore if $l \geq 1$, then $\pi^{(l')}$ is not contained in $\pi^{(l)} \otimes \pi^{(1)}$ if $l' \neq l+1, l$, and $l-1$ in which case

$$I_{jmm'}^{nn'll'} = 0 \quad \text{for all } j, m, m'. \tag{7.6.4}$$

If $l = 0$, then $\pi^{(l')}$ is not contained in $\pi^{(l)} \otimes \pi^{(1)}$ if $l' \neq 1$ in which case

$$I^{nn'0l'}_{jmm'} = 0 \quad \text{for all } j, m, m'. \tag{7.6.5}$$

Note that

$$I^{nn'll'}_{jmm'} = \int_0^\infty \int_0^\pi \int_0^{2\pi} p_j(g(r,\theta,\phi))\, \psi_{nlm}(g(r,\theta,\phi)) \cdot \overline{\psi}_{n'l'm'}(g(r,\theta,\phi)) r^2 \sin\phi \; d\phi d\theta dr \tag{7.6.6}$$

with

$$\begin{aligned} p_1(g(r,\theta,\phi)) &= r\cos\theta\sin\phi, \\ p_2(g(r,\theta,\phi)) &= r\sin\theta\cos\phi, \\ p_3(g(r,\theta,\phi)) &= r\cos\phi; \end{aligned} \tag{7.6.7}$$

see definition (4.A.2) in Appendix 4A.

Since Y_l^m is homogeneous of degree l, $Y_l^m(-x,-y,-z) = (-1)^l\, Y_l^m(x,y,z)$. By (5.2.12) it follows therefore that

$$\psi_{nlm}(-x,-y,-z) = (-1)^l \psi_{nlm}(x,y,z). \tag{7.6.8}$$

Since Lebesgue measure $dxdydz$ is invariant under the transformation $(x,y,z) \to (-x,-y,-z)$ we can use (7.6.8) to deduce from definition (7.6.2) that

$$\begin{aligned} I^{nn'll'}_{jmm'} &= \int_{\mathbb{R}^3} p_j(-x,-y,-z)\, (-1)^{l+l'}\, \psi_{nlm}(x,y,z)\, \overline{\psi}_{n'l'm'}(x,y,z)\, dxdydz \\ &= (-1)^{l+l'+1}\, I^{nn'll'}_{jmm'}, \end{aligned} \tag{7.6.9}$$

which shows that

$$I^{nn'll'}_{jmm'} = 0 \quad \text{if } l + l' \text{ is even (i.e., if } l, l' \text{ have the same parity).} \tag{7.6.10}$$

In particular if $l = l'$ (so that $l + l'$ is even), then

$$I^{nn'll'}_{jmm'} = 0. \tag{7.6.11}$$

Combining statements (7.6.4), (7.6.5), (7.6.11) one obtains

Theorem 7.3. *If $l' \neq l+1$ and $l-1$, then the integrals $I^{nn'll'}_{jmm'}$ defined in (7.6.2) (also see (7.6.6), (7.6.7)) vanish for all n, n', j, m, m'—i.e., for $1 \leq j \leq 3$, $-l \leq m \leq l$, $-l' \leq m' \leq l'$, $n, n' \geq 1$, $0 \leq l \leq n-1$, $0 \leq l' \leq n'-1$.*

By (5.3.9) and (7.6.6)

$$I^{nn'll'}_{jmm'} = \int_0^\infty \int_0^\pi \int_0^{2\pi} p_j(g(r,\theta,\phi))\, R^-_{nl}(r)\, \Phi_m(\theta)\, X_{lm}(\phi) \cdot R^-_{n'l'}(r)\, \overline{\Phi}_{m'}(\theta)\, X_{l'm'}(\phi)\, r^2 \sin\phi \; d\theta d\phi dr. \tag{7.6.12}$$

Consider the integration here with respect to θ:

$$\begin{aligned}
&\int_0^{2\pi} p_j(g(r,\theta,\phi))\ \Phi_m(\theta)\overline{\Phi_{m'}}(\theta)\ d\theta \\
&\stackrel{\text{def}}{=} \frac{1}{2\pi}\int_0^{2\pi} p_j(g(r,\theta,\phi))\ e^{i(m-m')\theta}\ d\theta \\
&\stackrel{\text{def}}{=} \frac{r}{2\pi}\begin{bmatrix} (sin\phi)\int_0^{2\pi}\cos\theta e^{i(m-m')\theta}\ d\theta & \text{for } j=1 \\ (cos\phi)\int_0^{2\pi}\sin\theta e^{i(m-m')\theta}\ d\theta & \text{for } j=2 \\ (cos\phi)\int_0^{2\pi} e^{i(m-m')\theta}\ d\theta & \text{for } j=3 \end{bmatrix}.
\end{aligned} \tag{7.6.13}$$

By the orthogonality relations

$$\frac{1}{2\pi}\int_0^{2\pi} e^{ip\theta}e^{-iq\theta}\ d\theta = \delta_{pq} \tag{7.6.14}$$

for integers p, q, and the fact that $\cos\theta = (e^{i\theta}+e^{-i\theta})/2$, $\sin\theta = (e^{i\theta}-e^{-i\theta})/2i$ we deduce by (7.6.12), (7.6.13) the following.

Theorem 7.4. *If* $m' \neq m+1, m-1$, *and* m *then the integrals* $I_{jmm'}^{nn'll'}$ *defined in* (7.6.2) *(also see* (7.6.6), (7.6.7)*) vanish for all* n, n', l, l', *and* j. *If* $m' \neq m$, *then* $I_{3mm'}^{nn'll'} = 0$ *for all* n, n', l, l'. *If* $m' \neq m+1, m-1$ *then both* $I_{1mm'}^{nn'll'}, I_{2mm'}^{nn'll'} = 0$ *for all* n, n', l, l'.

7.7 Time-Dependent Perturbations and Spectroscopic Selection Rules

Our discussions regarding selection rules so far have been largely mathematical in character. To connect them more directly to physics we indicate how integrals like the $I_{jmm'}^{nn'll'}$ in (7.6.2) arise in connection with first-order *time-dependent perturbation theory*. This theory serves as a platform from which *spectroscopic selection rules* can be derived—i.e., rules by which one determines whether a transition from one quantum state to another is allowed or forbidden. Once some more general results are established we can specialize them to determine some allowed and forbidden transitions of the hydrogenic atom, given Theorems 7.3 and 7.4.

In *quantum dynamics* one considers transitions of a quantum system from one energy state to another—a process which is obviously time-dependent, and which is governed by the time-dependent Schrödinger equation

$$i\hbar\frac{\partial\psi}{\partial t} = \frac{\hbar^2}{2m}\nabla^2\psi + V\psi = H_o\psi \quad \text{for} \tag{7.7.1}$$

$$\nabla^2 = \frac{\partial^2}{\partial x^2}+\frac{\partial^2}{\partial y^2}+\frac{\partial^2}{\partial z^2}, \qquad H_o = -\frac{\hbar^2}{2m}\nabla^2 + V,$$

where we work in $\mathbb{R}^3$ for the sake of specificity. Suppose we are given n stationary states $\psi_1, \ldots, \psi_n : \mathbb{R}^3 \to \mathbb{C}$ (or even a countably infinite number of such states, as we have had in previous examples):

$$H_o\psi_j = E_j\psi_j, \qquad E_j \in \mathbb{R}. \tag{7.7.2}$$

E_j is the energy of the state ψ_j. If we set

$$\Psi_j(x, y, z, t) = \psi_j(x, y, z)e^{-iE_jt/\hbar}, \tag{7.7.3}$$

then, as observed in Chapter 1, we obtain solutions Ψ_j of (7.7.1). The $\{\psi_j\}_j$ will be assumed to form an orthonormal set. Consider the case $n = 2$ for example. Here we have in mind a quantum system passing from the state $\Psi_1(x, y, z, t)$ to the state $\Psi_2(x, y, z, t)$. During this passage its wave function will be given by a superposition

$$\Psi(x, y, z, t) = a_1(t)\Psi_1(x, y, z, t) + a_2(t)\Psi_2(x, y, z, t) \tag{7.7.4}$$

of the wave functions Ψ_1, Ψ_2 where the coefficients a_1, a_2 are time-dependent. The passage, or transition, could be triggered for example by the interaction of the system with electromagnetic radiation. Mathematically, this means that the function Ψ will be subject to Schrödinger's equation of the form

$$i\hbar\frac{\partial\Psi}{\partial t} = (H_o + H_1)\Psi \tag{7.7.5}$$

where H_1 is a suitable real-valued energy function of position (x, y, z) and *time* t: H_1 represents the *perturbation* of the initial state $\Psi_1(x, y, z, t)$ of the system. Thus the (time-dependent) Hamiltonian of the system is given by $H_o + H_1$. Since $\Psi_1(x, y, z, t)$ is the initial state of the system we require that at $t = 0$ (before the perturbation is "switched on") that Ψ and Ψ_1 coincide; i.e.,

$$\Psi(x, y, z, 0) = \Psi_1(x, y, z, 0) = \psi_1(x, y, z), \tag{7.7.6}$$

by (7.7.3).

Let's consider more generally solutions Ψ of (7.7.5) of the form

$$\Psi(x, y, z, t) = \sum_{j=1}^{n} a_j(t)\Psi_j(x, y, z, t) \tag{7.7.7}$$

where the Ψ_j, ψ_j are as before (with $\{\psi_j\}_j$ an orthonormal set), where Ψ is subject to the initial condition (7.7.6). Now

$$\begin{aligned}\frac{\partial\Psi}{\partial t}(x, y, z, t)i\hbar &= \sum_{j=1}^{n}\left[a_j(t)\frac{\partial\Psi_j}{\partial t}(x, y, z, t) + a_j'(t)\Psi_j(x, y, z, t)\right] i\hbar \\ &= \sum_{j=1}^{n} a_j(t)H_o\Psi_j(x, y, z, t) + \sum_{j=1}^{n} a_j'(t)\Psi_j(x, y, z, t)i\hbar\end{aligned} \tag{7.7.8}$$

by (7.7.1), where by (7.7.5) the left-hand side here is

$$(H_1 + H_o)\Psi(x, y, z, t) = H_1(x, y, z, t)\sum_{j=1}^{n} a_j(t)\Psi_j(x, y, z, t)$$

$$+\sum_{j=1}^{n} a_j(t) H_o \Psi_j(x, y, z, t) \quad (7.7.9)$$

since the a_j do not involve x, y, z. Thus

$$\sum_{j=1}^{n} a_j'(t) i\hbar \Psi_j(x, y, z, t) = H_1(x, y, z, t) \sum_{j=1}^{n} a_j(t) \Psi_j(x, y, z, t). \quad (7.7.10)$$

Note that by definition (7.7.3)

$$\Psi_j(x, y, z, t)\overline{\psi_l}(x, y, z) = \psi_j(x, y, z)\overline{\psi_l}(x, y, z)e^{-iE_j t/\hbar}. \quad (7.7.11)$$

Therefore since $\{\psi_j\}_j$ is an orthonormal set we multiply both sides of equation (7.7.10) by $\overline{\psi_l}(x, y, z)$ and integrate over $\mathbb{R}^3$ to obtain

$$i\hbar a_l'(t)e^{-iE_l t/\hbar} = \sum_{j=1}^{n} a_j(t) \iiint_{\mathbb{R}^3} H_1(x, y, z, t)\, \psi_j(x, y, z) \cdot \overline{\psi_l}(x, y, z)\, dxdydz\, e^{-iE_j t/\hbar}. \quad (7.7.12)$$

Given the perturbation H_1, its *matrix* $H(t)$ with respect to the set $\{\psi_j\}$ is defined by

$$\begin{aligned} H_{jl}(t) &\stackrel{\text{def}}{=} \iiint_{\mathbb{R}^3} H_1(x, y, z, t)\psi_j(x, y, z)\overline{\psi_l}(x, y, z)\, dxdydz \\ &= \langle H_1(\ ;t)\psi_j, \psi_l\rangle. \end{aligned} \quad (7.7.13)$$

$H(t)$ is Hermitian as H_1 is real-valued: $\overline{H}_{jl}(t) = H_{lj}(t)$. By (7.7.12) the coefficients $a_l(t)$, $1 \le l \le n$, therefore satisfy

$$\begin{aligned} a_l'(t) &= -\frac{i}{\hbar}\sum_{j=1}^{n} a_j(t) H_{jl}(t) e^{i(E_l - E_j)t/\hbar} \\ &= -\frac{i}{\hbar}\left[a_l(t)H_{ll}(t) + \sum_{j\neq l} a_j(t)H_{jl}(t)e^{i(E_l-E_j)t/\hbar}\right]. \end{aligned} \quad (7.7.14)$$

Also by the initial condition (7.7.6)

$$\begin{aligned} \delta_{1l} &= \iiint_{\mathbb{R}^3} \psi_1(x, y, z)\overline{\psi_l}(x, y, z)\, dxdydz \\ &= \iiint_{\mathbb{R}^3} \Psi(x, y, z, 0)\, \overline{\psi_l}(x, y, z)\, dxdydz \\ &\stackrel{\text{def}}{=} \sum_{j=1}^{n} a_j(0) \iiint_{\mathbb{R}^3} \psi_j(x, y, z)\overline{\psi_l}(x, y, z)\, dxdydz \\ &= \sum_{j=1}^{n} a_j(0)\delta_{jl} = a_l(0). \end{aligned} \quad (7.7.15)$$

That is, $a_1(0) = 1$, $a_l(0) = 0$ for $l \neq 1$, which we use in (7.7.14) to obtain moreover that for $l \neq 1$

$$\begin{aligned} a_l'(0) &= -\frac{i}{\hbar} \sum_{j \neq l} a_j(0) H_{jl}(0) \\ &= -\frac{i}{\hbar} \left[\sum_{j \neq l,1} a_j(0) H_{jl}(0) + a_1(0) H_{1l}(0) \right] \\ &= -\frac{i}{\hbar} H_{1l}(0). \end{aligned} \tag{7.7.16}$$

On the other hand, for $l = 1$,

$$\begin{aligned} a_1'(0) &= -\frac{i}{\hbar} \left[a_1(0) H_{11}(t) + \sum_{j \neq 1} a_j(0) H_{j1}(0) \right] \\ &= -\frac{i}{\hbar} H_{11}(t). \end{aligned} \tag{7.7.17}$$

That is, condition (7.7.6) implies

$$a_1(0) = 1, \qquad a_l(0) = 0 \text{ for } l \neq 1, \qquad a_l'(0) = -\frac{i}{\hbar} H_{1l}(0) \text{ for all } l. \tag{7.7.18}$$

We assume Ψ is normalized:

$$\begin{aligned} 1 = \langle \Psi, \Psi \rangle &= \sum_{j,l} a_j(t) \bar{a}_l(t)\, e^{i(E_l - E_j)t/\hbar} \langle \psi_j, \psi_l \rangle \\ &= \sum_j |a_j(t)|^2 \end{aligned} \tag{7.7.19}$$

since

$$\Psi_j(x, y, z, t)\, \overline{\Psi_l}(x, y, z, t) = \psi_j(x, y, z) \overline{\psi_l}(x, y, z) e^{i(E_l - E_j)t/\hbar} \tag{7.7.20}$$

by (7.7.3). The moduli $|a_j(t)|$ in (7.7.19) are given as follows.

$$\begin{aligned} \langle \Psi(\ ;t), \psi_l \rangle &\stackrel{\text{def}}{=} \iiint_{\mathbb{R}^3} \Psi(x, y, z, t)\, \overline{\psi_l}(x, y, z)\, dxdydz \\ &= \sum_j a_j(t) \iiint_{\mathbb{R}^3} \Psi_j(x, y, z, t)\, \overline{\psi_l}(x, y, z)\, dxdydz \\ &= \sum_j a_j(t)\, e^{-iE_j t/\hbar}\, \delta_{jl} = a_l(t)\, e^{-iE_l t/\hbar} \end{aligned} \tag{7.7.21}$$

by (7.7.7), and (7.7.11), which implies that

$$|a_j(t)|^2 = |\langle \Psi(\ ;t), \psi_j \rangle|^2, \tag{7.7.22}$$

$\Psi(\ ;t)$ being the function $(x,y,z) \to \Psi(x,y,z,t)$. In (7.7.7) we may write

$$\Psi(\ ;t) = \sum_j a_j(t)e^{-iE_jt/\hbar}\psi_j \tag{7.7.23}$$

by (7.7.3), where by (7.7.2)

$$H_o(e^{-iE_jt/\hbar}\psi_j) = E_j(e^{-iE_jt/\hbar}\psi_j). \tag{7.7.24}$$

In other words (7.7.23) is a type of "Fourier expansion" of $\Psi(\ ;t)$ in terms of eigenfunctions $e^{-iE_jt/\hbar}\psi_j$ of H_o (with corresponding eigenvalue E_j), with the $a_j(t)$ appearing as Fourier coefficients. Using (7.7.19), (7.7.24) we can discover the physical meaning of the $|a_j(t)|^2$. To this end consider the average or expected value $\langle E\rangle$ of the energy of our system at time t. By definition (3.9.2)

$$\langle E\rangle = \langle H_o\Psi(\ ;t), \Psi(\ ;t)\rangle \tag{7.7.25}$$

as H_o is the operator corresponding to the "observable" E, and as Ψ is normalized; cf. definitions (6.2.6), (6.2.7). Therefore by (7.7.23), (7.7.24)

$$\begin{aligned}\langle E\rangle &= \left\langle \sum_j a_j(t)E_je^{-iE_jt/\hbar}\psi_j, \sum_l a_l(t)e^{-iE_lt/\hbar}\psi_l \right\rangle \\ &= \sum_{j,l} a_j(t)\overline{a_l}(t)E_je^{i(E_l-E_j)t/\hbar}\delta_{jl} \\ &= \sum_j |a_j(t)|^2 E_j.\end{aligned} \tag{7.7.26}$$

But if one is given a set of "scores" $E_1, E_2, \ldots$ with frequencies $f_1, f_2, \ldots$, then the average value $\overline{E}$ of these scores in the ordinary sense is

$$\overline{E} = \frac{\sum_j f_jE_j}{\sum_j f_j} = \sum_j p_jE_j \tag{7.7.27}$$

where $p_j = f_j/\sum_k f_k$ is the probability that a given score has the value E_j. Comparing (7.7.26) with (7.7.27) we are naturally lead to regard each $|a_j(t)|^2$ as a *probability* (= a frequency of the "occurrence of E_j," by (7.7.19)). Namely, *$|a_j(t)|^2$ is the probability that a measurement of the energy of the system at time t will yield the value E_j.* We also say (somewhat less precisely) that $|a_j(t)|^2$ is the probability of finding the system in the state ψ_j at time t. The conclusion on the meaning of the $|a_j(t)|^2$ also follows by general principles of quantum mechanics. Compare the remarks following equation (3.9.10). Rather than a systematic development of such principles we have chosen to illustrate the same in specific situations.

For physical reasons we see therefore that it is of great importance to compute the $|a_j(t)|^2$, say for $j \geq 2$. However there is no general procedure to obtain solutions of the system of linear differential equations in (7.7.14), even when $n = 2$.

Thus one has no choice but to look for approximate solutions. Here we apply the method of *first-order time-dependent perturbation theory*. The idea here is very simple: From the initial conditions $a_1(0) = 1$, $a_l(0) = 0$ for $l \neq 1$ in (7.7.18) we look for approximate solutions $a_j(t)$, say for very small t, by replacing the $a_j(t)$ in (7.7.14) by their initial values. Thus, were we to take $a_1(t) = 1$, $a_l(t) = 0$ for $l \neq 1$, for small t, equations (7.7.14) would assume the following simpler form.

$$a_1'(t) \simeq -\frac{i}{\hbar}[H_{11}(t) + 0] \tag{7.7.28}$$

and for $l \neq 1$

$$\begin{aligned} a_l'(t) &\simeq -\frac{i}{\hbar}\left[0 + \sum_{j \neq l} a_j(t) H_{jl}(t) e^{i(E_l - E_j)t/\hbar}\right] \\ &= -\frac{i}{\hbar}\left[\text{term for } j = 1 + \sum_{j \neq 1,l} a_j(t) H_{jl}(t) e^{i(E_l - E_j)t/\hbar}\right] \\ &= -\frac{i}{\hbar}\,(\text{term for } j=1) \\ &= -\frac{i}{\hbar} H_{1l}(t) e^{i(E_l - E_1)t/\hbar}. \end{aligned} \tag{7.7.29}$$

We take the approximations in (7.7.28), (7.7.29) as *motivation* for the following formal definitions of functions $a_l^{(1)}(t)$, $1 \leq l \leq n$, where the superscript $^{(1)}$ denotes first-order approximation, not the first derivative:

$$a_1^{(1)}(t) \stackrel{\text{def}}{=} -\frac{i}{\hbar}\int_0^t H_{11}(r)\,dr + 1, \qquad \text{and} \tag{7.7.30}$$

$$a_l^{(1)}(t) \stackrel{\text{def}}{=} -\frac{i}{\hbar}\int_0^t H_{1l}(r) e^{i(E_l - E_1)r/\hbar}\,dr \qquad \text{for } l \neq 1. \tag{7.7.31}$$

The constant 1 is added in the definition of $a_1^{(1)}(t)$ to secure the initial conditions $a_1^{(1)}(0) = 1$, $a_l^{(1)}(0) = 0$ for $l \neq 1$, as in (7.7.18). By definition

$$a_1^{(1)\prime}(t) = -\frac{i}{\hbar} H_{11}(t), \qquad \text{and} \tag{7.7.32}$$

$$a_l^{(1)\prime}(t) = -\frac{i}{\hbar} H_{1l}(t) e^{i\omega_{l1}t} \qquad \text{for } l \neq 1 \tag{7.7.33}$$

where we now write

$$\omega_{lj} \stackrel{\text{def}}{=} (E_l - E_j)/\hbar. \tag{7.7.34}$$

The ω_{lj} are called *Bohr frequencies*. In particular we also have

$$a_l^{(1)\prime}(0) = -\frac{i}{\hbar} H_{1l}(0) \text{ for all } l, \tag{7.7.35}$$

as in (7.7.18). Since we used a rough approximation argument to motivate the definition of the $a_l^{(1)}(t)$, and also since we have already noted the system in (7.7.14) cannot be solved in general, the $a_l^{(1)}(t)$ cannot be expected to solve (7.7.14). We would like to understand however to what extent they provide an approximate solution. Let

$$A_l(t) \stackrel{\text{def}}{=} a_l^{(1)\prime}(t) \quad + \quad \frac{i}{\hbar}\left[a_l^{(1)}(t)H_{ll}(t) + \sum_{j\neq l} a_j^{(1)}(t)H_{jl}(t)e^{i\omega_{lj}t}\right]. \quad (7.7.36)$$

Then $a_l^{(1)}(t)$ solves (7.7.14) $\iff$ $A_l(t) = 0$. The point is to give a bound for $A_l(t)$.

Proposition 7.2.

$$|A_l(t)| \leq \frac{1}{\hbar^2}\sum_j |H_{jl}(t)| \int_0^t |H_{1j}(r)|\,dr.$$

Hence $A_l(t) \to 0$ as $t \to 0$.

Proposition 7.2 is proved in Appendix 7B to this chapter. We will mainly be interested in the case when H_1 is a *harmonic* perturbation:

$$H_1(x, y, z, t) = V_1(x, y, z)\cos\omega t \quad (7.7.37)$$

for some potential energy function V_1. If

$$V_{jl} \stackrel{\text{def}}{=} \iiint_{\mathbb{R}^3} V_1(x, y, z)\,\psi_j(x, y, z)\,\overline{\psi_l}(x, y, z)\,dxdydz, \quad (7.7.38)$$

then by definition (7.7.13)

$$H_{jl}(t) = V_{jl}\cos\omega t \quad (7.7.39)$$

for H_1 in (7.7.37), and in this case

$$|A_l(t)| \leq \frac{1}{\hbar^2}\sum_j |V_{jl}||V_{1j}|t \quad (7.7.40)$$

by Proposition 7.2. We see that in some sense the $a_l^{(1)}$ are justifiably first-order approximate solutions of (7.7.14). One can obtain better approximations (second, third-order, etc.), which we will not go into since the first-order approximations (first-order perturbation theory) suffice indeed for a wide range of practical applications.

Having noted the physical importance of the $|a_l^{(1)}(t)|^2$ we come now to the main theorem of this section.

Theorem 7.5. *For the harmonic perturbation H_1 in (7.7.37), and the corresponding matrix entries V_{jl} in (7.7.38) one has for $l \neq 1$,*

$$|a_l^{(1)}(t)|^2 = \frac{|V_{1l}|^2}{\hbar^2}\left[\frac{\sin^2(\omega+\omega_{l1})t/2}{(\omega+\omega_{l1})^2} + \frac{\sin^2(\omega_{l1}-\omega)t/2}{(\omega_{l1}-\omega)^2} + \right.$$
$$\left. + 2(\cos\omega t)\left(\frac{\sin(\omega_{l1}-\omega)t/2}{\omega_{l1}-\omega}\right)\left(\frac{\sin(\omega_{l1}+\omega)t/2}{\omega_{l1}+\omega}\right)\right],$$

where the Bohr frequencies ω_{lj} are defined by (7.7.34).

Proof. For $l \neq 1$, $a_l^{(1)}(t)$ is given by definition (7.7.31) and equation (7.7.39):

$$\begin{aligned} a_l^{(1)}(t) &= -\frac{i}{\hbar}\int_0^t (\cos\omega r)V_{1l}e^{i\omega_{l1}r}\,dr \\ &= -\frac{V_{1l}}{2\hbar}\left[\frac{e^{i(\omega+\omega_{l1})t}-1}{\omega+\omega_{l1}} + \frac{e^{i(\omega_{l1}-\omega)t}-1}{\omega_{l1}-\omega}\right] \end{aligned} \tag{7.7.41}$$

where we use that

$$2\cos\omega r = e^{i\omega r} + e^{-i\omega r}. \tag{7.7.42}$$

Then multiplying $a_l^{(1)}(t)$ by $\overline{a_l^{(1)}(t)}$ and using (7.7.42) again we obtain

$$\begin{aligned} |a_l^{(1)}(t)|^2 &= \frac{|V_{1l}|^2}{4\hbar^2}\left[\frac{2-e^{-i(\omega+\omega_{l1})t}-e^{i(\omega+\omega_{l1})t}}{(\omega+\omega_{l1})^2}\right. \\ &+ \frac{e^{i(\omega_{l1}-\omega)t}e^{-i(\omega+\omega_{l1})t}-e^{-i(\omega+\omega_{l1})t}-e^{i(\omega_{l1}-\omega)t}+1}{(\omega_{l1}-\omega)(\omega+\omega_{l1})} \\ &+ \frac{e^{i(\omega_{l1}+\omega)t}e^{-i(\omega_{l1}-\omega)t}-e^{-i(\omega_{l1}-\omega)t}-e^{i(\omega_{l1}+\omega)t}+1}{(\omega_{l1}-\omega)(\omega+\omega_{l1})} \\ &+ \left.\frac{2-e^{-i(\omega_{l1}-\omega)t}-e^{i(\omega_{l1}-\omega)t}}{(\omega_{l1}-\omega)^2}\right] \\ &= \frac{|V_{1l}|^2}{4\hbar^2}\left[\frac{2-2\cos(\omega+\omega_{l1})t}{(\omega+\omega_{l1})^2} + (e^{-2i\omega t}-e^{-i(\omega+\omega_{l1})t}\right. \\ &- e^{i(\omega_{l1}-\omega)t}+e^{2i\omega t}-e^{-i(\omega_{l1}-\omega)t}-e^{i(\omega+\omega_{l1})t}+2)\frac{1}{w_{l1}^2-\omega^2} \\ &+ \left.\frac{2-2\cos(\omega_{l1}-\omega)t}{(\omega_{l1}-\omega)^2}\right] \\ &= \frac{|V_{1l}|^2}{4\hbar^2}\left[\frac{2-2\cos(\omega+\omega_{l1})t}{(\omega+\omega_{l1})^2} + \frac{2-2\cos(\omega_{l1}-\omega)t}{(\omega_{l1}-\omega)^2}\right. \\ &+ \left.\frac{2\cos 2\omega t - 2\cos(\omega+\omega_{l1})t - 2\cos(\omega_{l1}-\omega)t + 2}{\omega_{l1}^2-\omega^2}\right]. \end{aligned} \tag{7.7.43}$$

To simplify the latter expression we bring in some trigonometric identities. $2\sin^2\frac{\theta}{2} = 1-\cos\theta \implies$

$$\frac{2-2\cos(\omega+\omega_{l1})t}{(\omega+\omega_{l1})^2} = \frac{4\sin^2(\omega+\omega_{l1})t/2}{(\omega+\omega_{l1})^2},$$
$$\frac{2-2\cos(\omega_{l1}-\omega)t}{(\omega_{l1}-\omega)^2} = \frac{4\sin^2(\omega_{l1}-\omega)t/2}{(\omega_{l1}-\omega)^2}. \tag{7.7.44}$$

Also

$$2\cos 2\omega t + 2 = 2(1+\cos 2\omega t) = 4\cos^2 \omega t. \tag{7.7.45}$$

Next

$$\sin\left(\frac{a-b}{2}\right)\sin\left(\frac{a+b}{2}\right) = \frac{\cos b - \cos a}{2} \tag{7.7.46}$$

$$\Longrightarrow 2(\cos b)\sin\left(\frac{a-b}{2}\right)\sin\left(\frac{a+b}{2}\right) = \cos^2 b - \cos b\cos a \tag{7.7.47}$$

and

$$\cos x + \cos y = 2\cos\left(\frac{x+y}{2}\right)\cos\left(\frac{x-y}{2}\right) \tag{7.7.48}$$

$$\Longrightarrow 2\cos(\omega+\omega_{l1})t + 2\cos(\omega_{l1}-\omega)t = 4(\cos\omega_{l1}t)(\cos\omega t). \tag{7.7.49}$$

The numerator of the third term in (7.7.43) is therefore rewritten as follows:

$$\begin{aligned} &2\cos 2\omega t - 2\cos(\omega+\omega_{l1})t - 2\cos(\omega_{l1}-\omega)t + 2 \\ &\quad = 4\cos^2\omega t - 4(\cos\omega_{l1}t)(\cos\omega t) \quad \text{by (7.7.45) and by (7.7.49)} \\ &\quad = 8(\cos\omega t)\sin(\omega_{l1}-\omega)t/2\,\sin(\omega_{l1}+\omega)t/2, \end{aligned} \tag{7.7.50}$$

by (7.7.47). Theorem 7.5 now follows by (7.7.43), (7.7.44), and (7.7.50). □

The general harmonic perturbation H_1 in (7.7.37) will now be specialized. We consider an atom subject to an electromagnetic wave. It will experience a change in its energy. It is known that the magnetic component of the wave will bear less influence on the atom. Any appreciable influence will therefore be due to the electric component. As illustrated in Figure 7.1 we take, for example, the direction of propagation of the wave to be along the positive y-axis, with the electric vector $\vec{E}$ pointing in the z-direction.

The magnitude of $\vec{E}$ at time t and position (x, y, z) is determined by a traveling wave expression

$$\begin{aligned} E(x,y,z,t) &= E_o\cos(\kappa y - \omega t) \\ &= E_o\cos\left(\omega t - \frac{2\pi y}{\lambda}\right) \end{aligned} \tag{7.7.51}$$

where λ is the wavelength, $\kappa = \frac{2\pi}{\lambda}$ is the wave number, $\nu = \frac{\omega}{2\pi}$ is the frequency and E_o is the harmonic amplitude of the wave; compare equation (1.4.7) and the basic definitions given there. If y/λ is sufficiently small, $E(x, y, z, t)$ is well approximated via initial Maclaurin series terms:

$$E(x,y,z,t) \simeq E_o\left[\cos\omega t + \frac{2\pi y}{\lambda}\sin\omega t - \frac{4\pi^2 y^2}{\lambda^2}\cos\omega t\right]. \tag{7.7.52}$$

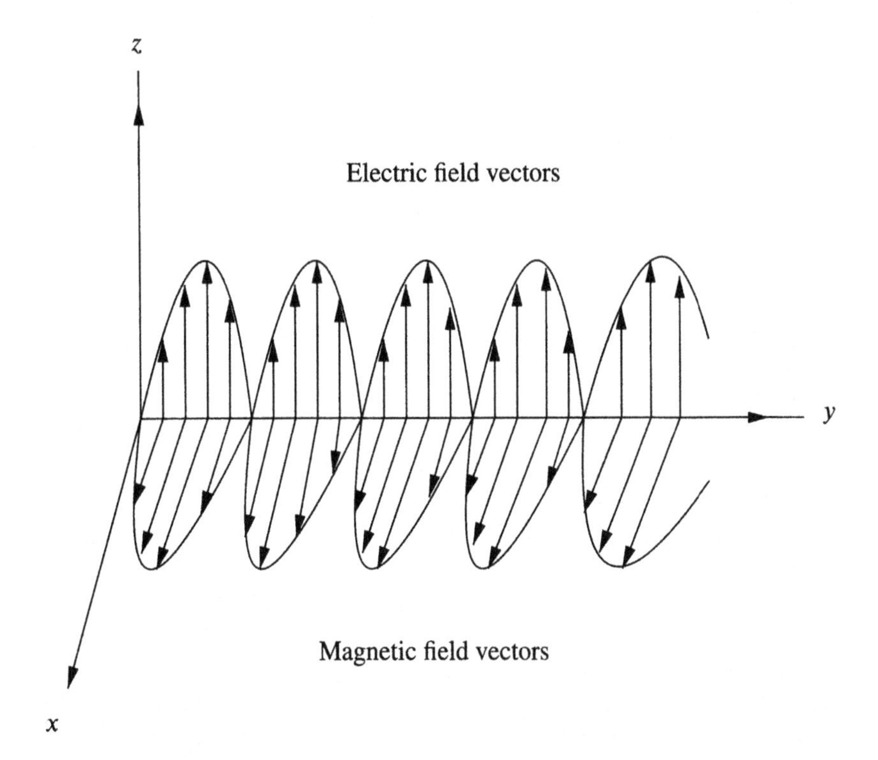

Figure 7.1: An electromagnetic wave propagated in the y-direction. The electric and magnetic field vectors are perpendicular to each other and to the direction in which the wave is propagated.

If for example λ is approximately of the order of a few thousand Ångströms ($\simeq 10^{-5}$ cm), as in the case of visible light, then since any distance y measured within the atom would roughly be of the order 10^{-8} cm = 1 Ångström, y/λ would be rather small, in which case one could reasonably choose the approximation

$$E(x, y, z, t) \simeq E_o \cos \omega t, \tag{7.7.53}$$

neglecting the powers $\frac{y}{\lambda}$, $\left(\frac{y}{\lambda}\right)^2$ in (7.7.52). Equation (7.7.53) means that there is effectively a uniform distribution of the electric field over the atom. One further point is to be made—a point which we will not attempt to develop here: the energy H_1, or Hamiltonian, of interaction of the atom and the electric field will be given by

$$H_1(x, y, z, t) = ezE(x, y, z, t) \tag{7.7.54}$$

where as before $-e$ is the electron charge. See remarks in Appendix 7D. In other words, in summary, an atom which is suddenly subjected to an electromagnetic wave (i.e., to interaction with electromagnetic radiation), with a large wavelength λ in comparison with the dimensions of that atom, will experience addition to its normal energy a perturbing energy of the amount $H_1(x, y, z, t) = ezE_o \cos \omega t$, approximately (by (7.7.53), (7.7.54)), where $\nu = \frac{\omega}{2\pi}$ is the wave frequency, and E_o is the harmonic amplitude of the electric component of the wave, whose electric vectors point in the z-direction (as in Figure 7.1). Thus, up to a reasonable approximation, $H_1(x, y, z, t)$ has the form (7.7.37), where

$$V_1(x, y, z) = ezE_o \tag{7.7.55}$$

in which case, by definition (7.7.38),

$$\begin{aligned} V_{jl} = V_{jl}^z &= eE_o \iiint_{\mathbb{R}^3} z\psi_j(x, y, z)\overline{\psi_l}(x, y, z)\, dxdydz \\ &= eE_o \iiint_{\mathbb{R}^3} p_3(x, y, z)\, \psi_j(x, y, z)\, \overline{\psi_l}(x, y, z)\, dxdydz, \end{aligned} \tag{7.7.56}$$

where again $p_1, p_2, p_3 : \mathbb{R}^3 \to \mathbb{R}$ are projections onto the first, second, third coordinates, respectively.

In the preceding discussion the electric field was assumed to be *polarized* in the z-direction: the electric vectors were assumed to point in the z-direction. We were thus led to the approximation $H_1(x, y, z, t) = ezE_o \cos \omega t$ (which is independent of y—i.e., of the originally assumed direction of propagation). If the electric vectors of the wave point in the x-direction, or in the y-direction we will have approximate perturbing energies $H_1(x, y, z, t) = exE_o \cos \omega t$ or $eyE_o \cos \omega t$ with corresponding matrix entries

$$V_{jl} = V_{jl}^x = eE_o \iiint_{\mathbb{R}^3} p_1(x, y, z)\, \psi_j(x, y, z)\, \overline{\psi_l}(x, y, z)\, dxdydz, \tag{7.7.57}$$

$$V_{jl} = V_{jl}^y = eE_o \iiint_{\mathbb{R}^3} p_2(x, y, z)\, \psi_j(x, y, z)\, \overline{\psi_l}(x, y, z)\, dxdydz, \tag{7.7.58}$$

respectively. We now see that the integrals $I_{jmm'}^{nn'll'}$ in (7.6.2) (which initially arose as examples of the integrals I_{sij} in (7.5.11)) appear as special cases of the integrals in (7.7.56), (7.7.57), (7.7.58).

From Theorem 7.5 various deductions can be made.

(i) The "transition probability" $|a_l^{(1)}(t)|^2$ as a function of ω attains its maximal value at the points $\omega = \pm\omega_{l1} \stackrel{\text{def}}{=} \pm(E_l - E_1)/\hbar$. This result has the following meaning. The atom will not likely absorb or emit radiation in the transition from the initial state ψ_1 to the state ψ_l unless $\omega = \pm\omega_{l1}$—i.e., unless $E_l = E_1 \pm \hbar\omega = E_1 \pm h\nu$ (as $\nu = \omega/2\pi$), which is the Bohr frequency rule; compare equation (1.2.1). In the absorption process the atom absorbs energy from the electromagnetic field in the amount $E_l - E_1 = h\nu$. In the emission process energy of the amount $E_1 - E_l = h\nu$ is added to the field.

(ii) $|a_l^{(1)}(t)|^2$ is proportional to the modulus squared $|V_{1l}|^2$ of the perturbation matrix entry V_{1l} (which actually follows by (7.7.43)), a fact which is taken as axiomatic in Heisenberg's version of quantum mechanics (in his matrix mechanics). In (7.7.56), (7.7.57), (7.7.58) $|V_{1l}|^2$ in turn is proportional to the modulus squared $|E_o|^2$ of the harmonic amplitude E_o of the electric vector. The transition probability $|a_l^{(1)}(t)|^2$ is therefore proportional to the intensity of the radiation.

(iii) If the matrix entry V_{1l} vanishes, then of course $|a_l^{(1)}(t)|^2$ vanishes which means that a transition from the initial state ψ_1 to the state ψ_l is essentially forbidden. We have employed certain approximations in the foregoing analysis. Experiments show that "forbidden transitions" might actually occur—but such occurrences are very rare, and such transitions produce very faint spectral lines. The term "forbidden" therefore is taken to mean most likely forbidden, by far, but not necessarily to mean that a transition is *absolutely* forbidden.

(iv) Theorem 7.5, up to first-order approximation (which is usually sufficient for many practical problems), provides a formula for the probability $|a_l^{(1)}(t)|^2$ of finding a quantum system, subject to a harmonic perturbation H_1, in the state ψ_l (different from the initial state ψ_1) at time t; that is, as pointed out earlier, we have a formula for the probability that a measurement of the energy of the system at time t will yield the energy value E_l.

(v) Embodied in Theorem 7.5 is not only a basis by which certain spectroscopic selection rules are derived, and a means for determining the probability of a transition, but the theorem also serves as a basis by which certain *polarization rules* might be derived. For example, suppose that both V_{1l}^z and V_{1l}^x in (7.7.56), (7.7.57) vanish. Then in the transition $\psi_1 \to \psi_l$ only radiation in the y-direction could be absorbed or emitted—i.e., radiation with a specific polarization.

7.8 Spectroscopic Selection Rules for the Hydrogenic Atom

We apply the results of Section 7.7 to a hydrogenic atom interacting with electromagnetic radiation. To begin with a specific example, we consider the case when the electric field is polarized in the z-direction. A matrix entry V_{ij}^z in (7.7.56) then has the form $eE_o I_{3mm'}^{nn'll'}$ by definition (7.6.2). By Theorem 7.3

$$I_{jmm'}^{nn'll'} = 0 \qquad \text{for all } n, n', j, m, m' \text{ unless } l' = l+1 \text{ or } l-1. \tag{7.8.1}$$

By Theorem 7.4, $I_{3mm'}^{nn'll'} = 0$ unless $m' = m$. The following conclusion (selection rule) is reached: For the electromagnetic wave polarized in the z-direction a transition $\psi_{nlm} \to \psi_{n'l'm'}$ (from the state ψ_{nlm} to the state $\psi_{n'l'm'}$) is forbidden unless $l' = l+1$ or $l-1$ and $m' = m$. For example, in the beginning of this chapter we asked whether the transition $(5, 2, -2) \to (5, 4, -2)$ (i.e., $\psi_{52(-2)} \to \psi_{54(-2)}$) was always physically possible—and we pointed out that the answer was *no*, as would be seen. We now see why, in the present context, the answer is "no": $l = 2, l' = 4 \implies l' \neq l+1$ and $l' \neq l-1$, even though $m' = m(= -2)$. It is customary to express the condition $l' = l+1$ or $l-1$ as $\Delta l = \pm 1$; similarly the condition $m' = m$ is expressed as $\Delta m = 0$. Thus for the wave polarized in the z-direction, the hydrogenic atom can absorb or emit radiation only if $\Delta l = \pm 1$ and $\Delta m = 0$.

Suppose next that the electric wave is polarized in either the x or y-direction. Then V_{ij}^x, V_{ij}^y in (7.7.57), (7.7.58) have the form $eE_o I_{1mm'}^{nn'll'}$, $eE_o I_{2mm'}^{nn'll'}$, respectively, where by Theorem 7.4 both $I_{1mm'}^{nn'll'}$, $I_{2mm'}^{nn'll'}$ vanish unless $m' = m+1$ or $m-1$—i.e., unless $\Delta m = \pm 1$. By (7.7.57), (7.7.58) we therefore obtain the selection rules $\Delta l = \pm 1$, $\Delta m = \pm 1$ for electric waves polarized in either the x or y direction. If $\Delta l = \pm 1$ and $\Delta m = \pm 1$, so that the transition $\psi_{nlm} \to \psi_{n'l'm'}$ is allowed, then by the remarks in (v) of Section 7.7 only radiation of a definite polarization can be absorbed or emitted—namely (in the present case) radiation whose electric vectors point in the z-direction.

7.9 A General Quantum Mechanical Selection Rule

The abstract selection rule of Theorem 7.2 was formulated in terms of a compact group G acting on a locally compact space X and the corresponding unitary representation π of G on $L^2(X)$, or on $C_2(X)$. It is possible, and useful, to formulate a selection rule even a bit more abstract (and simpler), where the Hilbert space $L^2(X)$ is replaced by a general Hilbert space $\mathcal{H}$. This we do in Theorem 7.6 below. The basic new idea needed for this formulation is that of a *tensor operator*, to use the terminology of physicists. In practice $\mathcal{H}$ will be the Hilbert space of some quantum mechanical system. Thus it is appropriate to refer to Theorem 7.6 as a quantum mechanical selection rule.

Again let G be a compact group as above, let π be a continuous unitary representation of G on a complex Hilbert space $\mathcal{H}$, and let $\pi^{(3)}$ be a continuous representation of G on a finite-dimensional vector space V_3. The use of the notation $\pi^{(3)}$ and use of later notation will be for the sake of easy comparison with that of Theorem 7.2. Let $\mathbb{B}(\mathcal{H})$ denote the space of bounded linear operators A on $\mathcal{H}$: $||A\alpha|| \leq M||\alpha||$ for some $M > 0$ where the norm $||\alpha||$ of a vector $\alpha \in \mathcal{H}$ is induced by the inner product $\langle\ ,\ \rangle$ on $\mathcal{H}$: $||\alpha||^2 = \langle \alpha, \alpha \rangle$. A *tensor operator* T, relative to π, $\pi^{(3)}$, is a linear map $T : V_3 \to \mathbb{B}(\mathcal{H})$ such that

$$T(\pi^{(3)}(g)v) = \pi(g)T(v)\pi(g^{-1}) \qquad \text{for } (g, v) \in G \times V_3. \tag{7.9.1}$$

In regards to such kind of operators, one has (for the above notation)

Theorem 7.6 (A General Quantum Mechanical Selection Rule). *Let $T : V_3 \to \mathbb{B}(\mathcal{H})$ be a tensor operator relative to π, $\pi^{(3)}$. Let V_1, $V_2 \subset \mathcal{H}$ be finite-dimensional π-invariant subspaces, and let $\pi^{(1)}$, $\pi^{(2)}$ be the corresponding continuous subrepresentations of G on V_1, V_2. If the trivial one-dimensional representation* **1** *of G is* not *contained in $(\pi^{(1)} \otimes \pi^{(3)}) \otimes \pi^{(2)*}$, then*

$$\langle T(v_3)v_1, v_2 \rangle = 0 \quad \textit{for all } (v_1, v_2, v_3) \in V_1 \times V_2 \times V_3. \tag{7.9.2}$$

If $\pi^{(2)}$ is actually irreducible the conclusion (7.9.2) *is still valid provided that $\pi^{(2)}$ is not contained in $\pi^{(1)} \otimes \pi^{(3)}$.*

It is clear by our choice of notation that Theorem 7.6 compares indeed with Theorem 7.2, where the integrals I_{sij} there (given by the inner products in (7.5.11)) are replaced by the inner products in (7.9.2).

As an example, suppose in fact that in Section 7.5 one has that for each $f \in V_3 \stackrel{\text{def}}{=} G \cdot \phi$, the corresponding multiplication operator M_f $(f_1 \to f f_1, f_1 \in$

$L^2(X, dx)$) belongs to $\mathbb{B}(L^2(X, dx))$. Then we have a linear map

$$T : V_3 \to \mathbb{B}(L^2(X, dx)) \text{ given by } T(f) \stackrel{\text{def}}{=} M_f. \tag{7.9.3}$$

The identity $g \cdot [f(g^{-1} \cdot f_1)] = (g \cdot f) f_1$ (which was also used to establish (7.5.9)) translates to the statement that T is a tensor operator relative to π, $\pi^{(3)}$. Regarding $V_1, V_2 \subset C_2(X)$ in Section 7.5 as finite-dimensional π-invariant subspaces of $\mathcal{H} = L^2(X, dx)$, we conclude from Theorem 7.6 that

$$\int_X f_3(x) f_1(x) \overline{f_2}(x)\, dx = \langle T(f_3) f_1, f_2 \rangle = 0 \tag{7.9.4}$$

for $(f_1, f_2, f_3) \in V_1 \times V_2 \times V_3$, *if* the trivial representation $\mathbf{1}$ of G is not contained in $(\pi^{(1)} \otimes \pi^{(3)}) \otimes \pi^{(2)*}$ (or if $\pi^{(2)}$ is not contained in $\pi^{(1)} \otimes \pi^{(3)}$ in case $\pi^{(2)}$ is irreducible), which provides the conclusion of Theorem 7.2. A proof of Theorem 7.6 is also given in Appendix 7A.

7.10 Group Invariance of Schrödinger's Equation

For the Hamiltonian

$$H \stackrel{\text{def}}{=} -\frac{\hbar^2}{2\mu}\nabla^2 - \frac{Ze^2}{\sqrt{x^2 + y^2 + z^2}}, \tag{7.10.1}$$

equation (5.1.27), which is the first equation in (5.1.9), can be written as

$$H f_1 = E_1 f_1, \tag{7.10.2}$$

which as we have seen is solved by the square-integrable hydrogenic wave functions ψ_{nlm} for $E_1 = E^{(n)} = -\mu Z^2 e^4 / 2\hbar^2 n^2$. Now since any $A \in \mathrm{O}(3)$ preserves the norm $\|\quad\|$, where $\|(x, y, z)\| = \sqrt{x^2 + y^2 + z^2}$, it follows from (7.1.10) that A also commutes with H: For suitable $f : \mathbb{R}^3 \to \mathbb{C}$

$$\begin{aligned}
[H(A \cdot f)](x, y, z) &\stackrel{\text{def}}{=} -\frac{\hbar^2}{2\mu}(\nabla^2(A \cdot f))(x, y, z) - Ze^2\, \|(x, y, z)\|^{-1} \\
&= -\frac{\hbar^2}{2\mu}(A \cdot \nabla^2 f)(x, y, z) - Ze^2\, \|(x, y, z)\|^{-1} \\
&\stackrel{\text{def}}{=} -\frac{\hbar^2}{2\mu}(\nabla^2 f)((x, y, z)A) - Ze^2\, \|(x, y, z)\|^{-1}, \qquad \text{whereas} \\
[A \cdot (Hf)](x, y, z) &\stackrel{\text{def}}{=} (Hf)((x, y, z)A) \\
&\stackrel{\text{def}}{=} -\frac{\hbar^2}{2\mu}(\nabla^2 f)(x, y, z)A) - Ze^2\, \|(x, y, z)A\|^{-1} \\
&= -\frac{\hbar^2}{2\mu}(\nabla^2 f)(x, y, z)A) - Ze^2\, \|(x, y, z)\|^{-1} :
\end{aligned} \tag{7.10.3}$$

Theorem 7.7. *Let H be the Hamiltonian in* (7.10.1). *Thus the hydrogenic wave functions ψ_{nlm} of Chapter 5 are solutions of the Schrödinger equation $H\psi = E\psi$ for $E = E^{(n)} = -\mu Z^2 e^4/2\hbar^2 n^2$, $n = 1, 2, 3, \ldots$. Given equation* (7.1.10) (*whose proof will be given in Appendix* 7C) *one also has that each $A \in O(3)$ commutes with H:*

$$H(A \cdot \psi) = A \cdot (H\psi), \qquad \textit{where}$$
$$(A \cdot \psi)(x, y, z) \overset{\text{def}}{=} \psi((x, y, z)A) \quad \textit{for } \psi : \mathbb{R}^3 \to \mathbb{C}. \tag{7.10.4}$$

In particular if ψ is a solution of the Schrödinger equation $H\psi = E\psi$, so is $A \cdot \psi$ (for the same *eigenvalue E, which is Wigner's* 1927 *observation).*

Many applications of group theory to quantum mechanics are based on the commutativity of the Hamiltonian of some system with the action of some symmetry group G, as illustrated in Theorem 7.7, for $G = \mathrm{O}(3)$. Thus Schrödinger's equation for the system is *invariant* under G. In Chapter 10 we shall consider the commutativity or invariance condition (7.10.4) for $A \in G = S_n$, the symmetric group on n letters, in connection with a system of *identical* particles. For that discussion it will be useful to have a description of the one-dimensional representations of S_n. This we take up in the next section.

7.11 The Representations $1^{\pm}$ of S_n, and the Irreducible Representations of S_3

Besides the symmetry group O(3), another one of great importance in quantum mechanics is the symmetric group $G = S_n$ of permutation of n letters, say $X_n = \{1, 2, \ldots, n\}$, as we shall note in Chapter 10. As usual, any $\sigma \in G$ (σ being a bijection of X_n) is written

$$\sigma = \begin{pmatrix} 1 & 2 & \cdots & n \\ \sigma(1) & \sigma(2) & \cdots & \sigma(n) \end{pmatrix}. \tag{7.11.1}$$

There is a natural representation π of G on $\mathbb{C}^n$. $\pi(\sigma) : \mathbb{C}^n \to \mathbb{C}^n$ is given by

$$\pi(\sigma)(v_1, \ldots, v_n) \overset{\text{def}}{=} (v_{\sigma(1)}, \ldots, v_{\sigma(n)}) \tag{7.11.2}$$

for $(v_1, \ldots, v_n) \in \mathbb{C}^n$. Indeed

$$\begin{aligned} \pi(\sigma_1\sigma_2)(v_1, \ldots, v_n) &\overset{\text{def}}{=} (v_{(\sigma_1\sigma_2)(1)}, \ldots, v_{(\sigma_1\sigma_2)(n)}) \\ &\overset{\text{def}}{=} (v_{\sigma_1(\sigma_2(1))}, \ldots, v_{\sigma_1(\sigma_2(n))}), \quad \text{and} \\ \pi(\sigma_1)(\pi(\sigma_2)(v_1, \ldots, v_n)) &\overset{\text{def}}{=} \pi(\sigma_1)(v_{\sigma_2(1)}, \ldots, v_{\sigma_2(n)}) \\ &\overset{\text{def}}{=} (v_{\sigma_1(\sigma_2(1))}, \ldots, v_{\sigma_1(\sigma_2(n))}) \\ \Longrightarrow \pi(\sigma_1\sigma_2) &= \pi(\sigma_1)\pi(\sigma_2) \quad \text{for } \sigma_1, \sigma_2 \in G, \end{aligned} \tag{7.11.3}$$

which shows that π is a representation, which is automatically continuous since the topology of a finite group is discrete. To find the matrix version of π, let

$\{e_j\}_{j=1}^n$ be the standard basis of $\mathbb{C}^n$: $e_j = (0, \ldots, 0, 1, 0, \ldots, 0)$ where the 1 appears in the jth place. Write $e_j = (v_1^{(j)}, \ldots, v_n^{(j)})$; i.e., $v_l^{(j)} = \delta_{jl}$. By (7.11.2)

$$\begin{aligned}\pi(\sigma)e_j &= (v_{\sigma(1)}^{(j)}, \ldots, v_{\sigma(n)}^{(j)}) \\ &= \sum_{l=1}^n v_{\sigma(l)}^{(j)} e_l = \sum_{l=1}^n \delta_{j\sigma(l)} e_l = e_{\sigma^{-1}(j)}.\end{aligned} \tag{7.11.4}$$

In particular the matrix of $\pi(\sigma)$ is given by

$$\pi(\sigma) = \begin{bmatrix} \delta_{\sigma(1)1} & \cdots & \delta_{\sigma(1)n} \\ \delta_{\sigma(2)1} & \cdots & \delta_{\sigma(2)n} \\ \vdots & & \vdots \\ \delta_{\sigma(n)1} & \cdots & \delta_{\sigma(n)n} \end{bmatrix}; \quad \text{i.e., } \pi(\sigma)_{ij} = \delta_{\sigma(i)j}. \tag{7.11.5}$$

Note that

$$\sum_{k=1}^n \delta_{\sigma(i)k}\delta_{\sigma(j)k} = \delta_{ij}. \tag{7.11.6}$$

For if the left-hand side of (7.11.6) is not zero we must have $\delta_{\sigma(i)k}\delta_{\sigma(j)k} \neq 0$ for some $k \implies \sigma(i) = k = \sigma(j) \implies i = j$. That is, $i \neq j \implies$ the left-hand side of (7.11.6) is zero. If $i = j$ however, then the left-hand side of (7.11.6) is

$$\sum_{k=1}^n \delta_{\sigma(i)k}\delta_{\sigma(i)k} = \delta_{\sigma(i)\sigma(i)}\delta_{\sigma(i)\sigma(i)} = 1, \tag{7.11.7}$$

which proves (7.11.6). But given (7.11.5), (7.11.6) is just the statement that

$$\begin{aligned}[\pi(\sigma)\pi(\sigma)^t]_{ij} &\overset{\text{def}}{=} \sum_{k=1}^n \pi(\sigma)_{ik}\pi(\sigma)_{jk} \\ &= \sum_{k=1}^n \delta_{\sigma(i)k}\delta_{\sigma(j)k} = \delta_{ij};\end{aligned} \tag{7.11.8}$$

i.e., $\pi(\sigma)\pi(\sigma)^t = 1 \implies$ each $\pi(\sigma) \in \mathrm{O}(n)$. In particular $[\det \pi(\sigma)]^2 = 1 \implies$

$$\det \pi(\sigma) = \pm 1 \qquad \text{for each } \sigma \in S_n. \tag{7.11.9}$$

It is easy to construct the one-dimensional trivial representation 1^+ of S_n as a subrepresentation of π. Namely define

$$V_1 = \mathbb{C}(1, 1, \ldots, 1) = \mathbb{C}(e_1 + e_2 + \cdots + e_n), \tag{7.11.10}$$

which is a one-dimensional subspace of $\mathbb{C}^n$ on which π acts trivially: By (7.11.4)

$$\begin{aligned}\pi(\sigma)(e_1 + e_2 + \cdots + e_n) &= e_{\sigma^{-1}(1)} + e_{\sigma^{-1}(2)} + \cdots + e_{\sigma^{-1}(n)} \\ &= e_1 + e_2 + \cdots + e_n.\end{aligned} \tag{7.11.11}$$

Thus $1^+ = \pi|_{V_1}$. Given $w = (w_1, w_2, \ldots, w_n) \in \mathbb{C}^n$ define

$$\lambda = \frac{1}{n}\sum_{j=1}^{n} w_j \in \mathbb{C}, \qquad v_j = w_j - \lambda \in \mathbb{C}. \tag{7.11.12}$$

Then

$$\sum_{j=1}^{n} v_j = \sum_{j=1}^{n} w_j - n\lambda = 0 \implies \tag{7.11.13}$$

$v \overset{\text{def}}{=} (v_1, \ldots, v_n)$ belongs to the π-invariant subspace

$$V_2 \overset{\text{def}}{=} \left\{ v = (v_1, \ldots, v_n) \in \mathbb{C}^n \mid \sum_{j=1}^{n} v_j = 0 \right\} \tag{7.11.14}$$

of $\mathbb{C}^n$. Also $v + \lambda(1, 1, \ldots, 1) = (v_1 + \lambda, v_2 + \lambda, \ldots, v_n + \lambda) \overset{\text{def}}{=} (w_1, \ldots, w_n) = w$ which shows that

$$\mathbb{C}^n = V_1 \oplus V_2, \quad \pi = 1^+ \oplus \pi_2 \quad \text{for } \pi_2 \overset{\text{def}}{=} \pi|_{V_2} \tag{7.11.15}$$

since clearly $V_1 \cap V_2 = \{0\} : v = (v_1, \ldots, v_n) = \lambda(1, 1, \ldots, 1)$ with $v \in V_2 \implies 0 = \sum_{j=1}^{n} v_j = n\lambda \implies \lambda = 0 \implies v = 0$. We also see in (7.11.15) that π is *not* irreducible of course.

The group S_n has another one-dimensional (non-trivial) representation 1^- which is constructed as follows. $1^-(\sigma) : \mathbb{C} \to \mathbb{C}$ is given by

$$1^-(\sigma)z \overset{\text{def}}{=} (\det \pi(\sigma))z \qquad \text{for } z \in \mathbb{C}. \tag{7.11.16}$$

Since $\det \pi(\sigma) \neq 0$ ($\det \pi(\sigma) = \pm 1$ by (7.11.9)), $1^-(\sigma)$ is indeed a non-singular linear operator on $\mathbb{C}$ such that for $\sigma_1, \sigma_2 \in S_n$, $1^-(\sigma_1\sigma_2)z = (\det \pi(\sigma_1\sigma_2))z = (\det(\pi(\sigma_1)\pi(\sigma_2)))z = (\det \pi(\sigma_1) \det \pi(\sigma_2))z = 1^-(\sigma_1)\, 1^-(\sigma_2)z$; i.e., 1^- is indeed a representation of S_n. The following result is known.

Theorem 7.8. *Up to equivalence* $1^{\pm}$ *are the* only *one-dimensional representations of* S_n.

An important special case, which arises often enough in chemistry, is the case $n = 3$. Here S_3 is isomorphic to D_3, where in general D_n is the dihedral group of order $2n$—the symmetry group of a regular n-gon. We use this case to illustrate some of the preceding remarks. Write $S_3 = \{I, A, B, C, D, E\}$, where

$$\begin{aligned} I &= \begin{pmatrix} 1 & 2 & 3 \\ 1 & 2 & 3 \end{pmatrix}, \quad A = \begin{pmatrix} 1 & 2 & 3 \\ 3 & 1 & 2 \end{pmatrix}, \quad B = \begin{pmatrix} 1 & 2 & 3 \\ 2 & 3 & 1 \end{pmatrix}, \\ C &= \begin{pmatrix} 1 & 2 & 3 \\ 1 & 3 & 2 \end{pmatrix}, \quad D = \begin{pmatrix} 1 & 2 & 3 \\ 3 & 2 & 1 \end{pmatrix}, \quad D = \begin{pmatrix} 1 & 2 & 3 \\ 2 & 1 & 3 \end{pmatrix}; \end{aligned} \tag{7.11.17}$$

cf. (7.11.1). The $\pi(\sigma)$ in (7.11.5) are given by

$$\pi(A) = \begin{bmatrix} 0 & 0 & 1 \\ 1 & 0 & 0 \\ 0 & 1 & 0 \end{bmatrix}, \quad \pi(B) = \begin{bmatrix} 0 & 1 & 0 \\ 0 & 0 & 1 \\ 1 & 0 & 0 \end{bmatrix}, \quad \pi(C) = \begin{bmatrix} 1 & 0 & 0 \\ 0 & 0 & 1 \\ 0 & 1 & 0 \end{bmatrix},$$
$$\pi(D) = \begin{bmatrix} 0 & 0 & 1 \\ 0 & 1 & 0 \\ 1 & 0 & 0 \end{bmatrix}, \quad \pi(E) = \begin{bmatrix} 0 & 1 & 0 \\ 1 & 0 & 0 \\ 0 & 0 & 1 \end{bmatrix}. \tag{7.11.18}$$

Of course $\pi(I)$ is the identity matrix. A basis for the subspace V_2 in (7.11.14) is $\{e_1 - e_3, e_2 - e_3\}$. Relative to this basis one has

$$\pi_2(A) = \begin{bmatrix} -1 & -1 \\ 1 & 0 \end{bmatrix}, \quad \pi_2(B) = \begin{bmatrix} 0 & 1 \\ -1 & -1 \end{bmatrix}, \quad \pi_2(C) = \begin{bmatrix} 1 & 0 \\ -1 & -1 \end{bmatrix},$$
$$\pi_2(D) = \begin{bmatrix} -1 & -1 \\ 0 & 1 \end{bmatrix}, \quad \pi_2(E) = \begin{bmatrix} 0 & 1 \\ 1 & 0 \end{bmatrix}. \tag{7.11.19}$$

In particular we obtain the following character values:

$$\begin{aligned} &\chi_{\pi_2}(I) = 2, \quad \chi_{\pi_2}(A) = -1, \quad \chi_{\pi_2}(B) = -1, \\ &\chi_{\pi_2}(C) = 0, \quad \chi_{\pi_2}(D) = 0, \quad \chi_{\pi_2}(E) = 0. \end{aligned} \tag{7.11.20}$$

Now apply the *Frobenius criterion*: For a representation τ of a finite group G, τ is irreducible if and only if its character χ_τ satisfies

$$\frac{1}{\text{order of } G} \sum_{g \in G} |\chi_\tau(g)|^2 = 1. \tag{7.11.21}$$

Given (7.11.20) we indeed have for $G = S_3$, $\tau = \pi_2$,

$$\frac{1}{6} \sum_{\sigma \in S_3} |\chi_{\pi_2}(\sigma)|^2 = \frac{1}{6}[4 + 1 + 1 + 0 + 0 + 0] = 1. \tag{7.11.22}$$

Thus π_2 is irreducible and in (7.11.15) we have the complete reduction of $\pi = 1^+ \oplus \pi_2$ as a direct sum of irreducible representations.

Note that 1^- in (7.11.16) is given as follows, where $1_{\mathbb{C}}$ is the identity operator on $\mathbb{C}$:

$$\begin{aligned} &1^-(A) = 1^-(B) = 1^-(I) = 1_{\mathbb{C}}, \\ &1^-(C) = 1^-(D) = 1^-(E) = -1_{\mathbb{C}}, \end{aligned} \tag{7.11.23}$$

since

$$\begin{aligned} &\det \pi(\sigma) = \det \pi_2(\sigma) = 1 \quad \text{for } \sigma = A, B, I, \\ &\det \pi(\sigma) = \det \pi_2(\sigma) = -1 \quad \text{for } \sigma = C, D, E \end{aligned} \tag{7.11.24}$$

by (7.11.18), (7.11.19). In other words $1^-(\sigma) = \pm 1_{\mathbb{C}}$ according as σ is an even (+) or odd (−) permutation; the same observation is in fact true for S_n in general. C for example is an odd permutation as it is the product of r transpositions with $r = 1$; namely C is the transposition $2 \to 3, 3 \to 2$ with 1 being fixed.

Finally one can deduce Theorem 7.8 in the present case via the *completeness relation* for a finite group G: If $\tau_1, \ldots, \tau_m$ is a list of its irreducible representations, up to equivalence, then

$$\sum_{j=1}^{m} [\text{dimension } \tau_j]^2 = \text{ the order of } G. \tag{7.11.25}$$

Since $1^{\pm}$, π_2 are irreducible of dimensions 1, 1, 2, and since $1^2 + 1^2 + 2^2 = 6 =$ the order of S_3, where no other solutions of equation (7.11.25) exist, we have in fact constructed *all* of the irreducible representations of S_3, up to equivalence, and have verified Theorem 7.8 in particular for $n = 3$.

Extended and systematic applications of group theory, more than that which we have attempted to undertake here, can be found in the references [79, 83, 87], for example.

Appendix 7A Proof of Theorems 7.2 and 7.6

Given the formula (7.5.17) we show how Theorem 7.2 follows from a more general result presented here. First we recall the notion of a sesquilinear form.

Let V_1, V_2 be complex vector spaces and let $B : V_1 \times V_2 \to \mathbb{C}$ be a function which satisfies $B(x, y_1 + y_2) = B(x, y_1) + B(x, y_2)$, $B(x_1 + x_2, y) = B(x_1, y) + B(x_2, y)$, $B(cx, y) = cB(x, y)$, $B(x, cy) = \bar{c}B(x, y)$ for $x, x_1, x_2 \in V_1$, $y, y_1, y_2 \in V_2$, $c \in \mathbb{C}$, where $\bar{c}$ is the complex conjugate of c. Then B is called a *sesquilinear* form on $V_1 \times V_2$. If $A : V_1 \to V_2$ is any linear map and $\langle\ ,\ \rangle$ is a complex inner product on V_2 then there is a corresponding sesquilinear form $B = B_{A,\langle\,,\,\rangle}$ on $V_1 \times V_2$ given by $B(x, y) \stackrel{\text{def}}{=} \langle Ax, y\rangle$ for $(x, y) \in V_1 \times V_2$. Conversely one has

Proposition 7.3. *Let $B : V_1 \times V_2 \to \mathbb{C}$ be a sesquilinear form where $(V_2, \langle\ ,\ \rangle)$ is a finite-dimensional inner product space. Then there is a unique linear map $A : V_1 \to V_2$ such that $B(x, y) = \langle Ax, y\rangle$ $\forall (x, y) \in V_1 \times V_2$.*

Proof. Given $x \in V_1$, define $L_x : V_2 \to \mathbb{C}$ by $L_x(y) = \overline{B(x, y,)}$ for $y \in V_2$. As B is sesquilinear, $L_x \in V_2^*$. Hence there exists a unique element $x^* \in V_2$ such that

$$L_x(y) = \langle y, x^*\rangle \quad \forall y \in V_2; \tag{7.A.1}$$

i.e., $\langle x^*, y\rangle = \overline{L_x(y)} = B(x, y)$ $\forall y \in V_2$. Define $A : V_1 \to V_2$ by $Ax = x^*$. Then for $x_1, x_2 \in V_1$, $y \in V_2$, $\langle (x_1 + x_2)^*, y\rangle \stackrel{\text{def}}{=} B(x_1 + x_2, y) = B(x_1, y) + B(x_2, y) \stackrel{\text{def}}{=} \langle x_1^*, y\rangle + \langle x_2^*, y\rangle = \langle x_1^* + x_2^*, y\rangle \implies (x_1 + x_2)^* = x_1^* + x_2^*$. Similarly $(cx_1)^* = cx_1^*$ for $c \in \mathbb{C} \implies A$ is linear. A is clearly uniquely determined by B. Thus Proposition 7.3 is established. □

Note that if $(V_2, \langle\ ,\ \rangle)$ is an infinite-dimensional Hilbert space (a case which we shall not need) and B, A are required to be *bounded* (i.e., $|B(x, y)| \le M_1 ||x||\, ||y||$, $||Ax|| \le M_2||x||$ for some $M_1, M_2 > 0$ and some norm $||\ ||$ also on V_1), then Proposition 7.3 remains valid since equation (7.A.1) would follow by the theorem of Riesz. In fact we would have $||x^*|| = ||L_x||$ and consequently $||A|| = ||B||$.

Now let π_1, π_2 be finite-dimensional continuous representations of G on complex vector spaces W, V_2, say that π_2 is unitary with respect to the inner product $\langle\ ,\ \rangle$ on V_2. Here we take G to be a compact topological group (with a Hausdorff topology), and we denote normalized Haar measure on G by dg. For fixed $(\omega, L) \in W^* \times V_2^*$ there is a natural sesquilinear form B on $W \times V_2$ given by

$$B(w, y) = B_{\omega,L}(w, y) \stackrel{\text{def}}{=} \int_G \omega(\pi_1(g)w)\, \overline{L(\pi_2(g)y)}\, dg. \tag{7.A.2}$$

Since π_1, π_2 are continuous the integrand in (7.A.2) is continuous; cf. the definition at the close of Section 7.1. By Proposition 7.3 there is a unique linear map $A = A_{\omega,L} : W \to V_2$ such that $B(w, y) = \langle Aw, y\rangle$ on $W \times V_2$. For an arbitrary element $g_1 \in G$ and $(w, y) \in W \times V_2$

$$\begin{aligned} B(\pi_1(g_1)w, \pi_2(g_1)y) &\stackrel{\text{def}}{=} \int_G \omega(\pi_1(gg_1)w)\, \overline{L(\pi_2(gg_1)y)}\, dg \\ &\quad \text{(since } \pi_j(g)\pi_j(g_1) = \pi_j(gg_1) \text{ for } j = 1, 2) \\ &= \int_G \omega(\pi_1(g)w)\overline{L(\pi_2(g)y)}\, dg \\ &\quad \text{(by the } G\text{-invariance of } dg) \\ &\stackrel{\text{def}}{=} B(w, y). \end{aligned} \tag{7.A.3}$$

That is, $\langle A\pi_1(g)w, \pi_2(g)y\rangle = \langle Aw, y\rangle = \langle \pi_2(g)Aw, \pi_2(g)y\rangle$ (since π_2 is unitary) $\forall (w, y) \implies$

$$A\pi_1(g) = \pi_2(g)A : \tag{7.A.4}$$

The diagram

$$\begin{array}{ccc} W & \xrightarrow{\pi_1(g)} & W \\ {\scriptstyle A}\downarrow & & \downarrow{\scriptstyle A} \\ V_2 & \xrightarrow[\pi_2(g)]{} & V_2 \end{array}$$

is commutative $\forall g \in G$; compare examples 2, 4, 5 in Section 7.1. That is, $A : W \to V_2$ is a G-map: $A \in \mathrm{Hom}_G(W, V_2) \stackrel{\text{def}}{=}$ the space of linear maps $A : W \to V_2$ such that (7.A.4) holds. Now we use the following standard fact from finite-dimensional representation theory, which amounts to the classical orthogonality relations for a compact group:

$$\begin{aligned}\dim \operatorname{Hom}_G(W, V_2) &= \int_G \chi_{\pi_1}(g)\overline{\chi_{\pi_2}(g)}\, dg \\ &= \int_G \chi_{\pi_2}(g)\overline{\chi_{\pi_1}(g)}\, dg = \dim \operatorname{Hom}_G(V_2, W),\end{aligned} \tag{7.A.5}$$

where as before χ_π denotes the character of the representation π. In particular if $A \neq 0$ then $\operatorname{Hom}_G(V_2, W) \neq \{0\}$: $\exists \phi : V_2 \to W$ a non-zero, linear G-map. The kernel of ϕ is π_2 invariant (since ϕ is a G-map) and differs from V_2 (since $\phi \neq 0$). Hence if π_2 is irreducible we must have the kernel of $\phi = \{0\}$; i.e., ϕ is one-to-one and thus, by way of ϕ, V_2 is equivalent to the subrepresentation $\pi_1|_{\phi(V_2)}$ of π_1, where the range $\phi(V_2)$ is indeed π_1-invariant by the G-invariance of ϕ; i.e., (as defined in Section 7.3),

$$\pi_2 \text{ is } \textit{contained} \text{ in } \pi_1 \text{ if } \operatorname{Hom}_G(V_2, W) \neq \{0\} \text{ and } \pi_2 \text{ is irreducible.} \tag{7.A.6}$$

In other words if π_2 is irreducible and π_2 is *not* contained in π_1, then we must have $A = 0 \implies B = 0 \implies$ (by definition (7.A.2))

$$\int_G \omega(\pi_1(g)w)\overline{L(\pi_2(g)y)}\, dg = 0 \tag{7.A.7}$$

for all $(w, y) \in W \times V_2$, $(\omega, L) \in W^* \times V_2^*$. If π_2 is not necessarily irreducible we use (7.A.5) to write

$$\begin{aligned}\dim \operatorname{Hom}_G(W, V_2) &= \int_G \chi_{\pi_1}(g)\overline{\chi_{\pi_2}(g)}\, dg \\ &= \int_G \chi_{\pi_1}(g)\chi_{\pi_2^*}(g)\, dg = \int_G \chi_{\pi_1\otimes\pi_2^*}(g)\, dg \\ &= \int_G \chi_{\pi_1\otimes\pi_2^*}(g)\overline{\chi_1(g)}\, dg \\ &= \dim \operatorname{Hom}_G(\mathbb{C}, W \otimes V_2^*)\end{aligned} \tag{7.A.8}$$

by standard elementary properties of characters, where $\mathbf{1}$ is the trivial representation of G on $\mathbb{C}$. As $\mathbf{1}$ is irreducible, statement (7.A.6) applies: If $\mathbf{1}$ is not contained in $\pi_1 \otimes \pi_2^*$, then $\operatorname{Hom}_G(W, V_2) = \{0\}$ by (7.A.8) and since $A \in \operatorname{Hom}_G(W, V_2)$ we must have $A = 0 \implies B = 0$ so that again we arrive at (7.A.7), which proves

Theorem 7.9. *Let π_1, π_2 be finite-dimensional, continuous representations of G on complex vector spaces W, V_2 with π_2^* equal to the contragredient representation of π_2 acting on V_2^*. If $\pi_1 \otimes \pi_2^*$ does not contain the one-dimensional trivial representation of G on $\mathbb{C}$ then for all $(w, y) \in W \times V_2$, $(\omega, L) \in W^* \times V_2^*$*

$$\int_G \omega(\pi_1(g)w)\overline{L(\pi_2(g)y)}\, dg = 0.$$

This equation also holds under the assumption that π_2 is irreducible and is not contained in π_1.

We have established Theorem 7.9 strictly speaking only for π_2 *unitary*. But as is well known this is essentially a non-assumption. Namely if $\langle\ ,\ \rangle_o$ is any inner product whatsoever on V_2, then the equation

$$\langle v_2, v_2' \rangle = \int_G \langle \pi_2(g)v_2, \pi_2(g)v_2' \rangle_o \, dg \tag{7.A.9}$$

defines an inner product $\langle\ ,\ \rangle$ on V_2 for which π_2 is unitary.

Theorem 7.2 follows from Theorem 7.9. Namely, assume that $(\pi^{(1)} \otimes \pi^{(3)}) \otimes \pi^{(2)*}$ does not contain the trivial one-dimensional representation of G; or if $\pi^{(2)}$ is actually irreducible, assume that $\pi^{(2)}$ is not contained in $\pi^{(1)} \otimes \pi^{(3)}$. In Theorem 7.9 choose $\pi_1 = \pi^{(1)} \otimes \pi^{(3)}$ and $\pi_2 = \pi^{(2)}$. Thus we may conclude that

$$\int_G \omega((\pi^{(1)} \otimes \pi^{(3)})(g)w)\overline{L(\pi^{(2)}(g)y)} \, dg = 0 \tag{7.A.10}$$

for all $(w, y) \in (V_1 \otimes V_3) \times V_2$, $(\omega, L) \in (V_1 \otimes V_3)^* \times V_2^*$. In particular by (7.5.17) we conclude that $I_{tjl} = 0 \; \forall t, j, l$, which proves Theorem 7.2.

For the proof of Theorem 7.6 we make use of the notation established in Section 7.9. Thus let $T : V_3 \to \mathbb{B}(\mathcal{H})$ be a tensor operator relative to π, $\pi^{(3)}$, and let $V_1, V_2 \subset \mathcal{H}$ be finite-dimensional π-invariant subspaces with $\pi^{(1)}, \pi^{(2)}$ denoting the corresponding subrepresentations. As $T : V_3 \to \mathbb{B}(\mathcal{H})$ is linear and satisfies (7.9.1), it is immediate that the equation

$$b(v_3, v_1) \stackrel{\text{def}}{=} T(v_3)v_1, \quad (v_3, v_1) \in V_3 \times V_1 \tag{7.A.11}$$

defines a bilinear form $b = b_T : V_3 \times V_1 \to \mathcal{H}$, which satisfies the G-invariance condition

$$b(\pi^{(3)}(g)v_3, \pi(g)v_1) = \pi(g)b(v_3, v_1). \tag{7.A.12}$$

By the definition of $V_3 \otimes V_1$ given in Section 7.2 (the tensor product being taken over $\mathbb{C}$) there exists a unique linear map $\widetilde{b} = \widetilde{b_T} : V_3 \otimes V_1 \to \mathcal{H}$ such that

$$\widetilde{b}(v_3 \otimes v_1) = b(v_3, v_1) = T(v_3)v_1 \quad \text{for } (v_3, v_1) \in V_3 \times V_1. \tag{7.A.13}$$

For the tensor product representation $\pi^{(3)} \otimes \pi^{(1)}$ one has by (7.A.12)

$$\begin{aligned}\widetilde{b}((\pi^{(3)} \otimes \pi^{(1)})(g)(v_3 \otimes v_1)) &\stackrel{\text{def}}{=} \widetilde{b}((\pi^{(3)}(g) \otimes \pi^{(1)}(g))(v_3 \otimes v_1)) \\ &\stackrel{\text{def}}{=} \widetilde{b}(\pi^{(3)}(g)v_3 \otimes \pi^{(1)}(g)v_1) \stackrel{\text{def}}{=} b(\pi^{(3)}(g)v_3, \pi(g)v_1) \\ &= \pi(g)b(v_3, v_1) \stackrel{\text{def}}{=} \pi(g)\widetilde{b}(v_3 \otimes v_1),\end{aligned} \tag{7.A.14}$$

which shows that $\widetilde{b}$ intertwines $\pi^{(3)} \otimes \pi^{(1)}$ and π. That is,

$$\widetilde{b}((\pi^{(3)} \otimes \pi^{(1)})(g)\, t) = \pi(g)\widetilde{b}(t) \quad \text{for } (g, t) \in G \times (V_3 \otimes V_1). \tag{7.A.15}$$

Next define $B = B_T : (V_3 \otimes V_1) \times V_2 \to \mathbb{C}$ by

$$B(t, v_2) \stackrel{\text{def}}{=} \langle \widetilde{b}(t), v_2 \rangle \quad \text{for } (t, v_2) \in (V_3 \otimes V_1) \times V_2. \tag{7.A.16}$$

As $\widetilde{b}$ is linear and the inner product $\langle \, , \, \rangle$ on $\mathcal{H}$ is sesquilinear it follows that B is a sesquilinear form, and therefore by Proposition 7.3 there is a unique linear map $A = A_T : V_3 \otimes V_1 \to V_2$ such that

$$B(t, v_2) = \langle At, v_2 \rangle_2 \quad \text{for } (t, v_2) \in (V_3 \otimes V_1) \times V_2, \tag{7.A.17}$$

where $\langle \, , \, \rangle_2$ is the inner product $\langle \, , \, \rangle$ restricted to V_2. Now for $g \in G$ $A_g \stackrel{\text{def}}{=} \pi^{(2)}(g^{-1})A(\pi^{(3)} \otimes \pi^{(1)})(g) : V_3 \otimes V_1 \to V_2$ is a linear map, and since π is unitary A_g also satisfies

$$\begin{aligned}
\langle A_g t, v_2 \rangle_2 &= \langle A(\pi^{(3)} \otimes \pi^{(1)})(g)t, \pi^{(2)}(g)v_2 \rangle_2 \\
&\stackrel{\text{def. (7.A.17)}}{=} B((\pi^{(3)} \otimes \pi^{(1)})(g)t, \pi^{(2)}(g)v_2) \\
&\stackrel{\text{def. (7.A.16)}}{=} \langle \widetilde{b}(\pi^{(3)} \otimes \pi^{(1)})(g)t, \pi^{(2)}(g)v_2 \rangle \\
&= \langle \pi(g)\widetilde{b}(t), \pi(g)v_2 \rangle \qquad \text{(by (7.A.15))} \\
&= \langle \widetilde{b}(t), v_2 \rangle \stackrel{\text{def. (7.A.16)}}{=} B(t, v_2) \\
&\quad \text{for every } (t, v_2) \in (V_3 \otimes V_1) \times V_2.
\end{aligned} \tag{7.A.18}$$

By the uniqueness of A we must therefore have $A_g = A : A(\pi^{(3)} \otimes \pi^{(1)})(g) = \pi^{(2)}(g)A$: The diagram

$$\begin{array}{ccc}
V_3 \otimes V_1 & \xrightarrow{(\pi^{(3)} \otimes \pi^{(1)})(g)} & V_3 \otimes V_1 \\
{\scriptstyle A}\downarrow & & \downarrow{\scriptstyle A} \\
V_2 & \xrightarrow[\pi^{(2)}(g)]{} & V_2
\end{array} \tag{7.A.19}$$

is commutative: $A \in \mathrm{Hom}_G(V_3 \otimes V_1, V_2)$. This puts us in position to apply exactly the arguments that followed (7.A.4), where π_1 there is taken as $\pi^{(3)} \otimes \pi^{(1)}$ here. We can therefore conclude the following. If the trivial one-dimensional representation $\mathbf{1}$ of G is not contained in $(\pi^{(3)} \otimes \pi^{(1)}) \otimes \pi^{(2)*}$, or if $\pi^{(2)}$ is actually irreducible and is not contained in $\pi^{(3)} \otimes \pi^{(1)}$, then $A = 0$ and hence (by (7.A.17)) $B = 0 \implies \langle \widetilde{b}(t), v_2 \rangle = 0$ for $(t, v_2) \in (V_3 \otimes V_1) \times V_2$ (by (7.A.16)). In particular $\langle T(v_3)v_1, v_2 \rangle = 0$ for $(v_1, v_2, v_3) \in V_1 \times V_2 \times V_3$ by (7.A.13), which proves Theorem 7.6 since of course the representations $\pi^{(1)} \otimes \pi^{(3)}$, $\pi^{(3)} \otimes \pi^{(1)}$ are equivalent (via the isomorphism $V_1 \otimes V_3 \to V_3 \otimes V_1$).

Appendix 7B Proof of Proposition 7.2

Here we establish the proof of Proposition 7.2. First take $l = 1$:

$$
\begin{aligned}
A_1(t) &\stackrel{\text{def}}{=} a_1^{(1)\prime}(t) + \frac{i}{\hbar}[a_1^{(1)}(t)H_{11}(t) + \sum_{j\neq 1} a_j^{(1)}(t)H_{j1}(t)e^{i(E_1-E_j)t/\hbar}] \\
&\stackrel{\text{def}}{=} -\frac{i}{\hbar}H_{11}(t) + \frac{i}{\hbar}\Bigg[\left(-\frac{i}{\hbar}\int_0^t H_{11}(r)\,dr + 1\right)H_{11}(t) \\
&\quad + \sum_{j\neq 1}\left(-\frac{i}{\hbar}\right)\int_0^t H_{1j}(r)e^{i(E_j-E_1)r/\hbar}\,dr\, H_{j1}(t)e^{i(E_1-E_j)t/\hbar}\Bigg] \qquad (7.B.1) \\
&= \frac{1}{\hbar^2}\int_0^t H_{11}(r)\,dr\, H_{11}(t) \\
&\quad + \frac{1}{\hbar^2}\sum_{j\neq 1}\int_0^t H_{1j}(r)e^{i(E_j-E_1)r/\hbar}\,dr\, H_{j1}(t)e^{i(E_1-E_j)t/\hbar}.
\end{aligned}
$$

For $l \neq 1$

$$
\begin{aligned}
A_l(t) &= -\frac{i}{\hbar}H_{1l}(t)e^{i(E_l-E_1)t/\hbar} \\
&\quad + \frac{i}{\hbar}\Bigg[\left(-\frac{i}{\hbar}\right)\int_0^t H_{1l}(r)e^{i(E_l-E_1)r/\hbar}\,dr\, H_{ll}(t) \\
&\quad + \sum_{j\neq l,1}\left(-\frac{i}{\hbar}\right)\int_0^t H_{1j}(r)e^{i(E_j-E_1)r/\hbar}\,dr\, H_{jl}(t)e^{i(E_l-E_j)t/\hbar} \\
&\quad + \left(-\frac{i}{\hbar}\int_0^t H_{11}(r)\,dr + 1\right)H_{1l}(t)e^{i(E_l-E_1)t/\hbar}\Bigg] \qquad (= \text{term} j = 1) \\
&= \frac{1}{\hbar^2}\int_0^t H_{1l}(r)e^{i(E_l-E_1)r/\hbar}\,dr\, H_{ll}(t) \\
&\quad + \frac{1}{\hbar^2}\sum_{j\neq l,1}\int_0^t H_{1j}(r)e^{i(E_j-E_1)r/\hbar}\,dr\, H_{jl}(t)e^{i(E_l-E_j)t/\hbar} \\
&\quad + \frac{1}{\hbar^2}\int_0^t H_{11}(r)\,dr\, H_{1l}(t)e^{i(E_l-E_1)t/\hbar}. \qquad (7.B.2)
\end{aligned}
$$

In particular by (7.B.1) and (7.B.2)

$$
\begin{aligned}
|A_1(t)| &\leq \frac{|H_{11}(t)|}{\hbar^2}\int_0^t |H_{11}(r)|\,dr \\
&\quad + \frac{1}{\hbar^2}\sum_{j\neq 1}\int_0^t |H_{1j}(r)|\,dr\,|H_{j1}(t)| \qquad (7.B.3) \\
&= \frac{1}{\hbar^2}\sum_j |H_{j1}(t)|\int_0^t |H_{1j}(r)|\,dr,
\end{aligned}
$$

and for $l \neq 1$

$$\begin{aligned} |A_l(t)| &\leq \frac{|H_{ll}(t)|}{\hbar^2} \int_0^t |H_{1l}(r)\, dr \\ &\quad + \frac{1}{\hbar^2} \sum_{j \neq l,1} |H_{jl}(t)| \int_0^t |H_{1j}(r)|\, dr \\ &\quad + \frac{1}{\hbar^2} |H_{1l}(t)| \int_0^t |H_{11}(r)|\, dr \\ &= \frac{1}{\hbar^2} \sum_j |H_{jl}(t)| \int_0^t |H_{1j}(r)\, dr. \end{aligned} \tag{7.B.4}$$

That is,

$$|A_l(t)| \leq \frac{1}{\hbar^2} \sum_j |H_{jl}(t)| \int_0^t |H_{1j}(r)|\, dr \tag{7.B.5}$$

for every l, as desired. □

Appendix 7C Proof of Equation (7.1.10)

For a fixed $n \times n$ real matrix $g = [g_{ij}]$ define $\psi = \psi_g : \mathbb{R}^n \to \mathbb{R}^n$ by $\psi = (\psi_1, \ldots, \psi_n)$ where $\psi_j : \mathbb{R}^n \to \mathbb{R}$ is given by

$$\psi_j(x) = \sum_{i=1}^n g_{ji} x_i \qquad \text{for } x = (x_1, \ldots, x_n) \in \mathbb{R}^n. \tag{7.C.1}$$

Let $f : \mathbb{R}^n \to \mathbb{R}$ or $\mathbb{C}$ have continuous first and second partial derivatives: f is a C^2-function. Define $F : \mathbb{R}^n \to \mathbb{R}$ or $\mathbb{C}$ by $F = f \circ \psi$: $F(x) = f(\psi_1(x), \ldots, \psi_n(x))$ for $x \in \mathbb{R}^n$. If D_i denotes partial differentiation with respect to the ith variable, then as f is of class C^1 the chain rule gives

$$\begin{aligned} (D_i F)(x) &= \sum_{j=1}^n (D_j f)(\psi(x))(D_i \psi_j)(x) \\ &= \sum_{j=1}^n g_{ji} (D_j f)(\psi(x)). \end{aligned} \tag{7.C.2}$$

That is,

$$D_i(f \circ \psi) = \sum_{j=1}^n g_{ji} (D_j f) \circ \psi. \tag{7.C.3}$$

As f is of class C^2 we may replace f by $D_j f$ here:

$$D_l[(D_j f) \circ \psi] = \sum_{k=1}^n g_{kl} (D_k D_j f) \circ \psi \tag{7.C.4}$$

$$\Longrightarrow \; D_l D_i(f \circ \psi) = \sum_{j=1}^{n} g_{ji} D_l[(D_j f) \circ \psi] \tag{7.C.5}$$
$$= \sum_{j=1}^{n} g_{ji} \sum_{k=1}^{n} g_{kl}(D_k D_j f) \circ \psi.$$

Then for

$$\nabla^2 = \sum_{i=1}^{n} \frac{\partial^2}{\partial x_i^2} = \sum_{i=1}^{n} D_i^2, \tag{7.C.6}$$

$$\nabla^2(f \circ \psi) = \sum_{i=1}^{n} D_i^2(f \circ \psi) = \sum_{i=1}^{n}\sum_{j=1}^{n}\sum_{k=1}^{n} g_{ji} g_{ki}(D_k D_j f) \circ \psi$$
$$= \sum_{j=1}^{n}\sum_{k=1}^{n}\sum_{i=1}^{n} g_{ji}(g^t)_{ik}(D_k D_j f) \circ \psi \tag{7.C.7}$$
$$= \sum_{j=1}^{n}\sum_{k=1}^{n} (gg^t)_{jk}(D_k D_j f) \circ \psi.$$

In particular if $g \in \mathrm{O}(n)$ (i.e., $gg^t = 1$)

$$\nabla^2(f \circ \psi_g) = \sum_{j=1}^{n}\sum_{k=1}^{n} \delta_{jk}(D_k D_j f) \circ \psi_g$$
$$= \sum_{j=1}^{n}(D_j^2 f) \circ \psi_g = \left(\sum_{j=1}^{n} D_j^2 f\right) \circ \psi_g = \left(\nabla^2 f\right) \circ \psi_g. \tag{7.C.8}$$

However $f \circ \psi_g \equiv g^t \cdot f$ by the definition given in example 5, compared with definition (7.C.1). (7.C.8) therefore reads $\nabla^2(g^{-1} \cdot f) = g^{-1} \cdot \nabla^2 f$ for $g \in \mathrm{O}(n)$, which proves (7.1.10).

Appendix 7D Remarks on Equation (7.7.54)

A reason for equation (7.7.54) is roughly the following. If $\vec{E}$ were a static field (which is manifestly *not* the case, as it is time-dependent) the energy V of a particle of charge q under its influence would be given by

$$V(p) = -q \int_0^p \vec{E} \cdot dr \tag{7.D.1}$$

at a point $p \in \mathbb{R}^3$. Compare equation (2.2.8) and Appendix 8A, equation (8.A.5). As $\vec{E}$ is directed along the z-axis, V assumes the specialized form

$$V(0,0,z) = -q \int_0^z E(0,0,z',t)\, dz'$$
$$\simeq -qzE(x,y,z,t) \tag{7.D.2}$$

since by (7.7.53) $E(x, y, z, t)$ is independent of (x, y, z), up to an approximation. For a hydrogenic atom, the case of main interest to us, we choose $q = -e$ to obtain

$$V(0,0,z) \simeq ezE(x,y,z,t). \tag{7.D.3}$$

Thus we obtain the right-hand side of (7.7.54), by admittedly rough considerations. The reader may wish to consult an appropriate physics text for further discussions of equation (7.7.54).

8

The Quantized Hamiltonian for a Charged Particle in an Electromagnetic Field

8.1 The Motion of a Charged Particle in an Electromagnetic Field

Electric fields and magnetic fields exert a manifest influence, or effect, on the energy levels of an atom. These effects, revealed for example by the *splitting* of spectral lines, are called Stark and Zeeman effects, respectively. The Stark effect (first observed by Johannes Stark in 1913) plays an important role, for example, in theoretical attempts to understand the formation of molecules from atoms. Our interest will be in the Zeeman effect—especially as we consider the notion of "spin," and *spin wave functions* in the next chapter, particularly in Section 9.3 there.

To prepare for that section we shall need to understand how to quantize the classical Hamiltonian that governs the motion of a charged particle in an electromagnetic field. The ideas and results here of course are of independent interest. We draw upon the resources of Chapters 2 and 3. Our basic goal (to be realized in the next chapter) is to compute energy levels (*Landau spectra*), and spin wave functions of hydrogenic atoms under the influence of a uniform magnetic field. Also we compute the Landau spectrum (or Zeeman effect) in another simpler example.

We shall employ the Gaussian system of units mentioned in Chapter 0: an electrostatic unit of charge (1 esu or 1 stat-Coulomb) is that amount of charge on a test object for which an equal object of like charge placed 1 cm away is repelled by a force of 1 dyne, both objects considered placed in a vacuum.

To begin the discussion, consider the (non-relativistic) motion of a particle with a charge q in an electromagnetic field (E, B). That is, E and B are functions:

$\mathbb{R}^3 \times \mathbb{R} \to \mathbb{R}^3$ depending on position (x, y, z) and time t, which represent the *electric* and *magnetic* components (respectively) of the field, and which in practical situations will be given by some *scalar* and *vector potential* $\phi, A = (A_1, A_2, A_3)$: $\mathbb{R}^3 \times \mathbb{R} \to \mathbb{R}, \mathbb{R}^3$:

$$\begin{aligned} E &= \left(-\frac{\partial \phi}{\partial x} - \frac{1}{c}\frac{\partial A_1}{\partial t}, -\frac{\partial \phi}{\partial y} - \frac{1}{c}\frac{\partial A_2}{\partial t}, -\frac{\partial \phi}{\partial z} - \frac{1}{c}\frac{\partial A_3}{\partial t} \right), \\ B &= \left(\frac{\partial A_3}{\partial y} - \frac{\partial A_2}{\partial z}, \frac{\partial A_1}{\partial z} - \frac{\partial A_3}{\partial x}, \frac{\partial A_2}{\partial x} - \frac{\partial A_1}{\partial y} \right) \end{aligned} \tag{8.1.1}$$

for c = the speed of light. ϕ reminds us of course of the role of the electrostatic Coulomb potential; compare definition (2.5.4) and the equations of (2.5.7) for example. The vector potential A allows for generalization of Oersted and Faraday magnetic induction laws. Namely, use (8.1.1) to directly compute

$$\operatorname{curl} E \stackrel{\text{def}}{=} \left(\frac{\partial E_3}{\partial y} - \frac{\partial E_2}{\partial z}, \frac{\partial E_1}{\partial z} - \frac{\partial E_3}{\partial x}, \frac{\partial E_2}{\partial x} - \frac{\partial E_1}{\partial y} \right). \tag{8.1.2}$$

Assuming that ϕ, A are of class C^2, so that one has equality of mixed partials, the result is independent of ϕ:

$$\begin{aligned} &\operatorname{curl} E \\ &= -\frac{1}{c}\left(\frac{\partial}{\partial t}\left(\frac{\partial A_3}{\partial y} - \frac{\partial A_2}{\partial z} \right), \frac{\partial}{\partial t}\left(\frac{\partial A_1}{\partial z} - \frac{\partial A_3}{\partial x} \right), \frac{\partial}{\partial t}\left(\frac{\partial A_2}{\partial x} - \frac{\partial A_1}{\partial y} \right) \right) \\ &\equiv -\frac{1}{c}\frac{\partial B}{\partial t}. \end{aligned} \tag{8.1.3}$$

It is customary to express (8.1.1) as

$$E = -\nabla \phi - \frac{1}{c}\frac{\partial A}{\partial t}, \qquad B = \operatorname{curl} A. \tag{8.1.4}$$

The computation leading to (8.1.3) just mentioned simply amounts to the fact, of course, that the curl of a gradient vanishes. Similarly, because of the equality of mixed partial derivatives, the divergence of a curl vanishes: for A of class C^2

$$\operatorname{div} B \stackrel{\text{def}}{=} \frac{\partial B_1}{\partial x} + \frac{\partial B_2}{\partial y} + \frac{\partial B_3}{\partial z} = 0 \tag{8.1.5}$$

by direct computation, by (8.1.1). Equation (8.1.3) (Faraday's law) and equation (8.1.5) (the *no magnetic charges* law) are part of *Maxwell's equations* which govern the propagation of an electromagnetic field. There are two other Maxwell equations (Gauss' law and Ampère's law) which we shall not venture to discuss.

Example

For $B \in \mathbb{R}$ fixed define a map $B : \mathbb{R}^3 \times \mathbb{R} \to \mathbb{R}^3$ (a constant magnetic field along the z-axis) by

$$B(x, y, z, t) \stackrel{\text{def}}{=} (0, 0, B). \tag{8.1.6}$$

It is trivial to construct a vector potential $A : \mathbb{R}^3 \times \mathbb{R} \to \mathbb{R}^3$ for B:

$$A(x, y, z, t) \stackrel{\text{def}}{=} \left(-\frac{By}{2}, \frac{Bx}{2}, 0\right) \tag{8.1.7}$$

$\Longrightarrow$ $B = \operatorname{curl} A$. Note that another choice for A, which will also be useful later, is

$$A(x, y, z, t) \stackrel{\text{def}}{=} (-By, 0, 0). \tag{8.1.8}$$

Let m denote the mass of our charged particle, whose equation of motion $\Phi = (\Phi_1, \Phi_2, \Phi_3) : \mathbb{R} \to \mathbb{R}^3$ we turn to. For $v = \Phi' : \mathbb{R} \to \mathbb{R}^3$ the velocity function, Newton's second law is $F = m dv/dt$, where in the present case F is given by the *Lorentz formula*

$$F = q\left(E + \frac{1}{c}\, v \times B\right); \tag{8.1.9}$$

$v \times B$ is the cross product $\mathbb{R}^3 \times \mathbb{R} \to \mathbb{R}^3$:

$$(v \times B)(x, y, z, t) \stackrel{\text{def}}{=} \begin{vmatrix} i & j & k \\ \Phi_1'(t) & \Phi_2'(t) & \Phi_3'(t) \\ B_1(x, y, z, t) & B_2(x, y, z, t) & B_3(x, y, z, t) \end{vmatrix}. \tag{8.1.10}$$

One expands the determinant in (8.1.10) to calculate explicitly the components F_1, F_2, F_3 of F. For example, by (8.1.9), (8.1.10),

$$\begin{aligned} F_1(x, y, z, t) = qE_1(x, y, z, t) + \frac{q}{c}\, &\big[\Phi_2'(t) B_3(x, y, z, t) \\ &-\Phi_3'(t) B_2(x, y, z, t)\big]; \end{aligned} \tag{8.1.11}$$

similar equations hold for F_2, F_3. In particular by Newton's law

$$q\left[E_1(\Phi(t), t) + \frac{1}{c}\left(\Phi_2'(t) B_3(\Phi(t), t) - \Phi_3'(t) B_2(\Phi(t), t)\right)\right] = m\Phi_1''(t). \tag{8.1.12}$$

On the other hand, by (8.1.1),

$$E_1 = -\frac{\partial \phi}{\partial x} - \frac{1}{c}\frac{\partial A_1}{\partial t}, \quad B_2 = \frac{\partial A_1}{\partial z} - \frac{\partial A_3}{\partial x}, \quad B_3 = \frac{\partial A_2}{\partial x} - \frac{\partial A_1}{\partial y}$$

which substituted in (8.1.12) yields

$$\begin{aligned} m\Phi_1''(t) = q\Bigg[&-\frac{\partial \phi}{\partial x}(\Phi(t), t) - \frac{1}{c}\frac{\partial A_1}{\partial t}(\Phi(t), t) \\ &+ \frac{1}{c}\left\{\Phi_2'(t)\left(\frac{\partial A_2}{\partial x} - \frac{\partial A_1}{\partial y}\right)(\Phi(t), t)\right. \\ &\left.- \Phi_3'(t)\left(\frac{\partial A_1}{\partial z} - \frac{\partial A_3}{\partial x}\right)(\Phi(t), t)\right\}\Bigg]. \end{aligned} \tag{8.1.13}$$

Similarly

$$
\begin{aligned}
m\Phi_2''(t) = q\Bigg[& -\frac{\partial \phi}{\partial y}(\Phi(t),t) - \frac{1}{c}\frac{\partial A_2}{\partial t}(\Phi(t),t) \\
& + \frac{1}{c}\left\{ \Phi_3'(t)\left(\frac{\partial A_3}{\partial y} - \frac{\partial A_2}{\partial z}\right)(\Phi(t),t) \right. \\
& \left. - \Phi_1'(t)\left(\frac{\partial A_2}{\partial x} - \frac{\partial A_1}{\partial y}\right)(\Phi(t),t) \right\}\Bigg], \\
m\Phi_3''(t) = q\Bigg[& -\frac{\partial \phi}{\partial z}(\Phi(t),t) - \frac{1}{c}\frac{\partial A_3}{\partial t}(\Phi(t),t) \\
& + \frac{1}{c}\left\{ \Phi_1'(t)\left(\frac{\partial A_1}{\partial y} - \frac{\partial A_3}{\partial x}\right)(\Phi(t),t) \right. \\
& \left. - \Phi_2'(t)\left(\frac{\partial A_3}{\partial y} - \frac{\partial A_2}{\partial z}\right)(\Phi(t),t) \right\}\Bigg].
\end{aligned} \tag{8.1.14}
$$

Equations (8.1.13), (8.1.14) are thus the equations of motion of the charged particle under the influence of the field (E, B), expressed in terms of the scalar and vector potential (ϕ, A).

Example

Choose the scalar potential $\phi = 0$ and choose the vector potential A to be that given by (8.1.8), where we take $B \in \mathbb{R}$ to be *non-zero*. By (8.1.1), $E = 0$ and by (8.1.8),

$$
\frac{\partial A_1}{\partial x} = 0, \qquad \frac{\partial A_1}{\partial y} = -B, \qquad \frac{\partial A_1}{\partial z} = 0,
$$

with all partial derivatives of A_2, A_3 equal to zero. By (8.1.13), (8.1.14) we obtain therefore the following equations of motion.

$$
m\Phi_1''(t) = \frac{qB}{c}\Phi_2'(t), \quad m\Phi_2''(t) = -\frac{qB}{c}\Phi_1'(t), \quad m\Phi_3''(t) = 0. \tag{8.1.15}
$$

Then

$$
\Phi_3(t) = \alpha_3 t + \gamma_3 \tag{8.1.16}
$$

for constants $\alpha_3, \gamma_3 \in \mathbb{R}$. Define $\omega = \frac{qB}{mc}$ and deduce from (8.1.15) that

$$
\frac{d^2}{dt^2}\left[\Phi_1'(t)\right] = \Phi_1'''(t) = \omega\Phi_2''(t) = -\omega^2\Phi_1'(t), \tag{8.1.17}
$$

and similarly that

$$
\frac{d^2}{dt^2}\left[\Phi_2'(t)\right] = \Phi_2'''(t) = -\omega\Phi_1''(t) = -\omega^2\Phi_2'(t), \tag{8.1.18}
$$

which are equations of simple harmonic motion for $\Phi_j'(t)$, $j = 1, 2$, with frequency $\omega/(2\pi)$, and with general solutions

$$\Phi_j'(t) = \alpha_j \cos(\omega t - \beta_j), \quad j = 1, 2, \tag{8.1.19}$$

for constants α_j, $\beta_j \in \mathbb{R}$; compare equation (2.2.17). Now (8.1.15) and (8.1.19) together imply that

$$\begin{aligned} m\omega\alpha_2 \cos(\omega t - \beta_2) &\overset{\text{def. of } \omega}{=} \frac{qB}{c}\alpha_2 \cos(\omega t - \beta_2) = \frac{qB}{c}\Phi_2'(t) \\ &= \quad m\Phi_1''(t) = -m\omega\alpha_1 \sin(\omega t - \beta_1), \end{aligned} \tag{8.1.20}$$

and since $B \neq 0 \implies \omega \neq 0$ we conclude that for all t

$$-\alpha_1 \sin(\omega t - \beta_1) = \alpha_2 \cos(\omega t - \beta_2), \tag{8.1.21}$$

which we differentiate to get (again as $\omega \neq 0$)

$$\alpha_1 \cos(\omega t - \beta_1) = \alpha_2 \sin(\omega t - \beta_2). \tag{8.1.22}$$

Square both sides of (8.1.21), (8.1.22) and add: $\alpha_1^2 = \alpha_2^2 \implies \alpha_1 = \pm\alpha_2$. From (8.1.19)

$$\Phi_j(t) = \frac{\alpha_j}{\omega} \sin(\omega t - \beta_j) + \gamma_j, \qquad j = 1, 2 \tag{8.1.23}$$

for some constants $\gamma_j \in \mathbb{R}$. That is, by (8.1.22),

$$\begin{aligned} \Phi_1(t) &= \frac{\alpha_1}{\omega} \sin(\omega t - \beta_1) + \gamma_1, \\ \Phi_2(t) &= \frac{\alpha_2}{\omega} \sin(\omega t - \beta_2) + \gamma_2 = \frac{\alpha_1}{\omega} \cos(\omega t - \beta_1) + \gamma_2, \end{aligned} \tag{8.1.24}$$

so that by (8.1.16) we obtain the equation of motion (for $B \neq 0$ in (8.1.8))

$$\Phi(t) = \left(\frac{\alpha_1}{\omega} \sin(\omega t - \beta_1), \frac{\alpha_1}{\omega} \cos(\omega t - \beta_1), \alpha_3 t\right) + \gamma \tag{8.1.25}$$

for constants α_1, $\beta_1 \in \mathbb{R}$, $\gamma = (\gamma_1, \gamma_2, \gamma_3) \in \mathbb{R}^3$, $\omega = \frac{qB}{mc}$. We see that the path of the particle is a helix confined to the circular cylinder $x^2 + y^2 = \alpha_1^2/\omega^2$ of radius $|\alpha_1/\omega|$. The cylinder is parallel to the direction of the magnetic field (the z-axis).

One can set up a suitable Lagrangian L for the motion of a charged particle in an electromagnetic field (E, B). In Section 2.7 of Chapter 2 the Lagrangian for the motion of n particles in $\mathbb{R}^3$ was a map from $\mathbb{R}^{3n} \times \mathbb{R}^{3n} \to \mathbb{R}$. In the present case $n = 1$. However as ϕ, A are *time-dependent* we take L to be a time-dependent map from $\mathbb{R}^3 \times \mathbb{R}^3 \times \mathbb{R} \to \mathbb{R}$. Namely

$$L(x, y, z, v_1, v_2, v_3, t) \overset{\text{def}}{=} \frac{m}{2} \sum_{j=1}^{3} v_j^2 \; - \; q\phi(x, y, z, t) \; + \; \frac{q}{c} \sum_{j=1}^{3} v_j A_j(x, y, z, t). \tag{8.1.26}$$

Let $\widetilde{q} = (x, y, z)$, $v = (v_1, v_2, v_2)$. Clearly

$$\frac{\partial L}{\partial v_k}(\widetilde{q}, v, t) = m v_k + \frac{q}{c} A_k(\widetilde{q}, t), \qquad \text{and} \tag{8.1.27}$$

$$\frac{\partial L}{\partial x}(\widetilde{q}, v, t) = -q \frac{\partial \phi}{\partial x}(\widetilde{q}, t) + \frac{q}{c} \sum_{j=1}^{3} v_j \frac{\partial A_j}{\partial x}(\widetilde{q}, t), \tag{8.1.28}$$

with similar formulas for $\frac{\partial L}{\partial y}$, $\frac{\partial L}{\partial z}$. Then, by the chain rule,

$$\begin{aligned}
&\frac{d}{dt}\frac{\partial L}{\partial v_1}\left(\Phi(t), \Phi'(t), t\right) - \frac{\partial L}{\partial x}\left(\Phi(t), \Phi'(t), t\right) \\
&= m\Phi_1''(t) + \frac{q}{c}\left[\frac{\partial A_1}{\partial x}(\Phi(t), t)\Phi_1'(t) + \frac{\partial A_1}{\partial y}(\Phi(t), t)\Phi_2'(t) + \frac{\partial A_1}{\partial z}(\Phi(t), t)\Phi_3'(t)\right. \\
&\quad \left. + \frac{\partial A_1}{\partial t}(\Phi(t), t)\right] + q\frac{\partial \phi}{\partial x}(\Phi(t), t) - \frac{q}{c}\sum_{j=1}^{3} \Phi_j'(t) \frac{\partial A_j}{\partial x}(\Phi(t), t) \\
&= m\Phi_1''(t) + q\left[\frac{\partial \phi}{\partial x}(\Phi(t), t) + \frac{1}{c}\frac{\partial A_1}{\partial t}(\Phi(t), t)\right] + \frac{q}{c}\left[\left(\frac{\partial A_1}{\partial y}(\Phi(t), t)\right.\right. \\
&\quad \left.\left. - \frac{\partial A_2}{\partial x}(\Phi(t), t)\right)\Phi_2'(t) + \left(\frac{\partial A_1}{\partial z}(\Phi(t), t) - \frac{\partial A_3}{\partial x}(\Phi(t), t)\right)\Phi_3'(t)\right].
\end{aligned} \tag{8.1.29}$$

Thus we see that

$$\frac{d}{dt}\frac{\partial L}{\partial v_1}\left(\Phi(t), \Phi'(t), t\right) - \frac{\partial L}{\partial x}\left(\Phi(t), \Phi'(t), t\right) = 0 \tag{8.1.30}$$

$\Longleftrightarrow$ equation (8.1.13) holds. Similarly

$$\frac{d}{dt}\frac{\partial L}{\partial v_2}\left(\Phi(t), \Phi'(t), t\right) - \frac{\partial L}{\partial y}\left(\Phi(t), \Phi'(t), t\right) = 0 \tag{8.1.31}$$

and

$$\frac{d}{dt}\frac{\partial L}{\partial v_3}\left(\Phi(t), \Phi'(t), t\right) - \frac{\partial L}{\partial z}\left(\Phi(t), \Phi'(t), t\right) = 0 \tag{8.1.32}$$

$\Longleftrightarrow$ equations (8.1.14) hold, which proves

Theorem 8.1. (Lagrangian form of the equation of motion of a charged particle in an electromagnetic field (E, B) with scalar and vector potential (ϕ, A)).

Let $L : \mathbb{R}^3 \times \mathbb{R}^3 \times \mathbb{R} \to \mathbb{R}$ be the Lagrangian constructed in (8.1.26). *Then the equations of motion* (8.1.13), (8.1.14) *are equivalent to the equations* (8.1.30), (8.1.31), (8.1.32).

One may compare Theorem 8.1 with Theorem 2.3.

8.2 Classical and Quantized Hamiltonian for a Charged Particle

The Lorentz force F in (8.1.9) is not the gradient of some potential energy function V. Thus one cannot set up a Hamiltonian function H as in definition (2.6.1). However, as pointed out in that Chapter 1 can use equation (2.7.12) there as a means of defining H, given L in the absence of V. Thus given (2.7.12), we are led to define $H : \mathbb{R}^3 \times \mathbb{R}^3 \times \mathbb{R} \to \mathbb{R}$ by

$$\begin{aligned} H(\widetilde{q}, p, t) \stackrel{\text{def}}{=} & \frac{\partial L}{\partial p_1}(\widetilde{q}, p, t)p_1 + \frac{\partial L}{\partial p_2}(\widetilde{q}, p, t)p_2 \\ & + \frac{\partial L}{\partial p_3}(\widetilde{q}, p, t)p_3 - L(\widetilde{q}, p, t) \end{aligned} \tag{8.2.1}$$

for $\widetilde{q} = (x, y, z)$, $p = (p_1, p_2, p_3) \in \mathbb{R}^3$, $t \in \mathbb{R}$. Using (8.1.26) and (8.1.27) one has

$$\begin{aligned} H(\widetilde{q}, p, t) &= \sum_{k=1}^{3} \left(mp_k + \frac{q}{c} A_k(\widetilde{q}, t) \right) p_k - \frac{m}{2} \sum_{k=1}^{3} p_k^2 + q\phi(\widetilde{q}, t) - \frac{q}{c} \sum_{k=1}^{3} p_k A_k(\widetilde{q}, t) \\ &= \frac{1}{2m} \sum_{k=1}^{3} (mp_k)^2 + q\phi(\widetilde{q}, t) \\ &= \frac{1}{2m} \sum_{k=1}^{3} \left[\frac{\partial L}{\partial p_k}(\widetilde{q}, v, t) - \frac{q}{c} A_k(\widetilde{q}, t) \right]^2 + q\phi(\widetilde{q}, t). \end{aligned} \tag{8.2.2}$$

On the other hand, in accord with remarks of Chapter 2, taking (2.7.13) there as a cue, we define *generalized momenta* p_x, p_y, p_z by

$$p_x = \frac{\partial L}{\partial p_1}, \qquad p_y = \frac{\partial L}{\partial p_2}, \qquad p_z = \frac{\partial L}{\partial p_3}. \tag{8.2.3}$$

Equation (8.2.2) then takes the form

$$H = \frac{1}{2m} \left[\left(p_x - \frac{q}{c} A_1 \right)^2 + \left(p_y - \frac{q}{c} A_2 \right)^2 + \left(p_z - \frac{q}{c} A_3 \right)^2 \right] + q\phi. \tag{8.2.4}$$

Apply the usual quantization rules

$$p_x \to \frac{\hbar}{i} \frac{\partial}{\partial x}, \qquad p_y \to \frac{\hbar}{i} \frac{\partial}{\partial y}, \qquad p_z \to \frac{\hbar}{i} \frac{\partial}{\partial z} \tag{8.2.5}$$

to quantize H: If M_{A_1} denotes multiplication by A_1 we see that

$$\begin{aligned} \left(p_x - \frac{q}{c} A_1 \right)^2 &\to \left(\frac{\hbar}{i} \frac{\partial}{\partial x} - \frac{q}{c} M_{A_1} \right)^2 \\ &= -\hbar^2 \frac{\partial^2}{\partial x^2} - \frac{\hbar}{i} \frac{q}{c} \frac{\partial}{\partial x} \circ M_{A_1} - \frac{\hbar}{i} \frac{q}{c} M_{A_1} \circ \frac{\partial}{\partial x} + \frac{q^2}{c^2} A_1^2, \end{aligned} \tag{8.2.6}$$

where

$$\frac{\partial}{\partial x} \circ M_{A_1} = M_{\frac{\partial A_1}{\partial x}} + A_1 \frac{\partial}{\partial x}$$

(by the product rule) $\Longrightarrow$

$$\left[p_x - \frac{q}{c} A_1\right]^2 \to -\hbar^2 \frac{\partial^2}{\partial x^2} - \frac{2q}{c} A_1 \frac{\hbar}{i} \frac{\partial}{\partial x} - \frac{q}{c} \frac{\hbar}{i} \frac{\partial A_1}{\partial x} + \frac{q^2}{c^2} A_1^2. \tag{8.2.7}$$

By similar quantization rules for $[p_y - \frac{q}{c} A_2]^2$, $[p_z - \frac{q}{c} A_3]^2$, we arrive at the quantized Hamiltonian $\widehat{H}$ given by

$$\begin{aligned} H \to \widehat{H} \overset{\text{def}}{=} \frac{1}{2m} \Bigg[& -\hbar^2 \nabla^2 - \frac{2q}{c} \frac{\hbar}{i} \left(A_1 \frac{\partial}{\partial x} + A_2 \frac{\partial}{\partial y} + A_3 \frac{\partial}{\partial z} \right) \\ & - \frac{q}{c} \frac{\hbar}{i} \left(\frac{\partial A_1}{\partial x} + \frac{\partial A_2}{\partial y} + \frac{\partial A_3}{\partial z} \right) + \frac{q^2}{c^2} \left(A_1^2 + A_2^2 + A_3^2 \right) \Bigg] + q\phi \end{aligned} \tag{8.2.8}$$

$$\text{for } \nabla^2 = \frac{\partial^2}{\partial x^2} + \frac{\partial^2}{\partial y^2} + \frac{\partial^2}{\partial z^2};$$

see (8.2.4).

Example

Let B be the constant magnetic field given by (8.1.6). For the choice of vector potential A given by (8.1.7),

$$\frac{\partial A_1}{\partial x} = \frac{\partial A_2}{\partial y} = \frac{\partial A_3}{\partial z} = 0, \qquad A_1^2 + A_2^2 + A_3^2 = \frac{B^2}{4}(x^2 + y^2),$$

$$\text{and } \frac{\hbar}{i} \left(A_1 \frac{\partial}{\partial x} + A_2 \frac{\partial}{\partial y} + A_3 \frac{\partial}{\partial z} \right) = \frac{\hbar}{i} \left(-\frac{By}{2} \frac{\partial}{\partial x} + \frac{Bx}{2} \frac{\partial}{\partial y} \right) = \frac{B}{2} M_z,$$

where $M_z = \frac{\hbar}{i} \left(x \frac{\partial}{\partial y} - y \frac{\partial}{\partial x} \right)$ is the quantized angular momentum operator in the z-direction; see (5.4.4). $\widehat{H}$ in (8.2.8) therefore reduces to

$$\widehat{H} = -\frac{\hbar^2 \nabla^2}{2m} - \frac{q}{2mc} B M_z + \frac{q^2 B^2}{8mc^2}(x^2 + y^2) + q\phi. \tag{8.2.9}$$

For the same B of (8.1.6), another choice of vector potential A is given by (8.1.8). For this choice

$$\frac{\partial A_1}{\partial x} = \frac{\partial A_2}{\partial y} = \frac{\partial A_3}{\partial z} = 0$$

again, but now $A_1^2 + A_2^2 + A_3^2 = B^2 y^2$. As $A_2, A_3 = 0$, $\widehat{H}$ in (8.2.8) reduces to

$$\widehat{H} = -\frac{\hbar^2 \nabla^2}{2m} + \frac{q\hbar}{mci} By \frac{\partial}{\partial x} + \frac{q^2 B^2 y^2}{2mc^2} + q\phi. \tag{8.2.10}$$

Patrick Shanahan's Chapter 18 will include material from this chapter with further discussion, especially in the direction of gauge theory. As pointed out in Chapter 0 (in remarks following equation (0.0.16)) the quantity $\omega \overset{\text{def}}{=} \frac{qB}{mc}$ in (8.1.19), for B representing a magnetic field strength, indeed does have units $\frac{1}{\sec}$; i.e., indeed $\omega/2\pi$ is a frequency. Compare also definition (9.3.23).

Appendix 8A The Electric Field Due to a Collection of Point Charges, Electric Potential, and Voltage

The notion of a *magnetic dipole moment* will arise in connection with *Pauli's equation* in the next chapter. It is appropriate, prior to then, to consider the corresponding notion of an *electric dipole moment*. This we take up in Appendix 8B, which we prepare for now. We find another application of Coulomb's law as we consider the electric field due to a collection of point charges. Some elementary material on electric potential and voltage is also presented.

Fix a charge q at some point $A = (x_1, y_1, z_1) \in \mathbb{R}^3$. This charge defines a vector field E on the domain of points $B = (x_2, y_2, z_2) \neq A$, as follows. The vector E_B at B is defined as the force that a unit positive charge at B experiences due to the charge q at A. E_B can be found by an application of Coulomb's law:

$$E_B = \frac{q\{(x_2, y_2, z_2) - (x_1, y_1, z_1)\}}{\left[(x_2 - x_1)^2 + (y_2 - y_1)^2 + (z_2 - z_1)^2\right]^{3/2}}; \tag{8.A.1}$$

thus $E_B = -F^{(1)}(= F^{(2)})$ in equation (2.5.1), with q_1 there equal to q, and $q_2 = 1$. The direction of E_B is as follows. If $q > 0$, then the charge q and the unit test charge have the same sign; the charges are opposite if $q < 0$; see Figure 8.1. Note that the magnitude $|E_B|$ of E_B is given by

$$\begin{aligned} |E_B| &= \frac{|q|}{(x_2 - x_1)^2 + (y_2 - y_1)^2 + (z_2 - z_1)^2} \\ &= \frac{|q|}{[\text{distance from } A \text{ to } B]^2}. \end{aligned} \tag{8.A.2}$$

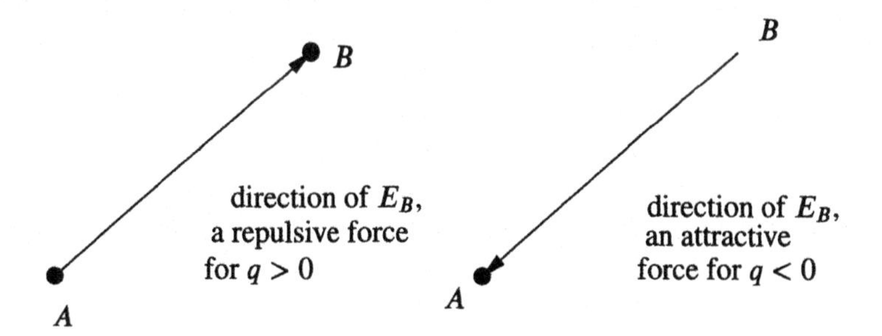

Figure 8.1: Direction of E_B

An *arbitrary* charge q' at B experiences a Coulomb force

$$\frac{qq'\,\{(x_2, y_2, z_2) - (x_1, y_1, z_1)\}}{\left[(x_2 - x_1)^2 + (y_2 - y_1)^2 + (z_2 - z_1)^2\right]^{3/2}} = q' E_B, \tag{8.A.3}$$

by (8.A.1). The vector field $E : B \to E_B$ is called the *electric field* due to the charge q. For $q' = 1$ esu in (8.A.3) one observes that E_B is given by the units dynes/esu, in the c.g.s. system.

Example

Find the electric field vector E_B 10 cm away from a charge $q = -1500$ esu. What force does a charge $q' = 30$ esu experience due to the charge q?

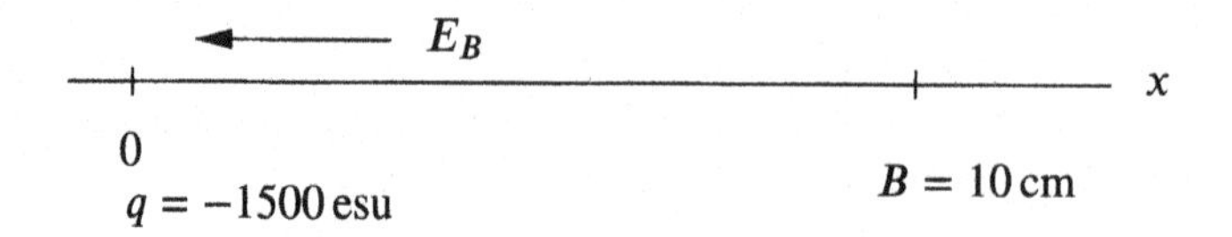

Figure 8.2: Example

We take $\kappa = 1$ is equation (0.0.4). By (8.A.2),

$$|E_B| = \frac{|q|}{10^2} = \frac{1500}{100} = 15 \text{ dynes/esu.}$$

By Figure 8.1, E_B is directed toward the origin, as illustrated in Figure 8.2. By (8.A.3) the charge q' at B experiences a force of 30 esu × 15 dynes/esu = 450 dynes due to the charge q at the origin.

So far we have considered the electric vector field generated by a *single* charge q at the point A. Suppose now there is a finite collection of charges $q_1, q_2, \ldots, q_n$ at distinct points $A_1, A_2, \ldots, A_n$. Experiments have well established the following *superposition principle*: If B is a point distinct from the A_j, and q' is a charge at B, then the Coulomb force exerted on the charge q' due to the charges q_j is the *sum* of the forces exerted individually on q' by the q_j. We can construct an electric vector field $E : B \to E_B$ away from the A_j in a manner similar to the case of a single charge. Namely, for a point B distinct from the A_j the vector E_B at B is defined as the force that a unit positive charge at B experiences due to the charges q_j at A_j. If E_j is the electric field due to the single charge q_j and E_{jB} is the corresponding electric field vector at B, then by the superposition principle

$$E_B = E_{1B} + E_{2B} + \cdots + E_{nB}. \tag{8.A.4}$$

By (8.A.3) an arbitrary charge q' at B experiences a force $q'E_{jB}$ due to the charge q_j and thus (by the superposition principle) a total force

$$F = q'E_{1B} + q'E_{2B} + \cdots + q'E_{nB} = q'E_B, \tag{8.A.5}$$

by (8.A.4), due to the field of charges.

Example

Charges $q_1 = .35$ esu and $q_2 = -(.55)$ esu, are situated at the points $A_1 = (0,4,0)$, $A_2 = (3,0,0)$ respectively. Taking the *dielectric constant* $1/\kappa = 1$ in

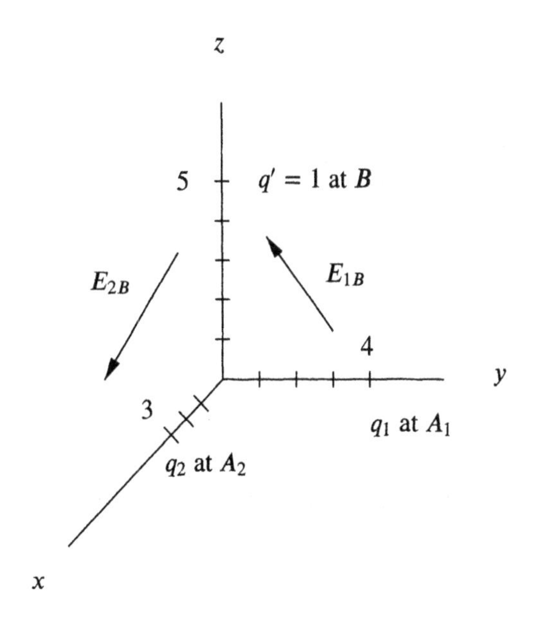

Figure 8.3: Multiple charges

equation (0.0.4) we wish to compute the electric field vector E_B at the point $B = (0, 0, 5)$. Let E_1, E_2 denote the electric fields due to the charges q_1, q_2. By (8.A.1) the unit positive charge at B experiences a force

$$\begin{aligned} E_{1B} &= \frac{q_1(0, -4, 5)}{[16 + 25]^{3/2}} = \frac{.35(0, -4, 5)}{262.52} \\ &= (.001333)(0, -4, 5) = (0, -.005333, .006665) \end{aligned} \tag{8.A.6}$$

due to q_1, and similarly a force

$$\begin{aligned} E_{2B} &= \frac{q_2(-3, 0, 5)}{[9 + 25]^{3/2}} = \frac{-(.55)(-3, 0, 5)}{198.25} \\ &= -(.00277)(-3, 0, 5) = (.00832, 0, -.01385). \end{aligned} \tag{8.A.7}$$

By (8.A.4) the desired electric vector E_B is given by

$$E_B = E_{1B} + E_{2B} = (.00832, -.00533, -.00718) \text{ dynes/esu.} \tag{8.A.8}$$

By (8.A.1) the force $E(x, y, z)$ on a *unit* charge at (x, y, z) due to the charge q at (x_1, y_1, z_1) is given by

$$E(x, y, z) = (E_1(x, y, z), E_2(x, y, z), E_3(x, y, z))$$

$$\begin{aligned} \text{for} \quad E_1(x, y, z) &= \frac{q(x - x_1)}{\left[(x - x_1)^2 + (y - y_1)^2 + (z - z_1)^2\right]^{3/2}}, \\ E_2(x, y, z) &= \frac{q(y - y_1)}{\left[(x - x_1)^2 + (y - y_1)^2 + (z - z_1)^2\right]^{3/2}}, \end{aligned} \tag{8.A.9}$$

$$E_3(x, y, z) = \frac{q(z - z_1)}{\left[(x - x_1)^2 + (y - y_1)^2 + (z - z_1)^2\right]^{3/2}},$$

where $(x, y, z) \in \mathbb{R}^3 - \{(x_1, y_1, z_1)\}$. It follows that we can immediately construct an *electric potential energy* function $V : \mathbb{R}^3 - \{(x_1, y_1, z_1)\} \to \mathbb{R}$ for the vector field E. Namely,

$$V(x, y, z) \stackrel{\text{def}}{=} \frac{q}{\left[(x - x_1)^2 + (y - y_1)^2 + (z - z_1)^2\right]^{1/2}}. \tag{8.A.10}$$

That is,

$$-\frac{\partial V}{\partial x} = E_1, \qquad -\frac{\partial V}{\partial y} = E_2, \qquad -\frac{\partial V}{\partial z} = E_3 \implies E = -\nabla V. \tag{8.A.11}$$

In particular the *work* W done by the field E in moving the unit test charge from point B_1 to point B_2 is *path-independent* as it is given by

$$W \stackrel{\text{def}}{=} \int_C E \cdot dr = -\int_C (\nabla V) \cdot dr = -[V(B_2) - V(B_1)], \tag{8.A.12}$$

where C is any curve from B_1 to B_2. By (8.A.3) the force on an arbitrary charge q' at $(x, y, z) \in \mathbb{R}^3 - \{(x_1, y_1, z_1)\}$ is $q'E(x, y, z) = -q'(\nabla V)(x, y, z) = -(\nabla q'V)(x, y, z)$, which means that the work $W_{q'}$ done by the field E in moving q' from B_1 to B_2 is given by

$$W_{q'} = -\int_{C \text{ in (8.A.12)}} (\nabla q'V) \cdot dr = -q'[V(B_2) - V(B_1)] = q'W. \tag{8.A.13}$$

A potential difference, $V(B_2) - V(B_1)$ as in (8.A.12), is called a *voltage difference*, or more simply a *voltage*, in honor of Alessandro Volta (1745–1827) who invented the electric battery which thus led to initial studies of electric currents. For $q' = 1$ esu in (8.A.13) one observes that voltage is measured in the unit (dyne×cm)/esu = erg/esu, in the c.g.s. system, which is also called a *stat-volt*. The more familiar unit, the *volt*, is given by 1 volt = 1/300 stat-volt. More precisely 1 volt= $\frac{1}{299.8}$ stat-volt.

Given a collection of charges $q_1, q_2, \ldots, q_n$ at distinct points $A_1, A_2, \ldots, A_n$, as before, and the electric fields $E_1, \ldots, E_n$ they generate (away from the points $A_1, A_2, \ldots, A_n$), the force $E(x, y, z)$ on a unit charge at $(x, y, z) \neq$ the A_j due to the q_j is given by (8.A.4):

$$E(x, y, z) = \sum_{j=1}^{n} E_j(x, y, z). \tag{8.A.14}$$

If V_j are electric potential energy functions for the E_j (say V_j is given by (8.A.10) with (x_1, y_1, z_1) there replaced by A_j), it follows that

$$E = -\nabla V \qquad \text{for } V \stackrel{\text{def}}{=} \sum_{j=1}^{n} V_j, \tag{8.A.15}$$

and thus if W, W_j is the work done by the fields E, E_j, respectively, in moving the unit test charge from B_1 to B_2, one has (as in (8.A.12))

$$W = -[V(B_2) - V(B_1)] = -\sum_{j=1}^{n}[V_j(B_2) - V_j(B_1)] = \sum_{j=1}^{n} W_j, \tag{8.A.16}$$

with $q'W =$ the work done by E

in moving an arbitrary charge from B_1 to B_2,

in comparison with (8.A.13). If the unit test charge is brought to the point B_2 from a distance infinitely far away ("$B_1 = \infty$," or each $V_j(B_1) = 0$ by comparison with (8.A.10)) then one has in (8.A.16)

$$W = -V(B_2), \tag{8.A.17}$$

which one can express as follows:

the *absolute* potential $V(B)$ at a point B equals $-W$ = the work done *against* the field E in bringing a unit positive test particle from infinitely far away to the point B. (8.A.18)

Example

In a container of oil with a dielectric constant $1/\kappa = 5.3$, two charges q_1, q_2 are separated 10 cm apart; see Figure 8.4. If $q_1 = 100$ esu and $q_2 = 50$ esu what is the absolute potential $V(B)$ at a point B midway between q_1 and q_2?

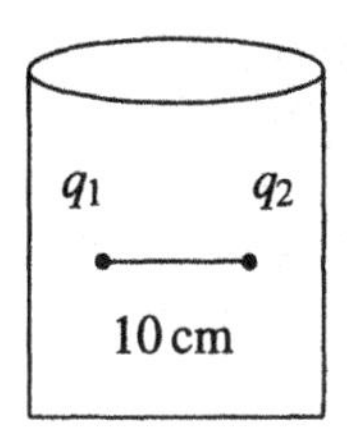

Figure 8.4: Two charges in oil

Solution: By (8.A.10),

$$V_1(B) = \frac{100}{(5.3)5} = 3.77 \text{ stat-volts}, \tag{8.A.19}$$

$$V_2(B) = \frac{50}{(5.3)5} = 1.89 \text{ stat-volts}.$$

By (8.A.15), $V(B) = V_1(B) + V_2(B) = 4.66$ stat-volts.

Another unit of energy is the *electron volt* (eV), which we have already encountered. It is defined as the amount of work W_{-e} required to move an electron through a potential difference of 1 volt. Thus in (8.A.13) and by the remarks following it, we require that $V(B_2) - V(B_1) = 1 \text{ volt} = \frac{1}{299.8}$ stat-volts $\stackrel{\text{def}}{=} \frac{1}{299.8}$ erg/esu, which by (0.0.3) gives $W_{-e} = \frac{-(-e)}{299.8}$ erg/esu $= \frac{4.80321\times10^{-10}}{299.8}$ erg $= 1.6021 \times 10^{-12}$ erg. Or 1 erg $= \frac{1}{1.60214} \times 10^{12}$ eV $= 6.2417 \times 10^{11}$ eV, as we have noted in Section 5.8 of Chapter 5, for example. Also see the computation in (5.6.5).

Appendix 8B Electric Dipole Moments

An *electric dipole* is a system of two opposite charges q, $-q$ separated by a distance d, as illustrated in Figure 8.5. There the *dipole moment* $\vec{m}$ of the system is also illustrated, $\vec{m}$ being the vector with magnitude qd and with direction that of an arrow from the negative charge towards the positive charge.

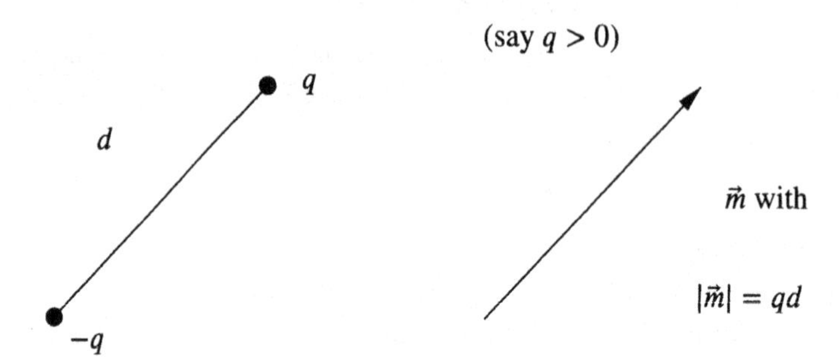

Figure 8.5: Dipole

We are interested in the behavior of the dipole in the presence of an *external* electric field E; that is, E is produced by other charges (independent of q, $-q$). We assume that E is *uniform*: every vector E_B has the same direction and magnitude. Take, for example, the direction of E_B to be that of the positive x-axis as indicated in Figure 8.6. The field E will cause the dipole to rotate to a direction parallel to E as illustrated in Figure 8.6. Thus the dipole tends to align itself with the field.

By (8.A.3) the Coulomb forces on the charges q, $-q$, given E, are given by $F_{\pm q} = \pm qE$. Their tangential components are denoted by $F_t^{\pm}$ in Figure 8.6, where $\theta + \gamma = \pi/2$. Now

$$|F_t^+| = |F_q| \cos\gamma = |F_q| \sin\theta, \tag{8.B.1}$$
$$|F_t^+| = |F_q| \cos\gamma = |F_q| \sin\theta,$$

and the distances from 0 to $\pm q$ are the same and equal to $d/2$. The work done by F_q as q undergoes a small displacement Δs along a circular arc of radius $d/2$ (see Figure 8.7) is (with the help of (8.B.1))

$$\Delta W^+ = |F_t^+|\Delta s = |F_q|(\sin\theta)\frac{d}{2}\Delta\theta. \tag{8.B.2}$$

The work done by F_{-q} as $-q$ undergoes the displacement Δs is, similarly,

$$\Delta W^- = |F_t^-|\Delta s = |F_q|(\sin\theta)\frac{d}{2}\Delta\theta. \tag{8.B.3}$$

We see that the total work W done by the field E as the dipole is rotated from some initial angle θ_o to the angle θ is given by

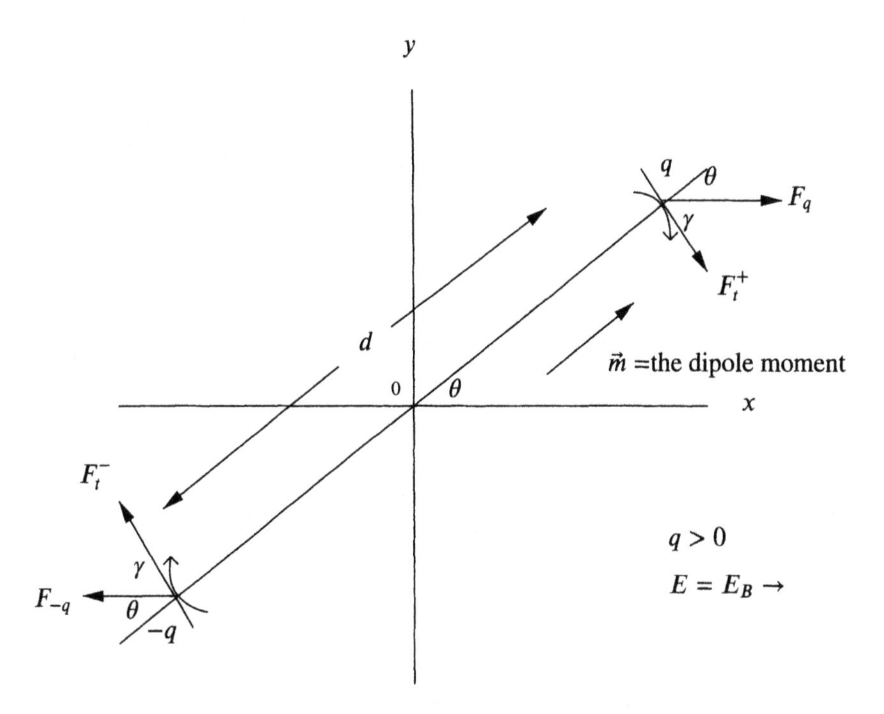

Figure 8.6: Dipole in a uniform field

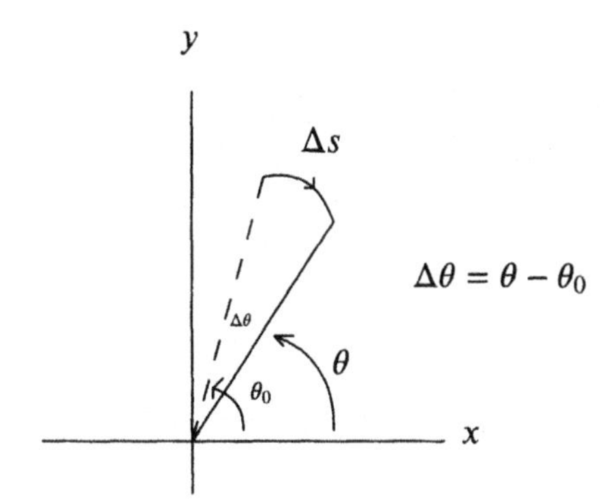

Figure 8.7: Circular displacement

$$\begin{aligned} W &= \int_{\theta_o}^{\theta} \left[|F_q|(\sin\theta)\frac{d}{2} + |F_q|(\sin\theta)\frac{d}{2} \right] d\theta \\ &= |F_q| d(\cos\theta_o - \cos\theta) \\ &= qd|E|(\cos\theta_o - \cos\theta). \end{aligned} \tag{8.B.4}$$

It is customary to choose $\theta_o = \pi/2$, in which case we see that

$$W = -qd|E|\cos\theta \equiv -\vec{m} \cdot E. \tag{8.B.5}$$

9

Spin Wave Functions

9.1 Spin Operators

In the introductory remarks to Chapter 5 mention was made of the 1925 proposal of Uhlenbeck and Goudsmit that to an electron there should be assigned an *intrinsic angular momentum* or *spin* $S = (S_x, S_y, S_z)$—this in addition to its quantum angular momentum (or *orbital* angular momentum) $M = (M_x, M_y, M_z)$ defined in (5.4.4). This is not to be taken to mean that an electron actually spins about an axis in some physical way. Experiments (in particular the noted Stern–Gerlach experiment) have provided adequate justification for the proposed concept of spin, a proposal also made by R. Bichowsky and H. Urey. Dirac has shown that the notion of spin is necessary if quantum mechanics and Einstein relativity are to be consistent theories. Chemical properties of atoms, and their ability to combine and form molecules, the splitting of spectral lines by a magnetic field applied to an atom (i.e., the *Zeeman effect*, which we discuss later), all depend on this important notion.

Proceeding in a purely formal manner, guided by analogies to orbital angular momentum, we show the mathematical existence of three *spin operators* (S_x, S_y, S_z), and the existence of three eigenvalues $\frac{1}{2}(\frac{1}{2}+1)\hbar^2$, $m_S\hbar$, where $m_S = \pm\frac{1}{2}$, results which are taken to have the following meaning in conformity with known physical facts: **(i)** $\sqrt{\frac{1}{2}(\frac{1}{2}+1)\hbar^2} = \frac{\sqrt{3}}{2}\hbar$ is the magnitude $|S|$ of the electron's intrinsic angular momentum (or spin) S, where S is a specific Hermitian operator which we construct with "components" S_x, S_y, S_z; **(ii)** the factor $\frac{1}{2}$ in the product $\frac{1}{2}(\frac{1}{2}+1)\hbar^2$ is the *spin quantum number* s of the electron $s = \frac{1}{2}$; **(iii)** as in the case of orbital angular momentum $M = (M_x, M_y, M_z)$ one also has spatial quantization of S: The z-component of S can assume only the two values $m_S\hbar = \pm\hbar/2$ (and, in particular, it cannot assume the value 0), which is in accord with the Stern–Gerlach experiment; recall Figure 5.4 and see Figure 9.1 below where $\cos\phi = \frac{S_z}{|S|} \stackrel{\text{def}}{=} \frac{m_s\hbar}{\frac{\sqrt{3}}{2}\hbar} = \pm\frac{1}{\sqrt{3}}$ restricts S to two orientations: $\phi = 54°45', 125°15'$. m_S is the electron's *magnetic spin quantum number*.

Starting with the commutation relations for orbital angular momentum

$$[M_x, M_y] = i\hbar M_z, \qquad [M_y, M_z] = i\hbar M_x, \qquad [M_z, M_x] = i\hbar M_y \tag{9.1.1}$$

(see Appendix 5A) and with the eigenvalues $l(l+1)\hbar^2$, $m_l\hbar$ of $M^2 \stackrel{\text{def}}{=} M_x^2 + M_y^2 + M_z^2$, M_z respectively (recalling that $M^2\psi_{nlm} = l(l+1)\hbar^2\psi_{nlm}$, $M_z\psi_{nlm} = m\hbar\psi_{nlm}$ by Theorems 5.2 and 5.3 where we write $m = m_l$) we seek to construct, analogously, mathematical spin operators S_x, S_y, S_z subject to the commutation relations

$$[S_x, S_y] = i\hbar S_z, \qquad [S_y, S_z] = i\hbar S_x, \qquad [S_z, S_x] = i\hbar S_y \tag{9.1.2}$$

such that $S^2 \stackrel{\text{def}}{=} S_x^2+S_y^2+S_z^2$ and S_z have eigenvalues $s(s+1)\hbar^2$, $m_s\hbar$ respectively. Whereas l, m_l are integers with $-l \le m_l \le l$, $l \ge 0$ we will want to have, for physical reasons, that in fact $s = 1/2$ and $m_s = \pm 1/2$, and that S^2 has the single eigenvalue $\frac{1}{2}(\frac{1}{2}+1)\hbar^2$; S_z will have $\pm\hbar/2$ only as eigenvalues. The construction of a triple $S = (S_x, S_y, S_z)$ which satisfies these requirements is a rather easy task. For this we make use of the *Pauli spin matrices*

$$\sigma_1 \stackrel{\text{def}}{=} \begin{bmatrix} 0 & 1 \\ 1 & 0 \end{bmatrix}, \qquad \sigma_2 \stackrel{\text{def}}{=} \begin{bmatrix} 0 & -i \\ i & 0 \end{bmatrix}, \qquad \sigma_3 \stackrel{\text{def}}{=} \begin{bmatrix} 1 & 0 \\ 0 & -1 \end{bmatrix}. \tag{9.1.3}$$

Let $h(n)$ denote the set of Hermitian matrices of degree n: $A \in h(n) \iff A$ is a complex $n \times n$ matrix such that $A = A^* \stackrel{\text{def}}{=} \overline{A}^t$ $(= \overline{A^t})$. Consider the case $n = 2$. For $A = \left[\begin{smallmatrix} a & b \\ c & d \end{smallmatrix}\right]$, $A^* = \left[\begin{smallmatrix} \bar{a} & \bar{c} \\ \bar{b} & \bar{d} \end{smallmatrix}\right] \implies A \in h(2) \iff a, d \in \mathbb{R}$ and $c = \bar{b}$. Writing $A = \left[\begin{smallmatrix} a & b_1+ib_2 \\ b_1-ib_2 & d \end{smallmatrix}\right]$ with $a, b_1, b_2, d \in \mathbb{R}$ we see that for $\mathbf{1} = \left[\begin{smallmatrix} 1 & 0 \\ 0 & 1 \end{smallmatrix}\right]$

$$A = \left(\frac{a+d}{2}\right)\mathbf{1} + b_1\sigma_1 + (-b_2)\sigma_2 + \left(\frac{a-d}{2}\right)\sigma_3, \tag{9.1.4}$$

where of course $\mathbf{1}, \sigma_1, \sigma_2, \sigma_3 \in h(2)$. By (9.1.4) we see that $h(2)$ is a 4-dimensional vector space with $\mathbb{R}$-basis $\mathbf{1}, \sigma_1, \sigma_2, \sigma_3$. One easily checks that

$$\begin{aligned} \sigma_1^2 = \sigma_2^2 = \sigma_3^2 &= \mathbf{1}, & \sigma_1\sigma_2 &= -\sigma_2\sigma_1 = i\sigma_3, \\ \sigma_2\sigma_3 = -\sigma_3\sigma_2 &= i\sigma_1, & \sigma_3\sigma_1 &= -\sigma_1\sigma_3 = i\sigma_2. \end{aligned} \tag{9.1.5}$$

Define $S_x, S_y, S_z \in h(2)$ by

$$S_x \stackrel{\text{def}}{=} \frac{\hbar}{2}\sigma_1, \qquad S_y \stackrel{\text{def}}{=} \frac{\hbar}{2}\sigma_2, \qquad S_z \stackrel{\text{def}}{=} \frac{\hbar}{2}\sigma_3. \tag{9.1.6}$$

Then by (9.1.5) the S_x, S_y, S_z indeed do satisfy the commutation relations

$$[S_x, S_y] = i\hbar S_z, \qquad [S_y, S_z] = i\hbar S_x, \qquad [S_z, S_x] = i\hbar S_y \tag{9.1.7}$$

postulated in (9.1.2). For example $S_xS_y - S_yS_x \stackrel{\text{def}}{=} \frac{\hbar^2}{4}\sigma_1\sigma_2 - \frac{\hbar^2}{4}\sigma_2\sigma_1 = \frac{\hbar^2}{4}2\sigma_1\sigma_2 = \frac{\hbar^2}{2}i\sigma_3 = \hbar i S_z$. As $S_z \stackrel{\text{def}}{=} \frac{\hbar}{2}\sigma_3 \stackrel{\text{def}}{=} \left[\begin{smallmatrix} \hbar/2 & 0 \\ 0 & -\hbar/2 \end{smallmatrix}\right]$, its eigenvalues are exactly $\pm\hbar/2$. Now $S^2 \stackrel{\text{def}}{=} S_x^2 + S_y^2 + S_z^2 \stackrel{\text{def}}{=} \frac{\hbar^2}{4}(\sigma_1^2 + \sigma_2^2 + \sigma_3^2) = \frac{3}{4}\hbar^2\mathbf{1}$, by (9.1.5), so that indeed S^2 has the single eigenvalue $\frac{3}{4}\hbar^2 = \frac{1}{2}(\frac{1}{2} + 1)\hbar^2$, as desired. We thus take $\frac{\sqrt{3}}{2}\hbar$ as the *magnitude* of S.

The eigenvectors $\chi_\pm$ corresponding to the eigenvalues $m_s\hbar = \pm\frac{\hbar}{2}$ of S_z are given by

$$\chi_+ \stackrel{\text{def}}{=} \begin{bmatrix} 1 \\ 0 \end{bmatrix}, \qquad \chi_- \stackrel{\text{def}}{=} \begin{bmatrix} 0 \\ 1 \end{bmatrix}; \qquad S_z\chi_\pm = \pm\frac{\hbar}{2}\chi_\pm. \tag{9.1.8}$$

We illustrate the spatial quantization of S as follows. Let ϕ denote the angle between the "vectors" S_z, S measured with respect to the z-axis indicated as in Figure 9.1. Note that $S_x = \left[\begin{smallmatrix} 0 & \hbar/2 \\ \hbar/2 & 0 \end{smallmatrix}\right]$, $S_y = \left[\begin{smallmatrix} 0 & -i\hbar/2 \\ i\hbar/2 & 0 \end{smallmatrix}\right] \Longrightarrow$

$$\begin{aligned} S_x \begin{bmatrix} 1 \\ 1 \end{bmatrix} &= \frac{\hbar}{2}\begin{bmatrix} 1 \\ 1 \end{bmatrix}, & S_x \begin{bmatrix} 1 \\ -1 \end{bmatrix} &= -\frac{\hbar}{2}\begin{bmatrix} 1 \\ -1 \end{bmatrix}, \\ S_y \begin{bmatrix} 1 \\ i \end{bmatrix} &= \frac{\hbar}{2}\begin{bmatrix} 1 \\ i \end{bmatrix}, & S_y \begin{bmatrix} 1 \\ -i \end{bmatrix} &= -\frac{\hbar}{2}\begin{bmatrix} 1 \\ -i \end{bmatrix}. \end{aligned} \tag{9.1.9}$$

That is, $\pm\hbar/2$ are also the eigenvalues of S_x, S_y. By general principles (principles which we have already employed; see (3.9.1)), the only possible values a physical observable can attain are the eigenvalues of the corresponding quantum operator assigned to that observable. We conclude therefore that the x, y, or z-component of the spin (of any 1/2-spin particle) can assume only the two values $\pm\hbar/2$, the relevant Hilbert space being $H = \mathbb{C}^2$.

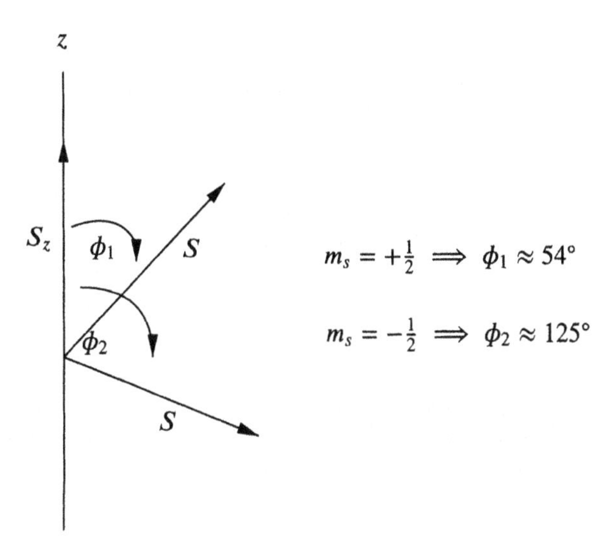

Figure 9.1: Spatial quantization of S

9.2 Pauli Spin Wave Functions

Taking into account various experimental results (such as alluded to earlier) which could not be explained by "prespin" quantum mechanics, Wolfgang Pauli in 1925 postulated the existence of two distinct states or "orientations" of an electron. We represent these states mathematically by the eigenvectors $\chi_\pm$ in (9.1.8). χ_+ is called the *spin-up* state and χ_- is called the *spin-down* state.

When the notion of spin is taken into account, one has in place of Schrödinger wave functions more general Pauli *spin wave functions* (also due to C.G. Darwin)

$$\psi_p(x, y, z, t) = \begin{bmatrix} \psi_+(x, y, z, t) \\ \psi_-(x, y, z, t) \end{bmatrix} = \psi_+(x, y, z, t)\chi_+ + \psi_-(x, y, z, t)\chi_- \quad (9.2.1)$$

satisfying the *Pauli equation*

$$i\hbar \frac{\partial \psi_p}{\partial t} = H\psi_p \quad (9.2.2)$$

for a suitable Hamiltonian H.

In general, the *state* of a $\frac{1}{2}$-spin particle (electron, proton, neutron, quark, neutrino, or muon, for example) is a *spinor*

$$\chi = \begin{bmatrix} z_1 \\ z_2 \end{bmatrix} = z_1\chi_+ + z_2\chi_- \in \mathbb{C}^2, \quad (9.2.3)$$

$\mathbb{C}^2$ being the Hilbert space with standard inner product structure

$$\left\langle \begin{bmatrix} z_1 \\ z_2 \end{bmatrix}, \begin{bmatrix} w_1 \\ w_2 \end{bmatrix} \right\rangle = z_1\overline{w_1} + z_2\overline{w_2}. \quad (9.2.4)$$

Hence

$$\langle \chi_+, \chi_+ \rangle = \langle \chi_-, \chi_- \rangle = 1, \qquad \langle \chi_+, \chi_- \rangle = 0 \quad (9.2.5)$$

which means that *normalized states* (χ with $||\chi||^2 = 1$) are given by

$$|z_1|^2 + |z_2|^2 = 1. \quad (9.2.6)$$

We concluded earlier that the only values the spin components S_x, S_y, S_z can assume are the eigenvalues $\pm\hbar/2$. Since a state

$$\chi = z_1\chi_+ + z_2\chi_-, \quad (9.2.7)$$

where (by (9.1.8)) $\chi_\pm$ are normalized eigenspinors of S_z with eigenvalues $\pm\hbar/2$, equation (9.2.7) being the analogue of the "Fourier expansion" of $\Psi(\,; t)$ in terms of the eigenvalues of H_0 in (7.7.23), we conclude the following for a normalized state: If a measurement of S_z were performed, the probability that the value $\hbar/2$ would be obtained is $|z_1|^2$; similarly $|z_2|^2$ is the probability that the value

$-\hbar/2$ would be obtained. Compare the remarks made in the discussion of equation (7.7.23), where now $\chi_\pm$ play the role of the states $\{\exp(-iE_jt/\hbar)\psi_j\}_{j=1}^n$ there and S_z plays the role of H_0. Also compare the remarks that followed equation (3.9.10).

As an example, take a $\frac{1}{2}$-spin particle in the normalized state

$$\chi = \frac{1}{\sqrt{62}}\begin{bmatrix} -3 \\ 2+7i \end{bmatrix}. \tag{9.2.8}$$

Then $|z_1|^2 = 9/62$ is the probability that a measurement of S_z will yield the value $\hbar/2$, and $|z_2|^2 = 53/62$ is the probability that a measurement of S_z will yield the value $-\hbar/2$.

To find the probability that a measurement of S_x, or of S_y, will yield the values $\hbar/2$ or $-\hbar/2$ we Fourier expand a state χ in terms of the normalized eigenspinors $\frac{1}{\sqrt{2}}\begin{bmatrix} 1 \\ 1 \end{bmatrix}, \frac{1}{\sqrt{2}}\begin{bmatrix} 1 \\ -1 \end{bmatrix}$ (or $\frac{1}{\sqrt{2}}\begin{bmatrix} 1 \\ i \end{bmatrix}, \frac{1}{\sqrt{2}}\begin{bmatrix} 1 \\ -i \end{bmatrix}$) for S_x, S_y respectively, similarly as we did in (9.2.7); see (9.1.9). For example, for S_x

$$\chi = \begin{bmatrix} z_1 \\ z_2 \end{bmatrix} = \frac{(z_1+z_2)}{\sqrt{2}}\begin{bmatrix} 1/\sqrt{2} \\ 1/\sqrt{2} \end{bmatrix} + \frac{(z_1-z_2)}{\sqrt{2}}\begin{bmatrix} 1/\sqrt{2} \\ -1/\sqrt{2} \end{bmatrix} \tag{9.2.9}$$

where, similar to (9.2.5),

$$\left\langle \begin{bmatrix} \frac{1}{\sqrt{2}} \\ \frac{1}{\sqrt{2}} \end{bmatrix}, \begin{bmatrix} \frac{1}{\sqrt{2}} \\ \frac{1}{\sqrt{2}} \end{bmatrix} \right\rangle = \left\langle \begin{bmatrix} \frac{1}{\sqrt{2}} \\ -\frac{1}{\sqrt{2}} \end{bmatrix}, \begin{bmatrix} \frac{1}{\sqrt{2}} \\ -\frac{1}{\sqrt{2}} \end{bmatrix} \right\rangle = 1, \quad \left\langle \begin{bmatrix} \frac{1}{\sqrt{2}} \\ \frac{1}{\sqrt{2}} \end{bmatrix}, \begin{bmatrix} \frac{1}{\sqrt{2}} \\ -\frac{1}{\sqrt{2}} \end{bmatrix} \right\rangle = 0 \tag{9.2.10}$$

$\Longrightarrow |\chi|^2 = \frac{|z_1+z_2|^2}{2} + \frac{|z_1-z_2|^2}{2} = |z_1|^2 + |z_2|^2$. If χ is normalized, then $\frac{|z_1 \pm z_2|^2}{2}$ is the probability, therefore, that a measurement of S_x will yield the value $\pm\hbar/2$. For the state χ in (9.2.8),

$$\chi = \frac{-1+7i}{\sqrt{2}\sqrt{62} = 2\sqrt{31}}\begin{bmatrix} \frac{1}{\sqrt{2}} \\ \frac{1}{\sqrt{2}} \end{bmatrix} + \frac{-5-7i}{2\sqrt{31}}\begin{bmatrix} \frac{1}{\sqrt{2}} \\ -\frac{1}{\sqrt{2}} \end{bmatrix}. \tag{9.2.11}$$

That is, the probability that a measurement of S_x (of a 1/2-spin particle in the state $\chi = 1/\sqrt{62}\begin{bmatrix} -3 \\ 2+7i \end{bmatrix}$) will yield the value $\hbar/2$ is $50/124$, and the probability that such a measurement will yield the value $-\hbar/2$ is $74/124$.

Let an electron in the presence of a magnetic field $\boldsymbol{B} = (B_1, B_2, B_3) : \mathbb{R}^3 \times \mathbb{R} \to \mathbb{R}^3$ (where each $\boldsymbol{B}_j : \mathbb{R}^3 \times \mathbb{R} \to \mathbb{R}$) be situated at some fixed point in $\mathbb{R}^3$. For this system Pauli's equation (9.2.2) assumes the form

$$i\hbar\frac{d}{dt}\begin{bmatrix} \psi_+(t) \\ \psi_-(t) \end{bmatrix} = H\begin{bmatrix} \psi_+(t) \\ \psi_-(t) \end{bmatrix}, \tag{9.2.12}$$

since the electron is at rest. Pauli postulated that in this case the Hamiltonian H is given by

$$H = H_B = (\mu\sigma) \cdot B \tag{9.2.13}$$

$$\stackrel{\text{i.e.,}}{=} \mu(B_1\sigma_1 + B_2\sigma_2 + B_3\sigma_3) \tag{9.2.14}$$

$$= \mu \begin{bmatrix} B_3 & B_1 - B_2 i \\ B_1 + B_2 i & -B_3 \end{bmatrix} \qquad \text{for } \mu \stackrel{\text{def}}{=} \frac{e\hbar}{2m_e c} \tag{9.2.15}$$

by (9.1.3), where m_e is the electron mass. The interesting factor $\mu = \frac{e\hbar}{2m_e c}$ in (9.2.15) (called the *Bohr magneton*—see (0.0.13), (0.0.15), in particular the ratio $\frac{e}{2m_e c}$, deserves some discussion. In Appendix 9A to this chapter, we provide some indication of how such a ratio (called a *gyromagnetic ratio*—of a charge to twice the mass of a particle times the speed of light) arises in connection with *magnetic dipole moments*, and we indicate why H_B is defined by (9.2.13). Later, in equation (9.3.14), it is indicated how a weak magnetic field produces an energy level shift proportional to the amount μ. Recall that μ also denotes reduced mass. This dual use of notation will not cause any problems.

Take for example B to be a constant magnetic field along the z-axis: $B = (0, 0, B_3)$ as in (8.1.6), where B_3 is a constant. Write B for B_3. Then as

$$H_B = \begin{bmatrix} \mu B & 0 \\ 0 & -\mu B \end{bmatrix} \tag{9.2.16}$$

equation (9.2.12) becomes the system

$$i\hbar\psi'_\pm(t) = \pm\mu B\psi'_\pm(t) \tag{9.2.17}$$

with solutions

$$\psi_\pm(t) = \psi_\pm(0)\exp\left(\pm\frac{\mu B}{i\hbar}t\right). \tag{9.2.18}$$

Example

Instead of a constant magnetic field, consider as a second example an electron at rest in a time-dependent oscillatory field given by

$$B(x, y, z, t) = (0, 0, B\cos\omega t) \tag{9.2.19}$$

for $B, \omega \in \mathbb{R}$ with $\omega \neq 0$. H_B in (9.2.15) is given by

$$H_B = \begin{bmatrix} \mu B\cos\omega t & 0 \\ 0 & -\mu B\cos\omega t \end{bmatrix} \tag{9.2.20}$$

and the corresponding Pauli equation (9.2.12) is equivalent to the differential equations

$$i\hbar\psi'_\pm(t) = \pm\mu B(\cos\omega t)\psi_\pm(t), \tag{9.2.21}$$

which are *separable* and which therefore can be immediately solved: For constants $C_\pm$

$$\psi_\pm(t) = C_\pm \exp\left(\pm\frac{\mu B}{i\hbar\omega}\sin\omega t\right). \tag{9.2.22}$$

The norm $||\psi(t)||$ of the state

$$\psi(t) \stackrel{\text{def}}{=} \begin{bmatrix} \psi_+(t) \\ \psi_-(t) \end{bmatrix} \tag{9.2.23}$$

is independent of t:

$$||\psi(t)||^2 \stackrel{\text{def}}{=} |\psi_+(t)|^2 + |\psi_-(t)|^2 = |C_+|^2 + |C_-|^2; \tag{9.2.24}$$

cf. (9.2.4). In particular, for any t, $\psi(t)$ is normalized $\iff |C_+|^2 + |C_-|^2 = 1$. Note that by (9.2.22), (9.2.23)

$$\psi(0) = \begin{bmatrix} C_+ \\ C_- \end{bmatrix}. \tag{9.2.25}$$

As an application of (9.2.22) suppose the electron (at rest in the oscillating magnetic field B given by (9.2.19)) at time $t = 0$ is in the *spin-up state* with respect to the x-axis—i.e., suppose $\psi(0) = \frac{1}{\sqrt{2}}\left[\begin{smallmatrix} 1 \\ 1 \end{smallmatrix}\right]$; compare remarks preceding (9.2.9). What is the probability $P(t)$ that a measurement of the x-component S_x of its spin at time t will yield the value $-\hbar/2$? To answer this question we note that first of all the electron initially (and therefore at all times, by (9.2.24)) is in a normalized state:

$$\begin{aligned} \psi(0) &= \frac{1}{\sqrt{2}} \begin{bmatrix} 1 \\ 1 \end{bmatrix} \implies \text{(by (9.2.25))} \\ C_+ &= C_- = \frac{1}{\sqrt{2}} \implies |C_+|^2 + |C_-|^2 = 1. \end{aligned} \tag{9.2.26}$$

Therefore, by remarks preceding (9.2.11), $P(t)$ is given by

$$P(t) = \frac{|z_1(t) - z_2(t)|^2}{2} \quad \text{where } \psi(t) = \begin{bmatrix} z_1(t) \\ z_2(t) \end{bmatrix} \stackrel{\text{i.e.,}}{=} \begin{bmatrix} \psi_+(t) \\ \psi_-(t) \end{bmatrix}.$$

Let $\lambda = \frac{\mu B}{\hbar\omega}$. Then by (9.2.22) (since $C_+ = C_- = \frac{1}{\sqrt{2}}$)

$$\begin{aligned} z_1(t) - z_2(t) &= \frac{1}{\sqrt{2}} \left[e^{-i\lambda \sin \omega t} - e^{i\lambda \sin \omega t} \right] \\ &= \frac{-2i}{\sqrt{2}} \sin(\lambda \sin \omega t) \\ \implies |z_1(t) - z_2(t)|^2 &= 2 \sin^2(\lambda \sin \omega t). \end{aligned} \tag{9.2.27}$$

That is, the desired probability $P(t)$ at time t is given by

$$P(t) = \sin^2 \left(\frac{\mu B}{\hbar\omega} \sin \omega t \right) = \sin^2 \left(\frac{eB}{2m_e c\omega} \sin \omega t \right), \tag{9.2.28}$$

by (9.2.15).

9.3 The Zeeman Effect for Weak Magnetic Fields

When an atom is placed in a magnetic field, the effect is that it experiences a shift of its energy levels. As indicated earlier, one observes a splitting of its spectral lines. This effect, called the *Zeeman effect*, was discovered by Charles Fievez around 1885, and ten years later by Pieter Zeeman. The field could be weak, intermediate, or strong. We shall confine our discussion to weak magnetic fields. One could also consider shifted energy levels of an atom placed in an electric field, the so-called *Stark effect*, after Johannes Stark.

As before we consider a constant magnetic field $\boldsymbol{B} \stackrel{\text{def}}{=} (0, 0, B)$, $B \in \mathbb{R}$, along the z-axis. For such a field we can choose a vector potential $A : \mathbb{R}^3 \times \mathbb{R} \to \mathbb{R}^3$ by $A(x, y, z, t) \stackrel{\text{def}}{=} (-\frac{By}{2}, \frac{Bx}{2}, 0) : \boldsymbol{B} = \operatorname{curl} A$. Recall (8.1.7) and also formula (8.2.9): The quantized Hamiltonian $\widehat{H}$ of a particle with charge q and mass m under the influence of the electromagnetic field $(E, \boldsymbol{B})$ is given by

$$\widehat{H} = -\frac{\hbar^2}{2m}\nabla^2 - \frac{qB}{2mc}M_z + \frac{q^2(A_1^2 + A_2^2 + A_3^2)}{2mc^2} + q\phi, \tag{9.3.1}$$

where

$$A_1^2 + A_2^2 + A_3^2 = \frac{B^2}{4}(x^2 + y^2), \text{ and}$$

$M_z = \frac{\hbar}{i}(x\frac{\partial}{\partial y} - y\frac{\partial}{\partial x})$ is the z-component of orbital angular momentum, and where ϕ is a scalar potential for E: $E = -\nabla\phi - \frac{1}{c}\frac{\partial A}{\partial t} = -\nabla\phi = 0$ for the choice $\phi = 0$. To treat the case of a hydrogenic atom influenced by $\boldsymbol{B}$, with a single electron of charge $q = -e$ and nucleus of charge Ze, we must consider (in addition to $\widehat{H}$) a Coulomb potential energy contribution V of the nucleus. To simplify matters we take the nuclear position to be fixed at the origin of the coordinates. Then away from the origin

$$V(x, y, z) = \frac{-Ze^2}{\sqrt{x^2 + y^2 + z^2}}, \tag{9.3.2}$$

which gives the electrostatic Coulomb force

$$F(x, y, z) = (-\nabla V)(x, y, z) = \frac{-Ze^2(x, y, z)}{|(x, y, z)|^3} \tag{9.3.3}$$

that the nucleus exerts on the electron; see Figure 5.1 where $F^{(2)}$ there is F here. To obtain Pauli's equation for our system we also add to $\widehat{H}$ Pauli's "magnetic" Hamiltonian H_B given in (9.2.15). Thus we obtain the Hamiltonian

$$H = \begin{bmatrix} \widehat{H} + V & 0 \\ 0 & \widehat{H} + V \end{bmatrix} + H_B \tag{9.3.4}$$

and the corresponding Pauli equation

$$i\hbar\frac{\partial\psi_p}{\partial t} = H\psi_p \tag{9.3.5}$$

for which we seek solutions; see (9.2.2). We should point out that one further term could be added to H, a term which takes into account so-called *spin-orbit interaction*, which we neglect here. Finally we assume that the field B is weak, since in this case it is well known, empirically, that the quadratic contribution $A_1^2 + A_2^2 + A_3^2$ to $\widehat{H}$ in (9.3.1) can be disregarded in many practical situations. Remembering the choice $\phi = 0$ for the scalar potential we therefore take

$$\widehat{H} = -\frac{\hbar^2}{2m_e}\nabla^2 + \frac{eBM_z}{2m_e c} \tag{9.3.6}$$

in (9.3.1). By (9.3.4) equation (9.3.5) becomes the system

$$i\hbar\frac{\partial\psi_\pm}{\partial t}(x,y,z,t) = \left[-\frac{\hbar^2\nabla^2}{2m_e} - \frac{Ze^2}{\sqrt{x^2+y^2+z^2}} + \frac{eB}{2m_e c}(M_z \pm \hbar)\right]\psi_\pm(x,y,z,t) \tag{9.3.7}$$

on $(\mathbb{R}^3 - \{(0,0,0)\}) \times \mathbb{R}$; see (9.2.1), (9.2.15), (9.2.16). If we have stationary state solutions $\psi_\pm(x,y,z)$, i.e.,

$$(H\psi_\pm)(x,y,z) = E\psi_\pm(x,y,z), \qquad E \in \mathbb{R}, \tag{9.3.8}$$

where H now denotes the operator on the right-hand side of (9.3.7), then the equation

$$\psi_\pm(x,y,z,t) \stackrel{\text{def}}{=} \psi_\pm(x,y,z)\exp\left(\frac{-iEt}{\hbar}\right) \tag{9.3.9}$$

provides for solutions $\psi_\pm$ of (9.3.7). Accordingly, we look for solutions of (9.3.8):

$$\left[-\frac{\hbar^2\nabla^2}{2m_e} - \frac{Ze^2}{\sqrt{x^2+y^2+z^2}} + \frac{eB}{2m_e c}(M_z \pm \hbar)\right]\psi_\pm = E\psi_\pm \tag{9.3.10}$$

on $\mathbb{R}^3 - \{(0,0,0)\}$ which may be written

$$(\nabla^2\psi_\pm)(x,y,z) + \frac{2m_e}{\hbar^2}\left[\frac{Ze^2}{\sqrt{x^2+y^2+z^2}} + E - \frac{eB}{2m_e c}(M_z \pm \hbar)\right]\psi_\pm(x,y,z) = 0 \tag{9.3.11}$$

on $\mathbb{R}^3 - \{(0,0,0)\}$. The point is that, modulo the term $-\frac{eB}{2m_e c}(M_z \pm \hbar)$, equation (9.3.11) is the first equation in (5.1.9), where the reduced mass $\mu \stackrel{\text{def}}{=} \frac{m_1 m_2}{m_1+m_2}$ there is replaced by m_e here. Now since the ψ_{nlm} in (5.2.12) are normalized solutions of the first equation in (5.1.9) for

$$E = E^{(n)} \stackrel{\text{def}}{=} \frac{-m_e Z^2 e^4}{2\hbar^2 n^2}, \qquad \alpha_n \stackrel{\text{def}}{=} \frac{2m_e Z e^2}{\hbar^2 n} \tag{9.3.12}$$

in (5.1.29), (5.1.30), and since

$$\frac{-eB}{2m_e c}(M_z \pm \hbar)\psi_{nlm} = \frac{-eB}{2m_e c}(m\hbar \pm \hbar)\psi_{nlm} \tag{9.3.13}$$

by Theorem 5.3, it follows that the normalized hydrogenic wave functions ψ_{nlm} are actually solutions of (9.3.11) for

$$E = E_{nm}^{\pm} \stackrel{\text{def}}{=} E^{(n)} + \frac{eB\hbar}{2m_e c}(m \pm 1). \tag{9.3.14}$$

This establishes the *normal* Zeeman effect for weak magnetic fields for the case at hand, where we see in fact that the field produces a shift in the energy level $E^{(n)}$ by the amount $\frac{eB\hbar}{2m_e c}(m \pm 1)$, $m = m_l$ being the magnetic quantum number of the hydrogenic atom under consideration in the state ψ_{nlm}. Now we see why m is called the magnetic quantum number. We also see, by (9.3.9), that we can obtain Pauli–Darwin spin wave functions ψ_p which solve equation (9.3.7) by setting

$$\psi_p(x, y, z, t) = \begin{bmatrix} \psi_{nlm}(x, y, z) \exp\left(-iE_{nm}^{+} t/\hbar\right) \\ \psi_{nlm}(x, y, z) \exp\left(-iE_{nm}^{-} t/\hbar\right) \end{bmatrix} \tag{9.3.15}$$

on $(\mathbb{R}^3 - \{(0,0,0)\}) \times \mathbb{R}$, for $E_{nm}^{\pm}$ given by (9.3.14).

For n, l fixed, with the field B "turned on," the spin-up states $\begin{bmatrix} \psi_{nlm} \\ 0 \end{bmatrix}$ have $2l + 1$ energy values $E_{nm}^{+} = E^{(n)} + \frac{eB\hbar}{2m_e c}(m + 1)$, $m = 0, \pm 1, \ldots, \pm l$. Similarly the spin-down states $\begin{bmatrix} 0 \\ \psi_{nlm} \end{bmatrix}$ have $2l + 1$ energy values $E_{nm}^{-} = E^{(n)} + \frac{eB\hbar}{2m_e c}(m - 1)$, $m = 0, \pm 1, \ldots, \pm l$. Note that for $m \neq 0$ there are $2l + 3$ distinct values $0, \pm 1, \pm 2, \ldots, \pm(l+1)$ of the $m \pm 1$ (and 2 distinct values ± 1, of course, if $m = 0$). With B turned on we have a *Zeeman splitting* of energy levels, illustrated for example for the states defined by $(n, l) = (1, 0), (2, 1)$ in Figure 9.2.

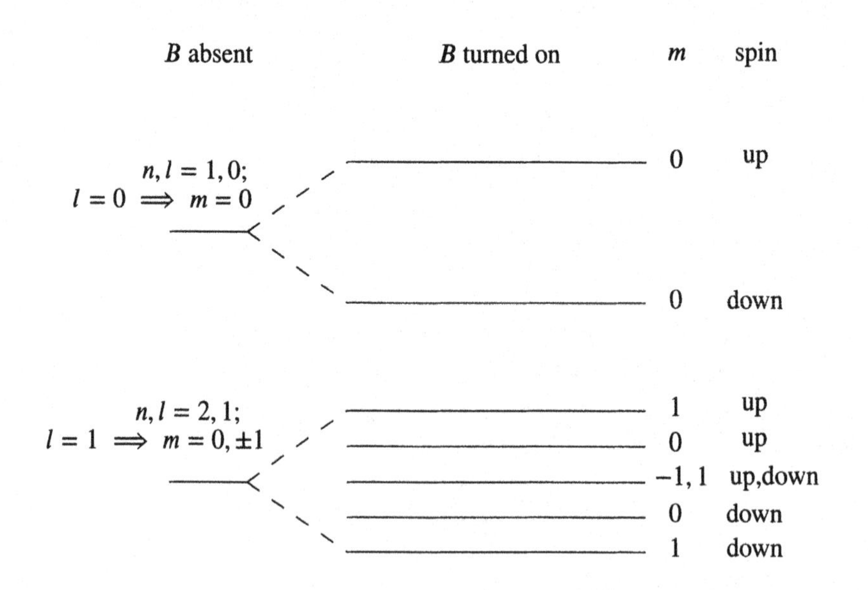

Figure 9.2: Zeeman splitting in a magnetic field

Example

As another example of the shift of energy levels by an imposed magnetic field B (the Zeeman effect) we consider a particle of charge q with the field constant as before along the z-axis. By (8.1.8) another choice of vector potential $A : \mathbb{R}^3 \times \mathbb{R} \to \mathbb{R}^3$ for $B = (0, 0, B)$ is given by $A(x, y, z, t) = (-By, 0, 0)$, in which case the quantized Hamiltonian $\widehat{H}$ is given by formula (8.2.10):

$$\widehat{H} = -\frac{\hbar^2 \nabla^2}{2m} + \frac{q\hbar By}{mci}\frac{\partial}{\partial x} + \frac{q^2 B^2 y^2}{2mc^2} + q\phi, \tag{9.3.16}$$

where m is the mass of the particle; we take the scalar potential $\phi = 0$. The present example is simpler than the preceding one of an electron bound to a nucleus. Here no Coulomb potential need be considered. We will consider moreover only Schrödinger's equation

$$i\hbar \frac{\partial \psi}{\partial t} = \widehat{H}\psi \tag{9.3.17}$$

where $\psi : \mathbb{R}^3 \times \mathbb{R} \to \mathbb{R}$. As usual it is enough to obtain solutions $\psi : \mathbb{R}^3 \to \mathbb{R}$ of

$$\widehat{H}\psi = E\psi, \qquad E \in \mathbb{R} \tag{9.3.18}$$

and set

$$\psi(x, y, z, t) = \psi(x, y, z)e^{Et/i\hbar} \tag{9.3.19}$$

to obtain solutions of (9.3.17).

To find solutions of (9.3.18) propose a separation of variables of the form

$$\psi(x, y, z) = e^{i(Ax+Cz)}\psi_o(y) \tag{9.3.20}$$

where $A, C \in \mathbb{R}$. Then equation (9.3.18) reduces to

$$-\frac{\hbar^2}{2m}\left[(-A^2 - C^2)\psi_o(y) + \psi_o''(y)\right] + \frac{AqB\hbar}{mc} y\psi_o(y) + \frac{q^2 B^2 y^2}{2mc^2}\psi_o(y) = E\psi_o(y) \tag{9.3.21}$$

or

$$\left[-\frac{\hbar^2}{2m}\frac{d^2}{dy^2} + \frac{q^2 B^2 y^2}{2mc^2} + \frac{A\hbar qBy}{mc}\right]\psi_o(y) = \left[E - \frac{\hbar^2}{2m}(A^2 + C^2)\right]\psi_o(y). \tag{9.3.22}$$

The trick for solving (9.3.22) is to use a change of variables. Namely, let

$$\psi(y) \stackrel{\text{def}}{=} \psi_o\left(y - \frac{\hbar A}{m\omega}\right) \qquad \text{where } \omega \stackrel{\text{def}}{=} \frac{qB}{mc}, \tag{9.3.23}$$

as in the second example of Chapter 8. Then

$$\hbar cAm\omega = \hbar AqB \implies \frac{qB\hbar A}{mc}\frac{\hbar A}{m\omega} = \frac{\hbar^2 A^2}{m}. \tag{9.3.24}$$

Now

$$
\begin{aligned}
&-\frac{\hbar^2}{2m}\psi''(y)+\frac{m}{2}\omega^2y^2\psi(y)\\
&\stackrel{\text{def}}{=} -\frac{\hbar^2}{2m}\psi_o''\left(y-\frac{\hbar A}{m\omega}\right)+\frac{m}{2}\omega^2\left[y-\frac{\hbar A}{m\omega}+\frac{\hbar A}{m\omega}\right]^2\psi_o\left(y-\frac{\hbar A}{m\omega}\right)\\
&=-\frac{\hbar^2}{2m}\psi_o''\left(y-\frac{\hbar A}{m\omega}\right)+\left[\frac{m}{2}\omega^2\left(y-\frac{\hbar A}{m\omega}\right)^2+m\omega^2\left(y-\frac{\hbar A}{m\omega}\right)\frac{\hbar A}{m\omega}\right.\\
&\quad\left.+\frac{m}{2}\omega^2\left(\frac{\hbar A}{m\omega}\right)^2\right]\psi_o\left(y-\frac{\hbar A}{m\omega}\right)\\
&=-\frac{\hbar^2}{2m}\psi_o''\left(y-\frac{\hbar A}{m\omega}\right)\\
&\quad+\left[\frac{m}{2}\omega^2\left(y-\frac{\hbar A}{m\omega}\right)^2+\omega y\hbar A-\frac{m}{2}\omega^2\left(\frac{\hbar A}{m\omega}\right)^2\right]\psi_o\left(y-\frac{\hbar A}{m\omega}\right),
\end{aligned}
\tag{9.3.25}
$$

where

$$
\begin{aligned}
-\frac{m\omega^2}{2}\left(\frac{\hbar A}{m\omega}\right)^2&=-\frac{\hbar^2A^2}{2m}=-\frac{\hbar^2A^2}{m}+\frac{\hbar^2A^2}{2m}\\
&=-\frac{qB\hbar A}{mc}\frac{\hbar A}{m\omega}+\frac{\hbar^2A^2}{2m},
\end{aligned}
\tag{9.3.26}
$$

by (9.3.24), $\Longrightarrow$ (by definition of ω)

$$
\omega y\hbar A-\frac{m\omega^2}{2}\left(\frac{\hbar A}{m\omega}\right)^2=\frac{qB\hbar A}{mc}\left(y-\frac{\hbar A}{m\omega}\right)+\frac{\hbar^2A^2}{2m}
\tag{9.3.27}
$$

$\Longrightarrow$ by (9.3.25)

$$
\begin{aligned}
&-\frac{\hbar^2}{2m}\psi''(y)+\frac{m}{2}\omega^2y^2\psi(y)\\
&=-\frac{\hbar^2}{2m}\psi_o''\left(y-\frac{\hbar A}{m\omega}\right)+\left[\frac{m}{2}\frac{q^2B^2}{m^2c^2}\left(y-\frac{\hbar A}{m\omega}\right)^2\right.\\
&\quad\left.+\frac{qB\hbar A}{mc}\left(y-\frac{\hbar A}{m\omega}\right)\right]\psi_o\left(y-\frac{\hbar A}{m\omega}\right)+\frac{\hbar^2A^2}{2m}\psi_o\left(y-\frac{\hbar A}{m\omega}\right)\\
&=\left[E-\frac{\hbar^2}{2m}(A^2+C^2)\right]\psi_o\left(y-\frac{\hbar A}{m\omega}\right)+\frac{\hbar^2A^2}{2m}\psi_o\left(y-\frac{\hbar A}{m\omega}\right)\\
&=\left[E-\frac{\hbar^2}{2m}C^2\right]\psi_o\left(y-\frac{\hbar A}{m\omega}\right)=\left[E-\frac{\hbar^2}{2m}C^2\right]\psi(y)
\end{aligned}
\tag{9.3.28}
$$

by definition (9.3.23)! That is, ψ satisfies Schrödinger's time independent equation for the simple harmonic oscillator with angular speed ω and frequency $\nu = \omega/(2\pi)$; see equation (3.2.3). We therefore obtain the *Landau energies* E given by

$$E - \frac{\hbar^2}{2m}C^2 = E_n \stackrel{\text{def}}{=} 2\pi\left(n + \frac{1}{2}\right)\nu\hbar = \left(n + \frac{1}{2}\right)\omega\hbar, \tag{9.3.29}$$

$n = 0, 1, 2, 3, \ldots$, and corresponding normalized L^2-solutions $\psi = \psi_n$ given in Section 4.8 of Chapter 4. Namely

$$\psi_n(y) = (-1)^n \left[\sqrt{\frac{b}{\pi}}\frac{1}{2^n\, n!}\right]^{1/2} \exp\left(-\frac{b}{2}y^2\right) H_n(\sqrt{b}y) \tag{9.3.30}$$

for

$$b \stackrel{\text{def}}{=} \frac{2\pi\nu m}{\hbar} = \frac{\omega m}{\hbar} \stackrel{\text{def}}{=} \frac{qB}{c\hbar}, \quad n = 0, 1, 2, 3, \ldots,$$

where H_n is the nth Hermite polynomial. Note that the Landau energies

$$E = E_n(C) \stackrel{\text{def}}{=} \frac{\hbar^2}{2m}C^2 + \left(n + \frac{1}{2}\right)\omega\hbar = \frac{\hbar^2}{2m}C^2 + \left(n + \frac{1}{2}\right)\frac{qB\hbar}{mc} \tag{9.3.31}$$

depend on the parameter C in (9.3.20) of course but are *independent* of the parameter A there. Thus infinitely many choices of A give rise to the *same* energy levels in (9.3.31), a phenomenon called *infinite degeneracy* in physics. By (9.3.20), (9.3.23) corresponding stationary Landau states ψ_n (i.e., solutions of (9.3.18), i.e., of (9.3.21) or (9.3.22) for $E = E_n(C)$) are given by

$$\psi_n(x, y, z) \stackrel{\text{def}}{=} e^{i(Ax+Cz)}\, \psi_n\left(y + \frac{\hbar A}{m\omega}\right), \quad n = 0, 1, 2, 3, \ldots \tag{9.3.32}$$

for ψ_n given in (9.3.30). Finally one obtains by (9.3.19) time-dependent solutions ψ_n of (9.3.17) (for $\widehat{H}$ given by (9.3.16) with $\phi = 0$ there) by setting

$$\psi_n(x, y, z, t) = e^{i(Ax+Cz)}\, e^{E_n(C)t/i\hbar}\, \psi_n\left(y + \frac{\hbar A}{m\omega}\right) \tag{9.3.33}$$

for $n = 0, 1, 2, 3, \ldots$, and for arbitrary choices of constants $A, C \in \mathbb{R}$.

We have considered the effect of a magnetic field on energy levels in this chapter. Another interesting study is the effect of a magnetic field on the Selberg trace formula. This is considered by the author in one of the papers in [16]. We take up applications of the trace formula in Chapters 16 and 17.

Appendix 9A Magnetic Dipole Moments

To have a better understanding of Pauli's magnetic Hamiltonian H_B (for example, how one arrives at definition (9.2.13)) we consider first how the ratio

$$\frac{e}{2m_e c} = \frac{\mu}{\hbar}$$

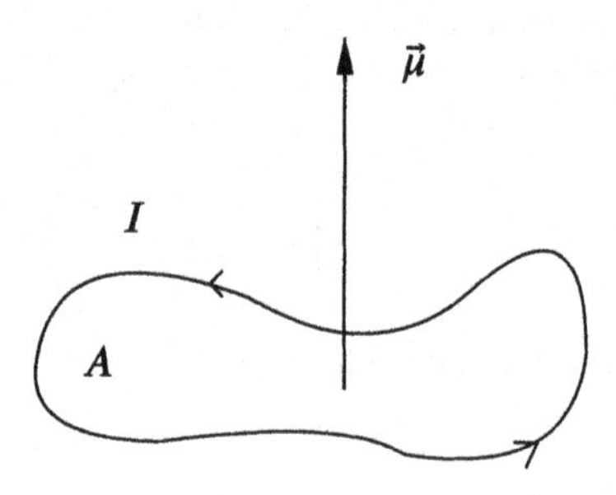

Figure 9.3: Right-hand rule

in (9.2.15) arises in a particular context, in regard to the notion of a magnetic dipole moment; cf. equation (9.A.3) below. First we replace the electron, of charge $-e$ and mass m_e, by an arbitrary charged particle.

Thus, suppose we have a particle with charge q and mass m. If the particle moves in a circular orbit of radius r and with a constant speed v, then as its displacement Δs along the circle over a period of time Δt is given by $\Delta s = r\Delta\theta$, $\Delta\theta$ being its angular displacement over the time period Δt (compare Figure 1.4), one has $\frac{\Delta s}{\Delta t} = r\frac{\Delta\theta}{\Delta t}$ or $v = \frac{ds}{dt} = r\frac{d\theta}{dt} = r\omega$ where ω is the (constant) angular speed of the particle. As before let T denote the period of the motion: given any point on the circle which the particle passes through at some instant, T is the amount of time required for the next passage through that point. Having noted in Section 1.4 of Chapter 1 that $T = 2\pi/\omega$ we can now use $v = r\omega$ to also write $T = 2\pi r/v$. The *current* i due to a charge is by definition the amount of charge passing through a given point per unit time. Thus in our case $i = q/T = qv/(2\pi r)$. In general if an amount of current I is flowing through a circuit which encloses an area A, then a *magnetic dipole moment* $\vec{\mu}$ is generated. Namely $\vec{\mu}$ is defined as a vector with magnitude $I\frac{A}{c}$, in the c.g.s. system, where c is the speed of light, with direction determined by the *right-hand rule*: if the curled fingers represent the direction of the current flow then the direction of $\vec{\mu}$ is that of the thumb, as illustrated in Figure 9.3, with $\vec{\mu}$ normal to the plane area A. In the case at hand $A = \pi r^2$ which means that for $|\vec{\mu}|$ the length of $\vec{\mu}$,

$$\vec{\mu} = \frac{iA}{c}\frac{\vec{\mu}}{|\vec{\mu}|} = \frac{qv\pi r^2}{2\pi rc}\frac{\vec{\mu}}{|\vec{\mu}|} = \frac{qvr}{2c}\frac{\vec{\mu}}{|\vec{\mu}|}. \tag{9.A.1}$$

On the other hand, suppose for example the circle is in the xy-plane and is centered at the origin, as in Figure 9.4. As discussed in Section 5.4 of Chapter 5, the angular momentum M of the particle is given by $M \stackrel{\text{def}}{=} \vec{r} \times m\vec{v}$. The angle between $\vec{r}, \vec{v}$ is $\pi/2$ ($\vec{r} \cdot \vec{v} = 0$). The length $|M|$ of M is therefore $|\vec{r}|m|\vec{v}| = mrv$. Also the direction of M is that of the positive z-axis, and thus is that of $\vec{\mu}$: $M = |M|\frac{\vec{\mu}}{|\vec{\mu}|} = mrv\frac{\vec{\mu}}{|\vec{\mu}|} \implies \frac{qM}{2mc} = \frac{qrv}{2c}\frac{\vec{\mu}}{|\vec{\mu}|}$; i.e., (by (9.A.1)),

$$\vec{\mu} = \frac{qM}{2mc}. \tag{9.A.2}$$

The ratio $\frac{|q|}{2mc}$ of the length of magnetic moment to angular momentum is called a *gyromagnetic ratio*.

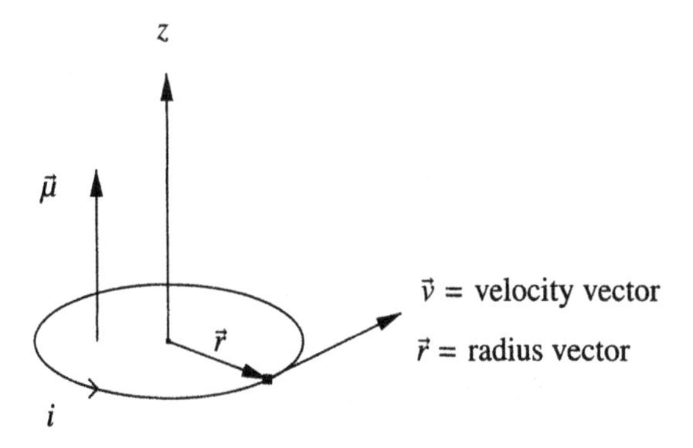

Figure 9.4: Magnetic dipole moment

In particular, for an electron in a circular orbit, moving classically with a constant speed, as in Figure 9.4, we see that the magnetic dipole moment $\vec{\mu}$ it generates (as a circulation charge about a circumference) is given by

$$\vec{\mu} = \frac{-e}{2m_e c} M = \frac{-\mu}{\hbar} M \quad \text{for } \mu \text{ in (9.2.15)}, \tag{9.A.3}$$

where M is its classical angular momentum. Thus we see the appearance of the ratio $\frac{e}{2m_e c}$ in (9.2.15).

Although we derived (9.A.3) by assuming a circular orbit for the electron, the same result would follow for an elliptic orbit or an orbit of any shape. Thus to an electron with classical orbital angular momentum M there is associated a magnetic dipole moment $\vec{\mu}_M = \vec{\mu}$ given by equation (9.A.3), which we can write as

$$\vec{\mu}_M = g_M \frac{\mu}{\hbar} M \qquad \text{for } g_M = -1. \tag{9.A.4}$$

There is, similarly, a magnetic moment μ_S associated to an electron with intrinsic spin angular momentum S:

$$\mu_S = g_S \frac{\mu}{\hbar} S \tag{9.A.5}$$

for a suitable factor g_S, called the *Landé (spin) g-factor*. Recall that $\frac{\sqrt{3}}{2}\hbar$ is the magnitude $|S|$ of μ_S. Thus

$$|\mu_S| = g_S \mu \frac{\sqrt{3}}{2}. \tag{9.A.6}$$

Now the value $|\mu_S|$ can be determined *experimentally*, and since μ is known (again see (0.0.15)) one finds that g_S is given by

$$g_S = -2.002319304\,. \tag{9.A.7}$$

That is $g_S \neq g_M$ but is about *twice* the value of g_M, a result which is known to follow somewhat naturally from Dirac's relativistic quantum theory. For practical purposes we will therefore express (9.A.5) as

$$\mu_S = -\frac{2\mu}{\hbar} S = -\frac{e}{m_e c} S, \tag{9.A.8}$$

which is in accord in fact with the result of the 1915 experiment of Einstein and de Haas.

Now we sketch how one arrives at definition (9.2.13). In Appendix 8B we indicated how an *electric* dipole with moment $\vec{m}$ placed in an electric field E acquires a potential energy given by $-\vec{m} \cdot E$. Similarly an electron placed in a magnetic field $B = (B_1, B_2, B_3)$ acquires an additional potential energy given by $-\mu_S \cdot B$. By (9.A.8) and (9.1.6)

$$\begin{aligned} -\mu_S \cdot B &= \frac{e}{m_e c} S \cdot B \\ &\stackrel{\text{def}}{=} \frac{e}{m_e c} \left(B_1 S_x + B_2 S_y + B_3 S_z\right) \\ &\stackrel{\text{def}}{=} \frac{e\hbar}{2m_e c} (B_1\sigma_1 + B_2\sigma_2 + B_3\sigma_3) \stackrel{\text{def}}{=} (\mu\sigma) \cdot B, \end{aligned} \tag{9.A.9}$$

which is exactly the Pauli definition (9.2.13) (or (9.2.14)).

Appendix 9B Quantization of the Magnetic Dipole Moment $\vec{\mu}$

Equation (9.A.3) of the preceding appendix provides the key to the question of how to assign to the observable $\vec{\mu}$ a corresponding operator. Namely, as we know how to quantize angular momentum M by Section 5.4 of Chapter 5, it is clear that we should make the following assignment:

$$\vec{\mu} \to \widehat{\mu} = (\mu_x, \mu_y, \mu_z) \stackrel{\text{def}}{=} \frac{-e}{2m_e c} \widehat{M} \tag{9.B.1}$$

where $\widehat{M} = (M_x, M_y, M_z)$ is given by (5.4.4). That is,

$$\begin{aligned} \mu_x &\stackrel{\text{def}}{=} \frac{-e}{2m_e c}\frac{\hbar}{i}\left(y\frac{\partial}{\partial z} - z\frac{\partial}{\partial y}\right), \\ \mu_y &\stackrel{\text{def}}{=} \frac{-e}{2m_e c}\frac{\hbar}{i}\left(z\frac{\partial}{\partial x} - x\frac{\partial}{\partial z}\right), \\ \mu_z &\stackrel{\text{def}}{=} \frac{-e}{2m_e c}\frac{\hbar}{i}\left(x\frac{\partial}{\partial y} - y\frac{\partial}{\partial x}\right). \end{aligned} \tag{9.B.2}$$

Appendix 9C Construction of Higher Spin Operators

The spin operators S_x, S_y, S_z constructed in Section 9.1 act on a two-dimensional vector space. The operator $S^2 \stackrel{\text{def}}{=} S_x^2 + S_y^2 + S_z^2$ was shown to have a single eigenvalue $s(s+1)\hbar^2$ where $s = 1/2$. Mathematically, this s-value represents the electron spin. There are other particles of course (protons, neutrons, quarks, neutrinos, for example) which also have spin $1/2$, as has been pointed out. There also exist in nature particles with higher spin. For example, photons have spin $s = 1$, gravitons have spin 2, "delta" particles have spin $3/2$. Another particle, the so-called h-meson, is known to have spin $s = 4$. Theoretically, particles can have arbitrarily high spin. Mathematically, one can construct spin operators S_x, S_y, S_z on an n-dimensional vector space where $n \geq 2$ is arbitrary. This we do in the present appendix. It should be noted however that at the present time particles with very high spin yet remain to be found. The results here therefore are of mathematical interest more than of physical interest.

To construct higher spin operators we apply the representation theory of the Lie algebra $g = sl(2, \mathbb{C})$ of 2×2 complex matrices with trace zero. A *representation* π of g on some complex vector space V is simply a Lie algebra homomorphism $\pi : g \to gl(V) \stackrel{\text{def}}{=}$ the Lie algebra of linear operators on V; i.e., π is a linear map such that

$$\pi([x, y]) = [\pi(x), \pi(y)] \qquad \text{for } x, y \in g \tag{9.C.1}$$

where $[x, y] = xy - yx$ (= matrix multiplication) and $[T, S] = TS - ST$ for the two linear operators on V. π is *irreducible* if $\{0\}$ and V are the only π-invariant subspaces of V; here we assume $V \neq \{0\}$. g has the standard $\mathbb{C}$-basis $\{H, E, F\}$ where

$$\begin{gathered} H = \begin{bmatrix} 1 & 0 \\ 0 & -1 \end{bmatrix}, \qquad E = \begin{bmatrix} 0 & 1 \\ 0 & 0 \end{bmatrix}, \qquad F = \begin{bmatrix} 0 & 0 \\ 1 & 0 \end{bmatrix}, \\ [H, E] = 2E, \qquad [H, F] = -2F, \qquad [E, F] = H. \end{gathered} \tag{9.C.2}$$

The following result is standard [51].

Theorem 9.1. *Let $m \geq 0$ be an integer, let V be a complex vector space of dimension $m+1$, and let $\{x_o, x_1, \ldots, x_m\}$ be a $\mathbb{C}$-basis of V. Then there exists an irreducible representation $\pi = \pi_m : g \to gl(V)$ of $g = sl(2, \mathbb{C})$ on V such that*

$$\begin{aligned} \pi(H)x_i &= (m - 2i)x_i, & \pi(E)x_i &= (m - i + 1)x_{i-1}, \\ \pi(F)x_i &= (i+1)x_{i+1} & &\textit{for } 0 \leq i \leq m \end{aligned} \tag{9.C.3}$$

where H, E, F are given in (9.C.2) *and where we set $x_{-1} = 0 = x_{m+1}$. Conversely any $m+1$-dimensional irreducible representation of g on a complex vector space is equivalent to π_m.*

Now let V be any complex vector space of dimension $n \geq 2$. We wish to construct linear operators (=*higher spin operators*) S_x, S_y, S_z on V subject to the following conditions (as in (9.1.2)):

$$
\begin{aligned}
&(i) \quad [S_x, S_y] = i\hbar S_z, \quad [S_y, S_z] = i\hbar S_x, \quad [S_z, S_x] = i\hbar S_y. \\
&(ii) \quad \text{the operator } S^2 \overset{\text{def}}{=} S_x^2 + S_y^2 + S_z^2 = s(s+1)\hbar^2 1_V \text{ where} \\
&\qquad \text{the } spin\ s \text{ is given by } s = \frac{n-1}{2};\ 1_V \text{ is the identity operator on } V. \\
&(iii) \quad \text{the eigenvalues of } S_z \text{ are } \left\{ \left(\frac{n-1-2j}{2} \right) \hbar \right\}_{j=0}^{n-1}.
\end{aligned}
\tag{9.C.4}
$$

Thus for $n = 2$, $s = 1/2$ and the eigenvalues of S_z are $\pm\hbar/2$, as for the case of the S_x, S_y, S_z of Section 9.1. To begin the construction apply Theorem 9.1 to V where m is chosen to be $n-1$: $m+1 = n$: there exists an irreducible representation π of g on V given by (9.C.3) for a choice of basis $\{x_o, x_1, \ldots, x_m = x_{n-1}\}$ of V:

$$
\begin{aligned}
&\pi(H)x_j = (n-1-2j)x_j \qquad 0 \le j \le n-1, \qquad \pi(E)x_o = 0, \\
&\pi(E)x_j = (n-j)x_{j-1}, \qquad 1 \le j \le n-1, \qquad \pi(F)x_{n-1} = 0, \\
&\pi(F)x_j = (j+1)x_{j+1}, \qquad 0 \le j \le n-2.
\end{aligned}
\tag{9.C.5}
$$

Using π we can define linear operators S_z, $S_\pm$ on V by

$$
S_z = \frac{\hbar}{2}\pi(H), \qquad S_+ = \hbar\pi(E), \qquad S_- = \hbar\pi(F). \tag{9.C.6}
$$

By (9.C.1) and (9.C.2) π satisfies

$$
\begin{aligned}
&[\pi(H), \pi(E)] = 2\pi(E), \quad [\pi(H), \pi(F)] = -2\pi(F), \\
&\qquad \text{and } [\pi(E), \pi(F)] = \pi(H)
\end{aligned}
\tag{9.C.7}
$$

which means that the operators S_z $S_\pm$ satisfy the commutation relations

$$
[S_z, S_\pm] = \pm\hbar S_\pm, \qquad [S_+, S_-] = 2\hbar S_z. \tag{9.C.8}
$$

Having defined S_z by (9.C.6) we now define S_x, S_y by

$$
S_\pm = S_x \pm iS_y; \quad \text{i.e., } S_x \overset{\text{def}}{=} \frac{S_+ + S_-}{2}, \quad S_y \overset{\text{def}}{=} \frac{S_+ - S_-}{2i} \text{ for } i^2 = -1. \tag{9.C.9}
$$

Then by (9.C.8) the operators S_x, S_y, S_z indeed do satisfy the commutation relation (i) in (9.C.4). For example

$$
\begin{aligned}
[S_z, S_x] &\overset{\text{def}}{=} \frac{[S_z, S_+] + [S_z, S_-]}{2} = \frac{\hbar S_+ - \hbar S_-}{2} \\
&= i\hbar \left(\frac{S_+ - S_-}{2i} \right) \overset{\text{def}}{=} i\hbar S_y.
\end{aligned}
\tag{9.C.10}
$$

Also by (9.C.5) and (9.C.6) we see that S_z is diagonalizable with eigenvalues $\frac{\hbar}{2}\times$ (the eigenvalues of $\pi(H)$)=the collection $\{\frac{\hbar}{2}(n-1-2j)\}_{j=0}^{n-1}$, as advertised in (iii) of (9.C.4). The final property to check is condition (ii) in (9.C.4). First note that for $S^2 \stackrel{\text{def}}{=} S_x^2 + S_y^2 + S_z^2$ one has

$$[S^2, S_x] = [S^2, S_y] = [S^2, S_z] = 0 \tag{9.C.11}$$

as in (5.A.6) of Appendix 5A, by the same proof of (5.A.6) there, given the relations (*i*) of (9.C.4). Equation (9.C.11) means that S^2 commutes with S_x, S_y, S_z, and therefore with S_+, S_-, S_z (by (9.C.9)), and therefore with $\pi(E)$, $\pi(F)$, $\pi(H)$ (by (9.C.6)), and therefore S^2 commutes with all $\pi(x)$, $x \in g$. But this means that S^2 must be a scalar operator $\lambda 1_V$, $\lambda \in \mathbb{C}$, since π is irreducible. Namely let λ be an eigenvalue of S^2 (since V is a finite-dimensional complex vector space $\neq \{0\}$). Then since S^2 commutes with $\pi(x)$, $x \in g$, the eigenspace V_λ of λ is π-invariant and thus by irreducibility of π, V_λ must coincide with V, as $V_\lambda \neq \{0\}$. Thus indeed $S^2 = \lambda 1_V$. To prove (9.C.4) (*ii*) we must therefore compute λ and show in fact that $\lambda = \frac{\hbar^2}{4}(n^2 - 1)$, which would indeed give

$$S^2 = \hbar^2 s(s+1)1_V \qquad \text{for } s = \frac{n-1}{2}. \tag{9.C.12}$$

Since S^2 is a scalar operator we can find λ by understanding the action of S^2 on a single non-zero vector $v \in V$. Namely we choose $v = x_o$, a so-called *highest-weight* vector for π, with highest weight $n - 1$; i.e., $\pi(H)x_o = (n-1)x_o$ and $\pi(E)x_o = 0$ (by (9.C.5)). We have by (9.C.5)

$$\begin{aligned}
S_+ x_o &\stackrel{\text{def}}{=} \hbar\pi(E)x_o = 0,\\
S_+ x_1 &\stackrel{\text{def}}{=} \hbar\pi(E)x_1 = \hbar(n-1)x_o,\\
S_- x_o &\stackrel{\text{def}}{=} \hbar\pi(F)x_o = \hbar x_1,\\
S_- x_1 &\stackrel{\text{def}}{=} \hbar\pi(F)x_1 = \begin{bmatrix} 2\hbar x_2 & \text{for } n \geq 3 \\ 0 & \text{for } n = 2 \end{bmatrix}.
\end{aligned} \tag{9.C.13}$$

Then

$$\begin{aligned}
S_x x_o &\stackrel{\text{def}}{=} \left(\frac{S_+ + S_-}{2}\right) x_o = \frac{\hbar}{2} x_1,\\
S_x x_1 &\stackrel{\text{def}}{=} \left(\frac{S_+ + S_-}{2}\right) x_1 = \begin{bmatrix} \frac{\hbar(n-1)x_o + 2\hbar x_2}{2} & \text{for } n \geq 3 \\ \frac{\hbar(n-1)x_o}{2} = \frac{\hbar x_o}{2} & \text{for } n = 2 \end{bmatrix}.
\end{aligned} \tag{9.C.14}$$

Similarly

$$\begin{aligned}
S_y x_o &= -\frac{\hbar}{2i} x_1,\\
S_y x_1 &= \begin{bmatrix} \frac{\hbar(n-1)x_o - 2\hbar x_2}{2i} & \text{for } n \geq 3 \\ \frac{\hbar x_o}{2i} & \text{for } n = 2 \end{bmatrix}.
\end{aligned} \tag{9.C.15}$$

Equations (9.C.14) and (9.C.15) imply that

$$S_x^2 x_o = \begin{bmatrix} \frac{\hbar^2}{4}(n-1)x_o + \frac{\hbar^2}{2}x_2 & \text{for } n \geq 3 \\ \frac{\hbar^2}{4}x_o & \text{for } n = 2 \end{bmatrix},$$
$$S_y^2 x_o = \begin{bmatrix} \frac{\hbar^2}{4}(n-1)x_o - \frac{\hbar^2}{2}x_2 & \text{for } n \geq 3 \\ \frac{\hbar^2}{4}x_o & \text{for } n = 2 \end{bmatrix}. \tag{9.C.16}$$

Finally since

$$S_z x_o \stackrel{\text{def}}{=} \frac{\hbar}{2}\pi(H)x_o = \frac{\hbar}{2}(n-1)x_o$$
$$\Longrightarrow \; S_z^2 x_o = \frac{\hbar}{4}(n-1)^2 x_o, \tag{9.C.17}$$

one obtains from (9.C.16), (9.C.17) that

$$S^2 x_o \stackrel{\text{def}}{=} (S_x^2 + S_y^2 + S_z^2)x_o \stackrel{\text{for } n \geq 3}{=}$$
$$\frac{\hbar^2}{4}(n-1)x_o + \frac{\hbar^2}{2}x_2 + \frac{\hbar^2(n-1)x_o}{4} - \frac{\hbar^2 x_2}{2} + \frac{\hbar^2}{4}(n-1)^2 x_o \tag{9.C.18}$$
$$= \frac{\hbar^2(n^2-1)}{4}x_o,$$

which also holds for $n = 2$.

10
Introduction to Multi-Electron Atoms

Some of the basic themes, or foundational concepts and principles of quantum mechanics (Planck's quantization, Schrödinger's equation, de Broglie's wavelength, Heisenberg's uncertainty principle, etc.) were introduced in Chapter 1. There is another basic principle, *Pauli's exclusion principle*, which was not introduced there but which will be considered in this chapter. This principle was actually postulated prior to the main developments of post-Bohr quantum mechanics, and has had a spectacular impact in regards to the understanding of why atoms exhibit different chemical properties. The principle resolves the question, among others, namely of what determines the arrangements of electrons about a nucleus, a question which we touch on briefly, or tangentially.

As our discussions up to this point have largely been confined to single electron atoms it is appropriate to now consider multi-electron atoms. We shall invade this vast and beautiful domain of study in only a minor and introductory way. Since no technology exists to solve Schrödinger's equation in a closed form when two or more electrons are present in an atomic system, one must resort to suitable approximation schemes. We introduce first-order *time-independent* perturbation theory and apply it to approximate the ground state energy of helium, which has two electrons—the simplest atom after hydrogen. We move on to some general remarks concerning multi-electron atoms, including some on *electron configurations* and the *principle of identity of particles.*

10.1 First-Order Correction to E_o

A discussion of first-order *time-dependent* perturbation theory was presented in Section 7.7 of Chapter 7. Here we consider an example of first-order *time-independent* perturbation theory. Our primary interest is an application of the latter theory to the computation of the ground state energy of the helium atom (or of helium-like atoms, more generally).

We begin with the problem of approximating E in the Schrödinger equation

$$H\psi = E\psi, \quad E \in \mathbb{R}, \tag{10.1.1}$$

where $H = H_o + H_1$ is the sum of two operators H_o, H_1 (a situation similar to that encountered in equation (7.7.5), where we assume that the equation

$$H_o\psi_o = E_o\psi_o, \quad E_o \in \mathbb{R}, \psi_o \neq 0, \tag{10.1.2}$$

has already been solved: ψ_o, E_o are known. In practice the perturbative effects of the operator H_1 will be considered as minimal. Write

$$\begin{aligned} \psi &= \psi_o + \Delta\psi, \\ E &= E_o + \Delta E, \end{aligned} \tag{10.1.3}$$

where, in some sense, the unknowns $\Delta\psi$, ΔE are first-order corrections to ψ_o, E_o, the idea being (very roughly) that if H_1 is sufficiently "small," then $H \simeq H_o$ in which case we think of equation (10.1.1) as $H_o\psi = E\psi$, and conclude by equation (10.1.2) that we should take $\psi \simeq \psi_o$, provided $E \simeq E_o$.

Proceeding formally, we in fact *define* the *first-order correction* ΔE to E_o by

$$\Delta E \overset{\text{def}}{=} \frac{\langle H_1\psi_o, \psi_o\rangle}{\langle \psi_o, \psi_o\rangle}. \tag{10.1.4}$$

The motivation for this definition is given in the following sketch. By (10.1.1), (10.1.2), (10.1.3), for $\psi_1 = \Delta\psi$, $E_1 = \Delta E$,

$$\begin{aligned} H_o\psi_o + H_1\psi_o + H_o\psi_1 + H_1\psi_1 &= (H_0 + H_1)(\psi_o + \psi_1) \\ &= H\psi = E\psi = (E_o + E_1)(\psi_o + \psi_1) \\ &= E_o\psi_o + E_1\psi_o + E_o\psi_1 + E_1\psi_1 \qquad (10.1.5) \\ \Longrightarrow H_1\psi_o + H_o\psi_1 &= E_1\psi_o + E_o\psi_1 + (E_1\psi_1 - H_1\psi_1) \\ &\quad (\text{as } H_o\psi_o = E_o\psi_o) \\ \Longrightarrow H_1\psi_o + H_0\psi_1 &\simeq E_1\psi_o + E_o\psi_1 \qquad (10.1.6) \end{aligned}$$

if we neglect the terms $E_1\psi_1$, $H_1\psi_1$ which are to be considered "small," or second order. Now take the inner product of both sides of (10.1.6) with ψ_o, and bear in mind that generically H_o is a Hermitian operator and E_o is real:

$$\begin{aligned} E_o\langle\psi_1, \psi_0\rangle = \langle\psi_1, E_o\psi_o\rangle = \langle\psi_1, H_o\psi_o\rangle &= \langle H_o\psi_1, \psi_o\rangle \Longrightarrow (\text{by } (10.1.6)) \\ \langle H_1\psi_o, \psi_o\rangle + E_o\langle\psi_1, \psi_o\rangle &= \langle H_1\psi_o, \psi_o\rangle + \langle H_o\psi_1, \psi_o\rangle \\ &= \langle H_1\psi_o + H_o\psi_1, \psi_o\rangle \simeq \langle E_1\psi_o + E_o\psi_1, \psi_o\rangle \\ &= E_1\langle\psi_o, \psi_o\rangle + E_o\langle\psi_1, \psi_o\rangle \\ \Longrightarrow \langle H_1\psi_o, \psi_o\rangle &\simeq E_1\langle\psi_o, \psi_o\rangle \\ \Longrightarrow E_1 \overset{\text{i.e.,}}{=} \Delta E &\simeq \frac{\langle H_1\psi_o, \psi_o\rangle}{\langle\psi_o, \psi_o\rangle}, \end{aligned} \tag{10.1.7}$$

which we take therefore as a reason for definition (10.1.4).

Example

A quantum system, an *anharmonic oscillator*, has potential V given by

$$V(x) = Ax^2 + Bx^3 + Cx^4 \quad \text{for } A, B, C \in \mathbb{R}, A > 0. \tag{10.1.8}$$

The quantized Hamiltonian H in the Schrödinger equation (10.1.1) in this case is given by (3.1.2):

$$\begin{aligned} H &= -\frac{\hbar^2}{2m}\frac{d^2}{dx^2} + V(x) = H_o + H_1 \text{ for} \\ H_o &\stackrel{\text{def}}{=} -\frac{\hbar^2}{2m}\frac{d^2}{dx^2} + Ax^2, \\ H_1 &\stackrel{\text{def}}{=} \text{multiplication by } Bx^3 + Cx^4. \end{aligned} \tag{10.1.9}$$

Define $\omega > 0$ by $A = \omega^2 m/2$ for $m > 0$ fixed. Thus the function $V_o(x) = Ax^2$ corresponds to the potential energy of a harmonic oscillator with mass m and frequency $\nu = \omega/(2\pi)$. Using first-order time-independent perturbation theory we approximate the ground state energy of the anharmonic oscillator. A normalized solution ψ_o of (10.1.2) with corresponding energy-eigenvalue E_o is given by equation (4.8.5) and by (4.8.1):

$$\psi_o(x) = \left(\frac{b}{\pi}\right)^{1/4} e^{-\frac{b}{2}x^2}, \qquad E_o = \pi\nu\hbar = \frac{\omega\hbar}{2} \tag{10.1.10}$$

for $b = 2\pi\nu m/\hbar = \omega m/\hbar$. The first-order correction ΔE to the ground state energy E_o of the harmonic oscillator is given by the key definition (10.1.4):

$$\Delta E \stackrel{\text{def}}{=} \langle H_1\psi_o, \psi_o\rangle \stackrel{\text{def}}{=} \left(\frac{b}{\pi}\right)^{1/2} \int_{\mathbb{R}} \left(Bx^3 + Cx^4\right) e^{-bx^2}\, dx \tag{10.1.11}$$

where

$$\int_{\mathbb{R}} Bx^3 e^{-bx^2}\, dx = 0 \tag{10.1.12}$$

as the integrand $Bx^3e^{-bx^2}$ is an odd function of x. Now by equation (4.7.59) (or by equation (4.7.60))

$$\int_{\mathbb{R}} x^4 e^{-bx^2}\, dx = 2\int_0^{\infty} x^4 e^{-bx^2}\, dx = \frac{3}{4b^2}\sqrt{\frac{\pi}{b}}. \tag{10.1.13}$$

Since $b^2 = \omega^2 m^2/\hbar^2 \stackrel{\text{def}}{=} 2Am^2/(m\hbar^2) = 2Am/\hbar^2$ we see that ΔE is independent of B and is given by

$$\Delta E = \frac{3C}{4b^2} = \frac{3C\hbar^2}{8Am}. \tag{10.1.14}$$

The ground state energy of the anharmonic oscillator is, by (10.1.3) and (10.1.14),

$$E = E_o + \Delta E = \frac{\omega\hbar}{2} + \frac{3C\hbar^2}{8Am} = \frac{1}{2}\sqrt{\frac{2A}{m}}\hbar + \frac{3C\hbar^2}{8Am}, \tag{10.1.15}$$

up to first-order approximation. From (10.1.15) it is intuitively clear that the "size" of the perturbative effect of H_1, in the present case, is controlled by the size of C (since H_1 is independent of A, which is regarded as fixed).

10.2 Helium-Like Atoms: Their Ground-State Energies

From equations (3.5.21), (3.5.22) the time-independent Schrödinger equation for the helium atom is given by

$$\begin{gathered} -\frac{\hbar^2}{2m_o}\nabla_o^2\psi - \frac{\hbar^2}{2m_e}\nabla_1^2\psi - \frac{\hbar^2}{2m_e}\nabla_2^2\psi + V\psi = E\psi \\ \text{for } V(x) = -\sum_{i=1}^{2}\frac{Ze^2}{|x^{(i)} - x^{(o)}|} + \frac{e^2}{|x^{(1)} - x^{(2)}|}, \end{gathered} \tag{10.2.1}$$

for m_o = the nuclear mass, m_e = the electron mass, $Z = 2$, $x^{(o)} = (x_o, y_o, z_o)$ = the nuclear position, $x^{(i)} = (x_i, y_i, z_i)$ = the position of the ith electron, $i = 1, 2$, $x = (x^{(o)}, x^{(1)}, x^{(2)})$, and for

$$\nabla_j^2 = \frac{\partial^2}{\partial x_j^2} + \frac{\partial^2}{\partial y_j^2} + \frac{\partial^2}{\partial z_j^2} \quad (j = 0, 1, 2) \tag{10.2.2}$$

as usual. Instead of fixing the atomic number $Z = 2$ we shall let Z vary, at no extra cost, in order to incorporate, more generally, helium-like atoms in the discussion. It is precisely because of the *inter-electronic repulsion term*

$$\frac{e^2}{|x^{(1)} - x^{(2)}|} \tag{10.2.3}$$

in (10.2.1) that an exact solution of (10.2.1) has never been found. It is known that no appreciable error is introduced if, as a first approximation, we consider the nucleus at rest, say at the origin $x^{(o)} = (0, 0, 0)$, given that $m_o \gg m_e$. That is, we replace equation (10.2.1) by the equation

$$\begin{aligned} -\frac{\hbar^2}{2m_e}&\left(\nabla_1^2 + \nabla_2^2\right)\psi(x^{(1)}, x^{(2)}) + V(x^{(1)}, x^{(2)})\psi(x^{(1)}, x^{(2)}) \\ &= E\psi(x^{(1)}, x^{(2)}) \qquad \text{for} \\ V(x^{(1)}, x^{(2)}) &= -\sum_{i=1}^{2}\frac{Ze^2}{|x^{(i)}|} + \frac{e^2}{|x^{(1)} - x^{(2)}|}, \end{aligned} \tag{10.2.4}$$

which still has never been solved exactly because the term (10.2.3) still appears.

Towards an approximate solution of (10.2.4), or at least an approximation of the ground energy $E = E_o$, we first ignore the term in (10.2.3). Then equation (10.2.4) reduces to

$$-\frac{\hbar^2}{2m_e}\left(\nabla_1^2 + \nabla_2^2\right)\ \psi(x^{(1)}, x^{(2)}) - \sum_{i=1}^{2} \frac{Ze^2}{|x^{(i)}|}\psi(x^{(1)}, x^{(2)}) = E\psi(x^{(1)}, x^{(2)}), \tag{10.2.5}$$

where $x^{(1)} \neq (0,0,0)$, $x^{(2)} \neq (0,0,0)$. Now equation (10.2.5) certainly has exact solutions. Namely, suppose we have solutions $\psi^{(i)}$ of the equations

$$-\frac{\hbar^2}{2m_e}\nabla_i^2\psi^{(i)}(x^{(i)}) - \frac{Ze^2\psi^{(i)}(x^{(i)})}{|x^{(i)}|} = E_i\psi^{(i)}(x^{(i)}), \qquad E_i \in \mathbb{R}, i = 1,2, \tag{10.2.6}$$

on $\mathbb{R}^3 - \{(0,0,0)\}$. Set $\psi = \psi^{(1)} \otimes \psi^{(2)}$; i.e.,

$$\psi(x_1, y_1, z_1, x_2, y_2, z_2) \stackrel{\text{def}}{=} \psi^{(1)}(x_1, y_1, z_1)\psi^{(2)}(x_2, y_2, z_2) \tag{10.2.7}$$

for $x^{(i)} = (x_i, y_i, z_i) \neq 0$. Then by straightforward differentiation (similar to the argument which lead to the converse of Theorem 3.3) one sees that indeed ψ is a solution of (10.2.5) for $E = E_1 + E_2$ there. On the other hand, (10.2.6) coincide with the first equation in (5.1.9), where the reduced mass μ there corresponds to m_e here. Therefore for internal *negative* energies E_1, E_2 we have normalized solutions $\psi^{(i)} = \psi_{nlm}^{(i)}$ of (10.2.6) given by (5.2.12) (or by (5.2.15)) for

$$E_i = E_i^{(n)} \stackrel{\text{def}}{=} -\frac{m_e Z^2 e^4}{2\hbar^2 n^2}, \tag{10.2.8}$$

as in definition (5.1.29). As usual n, l, m are integers with $n \geq 1, 0 \leq l \leq n-1$, $l \leq m \leq l$. Thus we obtain by (10.2.7) solutions

$$\psi = \psi_{nlm}^{(1)} \otimes \psi_{nlm}^{(2)} \tag{10.2.9}$$

of (10.2.5) for

$$\begin{aligned} E = E_1^{(n)} + E_2^{(n)} &= 2\left(-\frac{m_e Z^2 e^4}{2\hbar^2 n^2}\right) \\ &= -\frac{2Z^2}{n^2}E_H \end{aligned} \tag{10.2.10}$$

where

$$E_H = \frac{m_e e^4}{2\hbar^2}; \tag{10.2.11}$$

that is, $-E_H$ is the ground state energy of hydrogen, given by (5.8.2), where we have replaced $\mu = .99m_e$ there by m_e; recall (0.0.10).

Consider in particular the lowest state $\psi_o \stackrel{\text{def}}{=} \psi_{100}^{(1)} \otimes \psi_{100}^{(2)}$, as we are interested in approximating the ground state energy of a helium-like atom. Here of course we must compute the first-order correction term ΔE in (10.1.4), where the inner product there is given by an integral quite a bit more complicated than that given in (10.1.11). If

$$H \stackrel{\text{def}}{=} -\frac{\hbar^2}{2m_e}\left(\nabla_1^2 + \nabla_2^2\right) + V \tag{10.2.12}$$

is the Hamiltonian in (10.2.4), then clearly $H = H_o + H_1$ for

$$\begin{aligned} H_o &= -\frac{\hbar^2}{2m_e}\left(\nabla_1^2 + \nabla_2^2\right) - \sum_{i=1}^{2} \frac{Ze^2}{|x^{(i)}|}, \\ H_1 &\stackrel{\text{def}}{=} \text{multiplication by } \frac{e^2}{|x^{(1)} - x^{(2)}|}, \end{aligned} \tag{10.2.13}$$

where

$$H_o\psi_o = E_o\psi_o \quad \text{for } E_o = -2Z^2E_H \tag{10.2.14}$$

as we have just noted in (10.2.9), (10.2.10). Of course (10.2.4) is now the equation

$$H\psi = E\psi, \tag{10.2.15}$$

as in (10.1.1), and we are therefore in position to apply the perturbation theory of the preceding section. Note that the $\psi_{nlm}^{(1)} \otimes \psi_{nlm}^{(2)}$ in (10.2.9) (and in particular the state ψ_o) are all normalized, since by construction the variables (x_1, y_1, z_1), (x_2, y_2, z_2) are separated. To approximate E, to first-order, the main problem therefore is to compute the integral

$$\begin{aligned} \Delta E &\stackrel{\text{def}}{=} \langle H_1\psi_o, \psi_o\rangle \\ &= e^2 \int \cdots \int_{\mathbb{R}^6} \frac{|\psi_{100}^{(1)}(x_1, y_1, z_1)|^2 |\psi_{100}^{(2)}(x_2, y_2, z_2)|^2}{\sqrt{(x_1 - x_2)^2 + (y_1 - y_2)^2 + (z_1 - z_2)^2}} \, dx_1 dy_1 dz_1 dx_2 dy_2 dz_2, \end{aligned} \tag{10.2.16}$$

which is the first-order correction to the energy E_o in (10.1.4). Using spherical coordinates, we can express ΔE as

$$\begin{aligned} \Delta E = e^2 &\int_0^{2\pi}\int_0^{\pi}\int_0^{\infty}\int_0^{2\pi}\int_0^{\pi}\int_0^{\infty} |\psi_{100}^{(1)}(g(r_1, \theta_1, \phi_1))|^2 \\ &\cdot \frac{|\psi_{100}^{(2)}(g(r_2, \theta_2, \phi_2))|^2}{r_{12}} r_1^2 \sin\phi_1 d\phi_1 d\theta_1 dr_1 r_2^2 \sin\phi_2 d\phi_2 d\theta_2 dr_2, \end{aligned} \tag{10.2.17}$$

Atom	Z	$2Z^2 - 1.25Z$	$-E_t = (2Z^2 - 1.25Z)13.6\,\text{eV}$	$-E_e$ (eV)
He	2	5.5	74.8	79.02
Li^+	3	14.25	193.8	198.02
Be^{2+}	4	27.00	367.2	371.42
B^{3+}	5	43.75	595.0	599.35
C^{4+}	6	64.50	877.2	881.69
N^{5+}	7	89.25	1213.8	1218.56
O^{6+}	8	118.00	1604.8	1610.10

Table 10.1: Comparison of theoretical values E_t obtained in (10.2.22) with experimental values E_e of helium-like atoms

employing earlier notation, where r_{12} is the distance between the two electrons, and where by (5.3.12)

$$\psi_{100}^{(j)}(g(r_j, \theta_j, \phi_j)) = \left(\frac{Z}{a_o}\right)^{3/2} \frac{1}{\sqrt{\pi}} \exp\left(-\frac{Z}{a_o} r_j\right) \quad \text{for } j = 1, 2. \tag{10.2.18}$$

That is,

$$\begin{aligned} \Delta E = \frac{e^2}{\pi^2}\left(\frac{Z}{a_o}\right)^6 \int_0^{2\pi}\int_0^{\pi}\int_0^{\infty}\int_0^{2\pi}\int_0^{\pi}\int_0^{\infty} \exp\left(-\frac{2Z}{a_o}(r_1 + r_2)\right) \\ \cdot \frac{r_1^2 r_2^2}{r_{12}} \sin\phi_1 d\phi_1 d\theta_1 dr_1 \sin\phi_2 d\phi_2 d\theta_2 dr_2 \end{aligned} \tag{10.2.19}$$

for

$$a_o = \frac{\hbar^2}{m_e e^2}. \tag{10.2.20}$$

That is, in the definition of a_o in (5.3.1) we replace μ there by m_e, in accord with the remarks preceding (10.2.8). Although the iterated integral in (10.2.19) is not quite so simple it can be evaluated; see Appendix 10A to this chapter, and the General Appendix D following Chapter 18. The end result is that

$$\Delta E = \frac{5Ze^2}{8a_o} = \frac{5Zm_e e^4}{8\hbar^2} = \frac{5}{4} Z E_H, \tag{10.2.21}$$

by (10.2.11), (10.2.20).

By (10.1.3), (10.2.10), (10.2.21), we obtain the main result of this section: *the ground state energy E of a helium-like atom is given by*

$$\begin{aligned} E = E_1^{(1)} + E_2^{(1)} + \Delta E &= -(2Z^2 - \frac{5}{4}Z)E_H \\ &= -(2Z^2 - 1.25Z)13.6\,\text{eV}, \end{aligned} \tag{10.2.22}$$

up to a first-order approximation. Obviously it is of interest to compare the theoretical result (10.2.22) with known experimental results compiled by John Slater. This is done in Table 10.1; cf. [78, pp. 339–342].

Sometimes energy is expressed in terms of *Rydberg* units, where 1 Rydberg = 13.6 eV. Thus for helium, for example, $-E_e = \frac{79.02}{13.6} = 5.81$ Rydbergs. Even though the perturbation operator H_1 in (10.2.13) is not necessarily "small" the relative errors $(E_t - E_e)/E_e$ in Table 10.1 are small. For $Z = 2,3,4,5,6,7,8$ they are $-.05340$, $-.02131$, $-.01136$, $-.00725$, $-.00509$, $-.00390$, $-.00329$, respectively. Thus in absolute value the percentage errors range from about 5.3% for He to about $.33\%$ for O^{6+}, which shows fairly good agreement between theory and experiment.

10.3 Pauli's Exclusion Principle and Electron Configurations

We have seen that because of inter-electron interactions the calculation in closed form of energy levels and wave functions of atoms with Z electrons, $Z \geq 2$, is virtually a mathematical impossibility. In addition to electronic interaction Coulomb forces one must also deal with certain magnetic *spin-orbit* forces (Russell–Sanders L-S coupling).

Various approximation methods have been devised for multi-electron calculations. A simplifying assumption (one which indeed is subject to further refinements) is that each electron is considered to move independently under the Coulomb influence of the nucleus and under the *average* Coulomb field of the other $Z - 1$ electrons. In this *central field* approximation (introduced by John Slater in 1929) all magnetic interactions are neglected. The total wave function of the atomic system will be a product of Z one-particle wave functions (as the electrons move independently) corresponding to a spherically symmetric potential. As in Chapter 5 one can assign to each electron a set of four quantum numbers (n, l, m_l, m_s) which describes its state—the state being the corresponding individual wave function, which is also called an *orbital*. By calculations similar to those of Chapter 5 one finds that the angular momentum quantum number (or azimuthal quantum number) l and the magnetic quantum number m_l are the same as those for the hydrogenic atom. The electron spin quantum number m_s ($m_s = \frac{1}{2}$ or $-\frac{1}{2}$) also is the same as for the hydrogenic atom. However the (generalized) principal quantum number n will be *different* from the n in (5.2.6) which helped to label the radial wave function there. This is because now the potential in the radial equation is more general than the simple Coulomb potential. The energy levels will depend on *n and l*, and not just n alone as in the hydrogenic case.

Concerning the quantum numbers (n, l, m_l, m_s), Pauli in 1925 enunciated the following striking principle, after analyzing an array of spectroscopic data for multi-electron atoms:

No two electrons in an atom can have an identical set of quantum numbers. (10.3.1)

This *exclusion principle*, which is to be taken as axiomatic in quantum mechanics, has lead to a basic understanding, for example, of the *shell structure* of atoms, and to key facts concerning the chemical valence. Our discussion prior to its statement was a bit sketchy since in fact we plan to present a more general formulation of this fundamental principle.

Electrons with the same principal quantum number n are said to form a *shell*. With n fixed one can consider electrons with the same n and the same azimuthal number l. This collection of electrons forms a *subshell* of the shell defined by n. With n and l fixed there are $2l + 1$ possible values for m_l ($-l \le m_l \le l$), and two possible values for m_s. This means that the maximal number of electrons that can occupy a subshell, without violating Pauli's exclusion principle, is $2(2l + 1)$; this fact is known as *Stoner's rule*. For example for $n = 2$, $l = 1$, only the following states are possible: $(2, 1, -1, 1/2)$, $(2, 1, -1, -1/2)$, $(2, 1, 0, 1/2)$, $(2, 1, 0, -1/2)$, $(2, 1, -1, 1/2)$, $(2, 1, -1, -1/2)$. For $n = 2$ and $l = 0$ we must have $m_l = 0$. Then only the states $(2, 0, 0, 1/2)$, $(2, 0, 0, -1/2)$ are possible. Since each l gives rise to a maximum number of $2(2l + 1)$ states, for n fixed, there can be at most

$$\begin{aligned}\sum_{l=0}^{n-1} 2(2l+1) &= 2\left[2\sum_{l=0}^{n-1} l + \sum_{l=0}^{n-1} 1\right] \\ &= 2\left[2\left(0 + \frac{(n-1)n}{2}\right) + n\right] = 2n^2\end{aligned} \tag{10.3.2}$$

electrons occupying a shell. This number is 8 in the preceding example.

It is customary to employ, alternatively, x-ray spectroscopic notation: One writes

$$\begin{aligned}&\text{shell letters } K, L, M, N, O, P, \ldots, \\ &\text{for the } n \text{ numbers } 1, 2, 3, 4, 5, 6, \ldots, \\ &\text{and subshell letters } s, p, d, f, g, h, i, j, k, \ldots, \\ &\text{for the } l \text{ values } 0, 1, 2, 3, 4, 5, 6, 7, 8, \ldots,\end{aligned} \tag{10.3.3}$$

where the s stands for "sharp," p for "principal," d for "diffuse," and f stands for "fundamental." A d subshell for example (where $l = 2$) can accommodate $2(2l + 1) = 10$ electrons.

There is a tendency for a physical system to acquire its *lowest* energy state. Thus for the hydrogen atom, the lowest energy state, or ground state, is given by $n = 1$, as we have seen (the corresponding energy being about -13.59 electron volts). Chemists, and others use the notation in (10.3.3) to describe the *electron configuration* of atoms in their ground state. For nitrogen (N), with $Z = 7$, this configuration is $1s^2\ 2s^2\ 2p^3$ which means the following: reading pairs $1s^2$, $2s^2$, $2p^3$ from left to right with superscripts denoting the number of electrons with quantum numbers n, l, one takes $1s^2$ to mean there are two electrons with quantum numbers $n = 1$, $l = 0$ (since s corresponds to $l = 0$, by (10.3.3)); similarly $2s^2$ means there are two electrons with quantum numbers $n = 2$, $l = 0$, and $2p^3$ means there are three electrons with quantum numbers $n = 2$, $l = 1$ (since p corresponds to $l = 1$, by (10.3.3)). The $Z = 7$ ($= 2+2+3$) electrons thus are all accounted for.

As a second example consider oxygen (O), with $Z = 8$. Its ground state electron configuration is given by $1s^2\ 2s^2\ 2p^4$, which we translate to mean that there are two electrons with quantum numbers $n, l = 1, 0$, there are two electrons with quantum numbers $n, l = 2, 0$, and there are four electrons with quantum numbers $n, l = 2, 1$. Here we see that the K shell ($n = 1$) is filled or *closed* since the maximum number of electrons it can hold is $2n^2 = 2$. The L shell ($n = 2$) is not closed since it contains six (but not eight) electrons. For a third example, we take the sodium atom (Na) with $Z = 11$ electrons (and therefore 11 protons) and 12 neutrons (n). Its ground state electron configuration is $1s^2\ 2s^2\ 2p^6\ 3s^1$. One can picture this atom by way of the shell diagram in Figure 10.1.

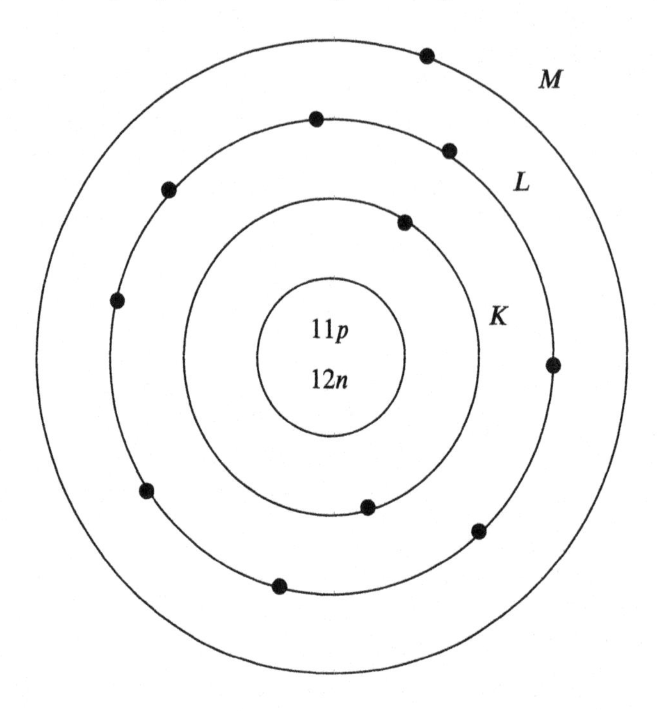

Figure 10.1: Shell diagram for Na

There are $2 + 6 = 8$ electrons in the L shell ($n = 2$), which is thus closed; two of these electrons have quantum numbers $n, l = 2, 0$ and six of them have quantum numbers $n, l = 2, 1$. There is one electron in the M shell with quantum numbers $n, l = 1, 0$, and there are two electrons in the K shell with quantum numbers $n, l = 1, 0$. The fact that the outer shell of sodium has a single electron (that is, its *valence number* is 1), whereas the outer shell of the chlorine atom lacks exactly one electron to close it (its valence number is -1) ultimately relates to the ability of these two atoms to combine to form sodium chloride (NaCl), ordinary table salt. The valence number is a measure of the number of electrons that an atom is willing to give up, accept, or share during a chemical reaction. For further details with regard to the example at hand we start with the electron configuration $1s^2\ 2s^2\ 2p^6\ 3s^2\ 3p^5$ of chlorine. Clearly $Z = 2+2+6+2+5 = 17$. One has the shell diagram depicted in Figure 10.2.

In the M shell there are two electrons with quantum numbers $n, l = 3, 0$ and five electrons with quantum numbers $n, l = 3, 1$. Recalling that the subshell defined by n, l can accommodate up to $2(2l + 1)$ electrons we see that the 3, 1 subshell of the chlorine atom (which can accommodate six electrons but only has five) is shy one electron, as remarked above. The sodium atom will readily give up (or lend) its one outer valence electron which the chlorine atom will readily accept to close its outer subshell. In the process sodium acquires a positive charge (having given up

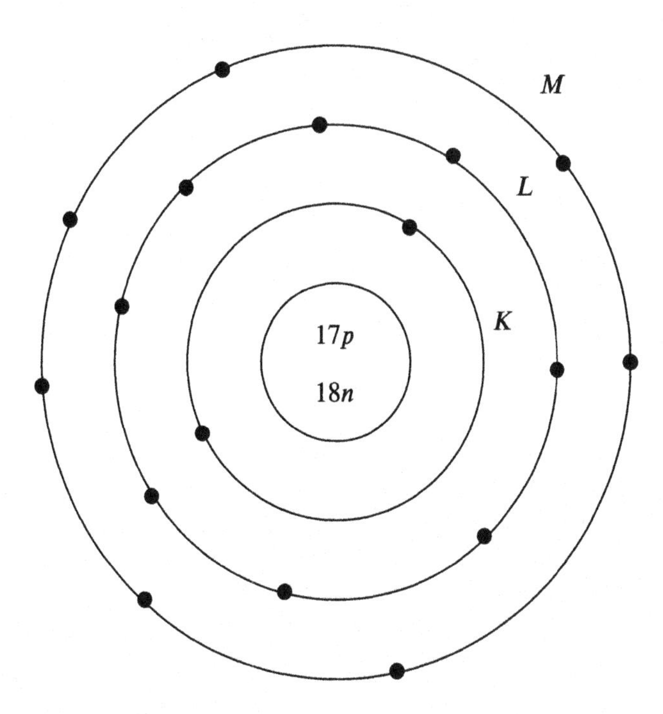

Figure 10.2: Shell diagram for Cl

an electron), and chlorine acquires a negative charge (having gained an electron). These oppositely charged particles attract each other to form sodium chloride.

In the electron configuration, again as one reads left to right, one notes the increase in electron energy (since n increases). Table 10.2 gives some additional electron configurations for a few light atoms.

The exclusion principle and the classification of electronic states by their quantum numbers (the latter being based on an approximation, as we have seen) provide quite adequate information and means for explanation of the Mendeléev *periodic table* (after Dimitri Ivanovitch Mendeléev (1834–1907)); see General Appendix B following Chapter 18. Elements in the table, apart from a central list of transition and rare earth elements, are organized into eight *groups*, a group being a column of elements with quite similar chemical properties, because they have similar electron configurations. Elements are also organized into *periods*, which are rows where the atomic number Z increases by one as one reads from left to right. Electrons first fill the lower energy shells. When these are filled (in the lighter atoms hydrogen, helium, lithium, etc.), higher energy shells start to be filled. One notes the outermost shells being filled as one reads across a row (again from left to right). By this *building-up principle* the table (with heavier elements being listed) is generated. The theoretical construction of the Mendeléev table, by Bohr in 1922 (also by Mainsmith and Stoner) was at the time one of the great and marvelous achievements of quantum theory.

At the quantum level, some particles (electrons, protons, neutrons, for example) are utterly *identical*. They are *indistinguishable* in the sense that they cannot be distinguished by some measurement. All electrons have the same mass, the same

Atom	Symbol	Electron Configuration
Hydrogen	H	$1s^1$
Helium	He	$1s^2$
Lithium	Li	$1s^2\,2s^1$
Beryllium	Be	$1s^2\,2s^2$
Boron	B	$1s^2\,2s^2\,2p^1$
Carbon	C	$1s^2\,2s^2\,2p^2$
Fluorine	F	$1s^2\,2s^2\,2p^5$
Neon	Ne	$1s^2\,2s^2\,2p^6$

Table 10.2: Ground state configurations

spin and charge, the same magnetic moment, etc.; "if you have seen one electron you have seen them all." There is a *principle of identity of particles* from which one can derive many fundamental and profound consequences. We consider it in relation to a more general formulation of Pauli's exclusion principle. As we shall see, there is imposed on the wave functions of systems of identical particles either a symmetry or asymmetry feature. A formal statement of the principle of identity of particles, one suitable for our purposes, is the following:

The state of a system of identical particles cannot change when two of them are interchanged. That is, if the coordinates and spin of two particles of the system are interchanged the wave function of the system can change only by some phase factor $e^{i\theta}$, $\theta \in \mathbb{R}$. (10.3.4)

We proceed mathematically to gain a more practical understanding of this principle. Suppose there is a system of n identical particles; i.e., they all have the same mass, the same charge, the same spin, etc. (assuming they have a charge or spin, for example). Let H denote the quantum Hamiltonian of the system. Then if P_{jk} denotes the *exchange* or *transposition* operator defined by

$$(P_{jk}f)(q_1,\ldots,q_j,\ldots,q_k,\ldots,q_n,t) \stackrel{\text{def}}{=} f(q_1,\ldots,q_k,\ldots,q_j,\ldots,q_n,t), \quad j \neq k \tag{10.3.5}$$

(here one may think of q_j as the position of the jth particle at time t), one requires that

$$HP_{jk} = P_{jk}H \qquad \text{for all } j,k,j \neq k. \tag{10.3.6}$$

Schrödinger's equation for the system is of course

$$i\hbar\frac{\partial\psi}{\partial t} = H\psi. \tag{10.3.7}$$

In (10.3.7) one could also take ψ to be a spinor. From (10.3.6) it is clear that *if* ψ *is a solution of* (10.3.7), then so is $P_{jk}\psi$, since

$$\frac{\partial}{\partial t}P_{jk}\psi = P_{jk}\frac{\partial\psi}{\partial t}. \tag{10.3.8}$$

That is, after the exchange of the jth and kth particle $P_{jk}\psi$ represents the new state of the system. (10.3.4) requires that

$$P_{jk}\psi = \lambda\psi \quad \text{for some } \lambda \in \mathbb{C} \text{ with } |\lambda| = 1. \tag{10.3.9}$$

As $P_{jk}^2 = 1$ we deduce that $\lambda^2 = 1$, or $\lambda = \pm 1$. That is, the wave functions of indistinguishable particles are either *symmetric* ($\lambda = 1$) or *anti-symmetric* ($\lambda = -1$). Particles with symmetric wave functions are called *bosons* (or Bose–Einstein particles, after Satyendranath Bose and Albert Einstein), and particles with anti-symmetric wave functions are called *fermions* (or Fermi–Dirac particles, after Enrico Fermi and Paul Dirac).

Pauli's exclusion principle, in a more general form, is the assertion:

> *The wave function for a system of identical particles having half-odd-integral intrinsic spin* (for example a multi-electron system) *must be anti-symmetric with respect to the interchange of any two of its particle's coordinates.* (10.3.10)

Here "coordinates" include spin parameters occurring in the wave function. Experiments have established that particles with half-odd-integral spin are indeed described by anti-symmetric wave functions, and moreover particles with integral spin are described by symmetric wave functions.

Consider, for example, the Hamiltonian H in (10.2.12) for helium-like atoms. If f is a function of six variables $(x^{(1)}, x^{(2)}) = (x_1, y_1, z_1, x_2, y_2, z_2)$, then by direct computation

$$\nabla_1^2 P_{12} f = P_{12} \nabla_2^2 f \implies \nabla_2^2 P_{12} f = P_{12} \nabla_1^2 f, \tag{10.3.11}$$

where the implication in (10.3.11) follows since $P_{12}^2 = 1$. Clearly $P_{12}V = V$ for V in (10.2.4). Thus indeed

$$HP_{12} = P_{12}H; \tag{10.3.12}$$

i.e., H satisfies (10.3.6). Also

$$H_o P_{12} = P_{12} H_o. \tag{10.3.13}$$

As in (10.2.9), (10.2.10) we have a solution

$$\psi_{12} \stackrel{\text{def}}{=} \psi_{n_1 l_1 m_1} \otimes \psi_{n_2 l_2 m_2} \tag{10.3.14}$$

of (10.2.5) (which is the equation $H_o\psi = E\psi$) for

$$E = E_1^{(n_1)} + E_2^{(n_2)}; \tag{10.3.15}$$

see (10.2.8). By (10.3.13)

$$\psi_{21} \stackrel{\text{def}}{=} P_{12}\psi_{12} \tag{10.3.16}$$

is also a solution of (10.2.5) for the same E as in (10.3.15). Of course the solutions ψ_{12}, ψ_{21} are *not* anti-symmetric, but an anti-symmetric solution ψ can immediately be constructed from them:

$$\begin{aligned} \psi &= \psi_{12} - \psi_{21}; \text{ i.e.,} \\ \psi(x^{(1)}, x^{(2)}) &= \psi_{n_1 l_1 m_1}(x^{(1)})\, \psi_{n_2 l_2 m_2}(x^{(2)}) \\ &\quad - \psi_{n_1 l_1 m_1}(x^{(2)})\, \psi_{n_2 l_2 m_2}(x^{(1)}). \end{aligned} \tag{10.3.17}$$

Now if $(n_1, l_1, m_1) = (n_2, l_2, m_2)$, then by (10.3.17) we see that $\psi \equiv 0$. That is, if the anti-symmetric wave function ψ is non-vanishing, then the states (n_1, l_1, m_1), (n_2, l_2, m_2) of the two electrons must differ: we must have $(n_1, l_1, m_1) \neq (n_2, l_2, m_2)$. Thus, disregarding spin considerations (for simplicity), we have presented an indication of why an anti-symmetry version of Pauli's exclusion (statement (10.3.10)) implies the version (10.3.1).

One can view the (somewhat mysterious) fact that nature allows only for symmetric or anti-symmetric wave functions in the description of indistinguishable particles as being related to the mathematical fact that the symmetric group S_n on n letters has *exactly two* one-dimensional representations $\pi^{\pm}$ (up to equivalence), where π^+ is the trivial representation and where π^- is the *sign* representation which is trivial on even permutations, but acts as -1 on odd permutations. Recall the discussion in Section 7.11 of Chapter 7. First note that since each $\sigma \in S_n$ is a product of transpositions one has that (10.3.6) holds for every $\sigma \in S_n$:

$$H\sigma = \sigma H \tag{10.3.18}$$

where

$$(\sigma f)(x_1, x_2, \ldots, x_n) \overset{\text{def}}{=} f(x_{\sigma^{-1}(1)}, x_{\sigma^{-1}(2)}, \ldots, x_{\sigma^{-1}(n)}). \tag{10.3.19}$$

Hence (by (10.3.8) again) if ψ is a (non-zero) solution of (10.3.7), so is $\sigma\psi$ and equation (10.3.9) (which followed by the principle of identity (10.3.4)) now becomes

$$\sigma\psi = \lambda_\sigma \psi \quad \text{for some } \lambda_\sigma \in \mathbb{C} \text{ with } |\lambda_\sigma| = 1. \tag{10.3.20}$$

By (10.3.19), $(\sigma_1\sigma_2)f = \sigma_1(\sigma_2 f)$ which means that equation (10.3.20) defines a one-dimensional representation π of S_n on the space $\mathbb{C}\psi$ spanned by ψ: $\pi(\sigma)\psi \overset{\text{def}}{=} \sigma\psi$. By the above remark, $\pi = \pi^+$ or π^-. If $\pi = \pi^+$, then $\psi = \sigma\psi\ \forall \sigma \in S_n \implies \psi$ is symmetric. If $\pi = \pi^-$, then in particular as every transposition is an odd permutation we have (by definition of π^-) $-\psi = P_{ij}\psi \implies \psi$ is anti-symmetric. Thus the one-dimensional *trivial* representation of the symmetric group S_n corresponds in some sense to bosonic particles of nature, whereas the one-dimensional *sign* representation π^- corresponds to fermionic particles.

Appendix 10A On the Integral in (10.2.19)

As was shown in Section 10.2 the first-order pertubative analysis of the ground state energy of a helium-like atom depends mainly on the computation of the integral I in (10.2.16) (or in (10.2.19)). There are various known methods of computing I, including one which uses an infinite series expansion of r_{12} in terms of spherical harmonics. Although the latter method is elegant, we shall indicate how the computation of I follows by some well-known Fourier transform computations. Since these Fourier transforms are known in an arbitrary dimension n we will in fact compute, more generally, an n-dimensional version of I, for the record. Namely, let

$$I_n \stackrel{\text{def}}{=} \int_{\mathbb{R}^n}\int_{\mathbb{R}^n} \frac{e^{-\alpha(|x|+|y|)}}{|x-y|^{n-\beta}}\,dxdy \text{ for } \alpha > 0, \quad 0 < \beta < n. \tag{10.A.1}$$

Then we can show

Theorem 10.1.

$$I_n = \frac{2^{n+\beta-1}\pi^{n-1}(n+\beta)\Gamma\left(\frac{\beta}{2}\right)\Gamma\left(\frac{n+\beta}{2}\right)\Gamma\left(\frac{n+1}{2}\right)^2}{n!\alpha^{n+\beta}\Gamma\left(\frac{n}{2}\right)}.$$

In particular for $n = 3$ and $\beta = 2$,

$$\begin{aligned} I_3 &\stackrel{\text{def}}{=} \int_{\mathbb{R}^3}\int_{\mathbb{R}^3} \frac{e^{-\alpha(|x|+|y|)}}{|x-y|}\,dxdy \\ &= \frac{2^4\pi^2 5\Gamma\left(\frac{5}{2}\right)}{3!\alpha^5\Gamma\left(\frac{3}{2}\right)} \end{aligned} \tag{10.A.2}$$

where $\Gamma(5/2)/\Gamma(3/2) = \Gamma(\frac{3}{2}+1)/\Gamma(3/2) = 3/2\,\Gamma(3/2)/\Gamma(3/2) = \frac{3}{2} \implies$

Corollary 10.1.

$$I_3 = \frac{20\pi^2}{\alpha^5} \quad \textit{for } \alpha > 0,\ \beta = 2.$$

Using Corollary 10.1 we indeed deduce the key result

$$\Delta E = \frac{5Ze^2}{8a_0} \tag{10.A.3}$$

in (10.2.21) by choosing $\alpha = \frac{2Z}{a_0}$. That is, by (10.2.19) and Corollary 10.1,

$$\begin{aligned} \Delta E &= \frac{e^2}{\pi^2}\left(\frac{Z}{a_0}\right)^6 I_3 = \frac{e^2}{\pi^2}\left(\frac{Z}{a_0}\right)^6 20\pi^2\left(\frac{a_0}{2Z}\right)^5 \\ &= \frac{5Ze^2}{8a_0}. \end{aligned} \tag{10.A.4}$$

Thus we can focus on Theorem 10.1. Its proof will be deferred for now. See the General Appendix D following Chapter 18.

Part II

Some Selected Topics

11

Fresnel Integrals and Feynman Integrals

In Part II of this book a transition is made as certain selected topics are considered. Among these are path integrals, zeta functions, and some applications of the Selberg trace formula. In some cases it will not be possible to develop all of the relevant background material. Rather, the intent will be to present some of the key ideas and, as in Part I, to present some concrete computational examples. Later we shall compute, for example, certain *Feynman path integrals*. For such computations one needs the value of certain *Fresnel* type integrals, integrals of independent interest. We begin with a discussion of the latter integrals.

11.1 Fresnel Integrals

The basic Fresnel integrals are the integrals

$$\int_0^\infty \sin x^2 \, dx, \qquad \int_0^\infty \cos x^2 \, dx. \tag{11.1.1}$$

We review the computation of these integrals, i.e., of the integral

$$\int_0^\infty \exp(i\lambda x^2) \, dx, \quad \lambda > 0, \tag{11.1.2}$$

and then show how to compute a much more complicated type of Fresnel integral of the form

$$\begin{aligned}\int_{-\infty}^\infty \cdots \int_{-\infty}^\infty &\exp\left(i\left\{\lambda\left[(x_1-a)^2+(x_2-x_1)^2+(x_3-x_2)^2+\cdots+(x_n-x_{n-1})^2\right.\right.\right.\\ &\left.\left.\left.+(b-x_n)^2\right]+\mu(x_1+\cdots+x_n)\right\}\right)\,dx_1dx_2\cdots dx_n \end{aligned}\tag{11.1.3}$$

for $\lambda > 0, a, b, \mu \in \mathbb{R}, n \geq 1$ an integer;

see Theorem 11.3.

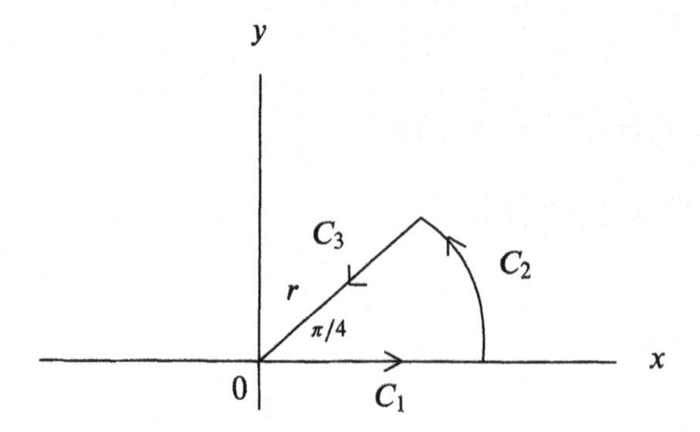

Figure 11.1: Closed contour

Let f be the holomorphic function $z \to \exp(i\lambda z^2)$ for $\lambda > 0$ fixed. Let $C = C_1 + C_2 + C_3$ be the closed contour of Figure 11.1, where C_2 is a portion of the circle about the origin of radius r. Parameterize C_1 by $x(t) = t$, $y(t) = 0$, $0 \le t \le r$, and parameterize $-C_3$ by $x(t) = t$, $y(t) = t$, $0 \le t \le r\cos(\pi/4) = r/\sqrt{2}$. Then $z(t) = x(t) + iy(t) = t$, $z'(t) = 1$ on C_1 and similarly $z(t) = t + it$, $z'(t) = 1 + i$ on $-C_3 \Longrightarrow$

$$\int_{C_1} f(z)\, dz = \int_0^r \exp(i\lambda t^2)\, dt, \tag{11.1.4}$$

$$\begin{aligned}\int_{C_3} f(z)\, dz = -\int_{-C_3} f(z)\, dz &= -\int_0^{r/\sqrt{2}} \exp(i\lambda t^2(1+i)^2)(1+i)\, dt \\ &= -(1+i)\int_0^{r/\sqrt{2}} \exp(-2\lambda t^2)\, dt.\end{aligned} \tag{11.1.5}$$

By Cauchy's theorem

$$\begin{aligned}0 &= \int_C f(z)\, dz \\ &= \int_0^r \exp(i\lambda t^2)\, dt + \int_{C_2} f(z)\, dz - (1+i)\int_0^{r/\sqrt{2}} \exp(-2\lambda t^2)\, dt.\end{aligned} \tag{11.1.6}$$

We will show that (for $\lambda > 0$)

$$\lim_{r\to\infty} \int_{C_2} f(z)\, dz = 0. \tag{11.1.7}$$

Assuming (11.1.7) for now we see that $\int_0^\infty \exp(i\lambda t^2)\, dt$ converges and has the value

$$\begin{aligned}(1+i)\int_0^\infty \exp(-2\lambda t^2)\, dt &= (1+i)\int_0^\infty \frac{\exp(-x^2)\, dx}{\sqrt{2\lambda}} \\ &= \frac{(1+i)}{\sqrt{2\lambda}}\frac{\sqrt{\pi}}{2} = \frac{\exp(i\pi/4)}{2}\sqrt{\frac{\pi}{\lambda}}.\end{aligned} \tag{11.1.8}$$

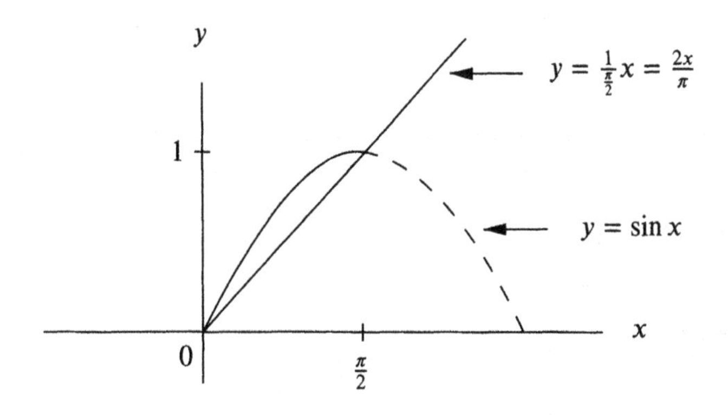

Figure 11.2: A lower bound for the sine

Parameterize C_2 by $x(\theta) = r\cos\theta$, $y(\theta) = r\sin\theta$, $0 \le \theta \le \pi/4$. Then

$$\int_{C_2} f(z)\,dz = \int_0^{\pi/4} \exp(i\lambda r^2 \exp(i2\theta)) \cdot ir\exp(i\theta)\,d\theta \tag{11.1.9}$$

$$\Longrightarrow \left|\int_{C_2} f(z)\,dz\right| \le \int_0^{\pi/4} \exp(\mathrm{Re}(i\lambda r^2\exp(i2\theta))) \cdot r\,d\theta$$

$$= r\int_0^{\pi/4} \exp(-\mathrm{Im}(\lambda r^2\exp(i2\theta)))\,d\theta \tag{11.1.10}$$

$$= r\int_0^{\pi/4} \exp(-\lambda r^2 \sin(2\theta))\,d\theta.$$

From Figure 11.2 we see that $\sin x \ge 2x/\pi$ for $0 \le x \le \pi/2$. Therefore for $0 \le \theta \le \pi/4$, $sin(2\theta) \ge 4\theta/\pi \implies -\lambda r^2\sin(2\theta) \le -\lambda r^2 4\theta/\pi$ (for $\lambda > 0$) $\Longrightarrow$

$$\int_0^{\pi/4} \exp(-\lambda r^2\sin(2\theta))\,d\theta \le \int_0^{\pi/4} \exp(-\lambda r^2 4\theta/\pi)\,d\theta$$
$$= [1-\exp(-\lambda r^2)]\frac{\pi}{4\lambda r^2} \to 0 \quad \text{as } r \to \infty, \text{ for } \lambda > 0. \tag{11.1.11}$$

Thus by (11.1.10), equation (11.1.7) is established, and by (11.1.8) we have proved

Theorem 11.1. *For $\lambda > 0$,*

$$\int_0^\infty \exp(i\lambda x^2)\,dx \quad \textit{converges}$$

and has the value

$$\frac{1}{2}\frac{(1+i)}{\sqrt{2}}\sqrt{\frac{\pi}{\lambda}} = \frac{e^{i\pi/4}}{2}\sqrt{\frac{\pi}{\lambda}}.$$

In particular the Fresnel integrals

$$\int_0^\infty \cos(\lambda x^2)\,dx, \quad \int_0^\infty \sin(\lambda x^2)\,dx \textit{ exist}$$

and have the common value

$$\frac{1}{2\sqrt{2}}\sqrt{\frac{\pi}{\lambda}}.$$

To make further progress we consider the process of *uncompleting the square.* Let $q(z) = -\alpha z^2 + \beta z + \gamma$ be a second degree polynomial function where $\alpha, \beta, \gamma \in \mathbb{C}$, $\alpha \neq 0$. Let z^* be the point where $q'(z)$ vanishes: $q'(z) = -2\alpha z + \beta \implies z^* = \beta/2\alpha$. Note that $q(z^*) = -\alpha(\beta/2\alpha)^2 + \beta(\beta/2\alpha) + \gamma = \beta^2/4\alpha + \gamma$. Then $\alpha(z-z^*)^2 = \alpha(z^2-2z\beta/2\alpha+\beta^2/4\alpha^2) = \alpha z^2 - z\beta + (q(z^*)-\gamma) = -q(z)+q(z^*)$; i.e.,

Proposition 11.1.

$$(z - z^*)^2 = \frac{-q(z) + q(z^*)}{\alpha}, \quad \textit{or} \tag{11.1.12}$$

$$q(z) = q(z^*) - \alpha(z - z^*)^2 \tag{11.1.13}$$

for

$$q(z) \stackrel{\text{def}}{=} -\alpha z^2 + \beta z + \gamma, \quad z^* \stackrel{\text{def}}{=} \frac{\beta}{2\alpha}, \quad \alpha \neq 0; \tag{11.1.14}$$

$$q(z^*) = \frac{\beta^2}{4\alpha} + \gamma; \tag{11.1.15}$$

here $\alpha, \beta, \gamma \in \mathbb{C}$.

In equation (11.1.13) of Proposition 11.1 we have "uncompleted the square." As an example, for $x, a, b \in \mathbb{R}$, $\lambda_1, \lambda_2 > 0$, $-\gamma \stackrel{\text{def}}{=} \lambda_1 a^2 + \lambda_2 b^2$, let

$$\begin{aligned} q(x) &\stackrel{\text{def}}{=} -(\lambda_1 + \lambda_2)x^2 + 2(a\lambda_1 + b\lambda_2)x + \gamma \\ &= -\lambda_1(x-a)^2 - \lambda_2(b-x)^2 = -\alpha x^2 + \beta x + \gamma \end{aligned} \tag{11.1.16}$$

for $\alpha = \lambda_1 + \lambda_2$, $\beta = 2(a\lambda_1 + b\lambda_2)$. Then

$$x^* \stackrel{\text{def}}{=} \frac{\beta}{2\alpha} = \frac{(a\lambda_1 + b\lambda_2)}{(\lambda_1 + \lambda_2)}. \tag{11.1.17}$$

By (11.1.13), (11.1.15), and (11.1.16)

$$\begin{aligned} \lambda_1(x-a)^2 + \lambda_2(b-x)^2 &= -q(x) = -q(x^*) + \alpha(x - x^*)^2 \\ &\stackrel{\text{def}}{=} -q(x^*) + (\lambda_1 + \lambda_2)(x - x^*)^2, \end{aligned} \tag{11.1.18}$$

where

$$\begin{aligned} q(x^*) &= \frac{\beta^2}{4\alpha} + \gamma \stackrel{\text{def}}{=} \frac{4(a\lambda_1 + b\lambda_2)^2}{4(\lambda_1 + \lambda_2)} - \lambda_1 a^2 - \lambda_2 b^2 \\ &= \frac{-\lambda_1\lambda_2(a-b)^2}{\lambda_1 + \lambda_2}. \end{aligned} \tag{11.1.19}$$

Therefore

$$\int_{-\infty}^{\infty} \exp(i[\lambda_1(x-a)^2 + \lambda_2(b-x)^2])\, dx$$
$$= \exp\left(\frac{i\lambda_1\lambda_2(a-b)^2}{\lambda_1+\lambda_2}\right) \int_{-\infty}^{\infty} \exp(i(\lambda_1+\lambda_2)(x-x^*)^2)\, dx. \quad (11.1.20)$$

For a continuous function f on $\mathbb{R}$ we take $I = \int_{-\infty}^{\infty} f(x)\, dx$ to mean

$$\begin{aligned}\int_{-\infty}^{\infty} f(x)\, dx &\stackrel{\text{def}}{=} \int_{-\infty}^{0} f(x)\, dx + \int_{0}^{\infty} f(x)\, dx \\ &\stackrel{\text{def}}{=} \lim_{r\to\infty} \int_{-r}^{0} f(x)\, dx + \lim_{r\to\infty} \int_{0}^{r} f(x)\, dx\end{aligned} \quad (11.1.21)$$

in this section. If both limits in (11.1.21) exist, i.e., if I converges, then of course I coincides with the Cauchy principal value (CPV):

$$I = \text{CPV} \int_{-\infty}^{\infty} f(x)\, dx \stackrel{\text{def}}{=} \lim_{r\to\infty} \int_{-r}^{r} f(x)\, dx. \quad (11.1.22)$$

Now for $r, \delta \in \mathbb{R}$, $\lambda > 0$

$$\begin{aligned}\int_{-r}^{0} \exp(i\lambda(x-\delta)^2)\, dx &= \int_{-r-\delta}^{-\delta} \exp(i\lambda u^2)\, du \\ &= \int_{-r-\delta}^{0} \exp(i\lambda u^2)\, du + \int_{0}^{-\delta} \exp(i\lambda u^2)\, du \\ &= \int_{0}^{r+\delta} \exp(i\lambda t^2)\, dt + \int_{0}^{-\delta} \exp(i\lambda u^2)\, du \quad (\text{for } t = -u).\end{aligned} \quad (11.1.23)$$

By Theorem 11.1 we see that

$$\int_{-\infty}^{0} \exp(i\lambda(x-\delta)^2)\, dx \quad \text{exists} \quad (11.1.24)$$

and has the value

$$\frac{e^{i\pi/4}}{2}\sqrt{\frac{\pi}{\lambda}} + \int_{0}^{-\delta} \exp(i\lambda u^2)\, du. \quad (11.1.25)$$

Similarly

$$\begin{aligned}\int_{0}^{r} \exp(i\lambda(x-\delta)^2)\, dx &= \int_{-\delta}^{r-\delta} \exp(i\lambda u^2)\, du \\ &= \int_{-\delta}^{0} \exp(i\lambda x^2)\, dx + \int_{0}^{r-\delta} \exp(i\lambda x^2)\, dx\end{aligned} \quad (11.1.26)$$

$$\Rightarrow \int_0^\infty \exp(i\lambda(x-\delta)^2)\,dx \quad \text{exists and has the value}$$

$$\int_{-\delta}^0 \exp(i\lambda x^2)\,dx + \frac{e^{i\pi/4}}{2}\sqrt{\frac{\pi}{\lambda}}. \tag{11.1.27}$$

Expressions (11.1.25), (11.1.27) taken together give

Corollary 11.1. *For $\delta \in \mathbb{R}$, $\lambda > 0$*

$$\int_{-\infty}^\infty \exp(i\lambda(x-\delta)^2)\,dx \quad \textit{converges}$$

(see (11.1.21)*) and has the value*

$$e^{i\pi/4}\sqrt{\frac{\pi}{\lambda}},$$

which is independent of δ.

By Corollary 11.1 and equation (11.1.20) we obtain

Corollary 11.2. *For $a, b \in \mathbb{R}$, $\lambda_1, \lambda_2 > 0$*

$$\int_{-\infty}^\infty \exp(i[\lambda_1(x-a)^2 + \lambda_2(b-x)^2])\,dx \quad \textit{converges}$$

(see (11.1.21)*) and has the value*

$$\exp\left(i\frac{\lambda_1\lambda_2(a-b)^2}{\lambda_1+\lambda_2}\right)e^{i\pi/4}\sqrt{\frac{\pi}{\lambda_1+\lambda_2}}.$$

Proposition 11.2. *For $a, b \in \mathbb{R}$, $a > 0$, $c \in \mathbb{C}$*

$$\int_{-\infty}^\infty \exp(i[ax^2+bx+c])\,dx \quad \textit{converges}$$

(see (11.1.21)*) and has the value*

$$e^{i\pi/4}\sqrt{\frac{\pi}{a}}\exp\left(-i\left(\frac{b^2}{4a}-c\right)\right).$$

Proof. Define $q(x) = -ax^2 - bx - c$ for $x \in \mathbb{R}$. By Proposition 11.1 $q(x) = q(x^*) - a(x-x^*)^2$ for $x^* = -b/2a$, where $q(x^*) = b^2/4a - c$. Then $i(ax^2+bx+c) = -iq(x) = -i[q(x^*) - a(x-x^*)^2] \Longrightarrow$

$$\int_{-\infty}^\infty \exp(i[ax^2+bx+c])\,dx = \exp(-iq(x^*))\int_{-\infty}^\infty \exp(ia(x-x^*)^2)\,dx. \tag{11.1.28}$$

Hence Proposition 11.2 follows by Corollary 11.1. □

The next theorem generalizes Corollary 11.2.

Theorem 11.2. *For $a, b, \mu \in \mathbb{R}$, $\lambda_1, \lambda_2 > 0$,*

$$\int_{-\infty}^{\infty} \exp(i[\lambda_1(x-a)^2 + \lambda_2(b-x)^2 + \mu x])\, dx \quad \textit{converges}$$

(see (11.1.21)*) and has the value*

$$e^{i\pi/4}\sqrt{\frac{\pi}{\lambda_1+\lambda_2}}\exp\left(i\left[\frac{-\mu^2}{4(\lambda_1+\lambda_2)} + \frac{\mu(\lambda_1 a + \lambda_2 b) + \lambda_1\lambda_2(a-b)^2}{\lambda_1+\lambda_2}\right]\right).$$

Proof. $\lambda_1(x-a)^2+\lambda_2(b-x)^2+\mu x = (\lambda_1+\lambda_2)x^2+(\mu-2\lambda_1 a-2\lambda_2 b)x+\lambda_1 a^2+\lambda_2 b^2$. Therefore by Proposition 11.2

$$\int_{-\infty}^{\infty} \exp(i[\lambda_1(x-a)^2 + \lambda_2(b-x)^2 + \mu x])\, dx \quad \text{converges}$$

and has the value

$$e^{i\pi/4}\sqrt{\frac{\pi}{\lambda_1+\lambda_2}}\exp\left(-i\left[\frac{(\mu-2\lambda_1 a-2\lambda_2 b)^2}{4(\lambda_1+\lambda_2)} - (\lambda_1 a^2 + \lambda_2 b^2)\right]\right),$$

which is the same value claimed in the theorem. □

Corollary 11.3. *For $\lambda > 0$, $a, b, \mu \in \mathbb{R}$*

$$\int_{-\infty}^{\infty} \exp(i\{\lambda[(x-a)^2 + (b-x)^2] + \mu x\})\, dx \quad \textit{converges}$$

and has the value

$$e^{i\pi/4}\sqrt{\frac{\pi}{2\lambda}}\exp\left(i\left[\frac{-\mu^2}{8\lambda} + \frac{\mu(a+b)}{2} + \frac{\lambda}{2}(a-b)^2\right]\right).$$

For $a, b, \mu \in \mathbb{R}$, $\lambda > 0$ we look at $J_2 \stackrel{\text{def}}{=}$

$$\int_{-\infty}^{\infty}\int_{-\infty}^{\infty} \exp\left(i\left\{\lambda\left[(x_1-a)^2 + (x_2-x_1)^2 + (b-x_2)^2\right] + \mu(x_1+x_2)\right\}\right) dx_1 dx_2$$

$$\begin{aligned} &= \int_{-\infty}^{\infty} \exp(i[\lambda(b-x_2)^2 + \mu x_2]) \\ &\quad \cdot \left[\int_{-\infty}^{\infty} \exp(i\{\lambda[(x_1-a)^2 + (x_2-x_1)^2] + \mu x_1\})\, dx_1\right] dx_2 \qquad (11.1.29) \\ &= \int_{-\infty}^{\infty} \exp(i[\lambda(b-x_2)^2 + \mu x_2]) e^{i\pi/4} \end{aligned}$$

$$\cdot \sqrt{\frac{\pi}{2\lambda}} \exp\left(i\left[-\frac{\mu^2}{8\lambda} + \frac{\mu(a+x_2)}{2} + \frac{\lambda}{2}(a-x_2)^2\right]\right) dx_2$$

by Corollary 11.3. The latter integral is

$$e^{i\pi/4}\sqrt{\frac{\pi}{2\lambda}} \exp\left(i\left(-\frac{\mu^2}{8\lambda}\right)\right) e^{i\mu a/2}$$
$$\cdot \int_{-\infty}^{\infty} \exp\left(i\left\{\lambda(b-x_2)^2 + \frac{\lambda}{2}(a-x_2)^2 + \frac{3\mu}{2}x_2\right\}\right) dx_2$$
$$= e^{i\pi/4}\sqrt{\frac{\pi}{2\lambda}} \exp\left(\frac{-i\mu^2}{8\lambda}\right) e^{i\mu a/2} e^{i\pi/4}\sqrt{\frac{\pi}{\lambda+\lambda/2}}$$
$$\cdot \exp\left(i\left[\frac{-(3\mu/2)^2}{4(\lambda+\lambda/2)} + \frac{\frac{3}{2}\mu(\frac{\lambda}{2}a+\lambda b) + \frac{\lambda}{2}\lambda(a-b)^2}{\lambda+\lambda/2}\right]\right)$$

(by Theorem 11.2). That is,

$$J_2 = e^{i2\pi/4}\sqrt{\frac{\pi^2}{3\lambda^2}} \exp\left(i\left\{\frac{-\mu^2}{2\lambda} + \frac{2}{2}\mu(a+b) + \frac{\lambda}{3}(a-b)^2\right\}\right), \tag{11.1.30}$$

which is a special case of the following more general result.

Theorem 11.3. *For $\lambda > 0$, $a, b, \mu \in \mathbb{R}$, $n \geq 1$ an integer the integral*

$$J_n = \int_{-\infty}^{\infty} \cdots \int_{-\infty}^{\infty} \exp\left(i\left\{\lambda\left[(x_1-a)^2 + (x_2-x_1)^2 + (x_3-x_2)^2\right.\right.\right.$$
$$\left.\left.\left. + \cdots + (x_n - x_{n-1})^2 + (b-x_n)^2\right] + \mu(x_1 + \cdots + x_n)\right\}\right)\, dx_1 dx_2 \cdots dx_n$$

converges (see (11.1.21)*) and has the value*

$$e^{in\pi/4}\sqrt{\frac{\pi^n}{(n+1)\lambda^n}} \exp\left(i\left\{\frac{-n(n+1)(n+2)}{48\lambda}\mu^2 + \frac{n\mu(a+b)}{2} + \frac{\lambda(a-b)^2}{n+1}\right\}\right),$$

where for $n = 1$ we take x_{n-1} to mean a.

Proof. The result is established for $n = 1$ in Corollary 11.3, and for $n = 2$ in (11.1.30). We proceed by induction:

$$J_{n+1} = \int_{-\infty}^{\infty} \cdots \int_{-\infty}^{\infty} \exp(i\{\lambda[(x_1-a)^2 + (x_2-x_1)^2$$
$$+ \cdots + (x_n - x_{n-1})^2 + (x_{n+1} - x_n)^2 + (b - x_{n+1})^2]$$
$$+ \mu(x_1 + \cdots x_n + x_{n+1})\})\, dx_1 \cdots dx_n dx_{n+1}$$

$$= \int_{-\infty}^{\infty} \exp(i\{\lambda(b-x_{n+1})^2 + \mu x_{n+1}\})$$
$$\cdot \left[\int_{-\infty}^{\infty} \cdots \int_{-\infty}^{\infty} \exp(i\{\lambda[(x_1-a)^2 + (x_2-x_1)^2\right.$$
$$\left. + \cdots + (x_n - x_{n-1})^2 + (x_{n+1}-x_n)^2] + \mu(x_1 + \cdots x_n)\}) \, dx_1 \cdots dx_n\right] dx_{n+1}$$
$$= \int_{-\infty}^{\infty} \exp(i\{\lambda(b-x_{n+1})^2 + \mu x_{n+1}\}) e^{in\pi/4} \sqrt{\frac{\pi^n}{(n+1)\lambda^n}}$$
$$\cdot \exp\left(i\left\{\frac{-n(n+1)(n+2)\mu^2}{48\lambda} + \frac{n}{2}\mu(a + x_{n+1}) + \frac{\lambda(a-x_{n+1})^2}{n+1}\right\}\right) dx_{n+1} \tag{11.1.31}$$

(by induction)

$$= e^{in\pi/4} \sqrt{\frac{\pi^n}{(n+1)\lambda^n}} \exp\left(\frac{-in(n+1)(n+2)\mu^2}{48\lambda}\right) e^{in\mu a/2}$$
$$\int_{-\infty}^{\infty} \exp\left(i\left\{\frac{\lambda}{n+1}(a-x_{n+1})^2 + \lambda(b-x_{n+1})^2 + \mu\left(1+\frac{n}{2}\right)x_{n+1}\right\}\right) dx_{n+1}$$
$$= e^{in\pi/4} \sqrt{\frac{\pi^n}{(n+1)\lambda^n}} \exp\left(\frac{-in(n+1)(n+2)\mu^2}{48\lambda}\right) e^{in\mu a/2} \quad \text{(by Theorem 11.2)}$$
$$e^{i\pi/4} \sqrt{\frac{\pi}{\frac{\lambda}{n+1} + \lambda}} \exp\left(i\left[\frac{-\mu^2(1+n/2)^2}{4(\frac{\lambda}{n+1} + \lambda)}\right.\right.$$
$$\left.\left. + \frac{\mu(1+n/2)(\frac{\lambda a}{n+1} + \lambda b)}{\frac{\lambda}{n+1} + \lambda} + \frac{\frac{\lambda}{n+1}\lambda(a-b)^2}{\frac{\lambda}{n+1} + \lambda}\right]\right)$$
$$= e^{i(n+1)\pi/4} \sqrt{\frac{\pi^{n+1}}{(n+2)\lambda^{n+1}}} \exp\left(i\left\{\frac{-n(n+1)(n+2)\mu^2}{48\lambda} - \frac{\mu^2(n+2)^2}{4\cdot 4\left(\frac{n+2}{n+1}\right)\lambda}\right.\right.$$
$$\left.\left. + \frac{\mu(n+2)\lambda(\frac{a}{n+1} + b)}{2\left(\frac{n+2}{n+1}\right)\lambda} + \frac{n\mu a}{2} + \frac{\lambda^2(a-b)^2}{(n+1)\left(\frac{n+2}{n+1}\right)}\right\}\right)$$
$$= a_n \exp\left(i\left\{-(n+1)(n+2)\frac{\mu^2}{\lambda}\left(\frac{n}{48} + \frac{1}{16}\right)\right.\right.$$
$$\left.\left. + \frac{\mu}{2}(n+1)(a+b) + \frac{\lambda^2(a-b)^2}{n+2}\right\}\right)$$
$$= a_n \exp\left(i\left\{\frac{-(n+1)(n+2)(n+3)\mu^2}{48\lambda} + \frac{\mu}{2}(n+1)(a+b) + \frac{\lambda^2(a-b)^2}{n+2}\right\}\right),$$

where

$$a_n = e^{i(n+1)\pi/4} \sqrt{\frac{\pi^{n+1}}{(n+2)\lambda^{n+1}}}, \tag{11.1.32}$$

which completes the induction. □

11.2 Feynman Path Integrals

In the year 1947 Richard Feynman completed a new formulation of quantum mechanics by means of his path integral. Seeds for this equivalent formulation were sown and developed in his 1942 Princeton thesis [31]. The original paper on the path integral, the 1948 paper [32], is considered by some to be his finest work. Feynman extended his path integral methods to cover quantum electrodynamics, quantum statistical mechanics, and other areas [33].

The path integral $I = I(t_a, t_b, x_a, x_b)$ in words is a "sum over histories"—a sum over all paths $x(t)$ of a particle from $x_a = x(t_a)$ to $x_b = x(t_b)$ for a time interval $[t_a, t_b]$, where each path is weighted by $e^{i S_c(x(t))/\hbar}$, $S_c(x(t))$ being the classical action for the path. A more analytic expression of such "words" will shortly follow. A quantum system can be equally well described by specifying its quantum Hamiltonian H (as we have done heretofore), or by specifying I. Physically, the modulus squared $|I|^2$ determines the probability of arrival of a particle at position x_b and time t_b from its initial position x_a at time t_a. Our purpose here is not to develope the path integral approach to quantum mechanics. We want to compute I in three concrete cases and then consider in Chapter 16 a more complicated type of path integral (a *Hawking* path integral [45]), whose meaning we should like to clarify using an appropriate zeta function. Applications to Kaluza–Klein space-times will be given. It should be pointed out that in contrast to *Wiener* integrals (or *Euclidean* path integrals which we take a look at in Chapter 13), Feynman integrals are usually plagued by difficult issues of mathematical rigor regarding their meaning—issues that we shall not attempt to resolve here.

In Schrödinger's formulation of quantum mechanics the underlying classical mechanics is Hamiltonian. Feynman's formulation has underlying ties to Lagrangian classical mechanics. We begin our discussion therefore of action integrals for a path; compare definition (2.7.20).

Consider a particle of mass m under the influence of a potential V, say in one dimension for simplicity. The motion during a time interval t_a to t_b ($t_a < t_b$) is represented by a path $x(t)$ from $x_a \stackrel{\text{def}}{=} x(t_a)$ to $x_b \stackrel{\text{def}}{=} x(t_b)$ where $x = x(t)$ is the equation of motion.

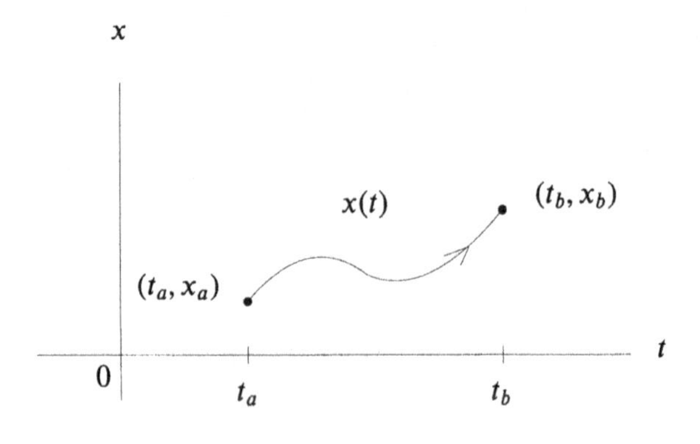

Figure 11.3: Path of particle

The *action integral* $S(x(t))$ for the path $x(t)$ is defined by

$$S(x(t)) \stackrel{\text{def}}{=} \int_{t_a}^{t_b} \left[\frac{m}{2}x'(t)^2 - V(x(t))\right] dt \tag{11.2.1}$$

where the integrand corresponds to the Lagrangian L of (2.7.1). One could think of $S(x(t))$ as evaluation at $x(t)$ of an action functional S on a space of paths. We shall have interest, in particular, in the case when $x(t)$ is a straight line. For a free particle, for example, $V = 0$ and Newton's equation $0 = ma = m\frac{d^2x}{dt^2} \Longrightarrow x = x(t) = ct + d$ is the equation of motion—a straight line. Here $x_a = ct_a + d$, $x_b = ct_b + d \Longrightarrow$

$$d = \frac{t_b x_a - t_a x_b}{t_b - t_a}, \qquad c = \frac{x_b - x_a}{t_b - t_a} \tag{11.2.2}$$

$$\Longrightarrow x(t) = \left(\frac{x_b - x_a}{t_b - t_a}\right)t + \frac{t_b x_a - t_a x_b}{t_b - t_a} \tag{11.2.3}$$

is the straight line path from x_a to x_b. For such a path the action integral in (11.2.1) is

$$S(x(t)) = \frac{m}{2}\frac{(x_b - x_a)^2}{t_b - t_a} - \int_{t_a}^{t_b} V(x(t))\, dt \tag{11.2.4}$$

where the second term in (11.2.4) is *zero* for a free particle.

The *Feynman path integral*, which we denote by

$$I(t_a, t_b, x_a, x_b) = \int_{x_a}^{x_b} \exp\left(\frac{i}{\hbar}S(x(t))\right) D[x(t)], \tag{11.2.5}$$

(the dependence on V being suppressed) is formally an integral over the space of all continuous paths $x(t)$ from x_a to x_b with each such path assigned a weight $\exp(\frac{i}{\hbar}S(x(t)))$. Here (t_a, x_a), (t_b, x_b) are now fixed points in $\mathbb{R}^2$ with $t_a < t_b$. As there is no suitable measure $D[x(t)]$ on the space of continuous paths we attempt to provide $I(t_a, t_b, x_a, x_b)$ a meaning by first taking polygonal paths (which may be considered as discrete approximations of continuous paths) and then by taking a suitable limit. We proceed as follows. Given an integer $n \geq 2$ choose a partition $P_n = \{t_o, t_1, t_2, \ldots, t_n\}$ of $[t_a, t_b]$ with each subinterval $[t_j, t_{j+1}]$, $0 \leq j \leq n-1$, of equal length $\epsilon = \epsilon_n \stackrel{\text{def}}{=} (t_b - t_a)/n$: $t_o = t_a$, $t_1 = t_a + \epsilon, \ldots, t_j = t_a + j\epsilon$, $0 \leq j \leq n$; $t_n = t_b$. Select real numbers x_j for each j, $0 \leq j \leq n$, such that $x_o = x_a$, $x_n = x_b$, and connect (t_j, x_j) to (t_{j+1}, x_{j+1}) by a straight line path $x_j(t)$ to obtain a polygonal path $x^{(n)}(t)$ from (t_a, x_a) to (t_b, x_b) as illustrated in Figure 11.4.

As

$$x_j(t) = x_j + \frac{x_{j+1} - x_j}{t_{j+1} - t_j}(t - t_j) \tag{11.2.6}$$

the action integral of $x_j(t)$ is

$$S(x_j(t)) = \frac{m}{2}\frac{(x_{j+1} - x_j)^2}{t_{j+1} - t_j} - \int_{t_j}^{t_{j+1}} V(x_j(t))\, dt, \tag{11.2.7}$$

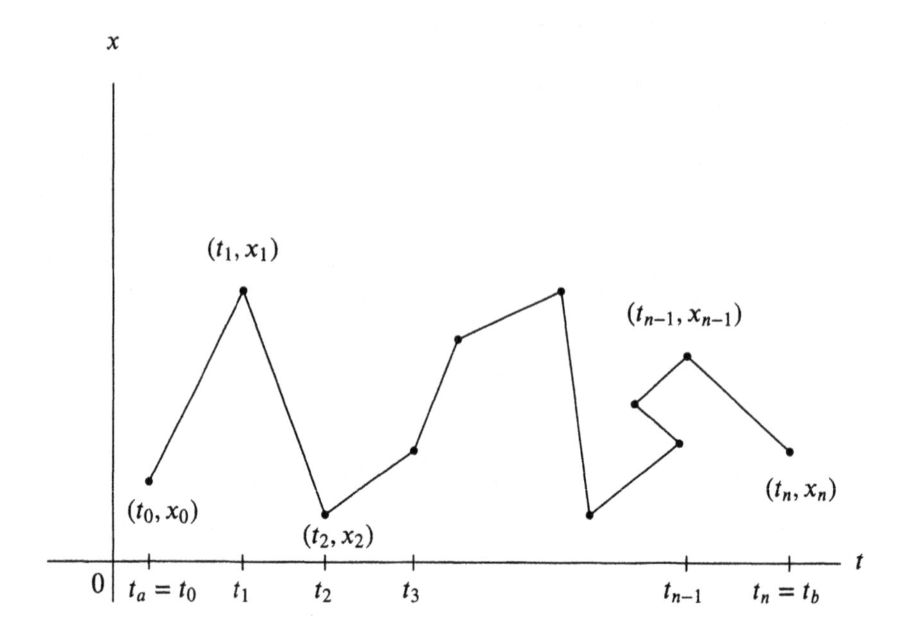

Figure 11.4: A polygonal path

similar to (11.2.4). For the full polygonal path $x^{(n)}(t)$

$$
\begin{aligned}
S(x^{(n)}(t)) &= \sum_{j=0}^{n-1} S(x_j(t)) \\
&= \frac{m}{2} \sum_{j=0}^{n-1} \frac{(x_{j+1} - x_j)^2}{t_{j+1} - t_j} - \sum_{j=0}^{n-1} \int_{t_j}^{t_{j+1}} V(x_j(t))\, dt.
\end{aligned}
\tag{11.2.8}
$$

Assuming that V is continuous, the mean value theorem for integrals gives

$$
\int_{t_j}^{t_{j+1}} V(x_j(t))\, dt = V(x_j(\xi_j))(t_{j+1} - t_j) \quad \text{for some } \xi_j \in (t_j, t_{j+1}). \tag{11.2.9}
$$

For large n, $t_{j+1} - t_j = \epsilon_n = (t_b - t_a)/n$ is small and consequently $V(x_j(\xi_j))$ is reasonably approximated by $V(x_j(t_{j+1})) = V(x_{j+1})$: in (11.2.9)

$$
\int_{t_j}^{t_{j+1}} V(x_j(t))\, dt \simeq V(x_{j+1})(t_{j+1} - t_j) \quad \text{for large } n. \tag{11.2.10}
$$

Of course one could consider other types of approximations as well. Given (11.2.8), (11.2.10),

$$
S(x^{(n)}(t)) \simeq \sum_{j=0}^{n-1} \left[\frac{m}{2\epsilon_n}(x_{j+1} - x_j)^2 - \epsilon_n V(x_{j+1}) \right]. \tag{11.2.11}
$$

At this point, engaging in a bit of pseudo-mathematics perhaps, we consider the right-hand side of (11.2.11) as a function of $n-1$ variables $x_1, x_2 \ldots, x_{n-1}$. Denote this function by $S(x^{(n)}(t))$ and form the Fresnel-type multiple integral

$$I_n(t_a, t_b, x_a, x_b) \stackrel{\text{def}}{=} c_n \int_{-\infty}^{\infty} \cdots \int_{-\infty}^{\infty} \exp\left(\frac{i}{\hbar} S(x^{(n)}(t))\right) dx_1 dx_2 \cdots dx_{n-1}, \tag{11.2.12}$$

for some suitable constant $c_n \neq 0$. Now define the Feynman path integral $I(t_a, t_b, x_a, x_b)$ in (11.2.5) by

$$I(t_a, t_b, x_a, x_b) \stackrel{\text{def}}{=} \lim_{n\to\infty} I_n(t_a, t_b, x_a, x_b), \tag{11.2.13}$$

provided the limit exists.

Let's analyze this definition in the case of a free particle: $V = 0 \implies$

$$\frac{iS(x^{(n)}(t))}{\hbar} \stackrel{\text{def}}{=} \frac{mi}{2\epsilon_n \hbar} \left[(x_1 - x_a)^2 + (x_2 - x_1)^2 + \cdots + \right.$$
$$\left. +(x_{n-1} - x_{n-2})^2 + (x_b - x_{n-1})^2\right]. \tag{11.2.14}$$

In Theorem 11.3 choose $\mu = 0$, $\lambda = \frac{m}{2\epsilon_n \hbar}$ and conclude from (11.2.12), (11.2.14) that

$$I_n(t_a, t_b, x_a, x_b)$$
$$= c_n \exp\left(\frac{i(n-1)\pi}{4}\right) \sqrt{\frac{\pi^{n-1}}{n}} \left[\frac{2\epsilon_n \hbar}{m}\right]^{\frac{n-1}{2}} \exp\left(i\frac{m}{2\epsilon_n \hbar n}(x_b - x_a)^2\right) \tag{11.2.15}$$
$$= c_n \left[\frac{2\pi i\hbar(t_b - t_a)}{nm}\right]^{n/2} \left[\frac{m}{2\pi i\hbar(t_b - t_a)}\right]^{1/2} \exp\left(\frac{im(x_b - x_a)^2}{2\hbar(t_b - t_a)}\right)$$

by the choice $i^{1/2} = \exp(i\pi/4)$ and the definition $\epsilon_n n = t_b - t_a$. It is now clear that if we make the choice

$$c_n \stackrel{\text{def}}{=} \left[\frac{mn}{2\pi i\hbar(t_b - t_a)}\right]^{n/2}, \tag{11.2.16}$$

for example, then

$$I_n(t_a, t_b, x_a, x_b) = \left[\frac{m}{2\pi i\hbar(t_b - t_a)}\right]^{1/2} \exp\left(\frac{im(x_b - x_a)^2}{2\hbar(t_b - t_a)}\right) \tag{11.2.17}$$

and in particular, by (11.2.13), we obtain

Theorem 11.4. *For the normalization constants c_n in* (11.2.16), *the Feynman path integral $I(t_a, t_b, x_a, x_b)$ in* (11.2.13) *for the free particle of mass m is given by*

$$I(t_a, t_b, x_a, x_b) = \left[\frac{m}{2\pi i\hbar(t_b - t_a)}\right]^{1/2} \exp\left(\frac{im(x_b - x_a)^2}{2\hbar(t_b - t_a)}\right). \tag{11.2.18}$$

Here $i^{1/2} = \exp(i\pi/4)$.

By (11.2.4) we note that equation (11.2.18) can be expressed as

$$I(t_a, t_b, x_a, x_b) = \left[\frac{m}{2\pi i\hbar(t_b - t_a)}\right]^{1/2} \exp\left(\frac{i}{\hbar}S(x(t))\right), \tag{11.2.19}$$

where $S(x(t))$ is the action integral of the classical path $x(t)$ from x_a to x_b of the free particle, with $x(t)$ given by (11.2.3).

To understand better the meaning of Theorem 11.4 consider Schrödinger's time-dependent equation

$$\frac{\partial\psi}{\partial t} = \frac{\hbar i}{2m}\frac{\partial^2\psi}{\partial x^2} \tag{11.2.20}$$

for the free particle. Define a *kernel function* G_o of (x, t), $x \in \mathbb{R}, t > 0$, by

$$G_o(x,t) \stackrel{\text{def}}{=} \frac{\exp(-\frac{mx^2}{2ti\hbar})}{\sqrt{\frac{2\pi ti\hbar}{m}}} = \frac{\exp(-\frac{x^2}{4t\alpha})}{\sqrt{4\pi t\alpha}} \quad \text{for } \alpha \stackrel{\text{def}}{=} \frac{i\hbar}{2m}, i^{1/2} = e^{i\pi/4}. \tag{11.2.21}$$

By direct computation:

Proposition 11.3. *G_o is a solution of Schrödinger's equation in* (11.2.20).

We see that the function G_o is the analogue of the *heat kernel*

$$H_o(x,t) \stackrel{\text{def}}{=} \frac{\exp(-\frac{x^2}{4t\alpha})}{\sqrt{4\pi t\alpha}}, \quad {\scriptstyle x\in\mathbb{R},\ t>0,\ \alpha>0} \tag{11.2.22}$$

which solves the *heat equation*

$$\frac{\partial\psi}{\partial t} = \alpha\frac{\partial^2\psi}{\partial x^2}. \tag{11.2.23}$$

Now define a *Green's function* G by

$$\begin{aligned} G(x,t;y,s) &\stackrel{\text{def}}{=} G_o(x-y, t-s) \quad \text{for } x, y, t, s \in \mathbb{R}, t > s \\ &\stackrel{\text{def}}{=} \left[\frac{m}{2\pi i\hbar(t-s)}\right]^{1/2} \exp\left(\frac{im(x-y)^2}{2\hbar(t-s)}\right). \end{aligned} \tag{11.2.24}$$

Then by (11.2.18) we see that for the free particle

$$I(t_a, t_b, x_a, x_b) \equiv G(x_b, t_b; x_a, t_a), \tag{11.2.25}$$

which relates the Feynman integral to a solution of Schrödinger's equation. A key point about the Feynman integral I (i.e., the Green's function G) is that (as can be shown) it relates a wave function at two different times:

$$\psi(x_b, t_b) = \int_{\mathbb{R}} G(x_b, t_b; x, t_a)\psi(x, t_a)\, dx, \tag{11.2.26}$$

which is *Huygens' principle*.

As a second example, a case with $V \neq 0$, choose $V(x) = mgx$, where g is a constant:

$$\begin{aligned}\frac{iS(x^{(n)}(t))}{\hbar} \\ &\overset{\text{def}}{=} \frac{mi}{\hbar 2\epsilon_n}\left[(x_1 - x_a)^2 + (x_2 - x_1)^2 + \cdots + (x_{n-1} - x_{n-2})^2 + (x_b - x_{n-1})^2\right] \\ &\quad - \frac{i}{\hbar}\epsilon_n mg(x_1 + \cdots + x_{n-1}) - \frac{i}{\hbar}\epsilon_n mgx_b.\end{aligned} \tag{11.2.27}$$

By definition (11.2.12) and Theorem 11.3,

$$\begin{aligned}I_n(t_a, t_b, x_a, x_b) \\ &= c_n \exp\left(-\frac{i}{\hbar}\epsilon_n mgx_b\right) \cdot \exp\left(\frac{i(n-1)\pi}{4}\right)\sqrt{\frac{\pi^{n-1}}{n\lambda^{n-1}}} \\ &\quad \cdot \exp\left(i\left\{-\frac{(n-1)n(n+1)}{48\lambda}\mu^2 + \frac{(n-1)\mu}{2}(x_a + x_b) + \frac{\lambda(x_b - x_a)^2}{n}\right\}\right) \\ &\quad \text{for } \lambda \overset{\text{def}}{=} \frac{m}{2\hbar\epsilon_n}, \quad \mu \overset{\text{def}}{=} -\frac{\epsilon_n mg}{\hbar}.\end{aligned} \tag{11.2.28}$$

Again use that $\epsilon_n n \overset{\text{def}}{=} t_b - t_a$:

$$\begin{aligned}-\frac{(n-1)n(n+1)\mu^2}{48\lambda} &= -\frac{(n^2-1)2\hbar\epsilon_n^2 m^2 g^2}{48m\hbar^2}(t_b - t_a) \\ &= -\frac{(n^2-1)\epsilon_n^2 mg^2(t_b - t_a)}{24\hbar} \\ &= -\frac{n^2\epsilon_n^2 mg^2(t_b - t_a)}{24\hbar} + \frac{\epsilon_n^2 mg^2(t_b - t_a)}{24\hbar} \\ &= -\frac{mg^2(t_b - t_a)^3}{24\hbar} + \frac{\epsilon_n^2 mg^2(t_b - t_a)}{24\hbar}.\end{aligned} \tag{11.2.29}$$

Also

$$\begin{aligned}\frac{(n-1)\mu}{2}(x_a + x_b) &\overset{\text{def}}{=} -\frac{(n-1)\epsilon_n mg}{2\hbar}(x_a + x_b) \\ &= -(t_b - t_a)\frac{mg}{2\hbar}(x_a + x_b) + \frac{\epsilon_n mg(x_a + x_b)}{2\hbar},\end{aligned} \tag{11.2.30}$$

and

$$\frac{\lambda}{n}(x_b - x_a)^2 \overset{\text{def}}{=} \frac{m}{2\hbar n\epsilon_n}(x_b - x_a)^2 = \frac{m(x_b - x_a)^2}{2\hbar(t_b - t_a)}. \tag{11.2.31}$$

Thus by (11.2.29), (11.2.30), and (11.2.31)

$$\exp\left(i\left\{-\frac{(n-1)n(n+1)\mu^2}{48\lambda}+\frac{(n-1)\mu}{2}(x_a+x_b)+\frac{\lambda(x_b-x_a)^2}{n}\right\}\right)$$
$$=\exp\left(i\left\{-\frac{mg^2(t_b-t_a)^3}{24\hbar}-(t_b-t_a)\frac{mg(x_a+x_b)}{2\hbar}+\frac{m(x_b-x_a)^2}{2\hbar(t_b-t_a)}\right\}\right)$$
$$\cdot\exp\left(i\left\{\frac{\epsilon_n^2 mg^2(t_b-t_a)}{24\hbar}+\frac{\epsilon_n mg(x_a+x_b)}{2\hbar}\right\}\right). \tag{11.2.32}$$

By definition of λ, ϵ_n

$$\exp\left(i\frac{(n-1)\pi}{4}\right)\sqrt{\frac{\pi^{n-1}}{n\lambda^{n-1}}}=\left[\frac{2\pi i\hbar(t_b-t_a)}{nm}\right]^{n/2}\left[\frac{m}{2\pi i\hbar(t_b-t_a)}\right]^{1/2} \tag{11.2.33}$$

for the choice $i^{1/2}=\exp(i\pi/4)$. For the same c_n defined in (11.2.16) we see by (11.2.28), (11.2.32), (11.2.33) that

$$I_n(t_a,t_b,x_a,x_b)$$
$$=\exp\left(-\frac{i}{\hbar}\epsilon_n mgx_b\right)\left[\frac{m}{2\pi i\hbar(t_b-t_a)}\right]^{1/2}$$
$$\cdot\exp\left(i\left\{-\frac{mg^2(t_b-t_a)^3}{24\hbar}-(t_b-t_a)\frac{mg(x_a+x_b)}{2\hbar}+\frac{m(x_b-x_a)^2}{2\hbar(t_b-t_a)}\right\}\right)$$
$$\cdot\exp\left(i\left\{\frac{\epsilon_n^2 mg^2(t_b-t_a)}{24\hbar}+\frac{\epsilon_n mg(x_a+x_b)}{2\hbar}\right\}\right). \tag{11.2.34}$$

As $n\to\infty$, $\epsilon_n\to 0$. Hence

Theorem 11.5. *For the choice of normalizing constants c_n given by* (11.2.16) *and for the potential $V(x)=mgx$, g a constant (such as the acceleration due to gravity) the Feynman path integral $I(t_a,t_b,x_a,x_b)$ is given by*

$$I(t_a,t_b,x_a,x_b)=\left[\frac{m}{2\pi i\hbar(t_b-t_a)}\right]^{1/2}\cdot\exp\left(\frac{i}{\hbar}\left\{-\frac{mg^2(t_b-t_a)^3}{24}\right.\right.$$
$$\left.\left.-(t_b-t_a)\frac{mg(x_a+x_b)}{2}+\frac{m(x_b-x_a)^2}{2(t_b-t_a)}\right\}\right). \tag{11.2.35}$$

Again we take $i^{1/2}=\exp(i\pi/4)$.

Suppose g is the acceleration due to gravity so that, as discussed in Chapter 2, a freely falling object of mass m experiences a force F of magnitude $-mg$. Label the vertical axis now by x. With F directed toward the origin $V(x)=mgx$ in

Theorem 11.5 is the corresponding potential: $-V'(x) = -mg$. The equation of motion is $-mg = m\frac{d^2x}{dt^2}$, or

$$x(t) = -\frac{gt^2}{2} + At + B. \tag{11.2.36}$$

Then for $x_a = x(t_a)$, $x_b = x(t_b)$

$$x_b - x_a = -\frac{g}{2}(t_b^2 - t_a^2) + A(t_b - t_a) \tag{11.2.37}$$

$$\Rightarrow A = \frac{x_b - x_a}{t_b - t_a} + \frac{g}{2}(t_b - t_a). \tag{11.2.38}$$

Similarly

$$t_b x_a = -\frac{gt_a^2 t_b}{2} + At_a t_b + Bt_b \text{ and}$$

$$t_a x_b = -\frac{gt_b^2 t_a}{2} + At_b t_a + Bt_a \tag{11.2.39}$$

$$\Rightarrow B = \frac{t_a x_b - t_b x_a}{t_a - t_b} - \frac{gt_a t_b}{2(t_a - t_b)}. \tag{11.2.40}$$

By definition (11.2.1)

$$\begin{aligned} S(x(t)) &= \int_{t_a}^{t_b} \left[\frac{m}{2}x'(t)^2 - V(x(t))\right] dt \\ &= \int_{t_a}^{t_b} \left[\frac{m}{2}(-gt + A)^2 - mg(-\frac{gt^2}{2} + At + B)\right] dt \\ &= \frac{mg^2}{3}(t_b^3 - t_a^3) - mgA(t_b^2 - t_a^2) + \left(\frac{mA^2}{2} - mgB\right)(t_b - t_a). \end{aligned} \tag{11.2.41}$$

Using the values of A, B given in (11.2.38), (11.2.40) one obtains, after some messy algebra,

Proposition 11.4. *For the potential $V(x) = mgx$ with g the acceleration due to gravity the classical path $x(t)$ from x_a to x_b is given by* (11.2.36) *with A, B given by* (11.2.38), (11.2.40), *and the corresponding action integral $S(x(t))$ of* (11.2.41) *is given by*

$$S(x(t)) = \frac{m}{2}\frac{(x_b - x_a)^2}{(t_b - t_a)} - \frac{mg}{2}(x_b + x_a)(t_b - t_a) - \frac{mg^2(t_b - t_a)^3}{24}. \tag{11.2.42}$$

Given (11.2.42) and Theorem 11.5 we can express the path integral $I(t_a, t_b, x_a, x_b)$ in (11.2.35) by

$$I(t_a, t_b, x_a, x_b) = \left[\frac{m}{2\pi i\hbar(t_b - t_a)}\right]^{1/2} \exp\left(\frac{i}{\hbar}S(x(t))\right). \tag{11.2.43}$$

This result is similar to the result (11.2.19) obtained for the free particle.

Appendix 11A Gaussian Integrals

We shall also need the following Gaussian version of Theorem 11.3.

Theorem 11.6. *For* $n = 1, 2, 3, \ldots, a, b \in \mathbb{R}$, $\lambda > 0$

$$I_n \stackrel{\text{def}}{=} \int_{-\infty}^{\infty} \cdots \int_{-\infty}^{\infty} e^{-\lambda[(x_1-a)^2+(x_2-x_1)^2+(x_3-x_2)^2+\cdots(x_n-x_{n-1})^2+(b-x_n)^2]} dx_1 dx_2 \cdots dx_n$$

converges and has the value $\sqrt{\frac{\pi^n}{(n+1)\lambda^n}}\, e^{\frac{-\lambda(a-b)^2}{n+1}}$. *For* $n = 1$, $x_{n-1} \stackrel{\text{def}}{=} a$.

Proof. The starting point is the fact that

$$\int_{-\infty}^{\infty} e^{-\lambda x^2} dx = \sqrt{\frac{\pi}{\lambda}}, \text{ or } \int_{-\infty}^{\infty} e^{-\lambda(x-\alpha)^2} dx = \sqrt{\frac{\pi}{\lambda}} \tag{11.A.1}$$

for $\alpha \in \mathbb{R}$, by the translational invariance of Lebesgue measure dx; cf. Corollary 11.1. By equations (11.A.1), (11.1.18), (11.1.19)

$$\lambda_1(x-a)^2 + \lambda_2(b-x)^2 = -q(x^*) + (\lambda_1+\lambda_2)(x-x^*)^2 \tag{11.A.2}$$

$$\Longrightarrow \int_{-\infty}^{\infty} e^{-[\lambda_1(x-a)^2+\lambda_2(b-x)^2]} dx = e^{q(x^*)} \int_{-\infty}^{\infty} e^{-(\lambda_1+\lambda_2)(x-x^*)^2} dx \tag{11.A.3}$$

$$= e^{\frac{-\lambda_1\lambda_2(a-b)^2}{\lambda_1+\lambda_2}} \sqrt{\frac{\pi}{\lambda_1+\lambda_2}}$$

for $\lambda_1, \lambda_2 > 0$. In particular for $\lambda_1 = \lambda_2 = \lambda > 0$

$$I_1 \stackrel{\text{def}}{=} \int_{-\infty}^{\infty} e^{-\lambda[(x-a)^2+(b-x)^2]} dx = \sqrt{\frac{\pi}{2\lambda}} e^{-\frac{\lambda(a-b)^2}{2}}. \tag{11.A.4}$$

That is, Theorem 11.6 is true for $n = 1$; we proceed by induction.

$$I_{n+1} = \int_{-\infty}^{\infty} \cdots \int_{-\infty}^{\infty} e^{-\lambda[(x_1-a)^2+(x_2-x_1)^2+\cdots+(x_n-x_{n-1})^2+(x_{n+1}-x_n)^2+(b-x_{n+1})^2]} dx_1 \cdots dx_n dx_{n+1}$$

$$\stackrel{\text{by induction}}{=} \sqrt{\frac{\pi^n}{(n+1)\lambda^n}} \int_{-\infty}^{\infty} e^{\frac{-\lambda}{n+1}(a-x_{n+1})-\lambda(b-x_{n+1})^2} dx_{n+1}$$

$$= \sqrt{\frac{\pi^n}{(n+1)\lambda^n}} \sqrt{\frac{\pi}{\frac{\lambda}{n+1}+\lambda}} e^{\frac{\frac{-\lambda}{n+1}\lambda(a-b)^2}{\frac{\lambda}{n+1}+\lambda}}, \tag{11.A.5}$$

by (11.A.3). That is,

$$I_{n+1} = \sqrt{\frac{\pi^{n+1}}{(n+1)\lambda^n\left(\frac{n+2}{n+1}\right)\lambda}}\, e^{\frac{-\lambda(a-b)^2}{n+2}} = \sqrt{\frac{\pi^{n+1}}{(n+2)\lambda^{n+1}}} \cdot e^{\frac{-\lambda(a-b)^2}{n+2}}, \tag{11.A.6}$$

which completes the induction. □

The next result generalizes (11.A.1).

Theorem 11.7. *Let A be an $n \times n$ real symmetric matrix with positive eigenvalues, and let $\alpha > 0$. Then*

$$\int_{\mathbb{R}^n} e^{-\alpha x A x^t} dx = \sqrt{\frac{\pi^n}{\alpha^n \det A}} \tag{11.A.7}$$

for Lebesgue measure dx on $\mathbb{R}^n$, $x =$ a row vector.

Proof. As A is symmetric we can choose an orthogonal matrix $B \ni$

$$BAB^{-1} = \begin{bmatrix} \lambda_1 & & 0 \\ & \ddots & \\ 0 & & \lambda_n \end{bmatrix} \tag{11.A.8}$$

where the λ_j are the eigenvalues of A. In general, if $\phi : \mathbb{R}^n \to \mathbb{R}^n$ is a non-singular linear operator on $\mathbb{R}^n$, then

$$\int_{\mathbb{R}^n} f \circ \phi dx = \frac{1}{|\det \phi|} \int_{\mathbb{R}^n} f dx \tag{11.A.9}$$

is a familiar property of Lebesgue measure dx, for any nice function f. Choose

$$\phi(x) = xB^t, f(x) = e^{-\alpha x \begin{bmatrix} \lambda_1 & & 0 \\ & \ddots & \\ 0 & & \lambda_n \end{bmatrix} x^t} = e^{-\alpha \sum_{j=1}^n \lambda_j x_j^2}$$

$$\text{for } x = (x_1, \ldots, x_n) \in \mathbb{R}^n. \tag{11.A.10}$$

Then $f(\phi(x)) = e^{-\alpha x A x^t}$ and $|\det \phi| = |\det B^t| = 1$ by (11.A.10), since B is orthogonal. (11.A.9) gives

$$\begin{aligned} \int_{\mathbb{R}^n} e^{-\alpha x A x^t} dx &= \int_{\mathbb{R}} \cdots \int_{\mathbb{R}} e^{-\alpha \lambda_1 x_1^2} \cdots e^{-\alpha \lambda_n x_n^2} dx_1 \cdots dx_n \\ &= \sqrt{\frac{\pi}{\lambda_1 \alpha}} \cdots \sqrt{\frac{\pi}{\lambda_n \alpha}} = \sqrt{\frac{\pi^n}{\alpha^n \det A}}, \end{aligned} \tag{11.A.11}$$

where we have used (11.A.1). □

We have noted that (11.A.1) is the Gaussian version of Corollary 11.1 where

$$\int_{-\infty}^{\infty} f(x) dx = \lim_{r \to \infty} \int_{-r}^{0} f(x) dx + \lim_{r \to \infty} \int_{0}^{r} f(x) dx \tag{11.A.12}$$

as in (11.1.21):

$$\int_{-\infty}^{\infty} e^{i\lambda x^2} dx = \int_{-\infty}^{\infty} e^{i\lambda (x-\delta)^2} dx = e^{i\pi/4} \sqrt{\frac{\pi}{\lambda}} \tag{11.A.13}$$

for $\lambda > 0$, $\delta \in \mathbb{R}$. Using (11.A.13) in place of (11.A.1) one can deduce that

$$\int_{\mathbb{R}^n} e^{\alpha i x A x^t} dx = e^{\frac{i\pi n}{4}} \sqrt{\frac{\pi^n}{\alpha^n \det A}} \tag{11.A.14}$$

for α, A in Theorem 11.7; $i^{1/2} = e^{i\pi/4}$.

12

Path Integral for the Harmonic Oscillator

So far we have computed path integrals for potentials V which are linear in the variable x: $V(x) = cx$ for some constant c. Consider now a *quadratic potential*: $V(x) = cx^2$. Specifically, consider $V(x) = \omega^2 mx^2/2$ where m is the mass of a simple harmonic oscillator with frequency $\nu = \omega/2\pi$. As discussed in Chapter 2 the equation of motion is

$$x(t) = A\cos(\omega t - \delta) = A\sin(\omega t + \phi), \tag{12.1.1}$$

where A is the amplitude of the oscillation and $\phi = \pi/2 - \delta$. δ is some phase constant. For an interval of time $t_a \leq t \leq t_b$ the action integral of the path (12.1.1) from $x_a = x(t_a)$ to $x_b = x(t_b)$ is given by

$$\begin{aligned} S(x(t)) &\stackrel{\text{def}}{=} \int_{t_a}^{t_b} \left[\frac{m}{2}x'(t)^2 - V(x(t))\right] dt \\ &= \int_{t_a}^{t_b} \left[\frac{m}{2}A^2\omega^2\cos^2(\omega t + \phi) - \frac{\omega^2 m}{2}A^2\sin^2(\omega t + \phi)\right] dt \\ &= \frac{m}{2}A^2\omega^2 \int_{t_a}^{t_b} [\cos^2(\omega t + \phi) - \sin^2(\omega t + \phi)]\, dt \\ &= \frac{mA^2\omega^2}{2} \int_{t_a}^{t_b} \cos 2(\omega t + \phi)\, dt \\ &= \frac{m\omega A^2}{4}[\sin(2\omega t_b + 2\phi) - \sin(2\omega t_a + 2\phi)]. \end{aligned} \tag{12.1.2}$$

On the other hand,

$$\begin{aligned} x_b &= A\sin(\omega t_b + \phi) = A(\sin\omega t_b \cos\phi + \cos\omega t_b \sin\phi), \\ x_a &= A\sin(\omega t_a + \phi) = A(\sin\omega t_a \cos\phi + \cos\omega t_a \sin\phi) \implies \end{aligned} \tag{12.1.3}$$

$$\begin{aligned} x_b\cos\omega t_a - x_a\cos\omega t_b &= (\sin\omega t_b\cos\omega t_a - \sin\omega t_a\cos\omega t_b)A\cos\phi \\ &= [\sin\omega(t_b - t_a)]A\cos\phi, \end{aligned} \tag{12.1.4}$$

and

$$\begin{aligned} x_b\sin\omega t_a - x_a\sin\omega t_b &= (\cos\omega t_b\sin\omega t_a - \cos\omega t_a\sin\omega t_b)A\sin\phi \\ &= [\sin\omega(t_a - t_b)]A\sin\phi \implies \end{aligned} \tag{12.1.5}$$

$$\begin{aligned} A\cos\phi &= \frac{x_b\cos\omega t_a - x_a\cos\omega t_b}{\sin\omega(t_b - t_a)} \quad \text{and} \\ A\sin\phi &= \frac{x_a\sin\omega t_b - x_b\sin\omega t_a}{\sin\omega(t_b - t_a)}, \quad \text{if} \ \sin\omega(t_b - t_a) \neq 0. \end{aligned} \tag{12.1.6}$$

Then using (12.1.3), (12.1.6) and $\sin 2\theta = 2\sin\theta\cos\theta$, one sees that

$$\begin{aligned} A^2\sin(2\omega t_b + 2\phi) &= A^2 2\sin(\omega t_b + \phi)\cos(\omega t_b + \phi) \\ &= 2Ax_b\cos(\omega t_b + \phi) = 2Ax_b[\cos\omega t_b\cos\phi - \sin\omega t_b\sin\phi] \\ &= 2x_b\cos\omega t_b\left[\frac{x_b\cos\omega t_a - x_a\cos\omega t_b}{\sin\omega(t_b - t_a)}\right] \\ &\quad - 2x_b\sin\omega t_b\left[\frac{x_a\sin\omega t_b - x_b\sin\omega t_a}{\sin\omega(t_b - t_a)}\right] \\ &= \frac{2x_b^2\cos\omega t_b\cos\omega t_a - 2x_a x_b + 2x_b^2\sin\omega t_b\sin\omega t_a}{\sin\omega(t_b - t_a)}. \end{aligned} \tag{12.1.7}$$

Similarly

$$\begin{aligned} A^2\sin(2\omega t_a + 2\phi) &= \frac{-2x_a^2\cos\omega t_a\cos\omega t_b + 2x_a x_b - 2x_a^2\sin\omega t_a\sin\omega t_b}{\sin\omega(t_b - t_a)} \\ &\implies A^2[\sin(2\omega t_b + 2\phi) - \sin(2\omega t_a + 2\phi)] \end{aligned} \tag{12.1.8}$$

$$\begin{aligned} &= \frac{2(x_b^2 + x_a^2)\cos\omega t_a\cos\omega t_b - 4x_a x_b + 2(x_b^2 + x_a^2)\sin\omega t_a\sin\omega t_b}{\sin\omega(t_b - t_a)} \\ &= \frac{2(x_b^2 + x_a^2)\cos\omega(t_b - t_a) - 4x_a x_b}{\sin\omega(t_b - t_a)} \quad \text{for} \ \sin\omega(t_b - t_a) \neq 0. \end{aligned} \tag{12.1.9}$$

Given (12.1.2), (12.1.9) we have thus shown

Proposition 12.1. *For the simple harmonic oscillator with frequency $\nu = \omega/2\pi$ the action integral $S(x(t))$ of (12.1.2) for its equation of motion $x(t)$ (see (12.1.1)) is given by*

$$S(x(t)) = \frac{m\omega}{2}\left[(x_b^2 + x_a^2)\cot\omega(t_b - t_a) - \frac{2x_a x_b}{\sin\omega(t_b - t_a)}\right] \tag{12.1.10}$$

for $\sin\omega(t_b - t_a) \neq 0$.

Given equations (11.2.19), (11.2.43) one might naturally wonder whether the path integral for the simple harmonic oscillator also has the form $c(t_a, t_b, m)\exp(\frac{i}{\hbar}S(x(t)))$ for $S(x(t))$ given by Proposition 12.1, where $c(t_a, t_b, m)$ is some constant depending on t_a, t_b, and m, and depending possibly on ω also. This indeed is the case, as was first shown by Feynman:

Theorem 12.1. *For the simple harmonic oscillator one has for* $\sin\omega(t_b - t_a) > 0$

$$I(t_a, t_b, x_a, x_b) = \sqrt{\frac{m\omega}{2\pi i\hbar\sin\omega(t_b - t_a)}}\exp\left(\frac{i}{\hbar}S(x(t))\right) \tag{12.1.11}$$

for $S(x(t))$ given by Proposition 12.1. The normalizing constants c_n in (11.2.12) *are again given by* (11.2.16).

A direct proof of Theorem 12.1 is long and cumbersome. A simpler, somewhat elegant, proof exists which is based on the consideration of the *deviation* $x_d(t)$ of a path $x(t)$ from the *classical path* $x_c(t)$ given by (12.1.1):

$$x_d(t) \overset{\text{def}}{=} x(t) - x_c(t) = x(t) - A\sin(\omega t + \phi) \quad \text{for } t_a \leq t \leq t_b. \tag{12.1.12}$$

As we always consider paths from x_a to x_b we have the boundary conditions

$$x_c(t_a) = x_a = x(t_a), \qquad x_c(t_b) = x_b = x(t_b) \tag{12.1.13}$$

$$\Longrightarrow x_d(t_a) = x_d(t_b) = 0. \tag{12.1.14}$$

Of course $x_c(t)$ satisfies

$$x_c''(t) = -\omega^2 x_c(t). \tag{12.1.15}$$

The next proposition is simple, but is a bit remarkable. From it one also sees why the factor $\exp(\frac{i}{\hbar}S(x(t)))$ occurs in (12.1.11) (which also occurs in (11.2.19), (11.2.43), as has been noted); see (12.1.19) below.

Proposition 12.2.

$$S(x(t)) = S(x_c(t)) + S(x_d(t)). \tag{12.1.16}$$

Proof. By definition

$$
\begin{aligned}
S(x(t)) &= \int_{t_a}^{t_b} \left[\frac{m}{2}x'(t)^2 - V(x(t))\right] dt \\
&= \int_{t_a}^{t_b} \left\{\frac{m}{2}[x_c'(t) + x_d'(t)]^2 - \frac{m\omega^2}{2}[x_c(t) + x_d(t)]^2\right\} dt \\
&= \int_{t_a}^{t_b} \left[\frac{m}{2}x_c'(t)^2 - \frac{m\omega^2}{2}x_c(t)^2\right] dt \\
&\quad + \int_{t_a}^{t_b} \left[\frac{m}{2}x_d'(t)^2 - \frac{m\omega^2}{2}x_d(t)^2\right] dt \\
&\quad + m\int_{t_a}^{t_b} [x_c'(t)x_d'(t) - \omega^2 x_c(t)x_d(t)]\, dt \\
&= S(x_c(t)) + S(x_d(t)) + mI,
\end{aligned}
\tag{12.1.17}
$$

where I denotes the latter integral which we shall see is zero. Namely, integration by parts provides an alternative expression of I:

$$
I = \left[x_c'(t)x_d(t)\right]_{t_a}^{t_b} - \int_{t_a}^{t_b} x_c''(t)x_d(t)\, dt - \int_{t_a}^{t_b} \omega^2 x_c(t)x_d(t)\, dt = 0 \tag{12.1.18}
$$

by the boundary conditions (12.1.14) and the equation of motion (12.1.15). □

Proposition 12.2 means that (as is intuitively clear) when computing the integral in (11.2.5) over paths from x_a to x_b (a sum over paths) one can factor out the contribution $\exp(\frac{i}{\hbar}S(x_c(t)))$ of the classical path and sum over paths of the form $x_d(t)$ with $x_d(t_a) = x_d(t_b) = 0$:

$$
I(t_a, t_b, x_a, x_b) = \exp\left(\frac{i}{\hbar}S(x_c(t))\right) \int_{x(t_a)=0}^{x(t_b)=0} \exp\left(\frac{i}{\hbar}S(x_d(t))\right) D[x(t)], \tag{12.1.19}
$$

which simplifies matters quite a bit. Here we take the integral on the right-hand side of (12.1.19) to mean the limit of $I_n(t_a, t_b, 0, 0)$ as $n \to \infty$, as in (11.2.13). Thus we must consider Fresnel-type multiple integrals in the special case $x_a = x_b = 0$. Again define $\epsilon_n = (t_b - t_a)/n$ for an integer $n \geq 2$. Define the $(n-1) \times (n-1)$ matrix A_n by

$$
A_n = \begin{pmatrix}
2 - \epsilon_n^2\omega^2 & -1 & & & 0 \\
-1 & 2 - \epsilon_n^2\omega^2 & & & \\
& & \ddots & & \\
& & & 2 - \epsilon_n^2\omega^2 & -1 \\
0 & & & -1 & 2 - \epsilon_n^2\omega^2
\end{pmatrix}. \tag{12.1.20}
$$

Recall that $S(x^{(n)}(t))$ defined as a function of $x_1, x_2, \ldots, x_{n-1}$ is given by

$$\begin{aligned} S(x^{(n)}(t)) &= \sum_{j=0}^{n-1}\left[\frac{m}{2\epsilon_n}(x_{j+1}-x_j)^2 - \epsilon_n V(x_{j+1})\right] \\ &= \sum_{j=0}^{n-1}\left[\frac{m}{2\epsilon_n}(x_{j+1}-x_j)^2 - \epsilon_n \frac{m\omega^2}{2}x_{j+1}^2\right] \end{aligned} \tag{12.1.21}$$

with $x_o = x_a, x_n = x_b$.

Proposition 12.3. *Given the boundary conditions $x_a = x_b = 0$ one has*

$$\begin{bmatrix} x_1, \dots, x_{n-1} \end{bmatrix} A_n \begin{bmatrix} x_1 \\ \vdots \\ x_{n-1} \end{bmatrix} = \sum_{j=0}^{n-1}[(x_{j+1}-x_j)^2 - \epsilon_n^2\omega^2 x_{j+1}^2]. \tag{12.1.22}$$

Proof. The left-hand side of the equation is

$$x_1^2(2-\epsilon_n^2\omega^2) - 2x_1x_2 + \cdots + x_{n-2}^2(2-\epsilon_n^2\omega^2) - 2x_{n-2}x_{n-1} + x_{n-1}^2(2-\epsilon_n^2\omega^2). \tag{12.1.23}$$

On the other hand, because $x_o = x_n = 0$ the right-hand side is

$$\begin{aligned} &\sum_{j=0}^{n-1}(x_{j+1}^2 - 2x_jx_{j+1} + x_j^2) - \epsilon_n^2\omega^2\sum_{j=0}^{n-1}x_{j+1}^2 = x_1^2 + x_2^2 + \cdots + x_{n-1}^2 \\ &\quad - 2\sum_{j=1}^{n-2}x_jx_{j+1} + x_1^2 + x_2^2 + \cdots + x_{n-1}^2 - \epsilon_n^2\omega^2(x_1^2 + x_2^2 + \cdots + x_{n-1}^2), \end{aligned} \tag{12.1.24}$$

which is the left-hand side, as desired. □

By (12.1.21), (12.1.22) we can write

$$S(x^{(n)}(t)) = \frac{m}{2\epsilon_n}xA_nx^t \text{ for } x = (x_1, \dots, x_{n-1}) \tag{12.1.25}$$

which puts us in position to apply (11.A.14) in Appendix 11A, where we take α there equal to $m/(2\epsilon_n\hbar)$; note that A_n is indeed symmetric; moreover we will have $\det A_n \neq 0$ at least for n sufficiently large:

$$\begin{aligned} I_n(t_a, t_b, 0, 0) &= c_n \exp\left(\frac{i\pi(n-1)}{4}\right)\left[\frac{\pi 2\epsilon_n\hbar}{m}\right]^{(n-1)/2}[\det A_n]^{-1/2} \\ &= \left[\frac{m}{2\pi\epsilon_n\hbar i}\right]^{1/2}[\det A_n]^{-1/2} \end{aligned} \tag{12.1.26}$$

for c_n given as usual. That is,

$$c_n \stackrel{\text{def}}{=} \left[\frac{mn}{2\pi i\hbar(t_b - t_a)}\right]^{n/2}. \tag{12.1.27}$$

To complete the proof of Theorem 12.1 we appeal to the following proposition.

Proposition 12.4. *For a real number c with $-1 < c < 1$ form the $n \times n$ matrix*

$$M_n(c) = \begin{pmatrix} 2c & -1 & & & \\ -1 & 2c & & 0 & \\ & & \ddots & & \\ & 0 & & 2c & -1 \\ & & & -1 & 2c \end{pmatrix}. \tag{12.1.28}$$

Write $c = \cos u$. Then we can write

$$\det M_n(c) = \frac{\sin(n+1)u}{\sin u}. \tag{12.1.29}$$

We assume this result for now. Noting that by definition (12.1.20) we can write $A_n = M_{n-1}(1 - \epsilon_n^2\omega^2/2)$, we conclude by Proposition 12.4 that

$$\det A_n = \frac{\sin nu_n}{\sin u_n} \quad \text{for } 1 - \frac{\epsilon_n^2\omega^2}{2} = \cos u_n \tag{12.1.30}$$

which by (12.1.26) gives

$$I_n(t_a, t_b, 0, 0) = \left[\frac{m}{2\pi\hbar i(t_b - t_a)}\right]^{1/2} \left[\frac{\sin nu_n}{n \sin u_n}\right]^{-1/2}. \tag{12.1.31}$$

Now

$$\sin^2 u_n = 1 - \cos^2 u_n = \epsilon_n^2\omega^2 - \frac{\epsilon_n^4\omega^4}{4} \text{ (by definition (12.1.30))} \tag{12.1.32}$$

$$\begin{aligned} \implies n^2 \sin^2 u_n &= (t_b - t_a)^2\omega^2 - \epsilon_n^2 \cdot \frac{(t_b - t_a)^2\omega^4}{4} \\ &\to (t_b - t_a)^2\omega^2 \quad \text{as } n \to \infty. \end{aligned} \tag{12.1.33}$$

Of course $0 \le u_n = \arccos(1 - \epsilon_n^2\omega^2/2) \to \arccos 1 = 0$ as $n \to \infty$. Thus if $f(x) \stackrel{\text{def}}{=} \sin x/x$ with $f(0) \stackrel{\text{def}}{=} 1$, then $f(u_n) \to 1$ so that we may choose a natural number N such that $n \ge N \implies |f(u_n) - 1| < 1/2$. In particular $n \ge N \implies |f(u_n)| > 1/2 \implies f(u_n) \ne 0$. Since $n^2 \sin^2 u_n = n^2 u_n^2 f(u_n)^2$ we see that for $n \ge N$, $n^2u_n^2 = n^2 \sin^2 u_n / f(u_n)^2$. Hence (by (12.1.33)) $n^2u_n^2 \to (t_b - t_a)^2\omega^2 \implies nu_n \to (t_b - t_a)\omega$ (since $u_n \ge 0$) $\implies$

$$\left[\frac{\sin nu_n}{n \sin u_n}\right]^{-1/2} \to \left[\frac{\sin(t_b - t_a)\omega}{(t_b - t_b)\omega}\right]^{-1/2} \quad \text{as } n \to \infty. \tag{12.1.34}$$

Actually $u_n > 0$ since otherwise $u_n = 0 \implies \epsilon_n^2\omega^2/2 = 0$ by (12.1.30). That is, by (12.1.31), (12.1.34)

$$\begin{aligned} I(t_a, t_b, 0, 0) &\stackrel{\text{def}}{=} \lim_{n\to\infty} I_n(t_a, t_b, 0, 0) \\ &= \left[\frac{m\omega}{2\pi\hbar i \sin(t_b - t_a)\omega}\right]^{1/2}, \end{aligned} \tag{12.1.35}$$

which together with (12.1.19) and Proposition 12.1 proves Theorem 12.1, modulo the proof of Proposition 12.4, which we now turn to.

For $n = 1$, $M_1(c) = 2c = 2\cos u$, whereas

$$\sin(n+1)u/\sin u = 2\sin u \cos u/\sin u = 2\cos u$$

(where $-1 < c < 1 \implies \sin u \neq 0$). Proposition 12.4 is also true for $n = 2$:

$$\det M_2(c) = \det \begin{bmatrix} 2c & -1 \\ -1 & 2c \end{bmatrix} = 4c^2 - 1 = 4\cos^2 u - 1 \tag{12.1.36}$$

whereas

$$\begin{aligned} \sin 3u/\sin u &= (3\sin u - 4\sin^3 u)/\sin u \\ &= 3 - 4(1-\cos^2 u) = 4\cos^2 u - 1. \end{aligned} \tag{12.1.37}$$

On the other hand, the following recursion formula holds:

$$\det M_{n+1}(c) + \det M_{n-1}(c) = 2c \det M_n(c). \tag{12.1.38}$$

Accordingly we may proceed by induction on n:

$$\begin{aligned} &\sin(x \pm y) = \sin x \cos y \pm \cos x \sin y \\ \implies &\sin(n+2)u = \sin((n+1)u + u) = \sin(n+1)u \cos u + \cos(n+1)u \sin u, \end{aligned} \tag{12.1.39}$$

and

$$\sin nu = \sin((n+1)u - u) = \sin(n+1)u\cos u - \cos(n+1)u \sin u; \tag{12.1.40}$$

i.e., $\cos(n+1)u \sin u = \sin(n+1)u \cos u - \sin nu \implies$

$$\sin(n+2)u = 2\sin(n+1)u\cos u - \sin nu. \tag{12.1.41}$$

By (12.1.38), with an induction hypothesis,

$$\begin{aligned} \det M_{n+1}(c) &= 2c \det M_n(c) - \det M_{n-1}(c) \\ &= 2\cos u \sin(n+1)u/\sin u - \sin nu/\sin u \\ &= \sin(n+2)u/\sin u \end{aligned} \tag{12.1.42}$$

by (12.1.41), which completes the induction.

Proposition 12.5. *Suppose in Proposition 12.4 that* $c > 1$. *Write* $c = \cosh u$. *Then we can write*

$$\det M_n(c) = \frac{\sinh(n+1)u}{\sinh u}. \tag{12.1.43}$$

The proof is analogous to that of Proposition 12.4 since the hyperbolic functions satisfy identities analogous to those of the trigonometric functions. Note that for $c = 1$, $M_n(c)$ is the *Cartan matrix* of the simple Lie algebra $sl(n+1, \mathbb{C})$ of rank n. One has $\det M_n(1) = n + 1$.

13

Euclidean Path Integrals

13.1 Euclidean Path Integrals

Besides Feynman's path integral formulation of quantum mechanics (and extended formulations of quantum electrodynamics and other areas, as mentioned earlier), his path integral formulation of statistical mechanics has also proved to be a very useful development. The latter theory however involves *Euclidean path integrals* or *Wiener* type integrals, which rest on a more steady mathematical foundation. Compare remarks in Section 14.2 of the next chapter, for example, and also in Section 16.1 of Chapter 16.

The Euclidean path integral is obtained formally from the Feynman path integral by analytically continuing real time t to imaginary time $-it$. Under this *Wick rotation* $t \to -it$ we declare that $\frac{d}{dt} \to -i\frac{d}{dt}$: $(\frac{dx}{dt})^2 \to (-i\frac{dx}{dt})^2 = -(\frac{dx}{dt})^2$. Then the action functional

$$S(x(t)) \to \int_{t_a}^{t_b} \left[-\frac{m}{2}x'(t)^2 - V(x(t)) \right] (-i)\, dt = iS_E(x(t)), \qquad (13.1.1)$$

where

$$S_E(x(t)) \stackrel{\text{def}}{=} \int_{t_a}^{t_b} \left[\frac{m}{2}x'(t)^2 + V(x(t)) \right] dt \qquad (13.1.2)$$

is the *Euclidean action integral* of the path $x(t)$; this corresponds to the action integral of our particle of mass m moving under the influence of the *inverted potential* $-V$. By (13.1.1)

$$\exp\left(\frac{iS(x(t))}{\hbar} \right) \to \exp\left(-\frac{S_E(x(t))}{\hbar} \right) \qquad (13.1.3)$$

and thus, in comparison with (11.2.5), we denote the Euclidean path integral $I_E(t_a, t_b, x_a, x_b)$ (not yet defined) by

$$I_E(t_a, t_b, x_a, x_b) = \int_{x_a}^{x_b} \exp\left(-\frac{S_E(x(t))}{\hbar}\right) D[x(t)]. \tag{13.1.4}$$

We also declare that $t_b - t_a \to -i(t_b - t_a)$ so that in (12.1.27)

$$c_n \to \tilde{c}_n \stackrel{\text{def}}{=} \left[\frac{mn}{2\pi\hbar(t_b - t_a)}\right]^{n/2}. \tag{13.1.5}$$

Given (13.1.2) we have the reasonable approximation (for large n)

$$S_E(x^{(n)}(t)) \simeq \sum_{j=0}^{n-1} \left[\frac{m}{2\epsilon_n}(x_{j+1} - x_j)^2 + \epsilon_n V(x_{j+1})\right], \tag{13.1.6}$$

as in (11.2.11). It is now clear by comparison with (11.2.12), (11.2.13), how to define $I_E(t_a, t_b, x_a, x_b)$ in (13.1.4). Namely

$$I_E(t_a, t_b, x_a, x_b) \stackrel{\text{def}}{=} \lim_{n\to\infty} I_{E,n}(t_a, t_b, x_a, x_b), \tag{13.1.7}$$

where for $\tilde{c}_n$ in (13.1.5)

$$I_{E,n}(t_a, t_b, x_a, x_b) \stackrel{\text{def}}{=} \tilde{c}_n \int_{-\infty}^{\infty} \cdots \int_{-\infty}^{\infty} \exp\left(-\frac{S_E(x(t))}{\hbar}\right) dx_1 dx_2 \cdots dx_{n-1} \tag{13.1.8}$$

with $S_E(x(t))$ regarded as a function of $x_1, x_2, \ldots, x_{n-1}$, given by the right-hand side of (13.1.6).

As a simple example consider again the free particle of mass m: $V = 0$. In this case we have

Proposition 13.1.

$$I_E(t_a, t_b, x_a, x_b) = \sqrt{\frac{m}{2\pi\hbar(t_b - t_a)}} \exp\left(-\frac{m}{2\hbar}\frac{(x_a - x_b)^2}{(t_b - t_a)}\right).$$

This compares with Theorem 11.4.

Proof. The Proposition follows directly from Theorem 11.6, just as Theorem 11.4 followed from Theorem 11.3. □

As a second example, which is non-trivial, consider again the simple harmonic oscillator with frequency ν and potential $V(x) = m\omega^2x^2/2$ where $\omega = 2\pi\nu$. To compute $I_E(t_a, t_b, x_a, x_b)$ we follow exactly the methods of Chapter 12, setting up the appropriate analogous results. First of all we need a classical path $x_c(t)$ analogous to the path $x(t) = A\sin(\omega t + \phi)$ in (12.1.1) there for V. Since the Euclidean action functional S_E corresponds to the inverted potential $-V$ (as noted earlier) the corresponding equation of motion $mx''(t) = F = -(-V)' = m\omega^2 x(t)$ reduces to $x''(t) = \omega^2 x(t)$, which has as a general solution $x(t) = A\sinh(\omega t + \phi)$. Thus in this section we take

$$x_c(t) \stackrel{\text{def}}{=} A\sinh(\omega t + \phi) \tag{13.1.9}$$

where $A, \phi \in \mathbb{R}$ are subject to the initial conditions $x_c(t_a) = x_a$, $x_c(t_b) = x_b$. Using analogous identities like $\cosh^2 x + \sinh^2 x = \cosh 2x$, $\sinh(x \pm y) = \sinh x \cosh y \pm \cosh x \sinh y$, $\cosh(x \pm y) = \cosh x \cosh y \pm \sinh x \sinh y$ one has, comparing the proof of Proposition 12.1,

Proposition 13.2. *For the classical path $x_c(t)$ given in* (13.1.9), *the Euclidean action integral $S_E(x_c(t))$ in* (13.1.2) *is given by*

$$S_E(x_c(t)) = \frac{m\omega}{2}\left[\frac{(x_b^2 + x_a^2)\cosh\omega(t_b - t_a) - 2x_a x_b}{\sinh\omega(t_b - t_a)}\right]. \tag{13.1.10}$$

Proposition 13.2 compares with Proposition 12.1.

Given the boundary conditions $x_c(t_a) = x_a = x(t_a)$, $x_c(t_b) = x_b = x(t_b)$, $x_d(t_a) = x_d(t_b) = 0$ (cf. (12.1.13), (12.1.14)) where again we set $x_d(t) = x(t) - x_c(t)$ for a path $x(t)$ (where $x_c(t)$ will continue to be given by (13.1.9)), and the equation of motion $x_c''(t) = \omega^2 x_c(t)$, one may proceed exactly as in the proof of Proposition 12.2 to establish

Proposition 13.3.

$$S_E(x(t)) = S_E(x_c(t)) + S_E(x_d(t)). \tag{13.1.11}$$

As in the case of Feynman path integrals, given Proposition 13.3, we can factor out the contribution of the classical path $x_c(t)$ to the Euclidean path integral $I_E(t_a, t_b, x_a, x_b)$ and write

$$I_E(t_a, t_b, x_a, x_b) = \exp\left(-S_E(x_c(t))/\hbar\right) \int\limits_{\substack{\text{paths } x(t) \\ \text{st. } x(t_a)=x(t_b)=0}} \exp\left(-S_E(x(t))/\hbar\right)\, D[x(t)] \tag{13.1.12}$$

(similar to (12.1.19)) which greatly simplifies matters. That is,

$$I_E(t_a, t_b, x_a, x_b) = \exp\left(-S_E(x_c(t))/\hbar\right) \lim_{n\to\infty} I_{E,n}(t_a, t_b, 0, 0) \tag{13.1.13}$$

where for $\widetilde{c_n}$ given by (13.1.5)

$$I_{E,n}(t_a, t_b, 0, 0) \stackrel{\text{def}}{=} \widetilde{c_n} \int_{-\infty}^{\infty} \cdots \int_{-\infty}^{\infty} \exp\left(-S_E(x^{(n)}(t))/\hbar\right)\, dx_1 dx_2 \cdots dx_{n-1} \tag{13.1.14}$$

with $S_E(x^{(n)}(t))$ as a function of $x_1, \dots, x_{n-1}$ given by

$$S_E(x^{(n)}(t)) = \sum_{j=0}^{n-1} \left[\frac{m}{2\epsilon_n}(x_{j+1} - x_j)^2 + \epsilon_n V(x_{j+1}) \right]$$
$$\text{for } \epsilon_n \stackrel{\text{def}}{=} \frac{t_b - t_a}{n}, \quad n \geq 2, \quad x_o = x_a = 0 = x_b = x_n. \tag{13.1.15}$$

In place of A_n defined in (12.1.20) we now set

$$A_n \stackrel{\text{def}}{=} \begin{pmatrix} 2+\epsilon_n^2\omega^2 & -1 & & & 0 \\ -1 & 2+\epsilon_n^2\omega^2 & & & \\ & & \ddots & & \\ & & & 2+\epsilon_n^2\omega^2 & -1 \\ 0 & & & -1 & 2+\epsilon_n^2\omega^2 \end{pmatrix}; \tag{13.1.16}$$

A_n is an $(n-1) \times (n-1)$ symmetric matrix. The analogue of Proposition 12.3 is the statement

$$\begin{bmatrix} x_1, \dots, x_{n-1} \end{bmatrix} A_n \begin{bmatrix} x_1 \\ \vdots \\ x_{n-1} \end{bmatrix} = \sum_{j=0}^{n-1} [(x_{j+1} - x_j)^2 + \epsilon_n^2 \omega^2 x_{j+1}^2], \tag{13.1.17}$$

given the boundary conditions $x_o = x_a = 0$, $x_n = x_b = 0$, which allows us to write

$$S_E(x^{(n)}(t)) = \frac{m}{2\epsilon_n} x A_n x^t \qquad \text{for } x = (x_1, \dots, x_{n-1}), \tag{13.1.18}$$

and thus to use the result (11.A.7) in Appendix 11A

$$\int_{\mathbb{R}^d} \exp\left(-\lambda x M x^t\right)\, dx = \left(\frac{\pi}{\lambda}\right)^{d/2} \frac{1}{\sqrt{\det M}} \tag{13.1.19}$$

for $\lambda = m/(2\epsilon_n \hbar)$ to conclude that

$$I_{E,n}(t_a, t_b, 0, 0) = \widetilde{c}_n \left(\frac{\pi}{\lambda}\right)^{(n-1)/2} \frac{1}{\sqrt{\det A_n}} = \left[\frac{m}{2\pi\hbar\epsilon_n}\right]^{1/2} \frac{1}{\sqrt{\det A_n}}. \tag{13.1.20}$$

By definition (13.1.16), $A_n = M_{n-1}(1 + \epsilon_n^2\omega^2/2)$ so that by Proposition 12.5 we can write

$$\det A_n = \frac{\sinh nu_n}{\sinh u_n} \qquad \text{for } 1 + \frac{\epsilon_n^2\omega^2}{2} = \cosh u_n. \tag{13.1.21}$$

Applying the arguments of Chapter 12, following the statement of (12.1.31), we deduce that (as $u_n > 0$)

$$\frac{\det A_n}{n} = \frac{\sinh nu_n}{n \sinh u_n} \to \frac{\sinh \omega(t_b - t_a)}{\omega(t_b - t_a)} \text{ as } n \to \infty \tag{13.1.22}$$

$$\Longrightarrow \; I_{E,n}(t_a, t_b, 0, 0) \to \left[\frac{m}{2\pi\hbar(t_b - t_a)}\right]^{1/2} \left[\frac{\omega(t_b - t_a)}{\sinh\omega(t_b - t_a)}\right]^{1/2} \tag{13.1.23}$$

$$= \left[\frac{m\omega}{2\pi\hbar\sinh(t_b - t_a)\omega}\right]^{1/2} \quad \text{as } n \to \infty, \text{ by (13.1.20).}$$

That is, by (13.1.13) and Proposition 13.2 we have established

Theorem 13.1. *For the simple harmonic oscillator with frequency $\nu = \omega/2\pi$ and potential $V(x) = \omega^2 mx^2/2$, the Euclidean path integral $I_E(t_a, t_b, x_a, x_b)$ of (13.1.7) is given by*

$$I_E(t_a, t_b, x_a, x_b) = \exp\left(-S_E(x_c(t))/\hbar\right) \left[\frac{m\omega}{2\pi\hbar\sinh\omega(t_b - t_a)}\right]^{1/2}, \tag{13.1.24}$$

where the Euclidean action $S_E(x_c(t))$ of the classical path $x_c(t) = A\sinh(\omega t + \phi)$, $A, \phi \in \mathbb{R}$ is given by

$$S_E(x_c(t)) = \frac{m\omega[(x_b^2 + x_a^2)\cosh\omega(t_b - t_a) - 2x_a x_b]}{2\sinh\omega(t_b - t_a)}; \tag{13.1.25}$$

see (13.1.2).

Theorem 13.1 compares with Theorem 12.1.

For $x, \theta \in \mathbb{R}$, $\sinh x = \sinh 2(\frac{x}{2}) = 2\sinh\frac{x}{2}\cosh\frac{x}{2} \implies \sinh x \tanh\frac{x}{2} = 2\sinh^2\frac{x}{2}$. Also $\cosh 2\theta = \cosh(\theta + \theta) = \cosh^2\theta + \sinh^2\theta = 1 + 2\sinh^2\theta \implies 2\sinh^2\frac{x}{2} = \cosh x - 1 \implies$

$$\sinh x \tanh\frac{x}{2} = 2\sinh^2\frac{x}{2} = \cosh x - 1. \tag{13.1.26}$$

Thus consider the special case $x_a = x_b = x$. By (13.1.25)

$$\begin{aligned} S_E(x_c(t)) &= m\omega x^2 \frac{\cosh\omega(t_b - t_a) - 1}{\sinh\omega(t_b - t_a)} \\ &= m\omega x^2 \tanh\frac{\omega}{2}(t_b - t_a) > 0. \end{aligned} \tag{13.1.27}$$

Corollary 13.1 (to Theorem 13.1).

$$I_E(t_a, t_b, x, x) = \left[\frac{m\omega}{2\pi\hbar\sinh\omega(t_b - t_a)}\right]^{1/2} \exp\left(-\frac{m\omega x^2}{\hbar}\tanh\frac{\omega}{2}(t_b - t_a)\right)$$

for $x \in \mathbb{R}$.

13.2 Wave Function Expansion of the Euclidean Path Integral

With a focus still on the example of the simple harmonic oscillator we relate the Euclidean path integral $I_E(t_a, t_b, x_a, x_b)$ computed in Theorem 13.1 to the normalized wave functions ψ_n and energy levels E_n of the oscillator. ψ_n and E_n are given by

$$\psi_n(x) = (-1)^n \sqrt{\sqrt{\frac{b}{\pi}} \frac{1}{2^n\, n!}} \exp\left(-\frac{b}{2}x^2\right) H_n(\sqrt{b}x) \quad \text{for } b = \frac{\omega m}{\hbar}, \tag{13.2.1}$$

$$E_n = \left(n + \frac{1}{2}\right) \nu h = \left(n + \frac{1}{2}\right) \omega\hbar \quad \text{for } n = 0, 1, 2, 3, \ldots; \tag{13.2.2}$$

see Section 4.8 of Chapter 4.

H_n is the nth Hermite polynomial. The following interesting expansion holds:

Theorem 13.2. $I_E(t_a, t_b, x_a, x_b) =$

$$\sum_{n=0}^{\infty} \psi_n(x_a)\psi_n(x_b) \exp(-E_n(t_b - t_a)/\hbar).$$

The meaning of Theorem 13.2 in regards to quantum statistical mechanics will be commented on in the next chapter; see remarks following the statement of Theorem 14.1 there. The key mathematical tool needed to prove Theorem 13.2 is the *Mehler formula*

$$\frac{1}{\sqrt{1-z^2}} \exp\left(\frac{2xyz - (x^2 + y^2)z^2}{1 - z^2}\right) = \sum_{n=0}^{\infty} \frac{H_n(x)H_n(y)}{n!} \left(\frac{z}{2}\right)^n, \tag{13.2.3}$$

or

$$\frac{1}{\sqrt{1-z^2}} \exp\left(\frac{4xyz - (x^2 + y^2)(1 + z^2)}{2(1 - z^2)}\right)$$
$$= \exp(-(x^2 + y^2)/2) \sum_{n=0}^{\infty} \frac{H_n(x)H_n(y)}{n!} \left(\frac{z}{2}\right)^n \tag{13.2.4}$$

for $x, y \in \mathbb{R}$, $z \in \mathbb{C}$ with $|z| < 1$.

To prove Theorem 13.2 we choose $x = \sqrt{m\omega/\hbar}\; x_a$, $y = \sqrt{m\omega/\hbar}\; x_b$, $z = \exp(-\omega(t_b - t_a))$. Then for $T \overset{\text{def}}{=} t_b - t_a$,

$$1 \pm z^2 = 1 \pm \exp(-2\omega T) = \frac{\exp(\omega T) \pm \exp(-\omega T)}{\exp(\omega T)} \tag{13.2.5}$$

$$\Rightarrow 1 - z^2 = \frac{2\sinh \omega T}{\exp(\omega T)}, \quad \text{or} \quad \frac{1}{\sqrt{1 - z^2}} = \frac{\exp(\omega T/2)}{\sqrt{2\sinh \omega T}}. \tag{13.2.6}$$

Also by (13.1.25)

$$\frac{4xyz - (x^2 + y^2)(1 + z^2)}{2(1 - z^2)}$$

$$= \frac{4\frac{m\omega}{\hbar}x_a x_b \exp(-\omega T) - \frac{m\omega}{\hbar}(x_a^2 + x_b^2)\frac{2\cosh\omega T}{\exp(\omega T)}}{2\left(\frac{2\sinh\omega T}{\exp(\omega T)}\right)} \tag{13.2.7}$$

$$= -\frac{m\omega}{2\hbar}\left[\frac{(x_a^2 + x_b^2)\cosh\omega T - 2x_a x_b}{\sinh\omega T}\right] \equiv -S_E(x_c(t))/\hbar.$$

By (13.2.1), (13.2.4), (13.2.6), and (13.2.7) we see that

$$\frac{\exp(\omega T/2)}{\sqrt{2}\sinh\omega T}\exp(-S_E(x_c(t))/\hbar) = \exp\left(-\frac{m\omega}{2\hbar}(x_a^2 + x_b^2)\right)$$
$$\cdot\sum_{n=0}^{\infty}\frac{H_n(\sqrt{\frac{m\omega}{\hbar}}x_a)H_n(\sqrt{\frac{m\omega}{\hbar}}x_b)}{2^n\, n!}\exp(-n\omega T) \tag{13.2.8}$$
$$= \sqrt{\frac{\pi\hbar}{m\omega}}\sum_{n=0}^{\infty}\psi_n(x_a)\psi_n(x_b)\exp(-n\omega T) \implies$$

$$\sum_{n=0}^{\infty}\psi_n(x_a)\psi_n(x_b)\exp(-n\omega T)$$
$$= \sqrt{\frac{m\omega}{2\pi\hbar\sinh\omega T}}\exp(\omega T/2)\exp(-S_E(x_c(t))/\hbar) \implies \tag{13.2.9}$$

$$\sqrt{\frac{m\omega}{2\pi\hbar\sinh\omega T}}\exp(-S_E(x_c(t))/\hbar)$$
$$= \sum_{n=0}^{\infty}\psi_n(x_a)\psi_n(x_b)\exp\left(-\left(n + \frac{1}{2}\right)\omega T\right) \tag{13.2.10}$$
$$= \sum_{n=0}^{\infty}\psi_n(x_a)\psi_n(x_b)\exp(-E_n T/\hbar),$$

which by Theorem 13.1 proves Theorem 13.2.

14

The Density Matrix and Partition Function in Quantum Statistical Mechanics

14.1 Helmholtz Free Energy, Entropy and Internal Energy

In quantum statistical mechanics one has the fundamental object $\rho(\beta; x_a, x_b)$, a matrix entry which defines the *density matrix* ρ, given in the Feynman formulation by the Euclidean path integral of the preceding chapter:

$$\begin{aligned}\rho(\beta; x_a, x_b) &\stackrel{\text{def}}{=} \int_{x_a=x(0)}^{x_b=x(\beta\hbar)} \exp(-S_E(x(t))/\hbar)\ D[x(t)] \\ &= I_E(0, \beta\hbar, x_a, x_b)\end{aligned} \tag{14.1.1}$$

where $\beta = \frac{1}{kT}$ is the inverse temperature $\frac{1}{T}$ up to a constant $\frac{1}{k}$, k being *Boltzmann's constant*. From ρ one obtains the all-important *partition function* Z as a trace:

$$\begin{aligned}Z &= \int_{-\infty}^{\infty} \rho(\beta; x, x)\ dx \\ &= \int_{-\infty}^{\infty} \int_{x(0)=x}^{x(\beta\hbar)=x} \exp(-S_E(x(t))/\hbar)\ D[x(t)]\ dx.\end{aligned} \tag{14.1.2}$$

One usually regards ρ and Z as functions of the temperature T of some system in thermal equilibrium. Generally speaking the Hamiltonian H of the system has continuous spectral contributions. In the definition (14.1.2) we assume, strictly speaking, that H has a purely discrete spectrum—which is the case in the example of interest to us.

From Z one derives the basic thermodynamic quantities. For example:

$$F = F(T) \stackrel{\text{def}}{=} -kT \log Z(T), \qquad \text{the } \textit{Helmholtz free energy} \tag{14.1.3}$$

$$\Longrightarrow\ Z = \exp(-\beta F), \tag{14.1.4}$$

$$S = S(T) \stackrel{\text{def}}{=} -\frac{\partial F}{\partial T}, \qquad \text{the } \textit{entropy} \tag{14.1.5}$$

$$U = U(T) \stackrel{\text{def}}{=} F(T) + TS(T), \qquad \text{the } \textit{internal energy} \tag{14.1.6}$$

$$= F(T) - T\frac{\partial F}{\partial T}.$$

Note that $Z = \exp(-\beta F) \implies$

$$\frac{\partial Z}{\partial T} = Z\left[\frac{F}{kT^2} - \frac{1}{kT}\frac{\partial F}{\partial T}\right] \implies U = \frac{kT^2}{Z}\frac{\partial Z}{\partial T}. \tag{14.1.7}$$

We compute F, S, and U in the example of continued interest, the harmonic oscillator. By Theorem 13.1 the density matrix as a path integral is given by

Theorem 14.1.

$$\rho(\beta; x_a, x_b) = \left[\frac{m\omega}{2\pi\hbar \sinh \beta\hbar\omega}\right]^{1/2} \exp(-S_E(x_c(t))/\hbar)$$

for

$$S_E(x_c(t)) = \frac{m\omega}{2}\left[\frac{(x_b^2 + x_a^2)\cosh \beta\hbar\omega - 2x_a x_b}{\sinh \beta\hbar\omega}\right].$$

On the other hand, by Theorem 13.2, we can express the density matrix in terms of the normalized wave functions ψ_n and energy levels E_n given in (13.2.1), (13.2.2) before its statement:

$$\rho(\beta; x_a, x_b) = \sum_{n=0}^{\infty} \psi_n(x_a)\psi_n(x_b)\exp(-E_n\beta). \tag{14.1.8}$$

In other words, the path integral definition (14.1.1) of the density matrix coincides with the more standard version expressed in (14.1.8), as formulated by von Neumann and Dirac. This is the meaning of Theorem 13.2 which we promised to comment on.

By definition (14.1.1) and Corollary 13.1

$$\rho(\beta; x, x) = \left[\frac{m\omega}{2\pi\hbar \sinh \omega\beta\hbar}\right]^{1/2} \exp(-\lambda x^2) \tag{14.1.9}$$

for $\lambda \overset{\text{def}}{=} \frac{m\omega}{\hbar} \tanh \frac{\omega}{2}\beta\hbar$. Hence, by (14.1.2),

$$\begin{aligned} Z &\overset{\text{def}}{=} \left[\frac{m\omega}{2\pi\hbar \sinh \omega\beta\hbar}\right]^{1/2} \int_{-\infty}^{\infty} \exp(-\lambda x^2)\, dx \\ &= \left[\frac{m\omega}{2\pi\hbar \sinh \omega\beta\hbar}\right]^{1/2} \sqrt{\frac{\pi}{\lambda}} \\ &= \left[\frac{1}{2 \sinh \omega\beta\hbar \tanh \frac{\omega}{2}\beta\hbar}\right]^{1/2} \\ &= \frac{1}{2 \sinh \frac{\omega}{2}\beta\hbar}, \end{aligned} \tag{14.1.10}$$

by (13.1.26). Using $E_n = \hbar\omega(n + 1/2)$, we note that for $t > 0$

$$\begin{aligned} \sum_{n=0}^{\infty} \exp(-E_n t) &= \exp(-\hbar\omega t/2) \sum_{n=0}^{\infty} (\exp(-\hbar\omega t))^n = \frac{\exp(-\hbar\omega t/2)}{1 - \exp(-\hbar\omega t)} \\ &= \frac{1}{\exp(\hbar\omega t/2) - \exp(-\hbar\omega t/2)} = \frac{1}{2 \sinh \hbar\omega t/2}. \end{aligned} \tag{14.1.11}$$

Hence the following well-known standard result holds:

Theorem 14.2.

$$Z = \frac{1}{2 \sinh \frac{\omega}{2}\beta\hbar} = \sum_{n=0}^{\infty} \exp(-E_n\beta). \tag{14.1.12}$$

The representation $\sum_{n=0}^{\infty} \exp(-E_n\beta)$ of Z also follows, of course, from (14.1.8) directly, since the ψ_n are normalized:

$$\int_{\mathbb{R}} \psi_n(x)\psi_n(x)\, dx = 1. \tag{14.1.13}$$

Corollary 14.1. *The free energy, entropy, and internal energy are given, respectively, by*

$$\begin{aligned} F(T) &= \hbar\omega/2 + kT \log(1 - \exp(-\beta\hbar\omega)), \\ S(T) &= k\left[\frac{\beta\hbar\omega}{\exp(\beta\hbar\omega) - 1} - \log(1 - \exp(-\beta\hbar\omega))\right], \textit{ and} \\ U(T) &= \hbar\omega\left[\frac{1}{2} + \frac{1}{\exp(\beta\hbar\omega) - 1}\right]. \end{aligned} \tag{14.1.14}$$

Proof. We use the definitions in (14.1.3)–(14.1.6).

$$\begin{aligned} F(T) &= -kT \log\left[\frac{1}{2 \sinh \frac{\omega}{2}\beta\hbar}\right] \\ &= -kT\left[-\frac{\hbar\omega\beta}{2} - \log(1 - \exp(-\hbar\omega\beta))\right] \end{aligned} \tag{14.1.15}$$

by (14.1.12), which gives the first formula in (14.1.14) since $kT\beta \overset{\text{def}}{=} 1$. By that first formula

$$\begin{aligned} -\frac{1}{k}\frac{\partial F}{\partial T} &= T\left[\frac{\exp(-\beta\hbar\omega)\frac{\hbar\omega}{kT^2}}{1-\exp(-\beta\hbar\omega)}\right] - \log(1-\exp(-\beta\hbar\omega)) \\ &= \frac{\hbar\omega\beta}{\exp(\beta\hbar\omega)-1} - \log(1-\exp(-\beta\hbar\omega)), \end{aligned} \tag{14.1.16}$$

which gives the second formula in (14.1.14), again as $\beta = \frac{1}{kT}$. Using $U(T) \overset{\text{def}}{=} F(T) + TS(T)$ one obtains the third formula of (14.1.14) immediately from the first two. □

By Corollary 14.1 we can understand, for example, the low and high temperature behavior of the internal energy. For low temperature (i.e., for β sufficiently large) the summand $\frac{\hbar\omega}{\exp(\beta\hbar\omega)-1}$ in (14.1.14) can be neglected. One obtains right away that $U(T) \simeq \hbar\omega/2$. To understand the high temperature behavior of $U(T)$, first recall the definition of the nth *Bernoulli number* B_n:

$$\frac{x}{e^x-1} \overset{\text{def}}{=} \sum_{n=0}^{\infty}\frac{B_n x^n}{n!}, \quad \text{where } \frac{x}{e^x-1} \overset{\text{def}}{=} 1, \text{ for } x = 0. \tag{14.1.17}$$

Thus one has, for example, $B_o = 1$, $B_1 = -\frac{1}{2}$, $B_2 = \frac{1}{6}$, $B_3 = 0$, $B_4 = -\frac{1}{30}$, $B_5 = 0$, $B_6 = \frac{1}{42}$, $B_7 = 0$, $B_8 = -\frac{1}{30}$, $B_9 = 0$, $B_{10} = \frac{5}{66}$; $B_{\text{odd}} = 0$. By (14.1.14), (14.1.17),

$$U(T) = \frac{\hbar\omega}{2} + \frac{1}{\beta}\left[\frac{\beta\hbar\omega}{\exp(\beta\hbar\omega)-1}\right] = \frac{\hbar\omega}{2} + \frac{1}{\beta}\sum_{n=0}^{\infty}\frac{B_n(\beta\hbar\omega)^n}{n!}. \tag{14.1.18}$$

For high temperature T (i.e., for β sufficiently small) the following first approximation is immediately obtained:

$$U(T) \simeq \frac{\hbar\omega}{2} + \frac{1}{\beta}[B_o + B_1\beta\hbar\omega] = \frac{1}{\beta} = kT. \tag{14.1.19}$$

14.2 A Zeta Function Representation of the Free Energy

The goal of the present section is to construct a zeta function $\zeta(s;T)$ such that the special value $\zeta'(0;T)$ of its derivative gives the free energy $F(T)$, for the harmonic oscillator. This provides another example of the occurrence of various zeta functions in quantum theory. The main result, Theorem 14.3, moreover is of a prototypical nature. It serves as an important simple example of the following type of situation that occurs often enough in quantum field theory, cosmology, and other areas—a more complicated situation that we will encounter in Chapter 16. Namely, one is given, intuitively, a suitable partition function Z, perhaps with some path integral representation; cf. (14.1.2). The problem is to give some precise meaning to $\log Z$ (or to the path integral itself); recall by (14.1.3), for example, $F(T) \stackrel{\text{def}}{=}$ constant $T \log Z(T)$. That is, one seeks a suitable *regularization* of the quantity $\log Z$. One approach to the problem is to construct a suitable zeta function ζ such that $\log Z$ may be appropriately defined in terms of the special values $\zeta'(0)$ and $\zeta(0)$. We consider this approach in Chapter 16 in the study of finite temperature effects of scalar (or spinor) quantum fields over certain negatively curved space-times. It must be mentioned however that the quantity $\log Z$ might have two or more regularizations which are rather non-related. The remarks here serve to provide some meaning of, and perspective for Theorem 14.3, whose statement we now turn to.

Our interest is in the following zeta function.

$$\zeta(s;T) \stackrel{\text{def}}{=} \sum_{n\in\mathbb{Z}} \frac{1}{(4\pi^2 n^2 + \hbar^2\omega^2\beta^2)^s}, \qquad \operatorname{Re} s > \frac{1}{2} \tag{14.2.1}$$

where again $\beta = \frac{1}{kT}$.

Theorem 14.3. *The zeta function $\zeta(s;T)$ of* (14.2.1) *admits a meromorphic continuation which in particular is holomorphic at $s = 0$. Moreover one has the relation*

$$F(T) = -\frac{kT}{2}\zeta'(0;T) \tag{14.2.2}$$

for the free energy.

The proof is based on a classical *Jacobi inversion formula* for the following theta function defined for $t > 0$:

$$\theta(t) \stackrel{\text{def}}{=} \sum_{n\in\mathbb{Z}} \exp(-t4\pi^2 n^2). \tag{14.2.3}$$

The inversion formula relates the value of θ at t with its value at the inverse $\frac{1}{t}$ of t:

$$\theta(t) = \frac{1}{\sqrt{4\pi t}}\,\theta\left(\tfrac{1}{16\pi^2 t}\right). \tag{14.2.4}$$

For a more general inversion formula, see Appendix 14A. It will be convenient to work with the slightly modified θ-function θ_a defined by

$$\theta_a(t) = \sum_{n\in\mathbb{Z}} \exp(-t(4\pi^2 n^2 + a)) = \exp(-ta)\,\theta(t) \tag{14.2.5}$$

for $a > 0$, a fixed. By the Jacobi formula (14.2.4)

$$\begin{aligned} \theta_a(t) &= \frac{\exp(-ta)}{\sqrt{4\pi t}} \sum_{n\in\mathbb{Z}} \exp(-n^2/4t) \\ &= \frac{\exp(-ta)}{\sqrt{4\pi t}} + 2\sum_{n=1}^{\infty} \frac{\exp(-(at + n^2/4t))}{\sqrt{4\pi t}}. \end{aligned} \tag{14.2.6}$$

Defining

$$\zeta_a(s) = \frac{1}{\Gamma(s)} \int_0^\infty \theta_a(t) t^{s-1}\, dt, \tag{14.2.7}$$

we can commute integration and summation and write

$$\begin{aligned} \zeta_a(s) &= \frac{1}{\Gamma(s)} \sum_{n\in\mathbb{Z}} \int_0^\infty \exp(-t(4\pi^2 n^2 + a)) t^{s-1}\, dt \\ &= \sum_{n\in\mathbb{Z}} \frac{1}{(4\pi^2 n^2 + a)^s} \qquad \text{for } \operatorname{Re} s > \frac{1}{2} \end{aligned} \tag{14.2.8}$$

since

$$\int_0^\infty \exp(-ta) t^{s-1}\, dt = a^{-s}\Gamma(s) \qquad \text{for } \operatorname{Re} s > 0. \tag{14.2.9}$$

We see that the zeta function in (14.2.1) coincides with ζ_a for the choice $a = \hbar^2\omega^2\beta^2$. Let us work therefore more generally with the zeta function ζ_a for an arbitrary $a > 0$.

To meromorphically continue ζ_a we use the Jacobi inversion formula (14.2.6). For $\operatorname{Re} s > \frac{1}{2}$

$$\begin{aligned} \zeta_a(s) &= \frac{1}{\sqrt{4\pi}\,\Gamma(s)} \int_0^\infty \exp(-ta) t^{s-1-1/2}\, dt + \frac{2}{\Gamma(s)} \sum_{n=1}^\infty \int_0^\infty \frac{\exp(-(at + \frac{n^2}{4t}))}{\sqrt{4\pi t}} t^{s-1}\, dt \\ &= \frac{a^{-(s-1/2)}\,\Gamma(s-1/2)}{\sqrt{4\pi}\,\Gamma(s)} + \frac{2}{\Gamma(s)} \sum_{n=1}^\infty \pi^{-1/2} \left(\frac{n}{2\sqrt{a}}\right)^{s-1/2} K_{-s+1/2}(\sqrt{a}n), \end{aligned} \tag{14.2.10}$$

where K_s is the K-Bessel function which is defined by

$$K_s(x) \stackrel{\text{def}}{=} \frac{1}{2} \int_0^\infty \exp\left(-\frac{x}{2}\left(t + \frac{1}{t}\right)\right) t^{s-1}\, dt \quad \text{for } x > 0, s \in \mathbb{C}, \tag{14.2.11}$$

and which satisfies

$$\int_0^\infty \frac{\exp\left(-\left(at + \frac{b}{4t}\right)\right)}{\sqrt{4\pi t}} t^{s-1}\, dt = \pi^{-1/2} \left(\frac{\sqrt{b}}{2\sqrt{a}}\right)^{s-1/2} K_{-s+1/2}(\sqrt{ab}) \tag{14.2.12}$$

for $a, b > 0$. On the other hand,

$$K_{-s+1/2}(\alpha) = \frac{(\alpha/2)^{-s+1/2}\sqrt{\pi}}{\Gamma(1-s)} \int_1^\infty \exp(-\alpha x)(x^2-1)^{-s}\, dx \tag{14.2.13}$$

for Re $s < 1$, $\alpha > 0$; see [40, p. 958]. Thus for Re $s < 1$ we can write the second term in (14.2.10) as

$$\frac{2}{\Gamma(s)}\sum_{n=1}^{\infty}\frac{\pi^{-1/2}n^{s-1/2}(\sqrt{a}n)^{-s+1/2}\sqrt{\pi}}{(2\sqrt{a})^{s-1/2}\,\Gamma(1-s)}\int_1^\infty \exp(-\sqrt{a}nx)(x^2-1)^{-s}\,dx$$

$$= \frac{2(\sqrt{a})^{1-2s}}{\Gamma(s)\,\Gamma(1-s)}\int_1^\infty \sum_{n=1}^{\infty}(\exp(-\sqrt{a}x))^n(x^2-1)^{-s}\,dx \tag{14.2.14}$$

$$= 2(\sqrt{a})^{1-2s}\frac{\sin \pi s}{\pi}\int_1^\infty \frac{\exp(-\sqrt{a}x)(x^2-1)^{-s}}{1-\exp(-\sqrt{a}x)}\,dx$$

since $\Gamma(1-s)\,\Gamma(s) = \pi(\sin \pi s)^{-1}$. That is,

Theorem 14.4. *The meromorphic continuation of $\zeta_a(s)$ to Re $s < 1$ is given by*

$$\zeta_a(s) = \frac{\Gamma(s-1/2)}{\sqrt{4\pi}\,\Gamma(s)a^{s-1/2}} + 2(\sqrt{a})^{1-2a}\frac{\sin \pi s}{\pi}\int_1^\infty \frac{(x^2-1)^{-s}}{\exp(\sqrt{a}x)-1}\,dx. \tag{14.2.15}$$

Corollary 14.2. *ζ_a is holomorphic at $s = 0$ and*

$$\zeta_a'(0) = -\sqrt{a} - 2\log(1-\exp(-\sqrt{a})). \tag{14.2.16}$$

Proof.

$$\begin{aligned}\zeta_a'(0) &= \left.\frac{\Gamma(s-1/2)}{\sqrt{4\pi}a^{s-1/2}}\right|_{s=0} + \left.2(\sqrt{a})^{1-2a}\frac{\sin \pi s}{\pi}\right|_{s=0}\frac{d}{ds}\int_1^\infty \frac{(x^2-1)^{-s}}{\exp(\sqrt{a}x)-1}\,dx \\ &\quad + \left.\frac{d}{ds}2(\sqrt{a})^{1-2s}\frac{\sin \pi s}{\pi}\right|_{s=0}\int_1^\infty \frac{dx}{\exp(\sqrt{a}x)-1} \\ &= -\frac{2\sqrt{\pi}}{\sqrt{4\pi}a^{-1/2}} + 2\sqrt{a}\int_1^\infty \frac{dx}{\exp(\sqrt{a}x)-1}\end{aligned} \tag{14.2.17}$$

since $\Gamma(-1/2) = -2\sqrt{\pi}$. By the transformation of variables $t = \exp(\sqrt{a}x) - 1$ one has

$$\begin{aligned}\int_1^b \frac{dx}{\exp(\sqrt{a}x)-1} &= \frac{1}{\sqrt{a}}\int_{\exp(\sqrt{a})-1}^{\exp(\sqrt{a}b)-1}\left[\frac{1}{t}-\frac{1}{t+1}\right]dt \\ &= \frac{1}{\sqrt{a}}\left[\log\left(1-\frac{1}{\exp(\sqrt{a}b)}\right) - \log(1-\exp(-\sqrt{a}))\right] \\ &\to -\frac{1}{\sqrt{a}}\log(1-\exp(-\sqrt{a}))\end{aligned} \tag{14.2.18}$$

as $b \to \infty$. That is,

$$\int_1^\infty \frac{dx}{\exp(\sqrt{a}x)-1} = -\frac{1}{\sqrt{a}}\log(1-\exp(-\sqrt{a})), \tag{14.2.19}$$

which proves the corollary. □

Having observed that the zeta function $\zeta(s;T)$ in (14.2.1) is given by $\zeta_a(s)$ for the choice $a = \hbar^2\omega^2\beta^2$, we conclude by Corollary 14.2 that

$$\begin{aligned} -\frac{kT}{2}\zeta'(0;T) &= \frac{kT}{2}\hbar\omega\beta + kT\log(1-\exp(-\hbar\omega\beta)) \\ &= \frac{\hbar\omega}{2} + kT\log(1-\exp(-\hbar\omega\beta)) \equiv F(T), \end{aligned} \tag{14.2.20}$$

by Corollary 14.1, which proves Theorem 14.3.

In Section 14.1 we considered the low and high temperature behavior of the internal energy $U(T)$. Similarly, in Section 16.2 of Chapter 16 we shall consider the low and high temperature behavior of the free energy $F(T)$, for $F(T)$ defined in a more complicated context.

Appendix 14A Jacobi Inversion

Formula (14.2.4) is a special case of a more general Jacobi inversion formula. Namely for $z \in \mathbb{C}$, $\tau > 0$ set

$$\theta_3(z|\tau) \stackrel{\text{def}}{=} \sum_{n\in\mathbb{Z}} e^{-\pi n^2\tau + 2\pi n z}. \tag{14.A.1}$$

Then

Theorem 14.5 (Jacobi Inversion Formula).

$$\theta_3(z|\tau) = \frac{e^{\pi z^2/\tau}}{\sqrt{\tau}}\theta_3\left(\frac{z}{i\tau}\Big|\frac{1}{\tau}\right). \tag{14.A.2}$$

In particular for $z = 0$, $t > 0$

$$\theta_3(0|4\pi t) = \sum_{n\in\mathbb{Z}} e^{-4\pi^2 n^2 t} \equiv \theta(t) \tag{14.A.3}$$

for $\theta(t)$ in (14.2.3). Thus (14.A.2) implies (14.2.4):

$$\theta(t) = \theta_3(0|4\pi t) = \frac{1}{\sqrt{4\pi t}}\theta_3\left(0\Big|\frac{1}{4\pi t}\right) = \frac{1}{\sqrt{4\pi t}}\theta\left(\frac{1}{16\pi^2 t}\right). \tag{14.A.4}$$

For a certain application in Chapter 16 the following version of Jacobi inversion is required.

Theorem 14.6. *For $t, \beta > 0$, $\mu \in \mathbb{C}$*

$$\begin{aligned} \sum_{n\in\mathbb{Z}} e^{-t(\frac{2\pi n}{\beta}+i\mu)^2} &= \frac{\beta}{\sqrt{4\pi t}}\sum_{n\in\mathbb{Z}} e^{\frac{-n^2\beta^2}{4t}-n\mu\beta} \\ &= \frac{\beta}{\sqrt{4\pi t}}\left[1 + 2\sum_{n=1}^{\infty}(\cosh n\mu\beta)e^{\frac{-n^2\beta^2}{4t}}\right]. \end{aligned} \tag{14.A.5}$$

Proof. In (14.A.2) choose $\tau = \frac{4\pi t}{\beta^2}$, $z = \frac{-2it\mu}{\beta}$. Then $\frac{\pi z^2}{\tau} = -t\mu^2$ and $\frac{z}{i\tau} = \frac{-\mu\beta}{2\pi} \Longrightarrow$

$$\theta_3\left(\frac{-2it\mu}{\beta} \mid \frac{4\pi t}{\beta^2}\right) = \frac{e^{-t\mu^2}\beta}{\sqrt{4\pi t}}\theta_3\left(\frac{-\mu\beta}{2\pi} \mid \frac{\beta^2}{4\pi t}\right). \tag{14.A.6}$$

If one multiplies both sides of (14.A.6) by $e^{t\mu^2}$, one obtains exactly the first equation in (14.A.5). The second equation there follows by writing

$$\begin{aligned}\sum_{n\in\mathbb{Z}} e^{\frac{-n^2\beta^2}{4t}-n\mu\beta} &= 1+\sum_{n=1}^{\infty} e^{\frac{-(-n)^2\beta^2}{4t}-(-n)\mu\beta}+\sum_{n=1}^{\infty} e^{\frac{-n^2\beta^2}{4t}-n\mu\beta} \\ &= 1+\sum_{n=1}^{\infty} e^{\frac{-n^2\beta^2}{4t}}\left[e^{n\mu\beta}+e^{-n\mu\beta}\right].\end{aligned} \tag{14.A.7}$$

□

In application the μ in Theorem 14.6 will be chosen to be the *chemical potential* of a quantum field over a suitable Kaluza–Klein space-time. β will be chosen as the inverse temperature of that field.

15

Zeta Regularization

Zeta functions play a very useful role in physics, especially as they provide for a natural means of regularizing certain quantities which a priori are infinite, or meaningless. In Chapter 14 we presented a zeta function representation of the free energy of a harmonic oscillator. In Chapter 16 we shall employ a zeta function, for example, to provide for a clear mathematical meaning of a certain partition function related to the Hawking path integral [45]. The *Vacuum energy* of certain quantum fields over various space-times is given by a special value of a suitable zeta function—the value at $s = -1/2$, for example; cf. [4, 11, 12, 14, 88]; also compare the term in (16.1.47), where the special value $s = -p/2$ determines the vacuum or *Casimir* energy of a certain *Kaluza–Klein* space-time. One uses zeta functions to define the *determinant* in certain cases of infinite-dimensional operators. This is done by a simple procedure called *zeta regularization.* If $D = y^2(\frac{\partial^2}{\partial x^2} + \frac{\partial^2}{\partial y^2})$ is the non-Euclidean Laplace operator on the upper $1/2$-plane, for example, then one can *define* and compute the determinant of the operator $-D + s(s-1)\mathbf{1}$. This determinant arises in Polyakov string theory [26, 36, 70, 74]. Zeta regularization without doubt has proven to be a powerful tool for dealing with many interesting phenomena in physics including certain *anomalies* (more on this in Chapter 17), and *ambiguities* which arise in one-loop or external field approximations in quantum field theory (so-called *ultraviolet divergences*), to name a few. Before plunging into examples that involve more sophisticated zeta functions (such will appear in later chapters) we use this chapter to illustrate the regularization procedure in a few simple cases. The following example illustrates the basic idea.

If $n \geq 1$ is an integer, then the factorial $n! = 1 \cdot 2 \cdot 3 \cdots n$ is of course well defined, whereas $\infty! = 1 \cdot 2 \cdot 3 \cdots n = \prod_{m=1}^{\infty} m$ is not, since the infinite product diverges. The question arises whether this infinite product can be *regularized*. That is, can one assign, in a somewhat natural way, a meaning to this infinite product, say by looking at it from some alternate point of view, that actually renders it *finite*? The answer is yes, and it comes by way of the Riemann zeta function

$$\zeta(s) \stackrel{\text{def}}{=} \sum_{n=1}^{\infty} \frac{1}{n^s}, \operatorname{Re} s > 1, \tag{15.1.1}$$

which we have already encountered in Chapter 3. Formally we have by (15.1.1)

$$\zeta'(s) = \sum_{m=1}^{\infty} \frac{-\log m}{n^s} \implies -\zeta'(0) = \sum_{m=1}^{\infty} \log m \tag{15.1.2}$$

where the latter series does *not* converge of course. Write, incorrectly,

$$e^{-\zeta'(0)} = e^{\sum_{m=1}^{\infty} \log m} = \prod_{m=1}^{\infty} e^{\log m} = \prod_{m=1}^{\infty} m. \tag{15.1.3}$$

That is, we can now think of the infinite product (which indeed is non-finite) as $e^{-\zeta'(0)}$ which is *finite* since the Riemann zeta function admits a meromorphic continuation to the full complex plane, which is in fact holomorphic at $s = 0$. In fact one knows that

$$\zeta'(0) = \frac{-\log 2\pi}{2}. \tag{15.1.4}$$

Thus it is natural to *define* $\infty!$ as $e^{-\zeta'(0)}$; i.e., $\infty! \stackrel{\text{def}}{=} \sqrt{2\pi}$. This simple example of zeta regularization provides the key to the method of regularizing an infinite-dimensional determinant.

Namely, suppose we are given a linear operator T on some infinite-dimensional space. Assume T has a discrete spectrum $\{\lambda_j, n_j\}_{j=0}^{\infty}$ where the multiplicity n_j of the eigenvalue λ_j is finite for each j. T could be an elliptic operator on a vector bundle over a compact manifold, for example. Somehow we should have

$$\det T = \prod_{j=0}^{\infty} \lambda_j^{n_j}. \tag{15.1.5}$$

But again the infinite product generally diverges. To regularize it, consider the *spectral zeta function*

$$\zeta_T(s) = \sum_{j=0}^{\infty} \frac{n_j}{\lambda_j^s} \quad \text{for } \operatorname{Re} s >> 0 \tag{15.1.6}$$

of T where we are interested in the case when each $\lambda_j > 0$, for example. Proceeding formally, as in (15.1.2), (15.1.3) we have (incorrectly)

$$\zeta_T'(s) = \sum_{j=0}^{\infty} \frac{-n_j \log \lambda_j}{\lambda_j^s} \implies -\zeta_T'(0) = \sum_{j=0}^{\infty} n_j \log \lambda_j \tag{15.1.7}$$

$$\Rightarrow e^{-\zeta_T'(0)} = \prod_{j=0}^{\infty} e^{n_j \log \lambda_j} = \prod_{j=0}^{\infty} \lambda_j^{n_j}. \tag{15.1.8}$$

Thus by comparison with (15.1.3) it is natural to define the *determinant* of T by

$$\det T = e^{-\zeta_T'(0)} \tag{15.1.9}$$

provided ζ_T admits a meromorphic continuation at least to a half-plane $\operatorname{Re} s > -\epsilon$ for some $\epsilon > 0$, such that ζ_T is holomorphic at 0. We see how all of this works out for the example of the operators

$$T_M \stackrel{\text{def}}{=} \frac{-d^2}{dx^2} + x^{2M}, \quad M = 1, 2, 3, \ldots \tag{15.1.10}$$

on $L^2(\mathbb{R}, dx)$, where $T_M \psi = E\psi$ is a Schrödinger equation.

The potential energy function V of a harmonic oscillator of mass m and frequency ν is given by $V(x) = 2\pi^2\nu^2 m x^2$, as we have seen several times. More generally, we considered in Section 3.7 of Chapter 3 potential energy functions V of the form $V(x) = Ax^{2M}$ for $A > 0$, $M = 1, 2, 3, \ldots$. The case $M = 2$ gave rise to the anharmonic oscillator, which was considered in Chapter 10. The corresponding Hamiltonian operators (up to scaling) $H = \frac{-d^2}{dx^2} + Ax^{2M}$ have a discrete spectrum $\{\lambda_j(M), n_j\}_{j=0}^{\infty}$ where the multiplicity n_j of the eigenvalue $\lambda_j = \lambda_j(M)$ is 1 for every j. By this result of Titchmarsh and the asymptotic behavior of the $\lambda_j(M)$ as $j \to \infty$, given in Theorem 3.2, we checked in Corollary 3.2 that the corresponding spectral zeta functions (compare (15.1.6))

$$\zeta_H(s) = \zeta_{H,M}(s) \stackrel{\text{def}}{=} \sum_{j=0}^{\infty} \frac{n_j}{\lambda_j^s} = \sum_{j=0}^{\infty} \frac{1}{\lambda_j^s} \tag{15.1.11}$$

converged absolutely for $\operatorname{Re} s > \frac{M+1}{2M}$, and defined holomorphic functions on the half-planes $\operatorname{Re} s > \frac{M+1}{2M}$; $0 < \lambda_0 < \lambda_1 < \cdots$, $\lim_{j\to\infty} \lambda_j = \infty$. It is natural to inquire whether ζ_H admits a meromorphic continuation to the full complex plane, and if so whether ζ_H is holomorphic at $s = 0$.

To illustrate the meromorphic continuation of ζ_H, take $A = 1$, $M = 1$, for example. Only for the case $M = 1$ are the eigenvalues λ_j *known*: $\lambda_j = 2j + 1$ and for $\operatorname{Re} s > \frac{M+1}{2M} = 1$ we noted (see (3.7.39)) that

$$\zeta_H(s) = \sum_{n=0}^{\infty} \frac{1}{(2n+1)^s} = (1 - \frac{1}{2^s})\zeta(s) \tag{15.1.12}$$

for ζ in (15.1.1). Equation (15.1.12) therefore provides for the meromorphic continuation of ζ_H to the full complex plane. Since ζ is holomorphic except for a simple pole at $s = 1$, with residue 1, ζ_H has only a simple pole at $s = 1$ with residue there equal to $\frac{1}{2}$. Also one knows that $\zeta(0) = -\frac{1}{2}$. Hence (for $M = 1$)

Corollary 15.1. $\zeta_H'(0) = -\frac{1}{2}\log 2$.

Now consider M arbitrary again. The zeta functions $\zeta_{H,M}(s)$ in (15.1.11) (defined for $\operatorname{Re} s > \frac{M+1}{2M}$) still admit a meromorphic continuation to $\mathbb{C}$, with explicit poles and residues—due to André Voros.

Theorem 15.1 (A. Voros [85]). *$\zeta_{H,M}$ is holomorphic at $s = 0$ and in fact $\zeta'_{H,M}(0) = \log \sin \frac{\pi}{2(M+1)}$; here take $A = 1$.*

For $M = 1$, $\log \sin \frac{\pi}{2(M+1)} = \log \frac{1}{\sqrt{2}} = -\frac{1}{2} \log 2$. That is, Theorem 15.1 indeed reduces to Corollary 15.1 in case $M = 1$. By definition (15.1.9) it follows that, for $A = 1$,

Corollary 15.2. $\det T_M = \csc \frac{\pi}{2(M+1)}$!

Brief mention was made in the introductory remarks of this chapter about the occurrence of the determinant of Laplace type operators in Polyakov string theory. We shall have a bit more to say on this matter in the General Appendix C (following Chapter 18), where a concrete computational example will be considered.

16

Helmholtz Free Energy for Certain Negatively Curved Space-Times, and the Selberg Trace Formula

16.1 A Zeta Function for Kaluza–Klein Space-Times Modeled on $SO_1(d,1)/SO(d)$, and a Trace Formula for Compact Quotients of $SO_1(d,1)$

It is very interesting to consider the computation of the quantum statistical quantities $F(T)$, $S(T)$, $U(T)$, for example, of Chapter 14, in a topological setting, given a suitable partition function $Z(T)$, and in particular to understand the dependence of such quantities on the given topology. We are particularly interested in the case when the topology is that of a product $\mathbb{R}^p \times M^d$, where M^d is a compact hyperbolic manifold of dimension $d \geq 2$, which we conveniently express as

$$M^d = \Gamma\backslash X^d \qquad \text{for } \Gamma = \pi_1(M^d) = \text{ the fundamental group of } M^d \qquad (16.1.1)$$
$$\text{and } X^d = \text{ its universal covering manifold.}$$

$\mathbb{R}^p \times M^d = \mathbb{R} \times (\mathbb{R}^{p-1} \times M^d)$ is called a *Kaluza–Klein space-time*. Kaluza extended Einstein's four-dimensional theory of gravity to five dimensions (which for us corresponds to $p = 1$, $d = 4$) to incorporate Maxwell's theory of electromagnetism. Although a number of studies have been made on finite-temperature effects of quantum fields over space-times $\mathbb{R}^p \times N^d$, where the spatial sections N^d have zero or positive curvature, only recent investigations have focussed on the case of negatively curved spatial sections M^d. The results we present here therefore are new. They are a special case of more general results obtained by the author in [91] and presented at the Fourth Alexander Friedmann International Seminar on Gravitation and Cosmology, held in St. Petersburg, Russia. The key mathematical tool needed to obtain these results is the *Selberg trace formula*, a non-commutative version of the Jacobi inversion formula; see Appendix 14A.

Although we could take M^d to be complex or quaternionic as well, we will assume for simplicity that M^d is real. We can express X^d as

$$X^d = \{(x_1, x_2, \ldots, x_d) \in \mathbb{R}^d | x_d > 0\}, \tag{16.1.2}$$

which is a manifold with a global chart $(X^d, \mathbf{1} = (x_1, \ldots, x_d))$, where x_i is the ith projection function on X^d. X^d has a canonical Riemannian structure given by

$$\left\langle \frac{\partial}{\partial x_i}, \frac{\partial}{\partial x_j} \right\rangle = g_{ij}, \quad \text{where} \tag{16.1.3}$$

$$g_{ij}(x_1, \ldots, x_d) \stackrel{\text{def}}{=} \delta_{ij}/x_d^2,$$

$\frac{\partial}{\partial x_i}$ being the (global) ith coordinate vector field on X^d. The *Laplace–Beltrami operator* Δ corresponding to a metric $g = [g_{ij}]$ is (by definition) given by

$$\Delta : f \to \Delta f = \frac{1}{|\det g|^{1/2}} \sum_{j=1}^{d} \frac{\partial}{\partial x_j} \left(\sum_{i=1}^{d} g^{ij} |\det g|^{1/2} \frac{\partial f}{\partial x_i} \right) \tag{16.1.4}$$

where $g^{-1} = [g^{ij}]$, $f \in C^\infty(X^d)$. Clearly in our case $g^{ij}(x_1, \ldots, x_d) = \delta_{ij} x_d^2$ by (16.1.3). Hence by (16.1.4) we immediately obtain

Proposition 16.1.

$$\Delta = x_d^2 \sum_{j=1}^{d} \frac{\partial^2}{\partial x_j^2} + (2 - d) x_d \frac{\partial}{\partial x_d}$$

for the metric given in (16.1.3). *In particular for* $d = 2$ *(in which case* $X^d = \{(x, y) \in \mathbb{R}^2 | y > 0\}$ *is simply the upper half-plane, by* (16.1.2)*)*

$$\Delta = y^2 \left(\frac{\partial^2}{\partial x^2} + \frac{\partial^2}{\partial y^2} \right).$$

Besides the concrete representation (16.1.2), X^d has a nice well-known group-theoretic realization as a homogeneous space:

$$X^d = \mathrm{SO}_1(d, 1)/\,\mathrm{SO}(d) \tag{16.1.5}$$

where $K \stackrel{\text{def}}{=} \mathrm{SO}(d)$ is the special orthogonal group (see Chapter 7), and where for $I_n =$ the nth order identity matrix and

$$I_{pq} = \begin{bmatrix} -I_p & 0 \\ 0 & I_q \end{bmatrix}, \quad \text{one has} \tag{16.1.6}$$

$$\mathrm{SO}(p, q) \stackrel{\text{def}}{=} \{g \in \mathrm{Gl}(p + q, \mathbb{R}) | \det g = 1, g^t I_{pq} g = I_{pq}\}, \text{ with} \tag{16.1.7}$$

$$\mathrm{SO}_1(p, q) \stackrel{\text{def}}{=} \text{the identity component of } \mathrm{SO}(p, q). \tag{16.1.8}$$

In the realization (16.1.5) we regard Γ in (16.1.1) as a torsion-free (i.e., only $1 \in \Gamma$ has finite order), discrete subgroup of $G \overset{\text{def}}{=} SO_1(d,1)$ such that $\Gamma\backslash G$ is compact, since M^d is compact. Since Δ is G-invariant, and thus is Γ-invariant in particular, Δ projects to a differential operator Δ_Γ on the quotient space

$$X_\Gamma \overset{\text{def}}{=} \Gamma\backslash X^d = \Gamma\backslash G/K, \tag{16.1.9}$$

which is a compact negatively curved manifold. For convenience we work with $-\Delta_\Gamma$ rather than Δ_Γ. As an elliptic operator on a compact manifold X_Γ, $-\Delta_\Gamma$ has a well defined *spectrum* $\{\lambda_j, n_j\}_{j=0}^{\infty}$ where n_j is the (finite) multiplicity of the eigenvalue λ_j of $-\Delta_\Gamma$. The minus sign preceding Δ_Γ provides for the non-negativity of the λ_j, which we may order as follows:

$$0 = \lambda_0 < \lambda_1 < \lambda_2 < \lambda_3 < \dots, \qquad \text{with } \lim_{j\to\infty} \lambda_j = \infty. \tag{16.1.10}$$

Note that in the important special case $d = 2$, X_Γ in (16.1.9) is the typical compact Riemann surface (with fundamental group Γ) of genus ≥ 2; $SO_1(2,1)$ is locally isomorphic to $Sl(2,\mathbb{R})$, the group of real 2×2 matrices with determinant equal to 1.

For a quantum field in thermal equilibrium at a finite temperature $T = 1/(k\beta)$ (we will take $k = 1$) on some curved background space-time, a suitable partition function Z exists with a path integral representation; cf. [2, 4, 11, 12, 17, 28, 33, 45]. Similar to (14.1.2), where the integration is taken over periodic paths $x(\beta\hbar) = x(0)$, one obtains Z by integrating over *periodic fields* $x(\tau) = x(\tau + \beta)$, where $\tau = it$ is imaginary time; recall the Wick rotation $t \to -it$ which we used earlier to transform Feynman integrals into Euclidean path integrals. A discussion of quantum fields is beyond the scope of these notes, of course. The notion of a measure on a space of fields over some space, moreover, is a bit too fuzzy to engage our time in discussion here. Fortunately, we can however achieve a clear mathematical meaning of the partition function Z by using a suitable zeta function ζ_Γ attached to the space-times $\mathbb{R}^p \times (\Gamma\backslash X^d)$, which we shall describe presently. Namely for a suitable constant c we may set

$$\log Z(\beta) = \frac{1}{2}\left[\zeta_\Gamma'(0) + c\zeta_\Gamma(0)\right]. \tag{16.1.11}$$

Actually, when p is odd (in particular when $p = 1$, a case of special importance) it will turn out that $\zeta_\Gamma(0) = 0$. Since our interest is in the case of odd p it will not be necessary to specify c. Of course definition (16.1.11) is quite analogous to Theorem 14.3 which says precisely that

$$\log Z(\beta) = \frac{1}{2}\zeta'(0), \tag{16.1.12}$$

for the harmonic oscillator, for the zeta function ζ given in (14.2.1).

To describe the zeta function ζ_Γ attached to $X_\Gamma^{p,d} \overset{\text{def}}{=} \mathbb{R}^p \times M^d = \mathbb{R} \times \mathbb{R}^{p-1} \times (\Gamma\backslash X^d)$ we employ not only the spectrum $\{\lambda_j, n_j\}_{j=0}^{\infty}$ of the Laplacian $-\Delta_\Gamma$ on

$\Gamma\backslash X^d$ as mentioned earlier, but also the spectrum of the differential operator

$$D_\mu \stackrel{\text{def}}{=} -\left(\frac{d}{dt} - \mu\right)^2 = -\frac{d^2}{dt^2} + 2\mu\frac{d}{dt} - \mu^2, \tag{16.1.13}$$

where μ is a fixed complex number. μ has a physical meaning: it serves to represent the *chemical potential* of a scalar quantum field over $X_\Gamma^{p,d}$. Although we could take $\mu = 0$ we prefer to carry out our computations, more generally, in the presence of a non-vanishing chemical potential. Note that the periodic functions

$$f_n(t+\beta) = f_n(t) \stackrel{\text{def}}{=} \exp\left(\frac{2\pi i n}{\beta}t\right), \quad n \in \mathbb{Z}, \tag{16.1.14}$$

are eigenfunctions of D_μ:

$$D_\mu f_n = \lambda_n(\beta,\mu) f_n \quad \text{for } \lambda_n(\beta,\mu) \stackrel{\text{def}}{=} \left[\frac{2\pi n}{\beta} + i\mu\right]^2. \tag{16.1.15}$$

We use the eigenvalues λ_j, $\lambda_n(\beta,\mu)$ to form θ-functions (defined for $t > 0$)

$$\theta_{\beta,\mu}(t) \stackrel{\text{def}}{=} \sum_{n\in\mathbb{Z}} \exp(-t\lambda_n(\beta,\mu)), \tag{16.1.16}$$

$$\Theta_\Gamma(t;b) \stackrel{\text{def}}{=} \sum_{j=0}^{\infty} n_j \exp(-t(\lambda_j + b)), \tag{16.1.17}$$

where $b > 0$ is fixed. In practice one takes $b = m^2$, where m is the scalar field mass, which we assume is non-zero; with a little care we could easily treat the case $m = 0$ as well. With $m \neq 0$ there are no *zero modes* present; i.e., $\lambda_n(\beta,\mu) + \lambda_j + m^2 \neq 0$ for all n, j. We come now to the *main definition* of this chapter. The zeta function ζ_Γ which provides for a clear meaning of the path integral $Z(\beta)$ in (16.1.11) is defined by the following Mellin transform:

$$\zeta_\Gamma(s) = \zeta_\Gamma(s;b,\beta,\mu) \stackrel{\text{def}}{=} \frac{1}{(4\pi)^{(p-1)/2}\,\Gamma(s)} \cdot \int_0^\infty \theta_{\beta,\mu}(t)\, \Theta_\Gamma(t;b)\, t^{[s-(p-1)/2]-1}\, dt \tag{16.1.18}$$

for Re $s > \dim X_\Gamma^{p,q}/2 = (p+d)/2$.

By the Jacobi inversion formula, Theorem 14.6 of Appendix 14A, we can write

$$\theta_{\beta,\mu}(t) = \frac{\beta}{\sqrt{4\pi t}} \sum_{n\in\mathbb{Z}} \exp\left(-\frac{n^2\beta^2}{4t} - n\mu\beta\right) \tag{16.1.19}$$

$$= \frac{\beta}{\sqrt{4\pi t}} \left[1 + 2\sum_{n=1}^{\infty} (\cosh n\mu\beta) \exp\left(-\frac{n^2\beta^2}{4t}\right)\right]. \tag{16.1.20}$$

Therefore

$$\theta_{\beta,\mu}(t)\Theta_\Gamma(t;b)\, t^{[s-(p-1)/2]-1} = \frac{\beta}{\sqrt{4\pi}}\Theta_\Gamma(t;b)t^{[s-p/2]-1}$$
$$+\frac{2\beta}{\sqrt{4\pi t}}\sum_{n=1}^{\infty}(\cosh n\mu\beta)\exp\left(-\frac{n^2\beta^2}{4t}\right)\Theta_\Gamma(t;b)\, t^{[s-(p-1)/2]-1}, \quad (16.1.21)$$

so that if we set

$$\zeta_\Gamma(s;b) \stackrel{\text{def}}{=} \frac{1}{\Gamma(s)}\int_0^\infty \Theta_\Gamma(t;b)\, t^{s-1}\, dt \quad \text{for } \operatorname{Re} s > \frac{d}{2},$$
$$M_{\beta,\mu}(s;b) \stackrel{\text{def}}{=} \int_0^\infty \frac{\beta}{\sqrt{\pi t}}\sum_{n=1}^{\infty}(\cosh n\mu\beta)\exp\left(-\frac{n^2\beta^2}{4t}\right)\Theta_\Gamma(t;b)\, t^{s-1}\, dt \quad (16.1.22)$$
$$\text{for any } s \in \mathbb{C}$$

we have the following.

Proposition 16.2.

$$\zeta_\Gamma(s;b,\beta,\mu) = \frac{\beta}{(4\pi)^{p/2}}\frac{\Gamma(s-p/2)}{\Gamma(s)}\zeta_\Gamma(s-p/2;b) + \frac{M_{\beta,\mu}(s-(p-1)/2;b)}{(4\pi)^{(p-1)/2}\,\Gamma(s)}.$$

Theorem 16.1. *$M_{\beta,\mu}(s;b)$ is an entire function of s. The zeta function $\zeta_\Gamma(s;b)$ admits an explicit meromorphic continuation in terms of Γ-structure and in terms of the spherical harmonic analysis on the symmetric space $G/K = SO_1(d,1)/SO(d)$. In particular for d even ζ_Γ has at most simple poles at $s = 1, 2, \ldots d/2$ with known residues.*

Indeed, using the Selberg trace formula, the author has worked out in [90] the complete, explicit meromorphic structure of $\zeta_\Gamma(s;b)$. Thus Proposition 16.2 and Theorem 16.1 provide for the meromorphic structure of the *finite temperature* zeta function $\zeta_\Gamma(s;b,\beta,\mu)$, which we see is decomposed into an entire function $M_{\beta,\mu}$ translated by $(p-1)/2$, and a simpler *zero temperature* zeta function ζ_Γ at zero chemical potential.

An important special case is $p = 1$, $d = 2$, as indicated earlier.

Proposition 16.3. *If p is odd and d is even (example $p = 1$, and $d = 2$), then*

$$\zeta_\Gamma(s;b,\beta,\mu) = 0 \qquad \text{for } s = 0, -1, -2, -3, \ldots.$$

Proof. We use Proposition 16.2. $\frac{1}{\Gamma(s)} = 0$ for $s = 0, -1, -2, -3, \ldots$ and $M_{\beta,\mu}$ is entire. Thus

$$\frac{M_{\beta,\mu}(s-(p-1)/2)}{(4\pi)^{(p-1)/2}\,\Gamma(s)} = 0 \qquad \text{for } s = 0, -1, -2, -3, \ldots.$$

If p is odd, then $\Gamma(s - p/2)$ is finite for $s = 0, -1, -2, -3, \ldots$, and similarly $\zeta_\Gamma(s - p/2; b)$ is finite for the latter values of s (by Theorem 16.1 as d is even).

Consequently

$$\frac{\beta}{(4\pi)^{p/2}}\frac{\Gamma(s-p/2)}{\Gamma(s)}\zeta_\Gamma(s-p/2;b)=0 \text{ for } s=0,-1,-2,-3,\ldots.$$

Proposition 16.3 therefore follows by Proposition 16.2. □

In particular $\zeta_\Gamma(0;b,\beta,\mu)=0$ by Proposition 16.3, for p odd and d even, in which case (16.1.11) reduces to

$$\log Z(\beta)=\frac{1}{2}\zeta'_\Gamma(0;b,\beta,\mu) \text{ for } p \text{ odd, } d \text{ even,} \tag{16.1.23}$$

which is the exact analogue of Theorem 14.3.

Further notation is introduced at this point, notation necessary for the statement of the trace formula and the main result. Let

$$A=\left\{\left[\begin{array}{ccc}\cosh t & 0 & \sinh t\\ 0 & I_{d-1} & 0\\ \sinh t & 0 & \cosh t\end{array}\right]\middle|\, t\in\mathbb{R}\right\}, \tag{16.1.24}$$

which is a closed Abelian subgroup of G with Lie algebra

$$a_o=\mathbb{R}H_o \quad \text{for } H_o \stackrel{\text{def}}{=} \left[\begin{array}{ccccc}0 & 0 & \cdots & 0 & 1\\ 0 & & & & 0\\ \vdots & & & & \vdots\\ 0 & & & & 0\\ 1 & 0 & \cdots & 0 & 0\end{array}\right]; \tag{16.1.25}$$

recall I_n denotes the nth order identity matrix. Let Ad denote as usual the adjoint representation of G on its Lie algebra g_o, and its complexified Lie algebra $g=g_o^{\mathbb{C}}$. Then for $\gamma\in\Gamma-\{1\}$ we can define $t_\gamma>0$ by

$$e^{t_\gamma} \stackrel{\text{def}}{=} \max\{|c| \mid c= \text{an eigenvalue of } \operatorname{Ad}(\gamma): g\to g\}. \tag{16.1.26}$$

Actually t_γ arises as follows. Let M denote the centralizer of A in K:

$$M \stackrel{\text{def}}{=} \{m\in K|ma=am\ \forall a\in A\}\simeq \mathrm{SO}(d-1). \tag{16.1.27}$$

Also let

$$A^+=\{\exp tH_o|t>0\}\subset A. \tag{16.1.28}$$

Then since Γ is torsion free, each $\gamma\in\Gamma-\{1\}$ is conjugate in G to an element of MA^+:

$$\exists x\in G \text{ such that } x\gamma x^{-1}=m_\gamma\exp(t_\gamma H_o)\in MA^+. \tag{16.1.29}$$

Moreover the element $m_\gamma \in M$ is unique up to conjugation in M. The group G admits an *Iwasawa decomposition*

$$G = KAN \tag{16.1.30}$$

where $N \stackrel{\text{def}}{=} \exp n_o$ for

$$n_o \stackrel{\text{def}}{=} \left\{ \begin{bmatrix} 0 & x^t & 0 \\ -x & 0 & x \\ 0 & x^t & 0 \end{bmatrix} \middle| \; x = \text{ column vector in } \mathbb{R}^{d-1} \right\}, \tag{16.1.31}$$

a subalgebra of g_o. Since m_γ is unique up to conjugation in M the following function C on $\Gamma - \{1\}$ is well defined:

$$\frac{1}{C(\gamma)} \stackrel{\text{def}}{=} \exp\left(t_\gamma \frac{d-1}{2}\right) \left|\det{}_{n_o}(\mathrm{Ad}(m_\gamma \exp(t_\gamma H_o))^{-1} - 1\right|. \tag{16.1.32}$$

Finally, again using that Γ is torsion free, we can associate to each $\gamma \in \Gamma - \{1\}$ an integer $j(\gamma) \geq 1$: γ has a unique representation

$$\gamma = \delta^{j(\gamma)} \qquad \text{where } j(\gamma) \geq 1 \text{ is an integer, } \delta \in \Gamma - \{1\} \text{ is a } \textit{prime}; \tag{16.1.33}$$

i.e., δ cannot be written as γ_1^j for $\gamma_1 \in \Gamma$ and $j > 1$ an integer. The definitions and notation from (16.1.24) to (16.1.33) obviously involve basis structure theory of the pair (G, Γ). Such structure theory is available for any pair (G, Γ) where G is a connected, split rank 1, non-compact semi-simple Lie group ($G = \mathrm{SO}_1(d, 1)$ being an example of such) and where Γ is a co-compact, torsion-free, discrete subgroup of G; the reader may consult [89] for details. The latter reference contains also a proof of the following statement. For choice of Haar measure dx on G there is a unique G-invariant measure $d\tilde{x}$ on $\Gamma\backslash G$ such that

$$\int_G f(x)\, dx = \int_{\Gamma\backslash G} \left[\sum_{\gamma\in\Gamma} f(\gamma x)\right] d\tilde{x}(\Gamma x) \tag{16.1.34}$$

for any continuous, compactly supported test function f on G. We set

$$\mathrm{vol}(\Gamma\backslash G) \stackrel{\text{def}}{=} \int_{\Gamma\backslash G} 1\, d\tilde{x}. \tag{16.1.35}$$

The Selberg trace formula will now be stated for the triple $(\mathrm{SO}_1(d, 1), \mathrm{SO}(d), \Gamma)$. Actually we give a special version of the formula, one suitable for our purposes, which results from a special choice of test function. The more general version, for a general test function, and for a larger class of Lie groups G, is discussed in [89], where proofs are available in detail; also see [38, 48, 64, 77, 86, 90].

Theorem 16.2 (Selberg Trace Formula for co-compact quotients of $\mathrm{SO}_1(d,1)$). *As above let $\Gamma \subset G = \mathrm{SO}_1(d,1)$ be a torsion-free, discrete subgroup such that $\Gamma\backslash G$ is compact. Let*

$$\Theta_\Gamma(t;b) = \sum_{j=0}^{\infty} n_j \exp(-t(\lambda_j + b))$$

be the θ-function of (16.1.17), *where $t, b > 0$. Let C_Γ be a complete set of representatives in Γ of its conjugacy classes (here recall that γ_1, γ_2 in Γ are* conjugate *if $\gamma_1 = \gamma\gamma_2\gamma^{-1}$ for some $\gamma \in \Gamma$), and form the θ-function*

$$\theta_\Gamma(t;b) \overset{\text{def}}{=} \frac{1}{\sqrt{4\pi t}} \sum_{\gamma \in C_\Gamma - \{1\}} t_\gamma j(\gamma)^{-1} C(\gamma) \exp\left(-\left[bt + \left(\frac{d-1}{2}\right)^2 t + \frac{t_\gamma^2}{4t}\right]\right),$$

where t_γ, $C(\gamma)$, $j(\gamma)$, for $\gamma \in \Gamma - \{1\}$, are given by (16.1.26), (16.1.32), (16.1.33). *Then $\theta_\Gamma(t;b)$ is finite, and one has*

$$\Theta_\Gamma(t;b) = \frac{\mathrm{vol}(\Gamma\backslash G)}{4\pi} \int_{\mathbb{R}} \exp\left(-\left[r^2 + b + \frac{d-1}{2}\right] t\right) |c(r)|^{-2}\, dr + \theta_\Gamma(t;b), \tag{16.1.36}$$

where for a suitable choice of Haar measure dx on G $\mathrm{vol}(\Gamma\backslash G)$ *is given by* (16.1.35), *and where $|c(r)|^{-2}\, dr$ is the spherical* Harish-Chandra–Plancherel measure *of the symmetric pair* $(\mathrm{SO}_1(d,1), \mathrm{SO}(d))$, *given by*

$$|c(r)|^{-2} = \begin{bmatrix} C_G \pi r\ P(r) \tanh \pi r & \text{for } d = 2n \\ C_G \pi\ P(r) & \text{for } d = 2n+1 \end{bmatrix} \tag{16.1.37}$$

with

$$\begin{aligned} C_G &= \frac{1}{2^{4n-4}\,\Gamma(n)^2}, \qquad & P(r) &= \prod_{j=0}^{n-2} \left[r^2 + \frac{(2j+1)^2}{4}\right] \\ & & & \text{for } d = 2n, n \geq 1, \\ C_G &= \frac{1}{2^{4n-2}\,\Gamma(n+1/2)^2}, \qquad & P(r) &= \prod_{j=1}^{n} \left[r^2 + (n-j)^2\right] \\ & & & \text{for } d = 2n+1, n \geq 1. \end{aligned} \tag{16.1.38}$$

We shall assume $d = 2n$ is even and p is odd. Thus definition (16.1.23) applies. If one applies the trace formula (16.1.36), one obtains an alternative expression for $M_{\beta,\mu}(s;b)$ in (16.1.22), which can be made more explicit by the commutation of integration with certain summations. To justify these commutations one needs an assumption on the chemical potential μ. We assume

$$|\mathrm{Re}\,\mu| < \sqrt{\lambda_1}, \frac{d-1}{2}, \sqrt{b}, \tag{16.1.39}$$

where λ_1 is the first non-zero eigenvalue of $-\Delta_\Gamma$; see (16.1.10). Then one can show that

$$\begin{aligned} &M_{\beta,\mu}(s-(p-1)/2;b) \\ &\quad= \frac{\beta\operatorname{vol}(\Gamma\backslash G)}{\sqrt{\pi}4\pi}2^{-s+1+p/2} \\ &\qquad\cdot\sum_{n=1}^{\infty}(\cosh n\mu\beta)(n\beta)^{s-p/2}\int_{\mathbb{R}}\frac{K_{-s+p/2}(n\beta\sqrt{r^2+b+\rho_o^2})\,|c(r)|^{-2}}{\left[\sqrt{r^2+b+\rho_o^2}\right]^{s-p/2}}\,dr \\ &\qquad+\frac{\beta}{\pi\left[2\sqrt{b+\rho_o^2}\right]^{s-(p+1)/2}}\sum_{n=1}^{\infty}(\cosh n\mu\beta)\sum_{\gamma\in C_\Gamma-\{1\}}t_\gamma j(\gamma)^{-1}C(\gamma) \\ &\qquad\cdot\left[\sqrt{n^2\beta^2+t_\gamma^2}\right]^{s-(p+1)/2}K_{-s+(p+1)/2}\left(\sqrt{b+\rho_o^2}\sqrt{n^2\beta^2+t_\gamma^2}\right) \end{aligned} \tag{16.1.40}$$

for μ subject to (16.1.36), where

$$\rho_o \overset{\text{def}}{=} \frac{d-1}{2} \tag{16.1.41}$$

and where the K-Bessel function K_ν is defined in (14.2.11). In particular, by Proposition 16.2

$$\zeta_\Gamma'(0;b,\beta,\mu) = \frac{\beta}{(4\pi)^{p/2}}\Gamma(-p/2)\zeta_\Gamma(-p/2;b) + \frac{M_{\beta,\mu}(-(p-1)/2;b)}{(4\pi)^{(p-1)/2}}, \tag{16.1.42}$$

where

$$\begin{aligned} &M_{\beta,\mu}(-(p-1)/2;b) \\ &\quad= \frac{\beta\operatorname{vol}(\Gamma\backslash G)}{\sqrt{\pi}4\pi}2^{1+p/2} \\ &\qquad\cdot\sum_{n=1}^{\infty}(\cosh n\mu\beta)(n\beta)^{-p/2}\int_{\mathbb{R}}\frac{K_{p/2}(n\beta\sqrt{r^2+b+\rho_o^2})\,|c(r)|^{-2}}{\left[\sqrt{r^2+b+\rho_o^2}\right]^{-p/2}}\,dr \\ &\qquad+\frac{\beta\left[2\sqrt{b+\rho_o^2}\right]^{(p+1)/2}}{\pi}\sum_{n=1}^{\infty}(\cosh n\mu\beta)\sum_{\gamma\in C_\Gamma-\{1\}}t_\gamma j(\gamma)^{-1}C(\gamma) \\ &\qquad\cdot\left[\sqrt{n^2\beta^2+t_\gamma^2}\right]^{-(p+1)/2}K_{(p+1)/2}\left(\sqrt{b+\rho_o^2}\sqrt{n^2\beta^2+t_\gamma^2}\right). \end{aligned} \tag{16.1.43}$$

By analogy with (14.2.2) we define the Helmholtz free energy $F(T)$ in the present context, of some quantum field in thermal equilibrium at finite temperature $T=\frac{1}{\beta}$ over our Kaluza–Klein space-time $\mathbb{R}^p\times(\Gamma\backslash\operatorname{SO}_1(d,1)/\operatorname{SO}(d))$:

$$F(T) \overset{\text{def}}{=} -\frac{1}{\beta}\log Z(\beta) = -\frac{1}{2\beta}\zeta_\Gamma'(0;b,\beta,\mu) \tag{16.1.44}$$

by (16.1.23), where $m = \sqrt{b}$ is the mass of the field. Then by (16.1.42), (16.1.43) we obtain the following main result.

Theorem 16.3. *Assume p is odd, d is even, and μ satisfies* (16.1.39). *Then the free energy $F(T)$ in* (16.1.44) *is given by*

$$\begin{aligned} F(T) &= -\frac{\Gamma(-p/2)}{2(4\pi)^{p/2}}\zeta_\Gamma(-p/2;m^2) \\ &\quad -\frac{\operatorname{vol}(\Gamma\backslash G)2^{p/2}}{\sqrt{\pi}(4\pi)^{(p+1)/2}}\sum_{n=1}^{\infty}(\cosh n\mu\beta)(n\beta)^{-p/2} \\ &\quad \cdot\int_{\mathbb{R}} K_{p/2}(n\beta\sqrt{r^2+m^2+\rho_o^2})\left[\sqrt{r^2+m^2+\rho_o^2}\right]^{p/2}|c(r)|^{-2}\,dr \\ &\quad -\frac{[2\sqrt{m^2+\rho_o^2}]^{(p+1)/2}}{2\pi(4\pi)^{(p-1)/2}}\sum_{n=1}^{\infty}(\cosh n\mu\beta)\sum_{\gamma\in C_\Gamma-\{1\}} t_\gamma j(\gamma)^{-1}C(\gamma) \\ &\quad \cdot\left[\sqrt{n^2\beta^2+t_\gamma^2}\right]^{-(p+1)/2} K_{(p+1)/2}(\sqrt{m^2+\rho_o^2}\sqrt{n^2\beta^2+t_\gamma^2}), \end{aligned} \tag{16.1.45}$$

where $\rho_o = (d-1)/2$, ζ_Γ is the zeta function in (16.1.22), $\operatorname{vol}(\Gamma\backslash G)$, t_γ, $j(\gamma)^{-1}$, *$C(\gamma)$ are given in definitions* (16.1.26) *through* (16.1.35), *K_ν is the K-Bessel function of* (14.2.11), *and $|c(r)|^{-2}$ is the spherical Harish-Chandra–Plancherel density of* $SO_1(d,1)/SO(d)$ *given by* (16.1.37), (16.1.38).

Some remarks concerning Theorem 16.3 are in order. From definitions (16.1.17), (16.1.22) one has (commuting integration and summation)

$$\zeta_\Gamma(s;b) = \sum_{j=0}^{\infty}\frac{n_j}{(\lambda_j+b)^s} \qquad \text{for Re } s > \frac{d}{2}, \tag{16.1.46}$$

which realizes ζ_Γ as a spectral zeta function of Minakshisundaram–Pleijel type [65, 85]. The first term

$$\frac{-\Gamma(-p/2)}{2(4\pi)^{p/2}}\zeta_\Gamma(-p/2;m^2) \tag{16.1.47}$$

in (16.1.45), which is *temperature independent*, is the *Casimir energy* of the space-time. In case $p = 1$, for example, the space-time $\mathbb{R}\times(\Gamma\backslash G/K)$ is called *Clifford–Klein*, and a beautiful formula exists which expresses the Casimir energy in terms of the logarithmic derivative ψ_Γ of the *Selberg zeta* function attached to $\Gamma\backslash G/K$ [11, 12, 14, 38, 89]. Such a formula is established in [88], for a general rank 1 space form $\Gamma\backslash G/K$, and is a consequence of the following formula (also proved in [88]) which provides a direct relationship between ζ_Γ and ψ_Γ:

$$\begin{aligned} \zeta_\Gamma(s;b) &= \frac{\operatorname{vol}(\Gamma\backslash G)}{4\pi}I(s;b) \\ &\quad +\frac{\sin\pi s}{\pi}\int_0^\infty \psi_\Gamma(\rho_o+t+a(b))(2a(b)t+t^2)^{-s}\,dt \quad \text{for Re} s < 1, \end{aligned} \tag{16.1.48}$$

where $a(b) = \sqrt{b+\rho_o^2}$, and where $I(s;b)$ is explicitly given in terms of the Plancherel density $|c(r)|^{-2}$. Formula (16.1.48) also holds for $b = 0$, but in this case the summation in (16.1.46) begins with $j = 1$, and we require Re $s < 0$. We can relate the Casimir energy with the logarithmic derivative ψ_Γ of Selberg's zeta function for any odd p, more generally. Of course formula (16.1.48) is of independent interest.

Consider the second and third terms in (16.1.45) which represent the non-trivial (temperature dependent) contribution to the free energy. The sum of the latter two terms corresponds to $\Omega_\beta^-(\beta,\mu)$ in [91], where the superscript "−" indicates that the field is *bosonic*, whereas the superscript "+" in [91] is reserved for *fermionic* fields. By [40, p. 967]

$$K_{q+1/2}(x) = \sqrt{\frac{\pi}{2x}}e^{-x}\sum_{k=0}^{q}\frac{(q+k)!}{k!(q-k)!(2x)^k}. \tag{16.1.49}$$

Write $p = 2q+1$ so that $p/2 = q+1/2$. The choice $x = n\beta\sqrt{r^2+m^2+\rho_o^2}$ then allows us to express the second term in (16.1.45) as

$$-\frac{\mathrm{vol}(\Gamma\backslash G)2^{(p-1)/2}}{(4\pi)^{(p+1)/2}\beta^{(p+1)/2}}\sum_{n=1}^{\infty}(\cosh n\mu\beta)n^{-(p+1)/2}\sum_{k=0}^{q}\frac{(q+k)!}{k!(q-k)!}$$
$$\cdot\int_{\mathbb{R}}(r^2+m^2+\rho_o^2)^{(p-1-2k)/4}\exp\left(-n\beta\sqrt{r^2+m^2+\rho_o^2}\right)(2n\beta)^{-k}\,|c(r)|^{-2}\,dr, \tag{16.1.50}$$

which with the third term in (16.1.45) provides for the content of Theorem 3.1 of [91]. For the bosonic case there where r^- is taken $= 1$, $\mathrm{vol}(M^d) = \mathrm{vol}(\Gamma\backslash G)/4\pi$, and $F(T)$ is referred to as the *one-loop effective potential* $\Omega^-(\beta,\mu)$ or *thermodynamic potential.*

16.2 Temperature Asymptotics

In Section 14.1 of Chapter 14 we obtained the low and high temperature behavior of $U(T)$. Similarly by Theorem 16.3 we can obtain a low and high temperature asymptotic expansion of $F(T)$. For example the following expansion is good for large x:

$$K_\nu(x) = \sqrt{\frac{\pi}{2x}}e^{-x}\left[\sum_{k=0}^{l-1}\frac{\Gamma(\nu+k+1/2)}{(2x)^k\,k!\,\Gamma(\nu-k-1/2)}+\frac{\theta\,\Gamma(\nu+l+1/2)}{(2x)^l\,l!\,\Gamma(\nu-l+1/2)}\right], \tag{16.2.1}$$

say for $\nu\in\mathbb{R}$, $x>0$, $l>\nu-1/2$ $(l=1,2,3,\dots)$, where $|\theta|\le 1$; see [40, p. 963]. To obtain an expression for $F(T)$ good for *low temperatures* T (i.e., for large β) we take $x = \sqrt{m^2+\rho_o^2}\sqrt{n^2\beta^2+t_\gamma^2}$, $\nu = (p+1)/2$.

Theorem 16.4. *Using* (16.2.1), *we can write the third term in* (16.1.45) *as*

$$-\frac{\left[2\sqrt{m^2+\rho_o^2}\right]^{p/2}}{2\sqrt{\pi}(4\pi)^{(p-1)/2}}\sum_{n=1}^{\infty}(\cosh n\mu\beta)\sum_{\gamma\in C_\Gamma-\{1\}} t_\gamma j(\gamma)^{-1}C(\gamma)$$

$$\cdot\frac{\exp\left(-\sqrt{m^2+\rho_o^2}\sqrt{n^2\beta^2+t_\gamma^2}\right)}{\left[n^2\beta^2+t_\gamma^2\right]^{(p+2)/4}}$$

$$\cdot\left[\sum_{k=0}^{l-1}\frac{1}{(2\sqrt{m^2+\rho_o^2}\sqrt{n^2\beta^2+t_\gamma^2})^k\, k!}\frac{\Gamma(q+k+3/2)}{\Gamma(q-k+3/2)}\right.$$

$$\left.+\theta\frac{\Gamma(q+l+3/2)}{\left[2\sqrt{m^2+\rho_o^2}\sqrt{n^2\beta^2+t_\gamma^2}\right]^l\, l!\,\Gamma(q-l+3/2)}\right]$$

for any integer $l > p/2$*; here* $|\theta| \leq 1$.

Using the expression (16.1.50) for the second term in (16.1.45) and [40, formula 8.446, p. 961] for the behavior of $K_{q+1}(x)$ for *small* $x > 0$ $(q = 0, 1, 2, \dots)$, we can rewrite the third term in (16.1.45) and obtain similarly high temperature asymptotics for $F(T)$, which with Theorems 16.3, 16.4 extend the results of [10] (which focussed on the important special case $p = 1$) to any odd dimension p of the Euclidean factor $\mathbb{R}^p$ of the space-time $\mathbb{R}^p \times \Gamma\backslash X^d$.

The results of this chapter, including the trace formula, can be extended to a vector bundle setting—namely to an automorphic vector bundle $V_\chi \to \Gamma\backslash X^d$ induced by a finite-dimensional unitary representation χ of Γ. For simplicity, we have taken χ to be the one-dimensional trivial representation in this chapter. We will take up another application of the trace formula in the next chapter. For further applications of this important formula (and of the Selberg zeta function) in quantum mechanics and in quantum field theory the reader may consult [1, 11, 12, 16, 37, 52].

The idea of using the trace formula to find a fairly explicit meromorphic continuation of the spectral zeta function in Theorem 16.1 goes back to Burton Randol [72], in the special (but important) case when $d = 2$. The extension of Randol's idea to an arbitrary split rank one setting is carried out in [90]. Spectral zeta functions are defined for any compact manifold ($\Gamma\backslash G/K$ being a special case), and their meromorphic continuation is carried out in general, but not explicitly, by S. Minakshisundaram and A. Pleijel [65].

17

The Zeta Function of a Product of Laplace Operators and the Multiplicative Anomaly for X_Γ^d

17.1 $\det L_1 L_2 = \det L_1 \det L_2$?

In Chapter 15 we defined the determinant $\det L$, in certain situations, of an operator L on an infinite-dimensional space by way of zeta regularization. Given two operators L_1, L_2 it is natural to inquire whether the familiar formula

$$\det L_1 L_2 = \det L_1 \det L_2 \tag{17.1.1}$$

remains valid. We shall see in general that it does not. The question of the validity of (17.1.1) is one not only of pure mathematical interest but is of practical interest as well, as it arises in various field theories. Namely, in certain settings [13, 18], it is important to compute the *multiplicative anomaly* $\alpha(L_1, L_2)$ defined by

$$\alpha(L_1, L_2) \stackrel{\text{def}}{=} \log\det L_1 L_2 - \log\det L_1 - \log\det L_2; \tag{17.1.2}$$

$$\text{i.e., } \det L_1 L_2 = \exp(\alpha(L_1, L_2)) \det L_1 \det L_2. \tag{17.1.3}$$

We see that (17.1.1) is valid if and only if the anomaly $\alpha(L_1, L_2)$ vanishes.

Of particular interest is the computation of $\alpha(L_1, L_2)$ when L_1, L_2 are Laplace type operators. Taking advantage of the notation set up in Chapter 16 we specifically choose

$$L_r = -\Delta_\Gamma + b_r, \qquad r = 1, 2, \tag{17.1.4}$$

where $b_1, b_2 > 0$ (b_1, b_2 may represent some field masses), and where Δ_Γ is the Laplace–Beltrami operator on $X_\Gamma^d = \Gamma \backslash \mathrm{SO}_1(d,1)/\mathrm{SO}(d)$; see Proposition 16.1 and remarks following it. The computation of $\alpha(L_1, L_2)$ for such L_r is non-trivial.

The key mathematical tool which we rely on is, again, the Selberg trace formula—formula (16.1.36) of the previous chapter. An attempt to provide a full array of details, as we have done in other discussions, will not be undertaken here. We will highlight some of the key points involved in the computation, again making free use of the notation of Chapter 16. The starting point is the study of the spectral zeta function of the product of Laplace operators $L_1 L_2$.

As $\{\lambda_j, n_j\}_{j=0}^{\infty}$ denotes the spectrum of $-\Delta_\Gamma$, the spectrum of $L_r \stackrel{\text{def}}{=} -\Delta_\Gamma + b_r$, $1 \leq r \leq 2$, is $\{\lambda_j + b_r, n_j\}_{j=0}^{\infty}$ and the corresponding spectral zeta function of L_r (compare (15.1.6)) is

$$\zeta_{L_r}(s) \stackrel{\text{def}}{=} \sum_{j=0}^{\infty} \frac{n_j}{(\lambda_j + b_r)^s} \qquad \text{for } \operatorname{Re} s > \frac{d}{2}. \tag{17.1.5}$$

That is, by (16.1.46)

$$\zeta_{L_r}(s) \equiv \zeta_\Gamma(s; b_r), \tag{17.1.6}$$

and by definition (15.1.9)

$$\det L_r \stackrel{\text{def}}{=} \exp(-\zeta_\Gamma'(0; b_r)). \tag{17.1.7}$$

The spectrum of $L_1 L_2$ is $\{(\lambda_j + b_1)(\lambda_j + b_2), n_j\}_{j=0}^{\infty}$ (which is consistent with finite-dimensional linear algebra), and the corresponding spectral zeta function of the product of Laplacians $L_1 L_2$ is given by

$$\zeta_\Gamma(s; b_1, b_2) \stackrel{\text{def}}{=} \sum_{j=0}^{\infty} \frac{n_j}{[(\lambda_j + b_1)(\lambda_j + b_2)]^s} \qquad \text{for } \operatorname{Re} s \text{ sufficiently large.} \tag{17.1.8}$$

Definition (17.1.8) is the key definition for the present chapter; see Proposition 17.1 below. If $b_1 = b_2$, then $\zeta_\Gamma(s; b_1, b_2) = \zeta_\Gamma(2s; b_1)$, whose meromorphic structure is explicitly known [90], as pointed out in Theorem 16.1. Therefore we always assume $b_1 \neq b_2$, say $b_1 > b_2$.

Proposition 17.1. *$\zeta_\Gamma(s; b_1, b_2)$ is a well-defined holomorphic function on $\operatorname{Re} s > \frac{d}{4}$.*

The problem before us is to show that $\zeta_\Gamma(s; b_1, b_2)$ admits a meromorphic continuation to $\mathbb{C}$ which in particular is holomorphic at $s = 0$. For then by definition (15.1.9) we would set

$$\det L_1 L_2 \stackrel{\text{def}}{=} \exp(-\zeta_\Gamma'(0; b_1, b_2)), \tag{17.1.9}$$

and by (17.1.2), (17.1.7), (17.1.9) the goal would be to compute

$$\alpha(L_1, L_2) \stackrel{\text{def}}{=} -\zeta_\Gamma'(0; b_1, b_2) + \zeta_\Gamma'(0; b_1) + \zeta_\Gamma'(0; b_2). \tag{17.1.10}$$

We would like to have a very explicit description of the meromorphic continuation of $\zeta_\Gamma(s; b_1, b_2)$—say in terms of Γ-structure and in terms of the spherical harmonic analysis for the symmetric space $G/K = \mathrm{SO}_1(d,1)/\,\mathrm{SO}(d)$, as we have for the zeta function $\zeta_\Gamma(s; b)$, $b \geq 0$. We can achieve such a description, as we now indicate, by use of the trace formula.

17.2 Explicit Meromorphic Structure of $\zeta_\Gamma(s; b_1, b_2)$

Define

$$b_\pm \overset{\text{def}}{=} \frac{b_1 \pm b_2}{2}; \text{ i.e., } b_1 = b_+ + b_-, b_2 = b_+ - b_- \text{ with } b_- > 0; \quad \text{i.e., } b_1 > b_2 \text{ by assumption.} \tag{17.2.1}$$

As the K-Bessel functions K_ν have been vital for Chapters 14, 16, the *modified* Bessel functions I_ν are vital for the present chapter:

$$I_\nu(t) \overset{\text{def}}{=} \left(\frac{t}{2}\right)^\nu \sum_{m=0}^{\infty} \frac{(t/2)^{2m}}{\Gamma(\nu+m+1)\, m!} \quad \text{for } t > 0, \nu \in \mathbb{C}; \tag{17.2.2}$$

the series in (17.2.2) converges absolutely, by the ratio test. We are interested in the following Laplace–Mellin transform of I_ν:

$$I(s; \lambda, b_1, b_2) \overset{\text{def}}{=} \int_0^\infty e^{-(\lambda+b_+)t}\, I_{s-1/2}(b_- t) t^{s-1/2}\, dt \quad \text{for } \mathrm{Re} s > 0, \lambda \geq 0. \tag{17.2.3}$$

We shall see that $I(s; \lambda, b_1, b_2)$ is well defined. Using some standard asymptotics of $I_\nu(t)$ for large and small $t > 0$ [34] one sees that

Proposition 17.2. *The integral*

$$\int_0^\infty e^{-\alpha t} I_\nu(t) t^\mu\, dt$$

converges absolutely for $\alpha > 1$, Re $(\mu + \nu) > -1$.

By the transformation $t \to t/b_-$

$$I(s; \lambda, b_1, b_2) = b_-^{-s-1/2} \int_0^\infty e^{-\alpha t} I_{s-1/2}(t) t^{s-1/2}\, dt \tag{17.2.4}$$

for $\alpha = \frac{\lambda}{b_-} + \frac{b_+}{b_-}$, where by (17.2.1) $b_+ - b_- = b_2 > 0 \implies b_+ > b_- \implies \frac{b_+}{b_-} > 1 \implies \alpha > 1$ for $\lambda \geq 0$. That is, by Proposition 17.2, $I(s; \lambda, b_1, b_2)$ is indeed well defined for $\lambda \geq 0$, Re $s > 0$.

Theorem 17.1. *For $b_1 > b_2 > 0$, Re $s > 0$, $\lambda \geq 0$*

$$I(s; \lambda, b_1, b_2) = (2b_-)^{s-1/2} \frac{\Gamma(s)}{\sqrt{\pi}} \frac{1}{(\lambda + b_1)^s (\lambda + b_2)^s},$$

where $b_- \overset{\text{def}}{=} (b_1 - b_2)/2$.

Theorem 17.1 is critical for our computation of $\alpha(L_1, L_2)$; we indicate its proof. By [40, formula 5, p. 713]

$$\int_0^\infty \exp(-t\beta(\beta^2-1)^{-1/2}) I_\mu(t) t^\nu \, dt = \frac{\Gamma(\nu+\mu+1)}{(\beta^2-1)^{-(\nu+1)/2}} P_\nu^{-\mu}(\beta) \tag{17.2.5}$$

for Re $(\nu+\mu) > -1$, where $P_\nu^{-\mu}$ is an *associated Legendre function* of the *first kind*. Concerning $P_\nu^{-\mu}$ we shall only need the special values formula

$$P_\nu^{-\nu}(\cosh t) = \frac{1}{\Gamma(\nu+1)} \left(\frac{\sinh t}{2}\right)^\nu \tag{17.2.6}$$

given on [40, p. 1008]. For $\alpha = (\lambda + b_+)/b_- > 1$ in (17.2.4), choose $\beta = \frac{\alpha}{+\sqrt{\alpha^2-1}}$. Then

$$\beta^2 - 1 = \frac{1}{\alpha^2-1} = \left(\frac{\beta}{\alpha}\right)^2 > 0 \tag{17.2.7}$$

$\implies \beta > 1 \implies \beta = \cosh t$ for some $t > 0 \implies \sinh^2 t = \beta^2 - 1 = \frac{1}{\alpha^2-1} \implies$ for Re$s > 0$

$$P_{s-1/2}^{-(s-1/2)}(\beta) = \frac{1}{\Gamma(s+1/2)} \left(\frac{1}{2\sqrt{\alpha^2-1}}\right)^{s-1/2}, \tag{17.2.8}$$

by (17.2.6). Since $\beta(\beta^2-1)^{-1/2} = \beta\frac{\alpha}{\beta} = \alpha$ (by (17.2.7)) we have by (17.2.4), (17.2.5), (17.2.8)

$$\begin{aligned} I(s; \lambda, b_1, b_2) &= \frac{b_-^{-s-1/2}\,\Gamma(2s)\,(\alpha^2-1)^{-(s+1/2)/2}}{\Gamma(s+1/2)} \left[\frac{1}{2\sqrt{\alpha^2-1}}\right]^{s-1/2} \\ &= \frac{b_-^{-s-1/2}}{2^{s-1/2}} (\alpha^2-1)^{-s} \frac{\Gamma(2s)}{\Gamma(s+1/2)}. \end{aligned} \tag{17.2.9}$$

Now

$$\begin{aligned} \alpha^2 - 1 &\stackrel{\text{def}}{=} \left(\frac{\lambda+b_+}{b_-}\right)^2 - 1 = \left(\frac{\lambda+b_+}{b_-} - 1\right)\left(\frac{\lambda+b_+}{b_-} + 1\right) \\ &= \frac{(\lambda+b_2)(\lambda+b_1)}{b_-^2} \end{aligned} \tag{17.2.10}$$

by (17.2.1). Also by the Legendre duplicating formula

$$\frac{\Gamma(2s)}{\Gamma(s+1/2)} = \frac{2^{2s-1}}{\sqrt{\pi}} \Gamma(s). \tag{17.2.11}$$

This with (17.2.10) plugged into (17.2.9) gives the desired formula for $I(s; \lambda, b_1, b_2)$. □

For the θ-function $\Theta_\Gamma(t; b)$ in (16.1.17) with the choice $b = b_+ = \frac{b_1+b_2}{2} > 0$ we consider, similarly, the transform

$$I_\Gamma(s; b_1, b_2) = \int_0^\infty \Theta_\Gamma(t; b_+)\, I_{s-1/2}(b_- t)\, t^{s-1/2}\, dt. \tag{17.2.12}$$

We state without proof the following.

Theorem 17.2. *The integral in* (17.2.12) *converges absolutely for Re* $s > \frac{d}{4}$, *and one may commute the integration in* (17.2.12) *with the summation in* Θ_Γ.

Thus by Theorem 17.2 and definition (17.2.3) one has for Re $s > \frac{d}{4}$

$$\begin{aligned} I_\Gamma(s; b_1, b_2) &= \sum_{j=0}^\infty n_j \int_0^\infty \exp(-t(\lambda_j + b_+))\, I_{s-1/2}(b_- t)\, t^{s-1/2}\, dt \\ &= \sum_{j=0}^\infty n_j I(s; \lambda_j, b_1, b_2). \end{aligned} \tag{17.2.13}$$

That is, by Theorem 17.1 and definitions (17.1.8), (17.2.12)

$$\begin{aligned} &\frac{(2b_-)^{s-1/2}\,\Gamma(s)}{\sqrt{\pi}} \zeta_\Gamma(s; b_1, b_2) \\ &= \int_0^\infty \Theta_\Gamma(t; b_+)\, I_{s-1/2}(b_- t)\, t^{s-1/2}\, dt \quad \text{for Re} s > \frac{d}{4}, \end{aligned} \tag{17.2.14}$$

where we recall that $\zeta_\Gamma(s; b_1, b_2)$ is holomorphic in s on Re$s > d/4$ by Proposition 17.1. By the way, equation (17.2.14) is the analogue of equation (16.1.22). The main point about equation (17.2.14) (as with equation (16.1.22)) is that one is placed exactly in position to apply the trace formula. Namely, proceeding as in [90], one essentially obtains the explicit meromorphic continuation of $\zeta_\Gamma(s; b_1, b_2)$ to $\mathbb{C}$ by plugging into (17.2.14) the right-hand side of (16.1.36) and applying Fubini's theorem. Because of the factor $I_{s-1/2}(b_- t)$ in (17.2.14), the arguments now are a bit more difficult than those in [90], but on the other hand they are also more interesting.

To state the main result we introduce further notation. For $a \in \mathbb{C}$, $n \geq 0$ an integer the corresponding *Pochhammer symbol* $(a)_n$ is defined by

$$(a)_n \stackrel{\text{def}}{=} a(a+1)(a+2)\cdots(a+n-1) = \frac{\Gamma(a+n)}{\Gamma(a)}; (a)_o \stackrel{\text{def}}{=} 1. \tag{17.2.15}$$

Then the Gauss hypergeometric function $F(a, b; c; z)$ is defined by

$$F(a, b; c; z) \stackrel{\text{def}}{=} \sum_{n=0}^\infty \frac{(a)_n (b)_n}{(c)_n n!} z^n \quad \text{for } |z| < 1. \tag{17.2.16}$$

For $j \geq 0$ an integer and $B_1, B_2 > 0$ set

$$E_j(s; B_1, B_2) = 2 \int_0^\infty \frac{r^{2j+1}\, dr}{(r^2 + B_1)^s (r^2 + B_2)^s (1 + \exp(2\pi r))} \quad \text{for } s \in \mathbb{C}. \tag{17.2.17}$$

For the case of $d = 2n$ even let

$$|c(r)|^{-2} = C_G \pi r P(r) \tanh \pi r \tag{17.2.18}$$

be the Harish-Chandra–Plancherel density for $G/K = \mathrm{SO}_1(d,1)/\mathrm{SO}(d)$ as before, where C_G, $P(r)$ are given by (16.1.38). Write

$$P(r) = \prod_{j=0}^{n-2}\left[r^2 + \frac{(2j+1)^2}{4}\right] = \sum_{j=0}^{n-1} a_{2j} r^{2j}; \quad a_{2j} \in \mathbb{R}. \tag{17.2.19}$$

For the θ-function $\theta_\Gamma(t; b)$ of Theorem 16.2 with the choice $b = b_+$, set

$$E_\Gamma(s; b_1, b_2) \overset{\text{def}}{=} (2b_-)^{1/2-s} \frac{\sqrt{\pi}}{\Gamma(s)} \int_0^\infty \theta_\Gamma(t; b_+)\, I_{s-1/2}(b_- t)\, t^{s-1/2}\, dt \quad \text{for } s \in \mathbb{C}. \tag{17.2.20}$$

Finally, set

$$B_j \overset{\text{def}}{=} \left(\frac{d-1}{2}\right)^2 + b_j \quad \text{for } j = 1, 2; \text{ then} \tag{17.2.21}$$

$$0 < B_1 - B_2 = b_1 - b_2 = 2b_-, \tag{17.2.22}$$

$$B_1 + B_2 = 2\left(\frac{d-1}{2}\right)^2 + b_1 + b_2 = 2\left[\left(\frac{d-1}{2}\right)^2 + b_+\right], \tag{17.2.23}$$

$$\frac{B_1 - B_2}{B_1 + B_2} = \frac{b_-}{b_+ + \left(\frac{d-1}{2}\right)^2} < 1. \tag{17.2.24}$$

The form of the meromorphic continuation of $\zeta_\Gamma(s; b_1, b_2)$ is determined by the parity of d. We will state the main theorem only in the case when $d = 2n$ is even; this is the more difficult and more interesting case. Again we always assume $b_1 > b_2 > 0$.

Theorem 17.3 (The meromorphic continuation of the zeta function ζ_Γ in (17.1.8) of a product of Laplacians). *Suppose the dimension d of the symmetric space $G/K = \mathrm{SO}_1(d,1)/\mathrm{SO}(d)$ is even: $d = 2n$. Then for Re $s > \frac{d}{4}$*

$$\begin{aligned}\zeta_\Gamma(s; b_1, b_2) = {} & C_G \frac{\mathrm{vol}(\Gamma\backslash G)}{4\pi} \pi (B_1 B_2)^{-s} \\ & \cdot \sum_{j=0}^{n-1} \frac{a_{2j} j! \left(\frac{2B_1B_2}{B_1+B_2}\right)^{j+1} F\left(\frac{j+1}{2}, \frac{j+2}{2}; s+1/2; \left(\frac{B_1-B_2}{B_1+B_2}\right)^2\right)}{(2s-1)(2s-2)\cdots(2s-j)(2s-(j+1))} \\ & - 2\pi C_G \frac{\mathrm{vol}(\Gamma\backslash G)}{4\pi} \sum_{j=0}^{n-1} a_{2j} E_j(s; B_1, B_2) + E_\Gamma(s; b_1, b_2),\end{aligned} \tag{17.2.25}$$

where, for a suitable normalization of Haar measure on G, $\mathrm{vol}(\Gamma\backslash G)$ *is given by* (16.1.35) *and* $C_G = (2^{4n-4}\ \Gamma(n)^2)^{-1}$. *The coefficients* a_{2j} *are determined by the spherical Harish-Chandra–Plancherel measure* $|c(r)|^{-2}\ dr$ *of* G/K; *see* (17.2.18), (17.2.19). B_1, B_2 *are given by* (17.2.21), F *is the hypergeometric function of* (17.2.16), *and* E_j, E_Γ *are entire functions given by* (17.2.17), (17.2.20).

Since $s \to F(a, b; s; c)/\Gamma(s)$ is an entire function, for $a, b, c \in \mathbb{C}$ fixed, $|c| < 1$, we multiply the first term in (17.2.25) by $\Gamma(s + 1/2)/\Gamma(s + 1/2)$ and conclude that the poles of $\zeta_\Gamma(s; b_1, b_2)$ are simple and are located at the points $s = -\frac{1}{2}, -\frac{3}{2}, -\frac{5}{2}, -\frac{7}{2}, \ldots, s = \frac{1}{2}, \frac{2}{2}, \frac{3}{2}, \ldots, \frac{n}{2}$. In particular $\zeta_\Gamma(s; b_1, b_2)$ is holomorphic at $s = 0$, and the determinant $\det L_1 L_2$ is well defined in (17.1.9). Using Theorem 17.3 and results in [90], we can compute $\zeta'_\Gamma(0; b_1, b_2)$ and $\zeta'_\Gamma(0; b_j)$, $j = 1, 2$, and thus the multiplicative anomaly $\alpha(L_1, L_2)$ in (17.1.10). After some tedious calculations one arrives at the following theorem.

Theorem 17.4. *Suppose* $d = 2n$ *is even, and* C_G, $\mathrm{vol}(\Gamma\backslash G)$, B_j, a_{2j} *are as in Theorem* 17.3. *Then*

$$\alpha(L_1, L_2)\left[C_G \frac{\mathrm{vol}(\Gamma\backslash G)}{4}\right]^{-1}$$
$$= \sum_{j=0}^{n-1} \frac{a_{2j}(-1)^{j+1}}{2} \left\{ \frac{j}{2}(B_1 - B_2)^2 B_2^{j-1} + \frac{j(j-1)}{4}(B_1 - B_2)^3 B_2^{j-2} \right.$$
$$\left. + \sum_{p=3}^{j} \frac{j!}{(p+1)\, p!\, (j-p)!} \left[\frac{1}{p} + \frac{1}{p-1} + \sum_{q=1}^{p-2} \frac{1}{p-q-1}\right] (B_1 - B_2)^{p+1} B_2^{j-p} \right\}. \tag{17.2.26}$$

In particular $\alpha(L_1, L_2) = 0$ *for* $d = 2$ *(and thus equation* (17.1.1) *is valid), but for* $d = 4$

$$\alpha(L_1, L_2) = \frac{C_G\, \mathrm{vol}(\Gamma\backslash G)}{16} a_2 (b_1 - b_2)^2 \neq 0. \tag{17.2.27}$$

If d *is* odd, $\alpha(L_1, L_2) = 0$.

We see that an explicit expression is obtained for the anomaly $\alpha(L_1, L_2)$ in the special case of a hyperbolic space form X_Γ^d, where L_1, L_2 are Laplace type operators. In general $\alpha(L_1, L_2)$, for a class of invertible, elliptic, pseudo-differential operators L_1, L_2, can be expressed in terms of Wodzicki's *non-commutative residue* [92, 93]; also compare [13, 18, 29, 59]. One knows that (consistent with Theorem 17.4) one always has $\alpha(L_1, L_2) = 0$ for *odd* dimensional manifolds [59]. The importance of the Wodzicki residue for physics has come to light recently. This residue has been considered, for example, in the approach via non-commutative geometry to the *standard model* of electroweak interactions [20, 21, 54, 55].

18

Schrödinger's Equation and Gauge Theory

Patrick Shanahan

18.1 A Brief Introduction to Gauge Theory

Our starting point is the Schrödinger equation for the wave function Ψ of a charged particle in an electromagnetic field (E, B) defined on space-time

$$i\hbar\frac{\partial \Psi}{\partial t} = \frac{-\hbar^2}{2m}\left(\nabla - \frac{i}{\hbar}qA\right)^2 \Psi + q\phi\Psi,$$

or equivalently,

$$i\hbar\left(\partial_t + \frac{i}{\hbar}q\phi\right)\Psi$$
$$= \frac{-\hbar^2}{2m}\left(\left(\frac{\partial}{\partial x_1} - \frac{i}{\hbar}qA_1\right)^2 + \left(\frac{\partial}{\partial x_2} - \frac{i}{\hbar}qA_2\right)^2 + \left(\frac{\partial}{\partial x_3} - \frac{i}{\hbar}qA_3\right)^2\right)\Psi. \tag{18.1.1}$$

Here q is the charge and A and ϕ are vector and scalar potentials for the field. Thus, $\operatorname{curl} A = B$ and $-\operatorname{grad}\phi = E + \partial A/\partial t$. That such potentials exist is a consequence of Maxwell's equations for the given electromagnetic field.

The potentials A and ϕ are far from unique. If (A, ϕ) is a potential pair for the field (E, B), then so is (A', ϕ'), where

$$A' = A + \operatorname{grad} f \qquad \text{and} \qquad \phi' = \phi - \frac{\partial f}{\partial t}$$

for some function $f = f(x, t)$, and conversely. This ambiguity is the beginning of the subject known as *gauge theory*.

Suppose that in (18.1.1) we replace A and ϕ by the equally valid potentials A' and ϕ'. With the aid of the integrating factor $e^{(i/\hbar)f}$ one may show that if Ψ is a solution of (18.1.1), then $\Psi' = e^{(i/\hbar)f}\Psi$ is a solution of the new equation. Of course, since $|e^{(i/\hbar)f(x,t)}| = 1$ for each (x, t), the probability $|\Psi(x, t)|^2$ of finding the particle at point x and at time t will remain unchanged if Ψ is replaced by Ψ'. Nevertheless, the fact that the complex numbers $\Psi(x, t)$ and $\Psi'(x, t)$ have different phases (i.e., arguments) cannot be disregarded. For example, if Ψ_1 and Ψ_2 are solutions to a particular Schrödinger equation, then so is the superposition $\Psi = \Psi_1 + \Psi_2$, and to compute $|\Psi(x, t)|^2$, it is necessary to know the phases of $\Psi_1(x, t)$ and $\Psi_2(x, t)$ (or at least the difference between these phases) as well as their moduli.

The question we will address here is the following: how can one deal efficiently with the "relativity of phase" of the solutions to the Schrödinger equation for a charged particle which stems from the ambiguity in the choice of the potentials A and ϕ?

The situation is a bit like the one that one encounters when working with vectors. In order to carry out computations involving vectors one must choose a coordinate system, i.e., a basis. When a different basis is chosen, the components of a given vector are replaced by new ones. Thus the components are relative entities which depend on a choice of basis, while the invariant object they represent is the vector itself.

There is another question, closely related to the problem of ambiguity of phase. For each choice (A, ϕ) of potentials for a given electromagnetic field we obtain a different Schrödinger equation, although the physical situation remains the same. To put this in terms of differential operators, for the potentials A and ϕ, the relevant partial differential operators are

$$\frac{\partial}{\partial t} + \frac{i}{\hbar}q\phi \qquad \text{and} \qquad \frac{\partial}{\partial x_i} - \frac{i}{\hbar}qA_i,$$

where A_i, $i = 1, 2, 3$, is the ith component of the vector potential A. (Modified operators such as these are an example of *covariant derivative operators*, which we will have more to say about later.) If one chooses instead the potentials A' and ϕ' for the given electromagnetic field, the relevant operators are

$$\frac{\partial}{\partial t} + \frac{i}{\hbar}q\phi' = \frac{\partial}{\partial t} + \frac{i}{\hbar}q\phi - \frac{i}{\hbar}q\frac{\partial f}{\partial t}$$

and

$$\frac{\partial}{\partial x_i} - \frac{i}{\hbar}qA_i' = \frac{\partial}{\partial x_i} - \frac{i}{\hbar}qA_i - \frac{i}{\hbar}q\frac{\partial f}{\partial x_i},$$

where $A' = A + \operatorname{grad} f$ and $\phi' = \phi - \partial f/\partial t$. In other words, the ambiguity in the choice of vector and scalar potentials leads to a sort of "relativity of covariant derivatives."

In the remainder of this chapter, this can be put into a unifying geometric context, one that has far-reaching consequences for the understanding not only of the electromagnetic force, but also for nuclear forces, such as the weak force and the strong force, and ultimately even the force which is most familiar to us in everyday life, the gravitational force. In order to do this it is necessary to discuss (briefly) the concepts of vector bundles, sections of a bundle, covariant derivatives of sections, and finally, the curvature associated to a given covariant derivative.

18.2

We briefly recall the concept of a fiber bundle. Let E and X be topological spaces, and let $\pi : E \to X$ be a mapping of E onto X. Suppose that for each $x \in X$ the inverse image $\pi^{-1}(x)$ is homeomorphic to a fixed space F. (See Figure 18.1).

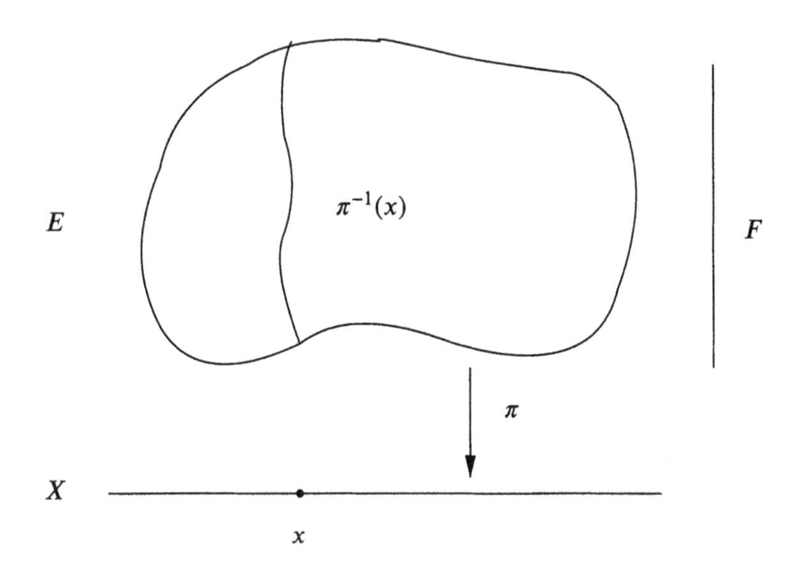

Figure 18.1: Basic set-up for a fiber bundle

The simplest example of such a system is the product space $X \times F$, where $\pi = p_1$ is the standard projection of $X \times F$ onto X. In general, however, E will not be homeomorphic to a product of two spaces.

Let us now assume that E is *locally* equivalent to a product, in the sense that for each x in X, there is a neighborhood U of x and a homeomorphism $h : \pi^{-1}(U) \to U \times F$ which is compatible with the projections $\pi : \pi^{-1}(U) \to U$ and $p_1 : U \times F \to U$. That is, we require for each $e \in \pi^{-1}(U)$ that

$$p_1(h(e)) = \pi(e).$$

In this situation we say that E is a *fiber bundle over X with fiber F*. Intuitively, a fiber bundle is a sort of "twisted product." The set X is called the *base space* of the bundle, E is called the *total space* of the bundle, and π is called the *projection*. The space $E_x = \pi^{-1}(x)$ is called the *fiber over x*.

A *section* s of a fiber bundle is a mapping $s : X \to E$ such that $\pi(s(x)) = x$ for each $x \in X$.

A simple example of a fiber bundle that is not a product bundle is the Möbius band, which is a fiber bundle over the circle S^1. Also, the set E of all tangent vectors to a smooth n-dimensional manifold M is a fiber bundle, with base space M and fiber $\mathbb{R}^n$; this bundle is called the *tangent bundle of M*, denoted TM. A section of TM is simply a field of tangent vectors on M.

A fiber bundle E is said to be *trivial*, or equivalent to a product bundle, if there exists a homeomorphism $h : E \to X \times F$ that is compatible with the projections π and p_1.

Let E be a fiber bundle and let $h_i : \pi^{-1}(U_i) \to U_i \times F$ be a local trivialization. For each $x \in U_i$, let $\phi_i(x) : E_x \to F$ be the restriction of h_i to E_x, followed by the projection p_2 of $U_i \times F$ onto F. If $h_j : \pi^{-1}(U_j) \to U_j \times F$ is another local trivialization, then for each $x \in U_i \cap U_j$ the composite $g_{ji}(x) = \phi_j(x) \circ \phi_i(x)^{-1}$ is a homeomorphism

$$g_{ji}(x) : F \to F.$$

(See Figure 18.2.)

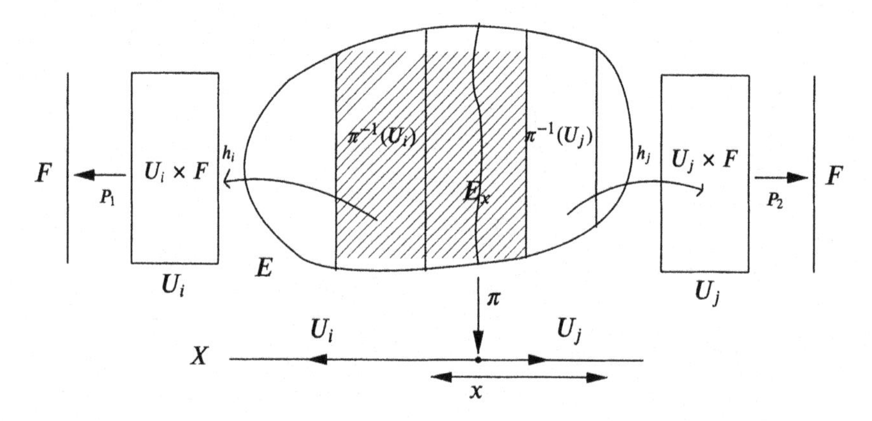

Figure 18.2: Transition mappings

We will refer to the mappings $g_{ji}(x)$ as "transition mappings" for the fiber E_x. It follows immediately from the definition of the transition mappings that for each i and each $x \in U_i$, g_{ii} is the identity mapping of F, and that if $x \in U_i \cap U_j \cap U_k$, then $g_{ki} = g_{kj} \circ g_{ji}$.

As an example, consider the Moebius band, regarded as a fiber bundle whose base is the circle S^1 and whose fiber is the interval $[-1, 1]$. For a trivializing covering we may take any two open arcs U_1 and U_2 which cover the circle. The local trivializations h_1 and h_2 may then be chosen in such a way that the transition functions $g_{21}(x)$ preserve or reverse the orientation of the fiber depending on which connected component of $U_1 \cap U_2$ the point x belongs to.

Let E be a fiber bundle whose fiber F is an n-dimensional vector space. If each of the fibers E_x is a vector space, such that the mappings $\phi_i(x) : E_x \to F$ are linear, then E is called a *vector bundle of dimension n*. When the fiber F is $\mathbb{R}^n$ (resp., $\mathbb{C}^n$), the bundle E is called a *real* (resp., *complex*) vector bundle. For example, the tangent bundle to a smooth manifold is a real vector bundle, whose transition mappings are given by the derivatives of the change-of-coordinate mappings associated to local coordinate charts.

Let E be an n-dimensional vector bundle, and let $h : \pi^{-1}(U) \to U \times F$ be a local trivialization. To each basis for the vector space F, there corresponds via h a family σ of sections $\sigma_1, \ldots, \sigma_n$ such that for each $x \in U$, the elements $\sigma_1(x), \ldots, \sigma_n(x)$ are a basis for E_x. Thus every section s defined over U may be expressed in the form

$$s = \sum_{i=1}^{n} a_i \sigma_i,$$

where the a_i are scalar-valued functions defined on U. The family σ is called a *local basis over* U.

If τ is another local basis over U, then for each x there exist scalars $c_{ij}(x)$ such that

$$\tau_j(x) = \sum_{i=1}^{n} c_{ij}(x)\sigma_i(x)$$

for $j = 1, \ldots, n$. We will refer to the matrix-valued function $C = (c_{ij})$ as the *transition matrix* from the local basis σ to the local basis τ.

Since our intention is to replace the wave functions Ψ of quantum mechanics by sections of a complex vector bundle over space-time $\mathbb{R}^4$, we will restrict our attention in what follows mainly to vector bundles whose fiber is complex n-dimensional space $\mathbb{C}^n$. In addition to the requirement that each fiber E_x be a complex vector space, we will further assume that each fiber E_x is supplied with a hermitian inner product $\langle\, , \rangle_x$, in such a way that the linear transformations $\phi_i(x) : E_x \to \mathbb{C}^n$ are hermitian with respect to the standard inner product on $\mathbb{C}^n$ defined by $\langle z, w\rangle = \sum_{i=1}^{n} z_i \overline{w}_i$. In particular, this will imply that the modulus $|s(x)|$ of a section s at a point $x \in X$ is a well-defined real number; this will prove to be important in connection with the interpretation of the modulus squared of a quantum wave function Ψ at a point of space-time in terms of probabilities.

When E is a hermitian vector bundle we will restrict our attention to *orthonormal* local bases $\sigma = \{\sigma_1, \ldots, \sigma_n\}$, i.e., to local bases over U which satisfy

$$\langle \sigma_i(x), \sigma_j(x)\rangle_x = \delta_{ij}$$

for each $x \in U$. If τ is another such local basis, then the transition matrices $C(x)$ from σ to τ will be unitary matrices for each $x \in U$.

At this point it will be useful to make the relation between sections s of vector bundles and quantum wave functions Ψ more explicit. Let E be a hermitian vector bundle of dimension 1 over space-time. Changing our notation slightly, we will denote a point in the base space $\mathbb{R}^4$ by (x, t). It is known that every vector bundle over $\mathbb{R}^n$ (or over any space which is contractible to a point, for that matter) is trivial, and hence we may consider local bases $\sigma = \{\sigma_1\}$ defined over the whole of $\mathbb{R}^4$. If σ is such a local basis, with $|\sigma_1| = 1$, and if s is a section of E, then $s = \Psi^\sigma \sigma_1$, where Ψ^σ is a complex-valued function on $\mathbb{R}^4$. Similarly, if τ is another such local basis, then $s = \Psi^\tau \tau_1$ for some function Ψ^τ. For each (x, t), the transition matrix $C(x, t)$ from $\sigma_1(x)$ to $\tau_1(x)$ is simply a complex number $c(x, t)$ of modulus 1. From

$$\Psi^\sigma \sigma_1 = s = \Psi^\tau \tau_1 = \Psi^\tau c \sigma_1$$

we see that the relation between the two representatives of s is

$$\Psi^{\tau} = c^{-1}\Psi^{\sigma}.$$

Moreover, since $|c^{-1}(x,t)| = 1$ for each $(x,t) \in \mathbb{R}^4$, it follows from the fact that $\mathbb{R}^4$ is simply connected that $c^{-1} = e^{if}$ for some real-valued function $f = f(x,t)$. Thus (except for the factor $1/\hbar$, which could be included by modifying f slightly) we have recaptured the relation $\Psi^{\tau} = e^{if}\Psi^{\sigma}$ of gauge equivalence of wave functions that was introduced in the first section.

Of course, there is no connection so far with the Schrödinger equation or with electromagnetic forces. In order to achieve such a connection we must first discuss the concept of covariant derivative operators on vector bundles.

18.3

A fiber bundle is *smooth* if the total space, the base space and the fiber are smooth manifolds, and the projection π and the local coordinate mappings $h_i : \pi^{-1}(U_i) \to U_i \times F$ are smooth (i.e., C^{∞}).

Let E be a smooth real or complex vector bundle over $\mathbb{R}^4$. In order to write down a "Schrödinger equation" in which the unknown is a section s of E, we need some kind of partial derivative operators which can act on sections. Somewhat more generally, we would like to be able to speak of "the derivative of a section s with respect to a vector $v \in \mathbb{R}^4$." One might be tempted to try to define such an operator just as one does for ordinary vector-valued functions, i.e., as the limit of a Newton quotient

$$\frac{s(x+hv)-s(x)}{h}$$

as h tends to zero. However, there is a problem: the numerator of this expression is not well defined, since $s(x)$ and $s(x+hv)$ lie in different fibers. If we had some sort of connection between the fibers E_x and E_{x+hv}, at least for h sufficiently small, then we could use this connection to make the numerator a well-defined element of E_x. Letting h go to zero, we would then have obtained an element (traditionally denoted $(\nabla_v s)(x)$) of E_x. This intuitive idea of a "connection" relating nearby (actually infinitesimally close) fibers is in fact one way of defining the so-called "covariant derivative operators" acting on sections of E. In fact, by abuse of language, covariant derivative operators are often referred to simply as "connections."

In practice, it is more convenient to approach the problem differently, by listing the properties which a first-order derivative operator on a smooth vector bundle ought to possess, and then using these properties to determine the form of all such operators.

A *covariant derivative operator* on a smooth vector bundle E is a rule ∇ which assigns to each vector field v on the base space X and each section s a section $\nabla_v s$, linear in v and s, and which satisfies

(i) $\nabla_{fv}s = f\nabla_v s$; and

(ii) $\nabla_v(fs) = (D_v f)s + f\nabla_v s$

for all smooth scalar-valued functions on X.

It follows from property (i) that for each $x \in X$, $(\nabla_v s)(x)$ depends only on the value of the vector field v at the point x.

Let ∇ be a covariant derivative on E, and let $\sigma = \{\sigma_1, \ldots, \sigma_n\}$ be a local basis for E, defined over $U \subset X$. If $s = f_1\sigma_1 + \cdots + f_n\sigma_n$, then by the linearity of ∇ and the Leibniz property (ii), $\nabla_v s$ is determined by the local sections $\nabla_v\sigma_1, \ldots, \nabla_v\sigma_n$. Moreover, for each $j = 1, \ldots, n$, the sections $\nabla_v\sigma_j$ are linear combinations of the sections σ_i, $i = 1, \ldots n$, and hence there exist scalars $A^{\sigma}_{ij}(v)$ such that

$$\nabla_v\sigma_j = \sum_{i=1}^{n} A^{\sigma}_{ij}(v)\sigma_i.$$

By the linearity in v, the scalars $A^{\sigma}_{ij}(v)$ define differential 1-forms A^{σ}_{ij} on U; the matrix A^{σ} may thus be regarded either as a matrix-valued 1-form or as a matrix of 1-forms.

If the section s is represented, relative to a local basis σ, by the vector-valued function $f = (f_1, \ldots, f_n)$, the section $\nabla_v s$ is represented by the function

$$D_v f + A^{\sigma}(v) f.$$

Let $\tau = \{\tau_1, \ldots \tau_n\}$ be another local basis for E, defined over $V \subset X$, and let $C = (c_{ij})$ be the transition matrix from σ to τ. Let A^{τ}_{ij} denote the 1-forms defined by $\nabla_v\tau_j = \sum_{i=1}^{n} A^{\tau}_{ij}\tau_i$. Then

$$\nabla_v\tau_j = \sum_{i=1}^{n} A^{\tau}_{ij}(v)\tau_i = \sum_{i=1}^{n} A^{\tau}_{ij}(v) \sum_{k=1}^{n} c_{ki}\sigma_k = \sum_{k=1}^{n} (CA(v))_{kj}\sigma_k.$$

On the other hand, we also have

$$\begin{aligned}\nabla_v\tau_j = \nabla_v \sum_{k=1}^{n} c_{kj}\sigma_k &= \sum_{k=1}^{n} (D_v c_{kj})\sigma_k + \sum_{k=1}^{n} c_{kj}\nabla_v\sigma_k \\ &= \sum_{k=1}^{n} (D_v c_{kj})\sigma_k + \sum_{k=1}^{n} c_{kj} \sum_{i=1}^{n} A^{\sigma}_{ik}(v)\sigma_i \\ &= \sum_{k=1}^{n} (D_v C + A^{\sigma}(v)C)_{kj}\sigma_k.\end{aligned}$$

It follows that on $U \cap V$ the matrix-valued 1-forms A^{σ} and A^{τ} are related by

$$CA^{\tau} = A^{\sigma}C + dC,$$

or finally,

$$A^{\tau} = C^{-1}A^{\sigma}C + C^{-1}dC,$$

where dC is the matrix-valued 1-form defined by $(dC)(v) = D_v C$.

In this way we have determined all covariant derivatives on E. Specifically, they correspond locally to operators of the form $D + A^{\sigma}$, where D is the ordinary derivative and A^{σ} is a matrix-valued 1-form, acting on column vectors by left multiplication, which transforms under a change of local basis according to the rule

$$A^{\tau} = C^{-1}A^{\sigma}C + C^{-1}dC.$$

The forms $\{A^{\sigma}\}$ are called the *connection forms* for the particular covariant derivative.

When E is a hermitian bundle (and this is the case with which we will be concerned) it is natural to restrict our attention to covariant derivatives ∇ that are compatible with the inner product on E, i.e., that satisfy

$$D_v\langle s_1, s_2\rangle = \langle \nabla_v s_1, s_2\rangle + \langle s_1, \nabla_v s_2\rangle$$

for all sections s_1 and s_2 of E. For such operators we have for any orthonormal local basis σ,

$$\begin{aligned} 0 = D_v\langle \sigma_i, \sigma_j\rangle &= \langle \nabla_v \sigma_i, \sigma_j\rangle + \langle \sigma_i, \nabla_v \sigma_j\rangle \\ &= \left\langle \sum_{\nu=1}^{n} A^{\sigma}_{\nu i}(v)\sigma_\nu, \sigma_j \right\rangle + \left\langle \sigma_i, \sum_{\mu=1}^{n} A^{\sigma}_{\mu j}(v)\sigma_\mu \right\rangle, \end{aligned}$$

which implies that $A^{\sigma}_{ji} = -\overline{A^{\sigma}_{ij}}$ for all i and j. Thus when ∇ is compatible with the hermitian inner product on E, the matrices $A^{\sigma}(v)$ are skew-hermitian; in particular, when $n = 1$ they are pure imaginary numbers.

We can use the above discussion to rewrite the Schrödinger equation for a charged particle in an electromagnetic field in terms of covariant derivative operators and sections.

Let E be a hermitian line bundle over space-time $\mathbb{R}^4$. Let $\{e_0, e_1, e_2, e_3,\}$ be the standard basis for $\mathbb{R}^4$, so that $(x, t) = x_1e_1 + x_2e_2 + x_3e_3 + te_0$. Let $\sigma = \{\sigma_1\}$ be a local (unitary) basis for E over $U \subset \mathbb{R}^4$; since E is trivial we may assume that $U = \mathbb{R}^4$.

Fix a covariant derivative with connection form A^{σ} on E, compatible with the hermitian structure, and consider the equation

$$i\nabla_{e_0} s = \frac{-1}{2m}(\nabla^2_{e_1} + \nabla^2_{e_2} + \nabla^2_{e_3})s, \qquad (18.3.1)$$

where the unknown s is a section of E. If we define real-valued functions A_1, A_2, A_3 and ϕ on $\mathbb{R}^4$ by

$$-iqA_1 = A^\sigma(e_1), \qquad -iqA_2 = A^\sigma(e_2),$$
$$-iqA_3 = A^\sigma(e_3), \quad \text{and} \qquad iq\phi = A^\sigma(e_0),$$

then the equation (18.3.1) goes over to the Schrödinger equation (18.1.1) (with $\hbar$ set equal to 1) for a particle with mass m and charge q, with Ψ replaced by Ψ^σ.

If we choose a different trivialization τ, related to σ by $\tau = c\sigma$, then since c and A^σ commute,

$$A^\tau = c^{-1}A^\sigma c + c^{-1}dc = A^\sigma + c^{-1}dc.$$

Letting f be a real-valued function such that $c = e^{-iqf}$, this relation becomes $A^\tau = A^\sigma - iq\, df$. In other words, the "potentials" A and ϕ are replaced by $A+\operatorname{grad} f$ and $\phi - \partial f/\partial t$, respectively. Moreover, the unknown Ψ^σ is replaced by $\Psi^\tau = e^{iqf}\Psi^\sigma$. Thus we see that the concepts of vector bundle, section and covariant derivative have led us to essentially the same relationships that arose in connection with the Schrödinger equation for a charged particle, but with connection forms and their transformation properties taking the place of vector and scalar potentials.

However, there is still one element missing. In the context of vector bundles and covariant derivatives, what could possibly correspond to the electromagnetic field (E, B) itself?

18.4

One of the great achievements of the first part of the 20th century was Einstein's development of the theory of general relativity, in which the force of gravity was understood to be an effect of the local curvature of space-time caused by the presence of matter (or more precisely, matter-energy). It turns out that the electromagnetic force can also be viewed as an effect of a kind of curvature. In order to understand this we must briefly recall the path by which one arrives at a useful and sufficiently general definition of the intrinsic (as opposed to extrinsic) curvature of a space at a point.

Let M be an oriented surface in Euclidean space $\mathbb{R}^3$. For each point $p \in M$, let $N(p)$ denote the unit normal vector at x determined by the right-hand rule with respect to the orientation, and let $N : M \to S^2$ be the mapping which assigns to each p the element $N(p) \in S^2$. The *Gaussian curvature* $K(p)$ of M at a point p is the Jacobian determinant of the mapping N at p. Thus the ratio of the variation of the normal vectors on a small neighborhood of p to the area of that neighborhood approximates $K(p)$. Although it might appear from this definition that the Gaussian curvature of a surface in space is an extrinsic concept, which depends on the relation of the surface to the surrounding space, it can also be defined in a purely intrinsic fashion, using only the metric properties inherited by the surface from the Euclidean metric on $\mathbb{R}^3$. One way to do this is to consider a curvilinear triangle on the surface, whose sides are geodesic arcs, and compute the sum of the angles of the triangle. If this sum does not equal 180 degrees, it indicates the presence of curvature in the interior of the triangle. In another approach one parallel transports a vector that is tangent to the surface around a closed loop; if the end result does not match the initial vector, this again indicates the presence of curvature inside the loop. (Of course, this presupposes that we can give an intrinsic meaning to the term "parallel transport" of a tangent vector on M.)

Riemann found a far-reaching generalization of the concept of the curvature of a surface, valid for what are now known as *Riemannian manifolds*. A Riemannian manifold M is a smooth n-dimensional manifold, each of whose tangent spaces $T_x M$ is supplied with an inner product $\langle\ ,\ \rangle_x$ which varies smoothly with x. A covariant derivative ∇ on TM is said to be "compatible with the metric" if $D_v\langle X,Y\rangle = \langle\nabla_v X, Y\rangle + \langle X, \nabla_v Y\rangle$ for all smooth vector fields X and Y on M.

One begins by showing that on the tangent bundle TM of a Riemannian manifold there is a unique covariant derivative ∇ such that

(i) ∇ is compatible with the metric on TM; and

(ii) $\nabla_X Y - \nabla_Y X = [X,Y]$ for all vector fields X and Y on M.

Here $[X,Y]$ is the usual commutator bracket of vector fields. Such a covariant derivative is often referred to as the "Levi–Civita connection" for the given Riemannian manifold. For example, for a surface embedded in $\mathbb{R}^3$, this unique covariant derivative can be shown to be the ordinary derivative D on $\mathbb{R}^3$, followed by orthogonal projection onto the surface.

The Riemann curvature tensor $R(p)$ is then defined by associating to each triple u, v and X of tangent vectors at p the tangent vector

$$R(u,v)X = \nabla_{\tilde{u}}(\nabla_{\tilde{v}}(\tilde{X})) - \nabla_{\tilde{v}}(\nabla_{\tilde{u}}(\tilde{X})) - \nabla_{[\tilde{u},\tilde{v}]}(X).$$

Here $\tilde{u}$, $\tilde{v}$, and $\tilde{X}$ are arbitrary extensions of u, v and X to smooth vector fields on a neighborhood of p. We will not explain how one arrives at this definition, except to say that, roughly speaking, the expression on the right-hand side represents the difference between X and the vector X' obtained by parallel transport of X around an "infinitesimally small parallelogram" spanned by u and v, where parallel transport of a tangent vector X along a curve $\gamma(t)$ is defined by requiring that $\nabla_{\gamma'(t)}X = 0$ for all t.

It follows immediately from the definition of $R(p)$ that it is linear in u, v and X, and it is not difficult to show with the aid of properties (i) and (ii) in the definition of a covariant derivative that it is independent of how we extend u, v and X to vector fields. Also, it is obvious that $R(u,v)(X) = -R(v,u)(X)$. In other words, one may regard $R(p)$ as an exterior 2-form on M whose values are linear transformations on T_pM.

The Riemann curvature tensor has an immediate generalization to vector bundles. Let E be a vector bundle over a smooth manifold M, and let ∇ be a covariant derivative on E. The *curvature* of ∇ is the exterior 2-form on M, whose values are linear transformations on the fibers of E, defined by

$$F(u,v)e = \nabla_{\tilde{u}}(\nabla_{\tilde{v}}(s)) - \nabla_{\tilde{v}}(\nabla_{\tilde{u}}(s)) - \nabla_{[\tilde{u},\tilde{v}]}(s).$$

Here e is an element of the fiber E_x, u and v are tangent vectors at $x \in M$, $\tilde{u}$, $\tilde{v}$ are vector fields which extend u and v, and s is a section of E such that $s(x) = e$. As before, this definition is independent of the choices of the extensions $\tilde{u}$, $\tilde{v}$ and s. The geometric interpretation is essentially the same as when E is the tangent bundle of M, except that now it is a measure of the effect on an element $e \in E_x$ of parallel transport around an "infinitesimal parallelogram" in M, and where a section s is said to be parallel over a path $\gamma(t)$ in M if $\nabla_{\gamma'(t)} s = 0$ for all t.

Note that in contrast to the Riemann curvature tensor on a Riemannian manifold M, which is determined by the metric on M, the curvature now depends on the connection one chooses.

Let σ be a local basis for E, and let F^σ be the matrix, relative to the local basis σ, of the linear transformation $e \mapsto F(u, v)(e)$. By using the connection 1-form A^σ of ∇ to compute $F(u, v)(\sigma_j(x))$ one finds that

$$F^\sigma(u, v) = dA^\sigma(u, v) + (A^\sigma \wedge A^\sigma)(u, v).$$

Here dA^σ is the usual exterior derivative of the 1-form A^σ, and $A^\sigma \wedge A^\sigma$ is the exterior product of A^σ with itself, defined by $(A^\sigma \wedge A^\sigma)(u, v) = A^\sigma(u)A^\sigma(v) - A^\sigma(v)A^\sigma(u)$. Note that in contrast to the case for scalar-valued forms, $A^\sigma \wedge A^\sigma$ is in general not zero, since matrix multiplication is non-commutative. However, for real or complex line bundles the connection forms have scalar values, so that in this case

$$F^\sigma = dA^\sigma.$$

Let τ be another local basis for E, and let C be the transition matrix from σ to τ. Then a rather tedious computation using the relation $A^\tau = C^{-1}A^\sigma C + C^{-1}dC$ shows that F^τ is related to F^σ by

$$F^\tau = C^{-1}F^\sigma C.$$

In particular, when E is a line bundle, this shows that the curvature 2-forms F^σ are independent of the choice of local basis σ.

Let us work out the curvature forms F^σ for the covariant derivative on a complex line bundle over $\mathbb{R}^4$ defined in the previous section. With respect to the basis $\{e_0, e_1, e_2, e_3\}$ we have

$$A^\sigma = -iq(A_1 dx_1 + A_2 dx_2 + A_3 dx_3 - \phi\, dt).$$

Therefore,

$$F^\sigma = dA^\sigma = -iq\left(\frac{\partial A_1}{\partial x_2} dx_2\, dx_1 + \frac{\partial A_1}{\partial x_3} dx_3\, dx_1 + \frac{\partial A_1}{\partial t} dt\, dx_1 \right.$$
$$\left. + \cdots - \frac{\partial \phi}{\partial x_1} dx_1\, dt + -\frac{\partial \phi}{\partial x_2} dx_2\, dt - \frac{\partial \phi}{\partial x_3} dx_3\, dt\right),$$

i.e.,

$$F^\sigma = -iq\,((\text{curl } A)_3 dx_1\, dx_2 + (\text{curl } A)_2 dx_3\, dx_1 + (\text{curl } A)_1 dx_2\, dx_3$$

$$- \left(\phi + \frac{\partial A_1}{\partial t} \right) dx_1\, dt - \left(\phi + \frac{\partial A_2}{\partial t} \right) dx_2\, dt - \left(\phi + \frac{\partial A_3}{\partial t} \right) dx_3\, dt \Bigg) .$$

Since A and ϕ are vector and scalar potentials for the field (E, B), this shows that

$$F^\sigma = -iq(E_1 dx_1\, dt + E_2 dx_2\, dt + E_3 dx_3\, dt + B_1 dx_2\, dx_3 \\ + B_2 dx_3\, dx_1 + B_3 dx_1\, dx_2).$$

Since F^σ does not depend on σ, this means that the coefficients of the curvature 2-form F of ∇ are just (except for the factor $-iq$) the components of the given electromagnetic field. Thus we have completed the geometric picture begun in the previous section: the vector and scalar potentials correspond to a covariant derivative, and the electromagnetic field corresponds to the curvature 2-form of the covariant derivative multiplied by iq.

As for Maxwell's equations, one finds by computing the exterior derivative

$$dF^\sigma = -iq \left(\frac{\partial E_1}{\partial x_2} dx_2\, dx_1\, dt + \frac{\partial E_1}{\partial x_3} dx_3\, dx_1\, dt + \frac{\partial E_2}{\partial x_1} dx_1\, dx_2\, dt \right. \\ \left. + \cdots + \frac{\partial B_2}{\partial t} dt\, dx_3\, dx_1 + \frac{\partial B_3}{\partial x_3} dx_3\, dx_1\, dx_2 + \frac{\partial B_3}{\partial t} dt\, dx_1\, dx_2 \right)$$

that the two homogeneous equations

$$\operatorname{div} B = 0 \qquad \text{and} \qquad \operatorname{curl} E = -\frac{\partial B}{dt}$$

are equivalent to the single equation $dF^\sigma = 0$.

The remaining two equations

$$\operatorname{div} E = \rho \qquad \text{and} \qquad \operatorname{curl} B = j + \frac{\partial E}{\partial t}$$

are equivalent to the equation $\delta F^\sigma = -iqJ^\sigma$, where δ is the adjoint of d with respect to the metric on forms on space-time $\mathbb{R}^4$ and $J^\sigma = j_1 dx_1 + j_2 dx_2 + j_3 dx_3 - \rho dt$ is the "four-current" density. This may be shown by using the fact that $\delta = *d*$, where $*$ is the Hodge star operator, defined for differential forms on space-time by

$$*dx_1 = -dx_2 dx_3 dt, \quad *dx_1 dx_2 = -dx_3 dt, \quad *dx_1 dx_2 dx_3 = -dt, \\ *dt = -dx_1 dx_2 dx_3, \quad *dt dx_1 = -dx_2 dx_3, \quad *dt dx_1 dx_2 = -dx_3,$$

with the remaining formulas being obtained by cyclic permutation of x, y, and z.

18.5

Let us reflect a bit on the results of the previous section. With the aid of the concepts of vector bundles and sections, we have found an interesting geometric interpretation of electromagnetic potentials and forces in terms of covariant derivatives and their curvature.

One implication of this is the following. Let ∇ be a covariant derivative on a hermitian line bundle E over $\mathbb{R}^4$. As we have seen, if s is a section of E which satisfies a Schrödinger equation $i\nabla_{e_0}s = (-1/2m)(\nabla^2_{e_1} + \nabla^2_{e_2} + \nabla^2_{e_3})s$, then for each local trivialization σ of E, the representative Ψ^σ of s satisfies the equation $i(\partial/\partial t + i\phi)\Psi^\sigma = (-1/2m)(\text{grad} - iA)^2\Psi^\sigma$, where $A_i = iA^\sigma(e_i)$, $i = 1, 2, 3$, and $\phi = -iA^\sigma(e_0)$. Since this equation is the Schrödinger equation for a particle of unit charge moving in an electromagnetic field with vector and scalar potentials A and ϕ, we see that *the geometric point of view contains within itself the necessity of electromagnetic potentials.*

Of course, the physical reality of electromagnetism can only be demonstrated by experiment; geometry and physics occupy different domains. Still, there is another consideration which should not be overlooked. As we have already noted, the gravitational force can be viewed as an effect of the curvature of space-time due to the presence of matter. There is a covariant derivative involved here also, namely, the Levi–Civita covariant derivative associated with the Lorentz metric on the space-time manifold M. In light of this, the fact that the other fundamental physical force which we experience directly can be expressed in terms of the curvature of a covariant derivative is rather startling. Among other things, it suggests the possibility that a unified theory of electromagnetism and gravity may someday be found.

Remark. It may be appropriate to mention at this point that the term "gauge," as in "choice of gauge," has its origin in a suggestion made in 1919 by Hermann Weyl. Weyl thought that perhaps one should relax the condition on the covariant derivative on the space-time manifold M slightly and allow the possibility that the lengths of tangent vectors could change under parallel transport, with the consequence that there would be no fixed gauge (*Eich*, in German) or standard of length. Einstein objected that if this were the case, then hydrogen atoms with different histories would not have identical spectra, and the idea was temporarily abandoned. A few years later, with the appearance of quantum wave functions Ψ, Weyl and others realized that his idea that arbitrary gauge changes be allowed could be applied instead to the *phases* $\theta(x, t)$ of a quantum wave function Ψ.

There are implications in the idea of allowing arbitrary gauge changes in quantum wave functions which go well beyond electromagnetism. Recall that there are two other fundamental forces in nature, the weak nuclear force, which is responsible for β-decay, and the strong nuclear force, which holds the nucleus together. Is it possible that these forces can be interpreted in terms of connections and curvature as well?

First, consider the weak nuclear force. This force does not "see" the difference between protons and neutrons. Heisenberg had suggested in 1932 that perhaps the proton and neutron were just different states of a single "nucleon," analogous to the "spin up" and "spin down" states of the electron. If this is the case, then the wave function Ψ for the nucleon could be regarded as a pair of complex-valued functions (Ψ_1, Ψ_2), with $|\Psi_1(x, t)|^2$ being the relative probability of finding the nucleon at (x, t) in the "proton state" and $|\Psi_2(x, t)|^2$ being the relative probability

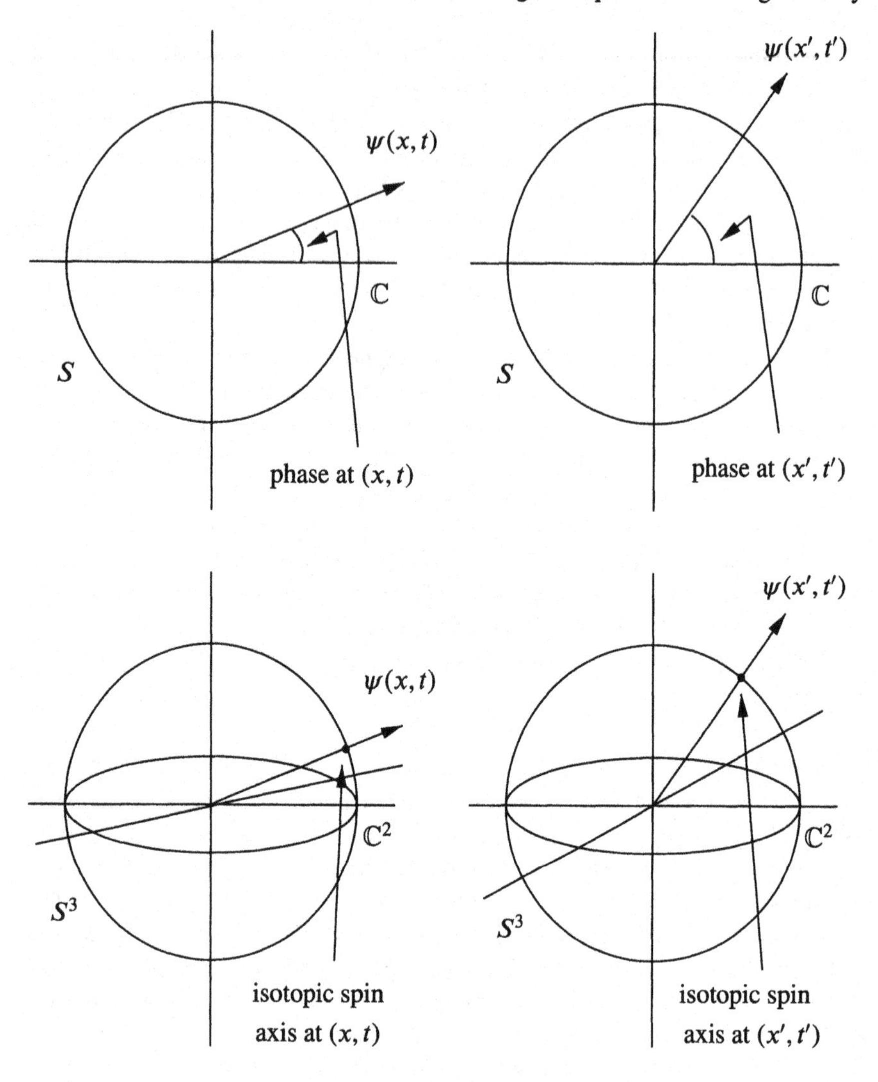

Figure 18.3: Isotopic spin

of finding it at (x, t) in the "neutron state." In this way Ψ could be viewed as a function on space-time with values in two-dimensional complex space $\mathbb{C}^2$. Of course, which state one chooses to call the proton state and which to call the neutron state is rather arbitrary, but it was assumed that once one had made such a choice at one point (x, t), then there would be no freedom of choice left for other points.

In 1954 Yang and Mills proposed that, in analogy with the situation in electrodynamics, one should be able make the choice of proton and neutron states in an arbitrary way, without being restricted by the choice made at one particular point (see Figure 18.3, in which the situations for wave functions with values in $\mathbb{C}$ and in $\mathbb{C}^2$ are illustrated). Thus from their viewpoint the "isotopic spin axes" in the space $\mathbb{C}^2$ could be chosen arbitrarily.

To put this another way, since the metric on $\mathbb{C}^2$ is invariant under $SU(2)$, the wave functions Ψ and $g\Psi$, where $g = g(x, t)$ is a function whose values are elements of the special unitary group $SU(2)$, should be regarded as representing the same physical state. This assumption leads, just as in electrodynamics, directly to the understanding that the wave function Ψ should be replaced by a section s of a trivial hermitian vector bundle E of dimension 2 over $\mathbb{R}^4$.

For each choice $\sigma = \{\sigma_1, \sigma_2\}$ of a basis for E, s will be represented by a function $\Psi^\sigma : \mathbb{R}^4 \to \mathbb{C}^2$; if τ is another trivialization, then $\Psi^\tau = C^{-1}\Psi^\sigma$, where $C = C(x,t)$ is the 2×2 transition matrix which relates τ to σ. The Schrödinger equation for s, which can be written in terms of a covariant derivative ∇ on E, corresponds via σ to a Schrödinger equation for Ψ^σ in which the ordinary derivative D is replaced by $D + A^\sigma$, the connection forms A^σ now being 1-forms on $\mathbb{R}^4$ whose values are 2×2 traceless skew-hemitian matrices, i.e., elements of the Lie algebra of $SU(2)$. The field A^σ, which may be thought of as a "generalized potential" for the weak force, may be written in the form $A^\sigma = \sum_{a=1}^{3} A_a^\sigma \xi_a$, where the A_a^σ are real-valued 1-forms on $\mathbb{R}^4$ and $\{\xi_1, \xi_2, \xi_3\}$ is a basis for the Lie algebra of $SU(2)$; one usually takes the ξ_i to be the Pauli matrices multiplied by i, so that

$$\xi_1 = i\begin{pmatrix} 0 & 1 \\ 1 & 0 \end{pmatrix} \qquad \xi_2 = i\begin{pmatrix} 0 & -i \\ i & 0 \end{pmatrix} \qquad \xi_3 = i\begin{pmatrix} 1 & 0 \\ 0 & -1 \end{pmatrix}.$$

Thus A^σ may be considered as a field on space-time with 12 components, in contrast to the four components of the potentials (A, ϕ) of electromagnetism. The transformation law which relates potentials relative to different trivializations of E is just as worked out in Section 18.3, namely

$$A^\tau = C^{-1}A^\sigma C + C^{-1}dC.$$

Here the transition matrix (or "gauge change") may be written as

$$C = \exp\left(\sum_{a=1}^{3} c_a\, \xi_a\right),$$

where the c_a are real-valued functions on $\mathbb{R}^4$.

The weak force then appears as the curvature 2-form F of ∇. Thus with respect to a trivialization σ of E, the weak force will be represented by a 2-form $F^\sigma = dF^\sigma + A^\sigma \wedge A^\sigma$ whose values are 2×2 traceless skew-hermitian matrices, with the transformation law

$$F^\tau = C^{-1}F^\sigma C.$$

We leave it to the reader to work out the components F_a^σ of F^σ in terms of the components A_a^σ of the potential. In terms of real-valued functions on $\mathbb{R}^4$, the weak force is a field with 18 components, as compared with the six components of electromagnetic fields.

One would also like to determine the field equations for this force, i.e., the analogue of Maxwell's equations. Unlike the case of hermitian line bundles, it is not true for vector bundles of dimension greater than 1 that $dF^\sigma = 0$. This equation must be replaced instead by the equation

$$dF^\sigma + A^\sigma \wedge F^\sigma - F^\sigma \wedge A^\sigma = 0,$$

which may be written as

$$dF^\sigma + [A^\sigma, F^\sigma] = 0, \tag{18.5.1}$$

where [,] is the commutator bracket of matrices. It is not difficult to verify that this equation is a simple consequence of the relation $F^\sigma = dA^\sigma + A^\sigma \wedge A^\sigma$. (To geometers it is known as the *Bianchi identity* for curvature.)

If we define the *covariant differential* associated to A^σ by $d^{A^\sigma} = d + [A^\sigma, \;]$, then (18.5.1) becomes simply

$$d^{A^\sigma} F^\sigma = 0. \tag{18.5.1$'$}$$

In addition to the Bianchi identity, the 2-form F^σ satisfies an equation analogous to the equation $\delta F^\sigma = J^\sigma$ which comprises the non-homogeneous pair of Maxwell's equations in the case of electromagnetism. However, in contrast to the situation in electromagnetic theory, where Maxwell's equations were known in advance, we have no such guide in the case of the weak force. The approach used by Yang and Mills was to appeal to a variational principle for the action density $|F^\sigma|^2 = F^\sigma \wedge *F^\sigma$ That this is a reasonable choice for an action density is based partly on the fact that when F^σ represents the electromagnetic force, then $F^\sigma \wedge *F^\sigma = (E^2 - B^2)dx_1dx_2dx_3dt$, which is the action density for the source-free electromagnetic field, and partly on the fact that there are few other choices that have the correct tensorial properties. By applying Hamilton's principle in the usual way, one arrives at the equation

$$\delta^{A^\sigma} F^\sigma = 0, \tag{18.5.2}$$

where $\delta^{A^\sigma} = * \, d^{A^\sigma} \, *$ is the adjoint of d^{A^σ}. The equation (18.5.2) is known as the *Yang–Mills equation.* (When a source term is included in the action density, the Yang–Mills equation takes the form $\delta^{A^\sigma} F^\sigma = J^\sigma$, where $J^\sigma = j_1dx_1 + j_2dx_2 + j_xdx_3 - \rho dt$ is a 1-form with values in the Lie algebra of $SU(2)$, with ρ being identified as the "charge" density and j as the "current density" associated to the source of the weak force field.)

Today it is understood that the proton and neutron are not fundamental particles, but are made up of quarks, as are other particles such as pions and kaons. The "isospin states" of quarks relative to the weak nuclear force are referred to as "flavors"; these occur in three different pairs, "up-down," "strange-charm" and "top-bottom." The quantum theory of each of these pairs is an $SU(2)$ gauge theory as described above.

Relative to the strong nuclear force, quarks have three isospin states, collectively referred to as "color." The color states are conventionally termed "red," "green" and "blue." The wave function for quarks which make up the proton and the neutron (each of which consist of three quarks) is regarded as a section of a hermitian bundle E of dimension 3. Thus, with respect to a trivialization σ of E, the wave function s is represented by a function $\Psi^\sigma = (\Psi_1^\sigma, \Psi_2^\sigma, \Psi_3^\sigma)$, where $|\Psi_1^\sigma(x,t)|^2$, $|\Psi_2^\sigma(x,t)|^2$ and $|\Psi_3^\sigma(x,t)|^2$ are the relative probabilities of finding the quark at (x, t) with color red, green or blue, respectively. However, except for the fact that the gauge group is now $SU(3)$, and the connection and curvature forms take their values in the Lie algebra of $SU(3)$, i.e, in the space of 3×3 traceless skew-hermitian matrices, the geometric formalism we have discussed in connection with the weak force goes over without any change to the strong force.

To conclude, the object of this talk has been to make the case that the geometric point of view, which makes use of the concepts of sections of vector bundles, connections and curvature, has a significant role to play in the development of the physical theory of the fundamental forces of nature.

About the Author

Patrick Shanahan is Professor Emeritus at Holy Cross College, where he taught for many years in the Mathematics Department. His many mathematical interests include topology, global analysis, gauge theory, quantum theory and string theory.

General Appendices

Appendix A

Some Further Electron Configurations

Electron configurations of a few light atoms in their ground states were presented in Table 10.2. We present here a few more ground state configurations.

The number of protons p and neutrons n in the nucleus, and the shell structures are presented in the following schematics for atoms with atomic numbers Z, where $1 \leq Z \leq 18$.

Atom	Symbol	Atomic Number Z	Electron Configuration
Nitrogen	N	7	$1s^2\ 2s^2\ 2p^3$
Oxygen	O	8	$1s^2\ 2s^2\ 2p^4$
Sodium	Na	11	$1s^2\ 2s^2\ 2p^6\ 3s^1$
Magnesium	Mg	12	$1s^2\ 2s^2\ 2p^6\ 3s^2$
Aluminum	Al	13	$1s^2\ 2s^2\ 2p^6\ 3s^2\ 3p^1$
Silicon	Si	14	$1s^2\ 2s^2\ 2p^6\ 3s^2\ 3p^2$
Phosphorous	P	15	$1s^2\ 2s^2\ 2p^6\ 3s^2\ 3p^3$
Sulfur	S	16	$1s^2\ 2s^2\ 2p^6\ 3s^2\ 3p^4$
Chlorine	Cl	17	$1s^2\ 2s^2\ 2p^6\ 3s^2\ 3p^5$
Argon	Ar	18	$1s^2\ 2s^2\ 2p^6\ 3s^2\ 3p^6$
Potassium	K	19	$1s^2\ 2s^2\ 2p^6\ 3s^2\ 3p^6\ 4s^1$
Calcium	Ca	20	$1s^2\ 2s^2\ 2p^6\ 3s^2\ 3p^6\ 4s^2$
Scandium	Sc	21	$1s^2\ 2s^2\ 2p^6\ 3s^2\ 3p^6\ 3d^1\ 4s^2$

Table A.1: More ground state configurations

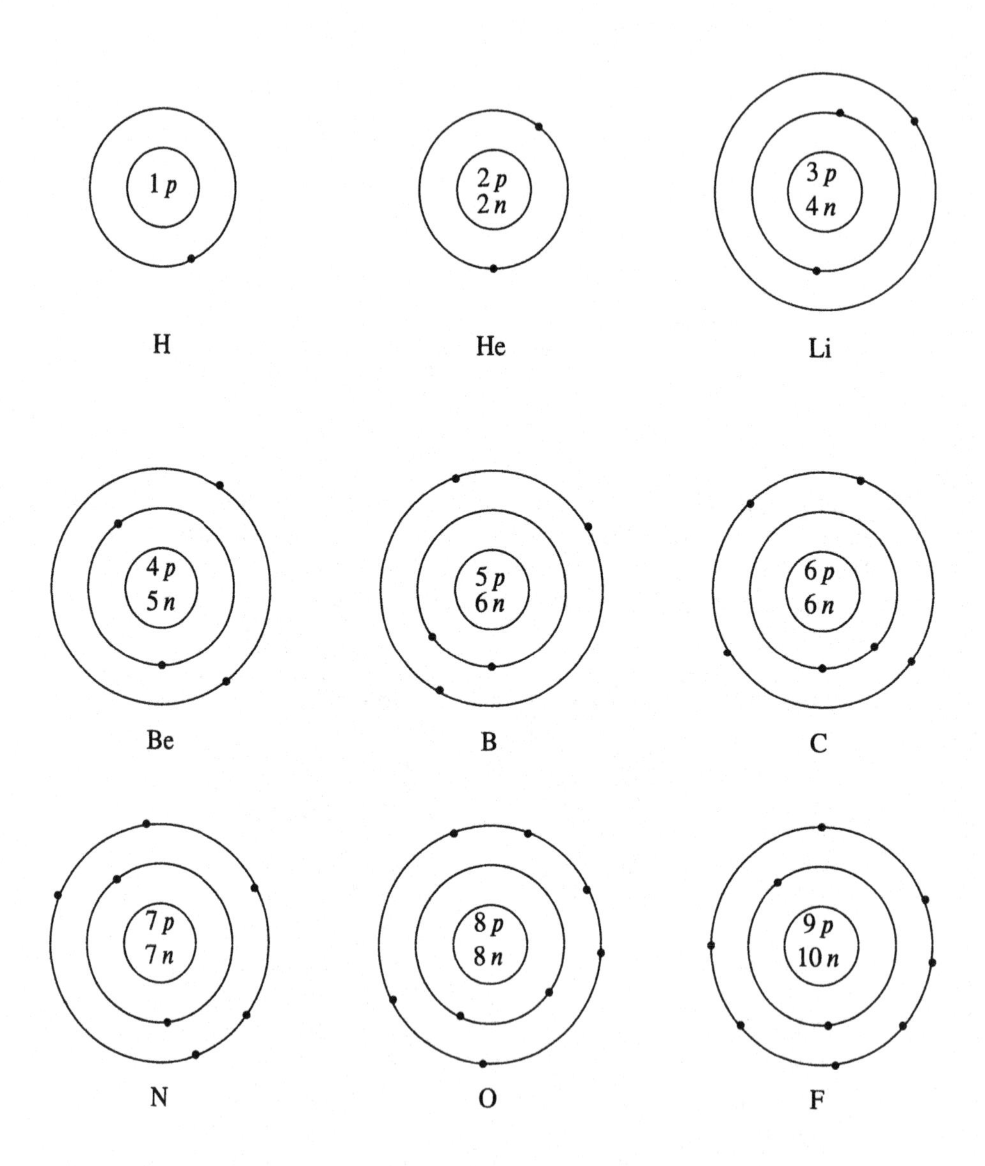
1 p
H
2 p
2 n
He
3 p
4 n
Li
4 p
5 n
Be
5 p
6 n
B
6 p
6 n
C
7 p
7 n
N
8 p
8 n
O
9 p
10 n
F

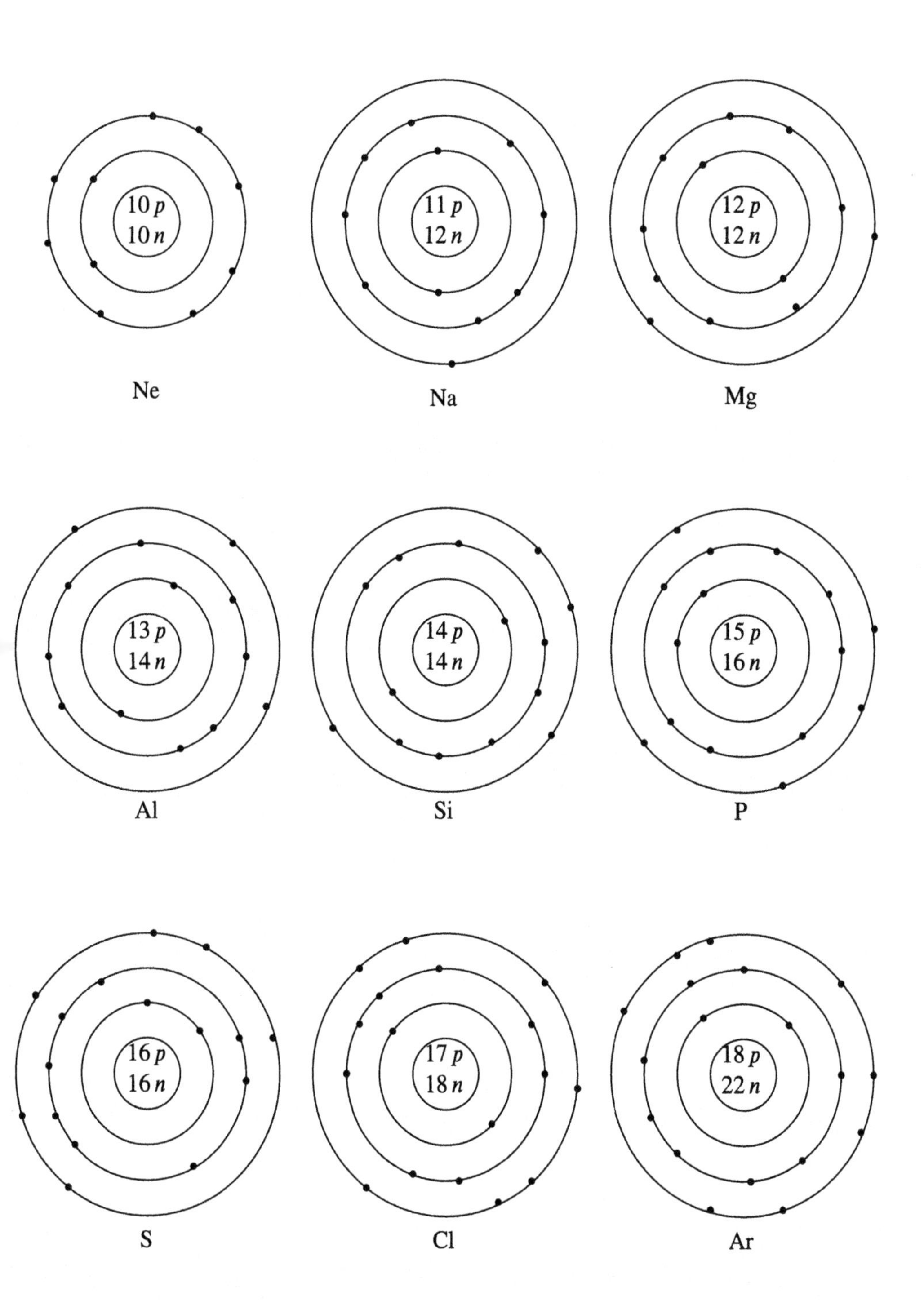

Figure A.1: Electron configurations

Appendix B

Mendeléev Periodic Table

Some brief remarks concerning the general set up of the periodic table appear in Section 10.3. Here the name, symbol, atomic number, and electron configuration of an element is listed. For example, in the table where the box

	2
chromium	8
Cr	13
24	1

appears, the element is chromium with symbol Cr, atomic number $Z = 24$, and which has 2 electrons in the K shell, 8 electrons in the L shell, 13 electrons in the M shell, and 1 electron in the N shell.

I	II	Transition elements						
lithium Li 3 2 1	beryllium Be 4 2 2							
sodium Na 11 2 8 1	magne-sium Mg 12 2 8 2							
potassium K 19 2 8 8 1	calcium Ca 20 2 8 8 2	scandium Sc 21 2 8 9 2	titanium Ti 22 2 8 10 2	vanadium V 23 2 8 11 2	chromium Cr 24 2 8 13 1	manganese Mn 25 2 8 13 2	iron Fe 26 2 8 14 2	cobalt Co 27 2 8 15 2
rubidium Rb 37 2 8 18 8 1	strontium Sr 38 2 8 18 8 2	yttrium Y 39 2 8 18 9 2	zirconium Zr 40 2 8 18 10 2	niobium Nb 41 2 8 18 12 2	molyb-denum Mo 42 2 8 18 13 1	techne-tium Tc 43 2 8 18 13 2	ruthenium Ru 44 2 8 18 15 1	rhodium Rh 45 2 8 18 16 1
cesium Cs 55 2 8 18 18 8 1	barium Ba 56 2 8 18 18 8 2	Lantha-nide series 174.97 lutetium Lu 71 2 8 18 32 9 2	hafnium Hf 72 2 8 18 32 10 2	tantalum Ta 73 2 8 18 32 12 2	tungsten W 74 2 8 18 32 12 2	rhenium Re 75 2 8 18 32 13 2	osmium Os 76 2 8 18 32 14 2	iridium Ir 77 2 8 18 32 15 2
francium Fr 87 2 8 18 32 18 8 1	radium Ra 88 2 8 18 32 18 8 2	Actinide series (257) lawren-cium Lw 103 2 8 18 32 32 9 2	kurchato-vium Ku 104 2 8 18 32 32 10 2	hahnium Ha 105 2 8 18 32 32 11 2	(263) 106 2 8 18 32 32 12 2	(261) 107 2 8 18 32 32 13 2		

Figure B.1: Metals

			III	IV	V	VI	VII	VIII
								helium He 2 (2)
			boron B 5 (2, 3)	carbon C 6 (2, 4)	nitrogen N 7 (2, 5)	oxygen O 8 (2, 6)	fluorine F 9 (2, 7)	neon Ne 10 (2, 8)
			aluminum Al 13 (2, 8, 3)	silicon Si 14 (2, 8, 4)	phosphorus P 15 (2, 8, 5)	sulfur S 16 (2, 8, 6)	chlorine Cl 17 (2, 8, 7)	argon Ar 18 (2, 8, 8)
nickel Ni 28 (2, 8, 16, 2)	copper Cu 29 (2, 8, 18, 1)	zinc Zn 30 (2, 8, 18, 2)	gallium Ga 31 (2, 8, 18, 3)	germanium Ge 32 (2, 8, 18, 4)	arsenic As 33 (2, 8, 18, 5)	selenium Se 34 (2, 8, 18, 6)	bromine Br 35 (2, 8, 18, 7)	krypton Kr 36 (2, 8, 18, 8)
palladium Pd 46 (2, 8, 18, 18, 0)	silver Ag 47 (2, 8, 18, 18, 1)	cadmium Cd 48 (2, 8, 18, 18, 2)	indium In 49 (2, 8, 18, 18, 3)	tin Sn 50 (2, 8, 18, 18, 4)	antimony Sb 51 (2, 8, 18, 18, 5)	tellurium Te 52 (2, 8, 18, 18, 6)	iodine I 53 (2, 8, 18, 18, 7)	xenon Xe 54 (2, 8, 18, 18, 8)
platinum Pt 78 (2, 8, 18, 32, 16, 2)	gold Au 79 (2, 8, 18, 32, 18, 1)	mercury Hg 80 (2, 8, 18, 32, 18, 2)	thallium Tl 81 (2, 8, 18, 32, 18, 3)	lead Pb 82 (2, 8, 18, 32, 18, 4)	bismuth Bi 83 (2, 8, 18, 32, 18, 5)	polonium Po 84 (2, 8, 18, 32, 18, 6)	astatine At 85 (2, 8, 18, 32, 18, 7)	radon Rn 86 (2, 8, 18, 32, 18, 8)

Figure B.2: Nonmetals

Lanthanide series

Element	Symbol	Atomic number	Electron shells
lanthanum	La	57	2 8 18 18 9 2
cerium	Ce	58	2 8 18 20 8 2
praseody-mium	Pr	59	2 8 18 21 8 2
neodymium	Nd	60	2 8 18 22 8 2
promethium	Pm	61	2 8 18 23 8 2
samarium	Sm	62	2 8 18 24 8 2
europium	Eu	63	2 8 18 25 8 2
gadolinium	Gd	64	2 8 18 25 9 2
terbium	Tb	65	2 8 18 27 8 2
dysprosium	Dy	66	2 8 18 28 8 2
holmium	Ho	67	2 8 18 29 8 2
erbium	Er	68	2 8 18 30 8 2
thulium	Tm	69	2 8 18 31 8 2
ytterbium	Yb	70	2 8 18 32 8 2

Actinide series

Element	Symbol	Atomic number	Electron shells
actinium	Ac	89	2 8 18 32 18 9 2
thorium	Th	90	2 8 18 32 18 10 2
protactinium	Pa	91	2 8 18 32 20 9 2
uranium	U	92	2 8 18 32 21 9 2
neptunium	Np	93	2 8 18 32 23 8 2
plutonium	Pu	94	2 8 18 32 24 8 2
americium	Am	95	2 8 18 32 24 9 2
curium	Cm	96	2 8 18 32 25 9 2
berkelium	Bk	97	2 8 18 32 27 8 2
californium	Cf	98	2 8 18 32 28 8 2
einsteinium	Es	99	2 8 18 32 29 8 2
fermium	Fm	100	2 8 18 32 30 8 2
mendelevium	Md	101	2 8 18 32 31 8 2
nobelium	No	102	2 8 18 32 32 8 2

Figure B.3: Rare earth elements

Appendix C

Determinants for String World-Sheets That Are Tori: Another Example of Zeta Regularization

In Chapter 11 (and other chapters) we have considered path integrals—"sums over histories or paths"—for particles. In string theory, a particle is replaced by a *string* and instead of a path integral one considers a "sum over surfaces" (a *Polyakov integral*), since a string moving in d-dimensional space-time $\mathbb{R}^d = \mathbb{R}^1 \times \mathbb{R}^{d-1}$ sweeps out a surface Σ, called a *world-sheet*. One also sums over Riemannian metrics γ on Σ and smooth maps $\Phi: \Sigma \to \mathbb{R}^d$. The sum over Φ, with (Σ, γ) fixed, is easy to deal with and can be given a mathematical meaning since it essentially reduces to $[\det(-\Delta(\gamma)')]^{-d/2}$, where $\Delta(\gamma)'$ is the restriction of the Laplace–Beltrami operator $\Delta(\gamma)$ of (Σ, γ) (see (16.1.4)) to the orthogonal complement of the kernel of $\Delta(\gamma)$. That is, formally, $\det(-\Delta(\gamma)')$ = product of *nonzero* eigenvalues of $-\Delta(\gamma)$, which can be made precise by the zeta-regularization method of Chapter 15. Here we take $d = 26$, the so-called *critical* dimension.

We show how to compute this determinant in a concrete example. Namely we consider a (Bosonic) string whose world-sheet is a complex torus: $\Sigma = \Sigma_\tau \stackrel{\text{def}}{=} \mathbb{C}/L_\tau$ where L_τ is the lattice $\{a + b\tau \mid a, b \in \mathbb{Z}\}$, where $\tau = \tau_1 + i\tau_2$ is a fixed number in the upper 1/2-plane $\pi^+ : \tau_2 > 0$. The Euclidean Laplacian $\Delta = \frac{\partial^2}{\partial x^2} + \frac{\partial^2}{\partial y^2}$ on $\mathbb{C} = \mathbb{R}^2$ commutes with translations by points of $\mathbb{R}^2$ (by the chain rule), and in particular with translations by points of L_τ. Thus Δ projects to an elliptic differential operator Δ_τ on Σ_τ, in the same way that the Laplacian Δ on G/K projected to the operator Δ_Γ on $\Gamma\backslash G/K$ in Chapter 16. One knows that the canonical metric on $\mathbb{R}^2$ is the pull-back of a metric γ on Σ_τ whose Laplace–Beltrami operator $\Delta(\gamma)$ is exactly Δ_τ. The spectrum of Δ_τ is known and is given as follows. For integers

m, n define ϕ_{mn} on $\mathbb{C}$ by

$$\begin{aligned}\phi_{mn}(z) &\stackrel{\text{def}}{=} e^{\frac{\pi}{\tau_2}[z(m+n\bar{\tau})-\bar{z}(m+n\tau)]} \\ &= e^{-2\pi i\, n\, x} e^{\frac{2\pi i y}{\tau_2}(m+n\tau_1)},\end{aligned} \tag{C.1.1}$$

for $z = x + iy$, $\tau = \tau_1 + i\tau_2$. Then one has immediately that

$$\begin{aligned}-\Delta\phi_{mn} &= \lambda_{mn}\phi_{mn} \quad \text{for} \\ \lambda_{mn} &\stackrel{\text{def}}{=} \frac{4\pi^2}{\tau_2^2}\,|m+n\tau|^2 \geq 0.\end{aligned} \tag{C.1.2}$$

Now $\phi_{mn} : \mathbb{C} \to T \stackrel{\text{def}}{=} \{z \in \mathbb{C} \mid |z| = 1\}$ is a homomorphism which is trivial on L_τ. Therefore ϕ_{mn} projects to a homomorphism $\widetilde{\phi}_{mn} : \Sigma_\tau \to T$ which satisfies

$$-\Delta_\tau\widetilde{\phi}_{mn} = \lambda_{mn}\widetilde{\phi}_{mn}. \tag{C.1.3}$$

It is known that $\{\lambda_{mn}\}_{m,n\in\mathbb{Z}}$ is the (multiplicity-free) spectrum of $-\Delta_\tau$ on $C^\infty(\Sigma_\tau)$. By (C.1.2), $\lambda_{mn} = 0 \iff m, n = 0$. Therefore the spectral zeta function $\zeta_{-\Delta(\gamma)'}$ of $-\Delta(\gamma)' = -\Delta'_\tau$ (cf. (15.1.6)) is given by

$$\begin{aligned}\zeta_{-\Delta(\gamma)'}(s) &\stackrel{\text{def}}{=} \sum_{(m,n)\neq(0,0)} \frac{1}{\lambda_{mn}^s} \\ &\stackrel{\text{by (C.1.2)}}{=} \frac{\tau_2^s}{(4\pi^2)^s} \sum_{(m,n)\neq(0,0)} \frac{\tau^s}{|m+n\tau|^{2s}},\end{aligned} \tag{C.1.4}$$

where

$$\sum_{(m,n)\neq(0,0)} \frac{\tau^s}{|m+n\tau|^{2s}} \stackrel{\text{def}}{=} E^*(s,\tau) \tag{C.1.5}$$

is a standard *Eisenstein series* which is holomorphic in s for $\text{Re}\, s > 1$ and which admits an explicit meromorphic continuation to $\mathbb{C}$; cf. [89], Chapter 15 for example. The only singularity of $s \to E^*(s,\tau)$ is a simple pole at $s = 1$ (with residue $\frac{3}{\pi}$). Thus by (C.1.4), (C.1.5) $\zeta_{-\Delta(\gamma)'}$, defined initially for $\text{Re}\, s > 1$, extends meromorphically (and explicitly) to $\mathbb{C}$ and, in particular, is holomorphic at $s = 0$, which means that definition (15.1.9) applies:

$$\det(-\Delta'_\tau) \stackrel{\text{def}}{=} e^{-\zeta'_{-\Delta'_\tau=-\Delta(\gamma)'}(0)} \tag{C.1.6}$$

By (C.1.4), (C.1.5) one therefore needs the value $\frac{\partial E^*}{\partial s}(0,\tau)$. But this value is indeed known, a result which amounts to the classical *Kronecker limit formula.* Namely, let $\eta(\tau)$ denote the *Dedekind eta function* on π^+:

$$\eta(\tau) \stackrel{\text{def}}{=} e^{i\pi\tau/12} \prod_{n=1}^{\infty} (1 - e^{2\pi i\, n\,\tau}). \tag{C.1.7}$$

Then

Theorem C.1.

$$\frac{\partial E^*}{\partial s}(s,\tau)\mid_{s=0}= -\log 4\pi^2 Im\tau|\eta(\tau)|^4.$$

It follows that (again for $\tau_2 = \text{Im}\tau$)

$$\zeta'_{-\Delta'_\tau}(0) = -\log \tau_2^2|\eta(\tau)|^4 \implies \det(-\Delta'_\tau) = \tau_2^2|\eta(\tau)|^4. \tag{C.1.8}$$

Thus we obtain the contribution $[\det(-\Delta(\gamma)')]^{-d/2} = \tau_2^{-d}|\eta(\tau)|^{-2d}$, for $d = 26$, of the sum over "embeddings" $\Phi : \Sigma \to \mathbb{R}^{26}$ in the Polyakov integral, in case Σ is the complex torus $\Sigma_\tau = \mathbb{C}/L_\tau$. This *one-loop* contribution was found by Joseph Polchinski (and others). One can view 26 as $24 + 2$ where $\eta(\tau)^{24}$ is a *cusp form* of weight 12 (in the theory of automorphic forms), the smallest weight possible for the existence of cusp forms (for $SL(2,\mathbb{Z})$). The "2" represents two degrees of freedom in a certain context. One also arrives at the dimension $d = 26 = 24 + 2$ via the spectral analysis of a Bosonic string (and by many other ways) where the special value $\zeta(-1) = -\frac{1}{12}$ of the Riemann zeta function ζ comes into play. That is, it can be shown that Lorentz invariance of the string requires that $\frac{d-2}{2}\zeta(-1) = -1$, which forces $d = 26$.

Appendix D

Evaluation of the Integral $I_n = \int_{\mathbb{R}^n}\int_{\mathbb{R}^n} \frac{e^{-\alpha(|x|+|y|)}}{|x-y|^{n-\beta}} dxdy$

Although we need only the value of the integral I_n in case $n = 3$ and $\beta = 2$, to complete the analysis in Chapter 10, we indicate how to compute I_n in general. Here $\alpha, \beta > 0$ with $\beta < n$.

Consider the function

$$\Phi(x) = \frac{1}{|x|^\beta(|x|^2 + A)^{\frac{n+1}{2}}} \quad \text{on } \mathbb{R}^n \text{ for } A > 0, \beta \in \mathbb{R}. \tag{D.1.1}$$

We claim that $\Phi \in L^1(\mathbb{R}^n, dx)$ for $-1 < \beta < n$. To see this we use polar coordinates. For $x \in \mathbb{R}^n - \{0\}$, write $x = r\sigma$, where $r = |x| > 0$, $\sigma = \frac{x}{|x|} \in S_1(0) \overset{\text{def}}{=} \{y \in \mathbb{R}^n \mid |y| = 1\}$. Then Lebesgue measure dx assumes the form $dx = r^{n-1}drd\sigma$. We need also that the surface area σ_n of $S_1(0)$ is given by

$$\sigma_n = \frac{2\pi^{\frac{n}{2}}}{\Gamma(\frac{n}{2})} \tag{D.1.2}$$

and that (by [40, p. 295])

$$\int_0^\infty r^{\mu-1}(A + r^2)^{\nu-1}dr < \infty \quad \text{for } \mu > 0,\ \nu + \frac{\mu}{2} < 1,\ A > 0; \tag{D.1.3}$$

the value of the integral in (D.1.3) is $\frac{A^{\frac{\mu+2\nu-2}{2}}}{2} B\left(\frac{\mu}{2}, 1 - \nu - \frac{\mu}{2}\right)$. Now

$$\begin{aligned}\int_{\mathbb{R}^n} \Phi(x)dx &= \int_{S_1(0)}\int_0^\infty \frac{r^{n-1}drd\sigma}{r^\beta(r^2 + A)^{\frac{n+1}{2}}} \\ &= \sigma_n \int_0^\infty r^{n-\beta-1}(A + r^2)^{-(\frac{n+1}{2})}dr < \infty\end{aligned} \tag{D.1.4}$$

by (D.1.3) for $n > \beta > -1$, which proves the claim.

For $f \in L^1(\mathbb{R}^n, dx)$, we take the following definition for its Fourier transform $\widehat{f}$:

$$\widehat{f}(x) \stackrel{\text{def}}{=} \int_{\mathbb{R}^n} e^{i x \cdot y} f(y) dy; \tag{D.1.5}$$

cf. definition (6.2.20). The following result is standard; see [34] for example.

Theorem D.1. (i) *For $f(x) \stackrel{\text{def}}{=} e^{-\alpha|x|}$, $\alpha > 0$, $\widehat{f}(x)$ is given by*

$$\widehat{f}(x) = \frac{\alpha c_n}{\left(\alpha^2 + |x|^2\right)^{\frac{n+1}{2}}} \quad \textit{for } c_n = \frac{(2\pi)^n \Gamma(\frac{n+1}{2})}{\pi^{\frac{n+1}{2}}}. \tag{D.1.6}$$

(ii) *For $g(x) \stackrel{\text{def}}{=} e^{-B|x|^2}$, $B > 0$*

$$\widehat{g}(x) = \left(\frac{\pi}{B}\right)^{\frac{n}{2}} e^{\frac{-|x|^2}{4B}}. \tag{D.1.7}$$

(iii) *For any $f, g \in L^1(\mathbb{R}^n, dx)$*

$$\int_{\mathbb{R}^n} \widehat{f}(x) g(x) dx = \int_{\mathbb{R}^n} f(x) \widehat{g}(x) dx. \tag{D.1.8}$$

(D.1.8) follows immediately by Fubini's theorem.

Define

$$\begin{aligned} f(x) &= e^{-\alpha|x|} \quad \text{for } \alpha > 0 \text{ (as in (i) of Theorem D.1), and} \\ h(x) &= \frac{1}{|x|^\beta} \widehat{f}(x) \quad \text{for } -1 < \beta < n. \end{aligned} \tag{D.1.9}$$

The main computation of this appendix is a Fourier transform formula for h, for $0 < \beta < n$. Note first that for the choice $A = \alpha^2$ we have that $h = \alpha c_n \Phi$ by definition (D.1.1) and Theorem D.1. As we have shown in equation (D.1.4) that $\Phi \in L^1(\mathbb{R}^n, dx)$ for $-1 < \beta < n$ we see that $\widehat{h}$ is indeed well defined: $h \in L^1(\mathbb{R}^n, dx)$:

Theorem D.2. *For h given by definition* (D.1.9) *one has that $h \in L^1(\mathbb{R}^n, dx)$. Moreover, the following formula holds for $0 < \beta < n$:*

$$\Gamma\left(\frac{\beta}{2}\right) \widehat{h}(x) = \pi^{\frac{n}{2}} 2^{n-\beta} \Gamma\left(\frac{n-\beta}{2}\right) \int_{\mathbb{R}^n} \frac{e^{-\alpha|x-y|}}{|y|^{n-\beta}} dy \tag{D.1.10}$$

Indeed the integral in (D.1.10) *exists for $\alpha, \beta > 0$.*

Proof. Apply the formula for the gamma function

$$\int_0^\infty e^{-\mu t} t^{\nu-1} dt = \frac{\Gamma(\nu)}{\mu^\nu} \quad \text{for } \mathrm{Re}\mu, \nu > 0: \tag{D.1.11}$$

$$\int_0^\infty e^{-|y|^2 t} t^{\frac{\beta}{2}-1} dt = \frac{\Gamma(\frac{\beta}{2})}{|y|^\beta} \quad \text{for } y \neq 0, \beta > 0. \tag{D.1.12}$$

For $c(\beta) \stackrel{\text{def}}{=} \Gamma(\beta/2)$, $0 < \beta < n$

$$\begin{aligned} c(\beta)\widehat{h}(x) &\stackrel{\text{def}}{=} c(\beta) \int_{\mathbb{R}^n} e^{i\,x\cdot y} h(y) dy \\ &\stackrel{\text{def}}{=} c(\beta) \int_{\mathbb{R}^n} e^{i\,x\cdot y} \frac{\widehat{f}(y)}{|y|^\beta} dy \\ &\stackrel{\text{by (D.1.12)}}{=} \int_{\mathbb{R}^n} \int_0^\infty e^{i\,x\cdot y} \widehat{f}(y) e^{-|y|^2 t} t^{\frac{\beta}{2}-1} dt dy. \end{aligned} \tag{D.1.13}$$

Here we wish to apply Fubini's theorem. Thus note that

$$\begin{aligned} &\int_{\mathbb{R}^n} \int_0^\infty \left| e^{i\,x\cdot y} \widehat{f}(y) e^{-|y|^2 t} t^{\frac{\beta}{2}} \right| \frac{dt}{t} dy \\ &= \int_{\mathbb{R}^n} \int_0^\infty e^{-|y|^2 t} t^{\frac{\beta}{2}} \frac{dt}{t} |\widehat{f}(y)| dy \\ &\stackrel{\text{by (D.1.12)}}{=} \int_{\mathbb{R}^n} c(\beta) \frac{|\widehat{f}(y)|}{|y|^\beta} dy = c(\beta) \int_{\mathbb{R}^n} |h(y)| dy < \infty, \end{aligned} \tag{D.1.14}$$

where dt/t = Haar measure on $(0, \infty)$, so that Fubini's theorem does apply:

$$c(\beta)\widehat{h}(x) = \int_0^\infty \int_{\mathbb{R}^n} e^{-|y|^2 t} e^{i\,x\cdot y} \widehat{f}(y) dy\, t^{\frac{\beta}{2}-1} dt. \tag{D.1.15}$$

For $f_b(x) \stackrel{\text{def}}{=} f(x+b)$, $b \in \mathbb{R}^n$, $\widehat{f_b}(x) = e^{-ib\cdot x} \widehat{f}(x)$, as in (iii) of Theorem 6.4, by the translation invariance of dy:

$$\begin{aligned} \widehat{f_b}(x) &= \int_{\mathbb{R}^n} e^{i\,x\cdot y} f(y+b) dy = \int_{\mathbb{R}^n} e^{ix\cdot(y-b)} f(y) dy \\ &= e^{-ix\cdot b} \widehat{f}(x). \end{aligned} \tag{D.1.16}$$

We can therefore write, for $g_t(y) \stackrel{\text{def}}{=} e^{-t|y|^2}$,

$$\begin{aligned} &\int_{\mathbb{R}^n} e^{-|y|^2 t} e^{i\,x\cdot y} \widehat{f}(y) dy \\ &= \int_{\mathbb{R}^n} g_t(y) \widehat{f}_{-x}(y) dy \stackrel{\text{by Theorem } D.1\text{(iii)}}{=} \int_{\mathbb{R}^n} \widehat{g}_t(y) f_{-x}(y) dy \\ &\stackrel{\text{by Theorem} D.1\text{(ii)}}{=} \left(\frac{\pi}{t}\right)^{n/2} \int_{\mathbb{R}^n} e^{-\frac{|y|^2}{4t}} f_{-x}(y) dy, \end{aligned} \tag{D.1.17}$$

which plugged into D.1.15 gives

$$c(\beta)\widehat{h}(x) = \int_0^\infty \int_{\mathbb{R}^n} e^{-\frac{|y|^2}{4t}} f(y-x) dy \left(\frac{\pi}{t}\right)^{n/2} t^{\frac{\beta}{2}-1} dt. \tag{D.1.18}$$

Here we wish to apply Fubini's theorem a second time:

$$\begin{aligned}
&\int_{\mathbb{R}^n}\int_0^\infty \left| e^{\frac{-|y|^2}{4t}} f(y-x)\left(\frac{\pi}{t}\right)^{\frac{n}{2}} t^{\frac{\beta}{2}} \right| \frac{dt}{t} dy \\
&\quad = \pi^{\frac{n}{2}} \int_{\mathbb{R}^n} |f(y-x)| \int_0^\infty e^{\frac{-|y|^2}{4t}} t^{\frac{\beta}{2}-\frac{n}{2}} \frac{dt}{t} dy \\
&\quad \overset{t\to\frac{1}{t}}{=} \pi^{\frac{n}{2}} \int_{\mathbb{R}^n} |f(y-x)| \int_0^\infty e^{\frac{-|y|^2 t}{4}} t^{\frac{n}{2}-\frac{\beta}{2}} \frac{dt}{t} dy \\
&\quad \overset{\text{by (D.1.11)}}{=} \pi^{\frac{n}{2}} \Gamma\left(\frac{n-\beta}{2}\right) 2^{n-\beta} \int_{\mathbb{R}^n} \frac{|f(y-x)|}{|y|^{n-\beta}} dy.
\end{aligned} \tag{D.1.19}$$

On the other hand, $|y| = |y-x+x| \le |y-x| + |x| \implies e^{-\alpha|y|} \ge e^{-\alpha|y-x|}e^{-\alpha|x|} \overset{\text{def(D.1.9)}}{=} f(y-x)e^{-\alpha|x|}$. That is, $|f(y-x)| = f(y-x)$ (since $f \ge 0$) $\le e^{\alpha|x|}e^{-\alpha|y|} \implies$ (by polar coordinates again)

$$\begin{aligned}
\int_{\mathbb{R}^n} \frac{|f(y-x)|dy}{|y|^{n-\beta}} &\le e^{\alpha|x|} \int_{\mathbb{R}^n} \frac{e^{-\alpha|y|}}{|y|^{n-\beta}} dy \\
&= e^{\alpha|x|} \int_{S_1(0)} \int_0^\infty \frac{e^{-\alpha r} r^{n-1}}{r^{n-\beta}} dr\, d\sigma \\
&= e^{\alpha|x|} \sigma_n \int_0^\infty e^{-\alpha r} r^{\beta-1} dr < \infty
\end{aligned} \tag{D.1.20}$$

for $\alpha, \beta > 0$; cf. D.1.11 again.

That is,

$$\int_{\mathbb{R}^n} \frac{e^{-\alpha|y-x|}}{|y|^{n-\beta}} dy < \infty \quad \text{for } \alpha, \beta > 0, \tag{D.1.21}$$

which establishes the last sentence in the statement of Theorem D.2, and by (D.1.19), (D.1.20) we can legitimately apply Fubini's theorem:

$$\begin{aligned}
c(\beta)\widehat{h}(x) &= \int_{\mathbb{R}^n} f(y-x)\pi^{\frac{n}{2}} \int_0^\infty e^{\frac{-|y|^2}{4t}} t^{\frac{\beta}{2}-\frac{n}{2}-1} dt\, dy \\
&\overset{\text{by (D.1.19)}}{=} \pi^{\frac{n}{2}} \Gamma\left(\frac{n-\beta}{2}\right) 2^{n-\beta} \int_{\mathbb{R}^n} \frac{e^{-\alpha|x-y|}}{|y|^{n-\beta}} dy,
\end{aligned} \tag{D.1.22}$$

which proves Theorem D.2 □

Corollary D.1. *For $\alpha > 0$, $0 < \beta < n$,*

$$x \to \int_{\mathbb{R}^n} \frac{e^{-\alpha|x-y|}}{|y|^{n-\beta}} dy \tag{D.1.23}$$

is a continuous function on $\mathbb{R}^n$ which vanishes at infinity.

Corollary D.1 follows since the function in (D.1.23) is the Fourier transform of an integrable function on $\mathbb{R}^n$, namely the function $c(n,\beta)h$ for a suitable constant $c(n,\beta)$.

As

$$\widehat{f}(x) = \frac{\alpha c_n}{\left(\alpha^2 + |x|^2\right)^{\frac{n+1}{2}}} \tag{D.1.24}$$

(by Theorem D.1) we can write (by definition (D.1.9))

$$h(y) = \frac{\alpha c_n}{|y|^\beta(\alpha^2 + |y|^2)^{\frac{n+1}{2}}} \tag{D.1.25}$$

and then express Theorem D.2 by the equation

$$\begin{aligned} &\Gamma\left(\frac{\beta}{2}\right)\int_{\mathbb{R}^n} e^{i\,x\cdot y}\frac{\alpha c_n dy}{|y|^\beta(\alpha^2 + |y|^2)^{\frac{n+1}{2}}} \\ &\quad\overset{y\to -y}{=} d(\beta)\int_{\mathbb{R}^n}\frac{e^{-\alpha|x+y|}}{|y|^{n-\beta}}dy \overset{y\to y-x}{=} d(\beta)\int_{\mathbb{R}^n}\frac{e^{-\alpha|y|}}{|y-x|^{n-\beta}}dy \end{aligned} \tag{D.1.26}$$

for $d(\beta) \overset{\text{def}}{=} \pi^{\frac{n}{2}} 2^{n-\beta}\Gamma(\frac{n-\beta}{2})$. Multiply both sides of (D.1.26) by $e^{-\alpha|x|}$ and integrate with respect to x:

$$d(\beta)\int_{\mathbb{R}^n}\int_{\mathbb{R}^n}\frac{e^{-\alpha(|x|+|y|)}}{|y-x|^{n-\beta}}dy\,dx = \alpha c_n\Gamma\left(\frac{\beta}{2}\right)\int_{\mathbb{R}^n}\int_{\mathbb{R}^n}\frac{e^{ix\cdot y}e^{-\alpha|x|}}{|y|^\beta(\alpha^2+|y|^2)^{\frac{n+1}{2}}}dy\,dx, \tag{D.1.27}$$

where

$$\begin{aligned} &\int_{\mathbb{R}^n}\int_{\mathbb{R}^n}\left|\frac{e^{ix\cdot y}e^{-\alpha|x|}}{|y|^\beta\left(\alpha^2+|y|^2\right)^{\frac{n+1}{2}}}\right|dx\,dy \\ &\quad= \int_{\mathbb{R}^n}\int_{\mathbb{R}^n}\frac{e^{-\alpha|x|}}{|y|^\beta\left(\alpha^2+|y|^2\right)^{\frac{n+1}{2}}}dx\,dy \\ &\quad= \int_{\mathbb{R}^n} e^{-\alpha|x|}dx\int_{\mathbb{R}^n}\Phi(y)dy < \infty \end{aligned} \tag{D.1.28}$$

by (D.1.4), for the choice $A = \alpha^2$ in (D.1.1). That is, by Fubini's theorem,

$$
\begin{aligned}
d(\beta) & \int_{\mathbb{R}^n}\int_{\mathbb{R}^n} \frac{e^{-\alpha(|x|+|y|)}}{|y-x|^{n-\beta}} dy\, dx \\
&= \alpha c_n \Gamma\left(\frac{\beta}{2}\right) \int_{\mathbb{R}^n}\int_{\mathbb{R}^n} \frac{e^{i x\cdot y} e^{-\alpha|x|}}{|y|^{\beta}\left(\alpha^2+|y|^2\right)^{\frac{n+1}{2}}} dx\, dy \\
&\overset{\text{by def (D.1.9)}}{=} \alpha c_n \Gamma\left(\frac{\beta}{2}\right) \int_{\mathbb{R}^n} \frac{\widehat{f}(y) dy}{|y|^{\beta}\left(\alpha^2+|y|^2\right)^{\frac{n+1}{2}}} \\
&\overset{\text{by (D.1.24)}}{=} \alpha^2 c_n^2 \Gamma\left(\frac{\beta}{2}\right) \int_{\mathbb{R}^n} \frac{dy}{|y|^{\beta}(\alpha^2+|y|^2)^{n+1}} \\
&\overset{\text{by polar coordinates}}{=} \alpha^2 c_n^2 \Gamma\left(\frac{\beta}{2}\right) \sigma_n \int_0^{\infty} \frac{r^{n-1} dr}{r^{\beta}(\alpha^2+r^2)^{n+1}} \\
&= \alpha^2 c_n^2 \Gamma\left(\frac{\beta}{2}\right) \sigma_n \frac{(\alpha^2)^{\frac{-n-\beta-2}{2}}}{2} B\left(\frac{n-\beta}{2}, \frac{n+\beta+2}{2}\right),
\end{aligned}
\tag{D.1.29}
$$

by the remark following (D.1.3). As $B(x, y) = \Gamma(x)\Gamma(y)/\Gamma(x+y)$ for $x, y > 0$ we get by definition of $d(\beta)$, c_n, and I_n, and by (D.1.29) (see (D.1.2), (D.1.6))

$$
\pi^{\frac{n}{2}} 2^{n-\beta} \Gamma\left(\frac{n-\beta}{2}\right) I_n = \frac{\alpha^{-n-\beta}(2\pi)^{2n} \Gamma\left(\frac{n+1}{2}\right)^2 \Gamma\left(\frac{\beta}{2}\right) 2\pi^{\frac{n}{2}} \Gamma\left(\frac{n-\beta}{2}\right) \Gamma\left(\frac{n+\beta+2}{2}\right)}{2\pi^{n+1}\Gamma(\frac{n}{2})\Gamma(n+1)},
\tag{D.1.30}
$$

or

$$
I_n = \frac{2^{\beta+n}\, \pi^{n-1} \Gamma\left(\frac{\beta}{2}\right) \Gamma\left(\frac{n+1}{2}\right)^2 \Gamma\left(\frac{n+\beta+2}{2}\right)}{\alpha^{n+\beta}\, \Gamma(\frac{n}{2}) n!}.
$$

Finally as $\Gamma(z+1) = z\Gamma(z)$ we can write $\Gamma\left(\frac{n+\beta+2}{2}\right) = \frac{(n+\beta)}{2}\Gamma\left(\frac{n+\beta}{2}\right)$, and conclude that

$$
I_n = \frac{2^{n+\beta-1}\pi^{n-1}(n+\beta)\Gamma\left(\frac{\beta}{2}\right)\Gamma\left(\frac{n+\beta}{2}\right)\Gamma\left(\frac{n+1}{2}\right)^2}{\alpha^{n+\beta}\Gamma(\frac{n}{2}) n!}
\tag{D.1.31}
$$

for $\alpha > 0$, $0 < \beta < n$, which proves Theorem 10.1 in Appendix 10A.

Appendix E

Some Informal Comments on QFT

Remarks in the Preface and at the beginning of Chapter 11 point to the contrasting nature of Parts I and II. As the latter part involves more technicalities, in some sense, with less exposition, and some casual mention of quantum fields, we offer here a few informal comments which might serve to ease the progression, at least slightly, to those latter chapters.

The wave function is a central notion in the quantum mechanical description of a single particle (or multi-particle) system. The integral of its modulus squared over a region is interpreted as the probability of finding the particle in that region. We have also considered Feynman's "sum over histories," path integral approach to quantum mechanics and its probabilistic meaning. In the discussion of Heisenberg's uncertainty principle in Chapter 6, the decline of determinism with the rise of quantum mechanics was noted: the behavior of particles at the fundamental level is unpredictable and probabilistic.

In *quantum field theory* (QFT), the basic entities are *fields* rather than particles. As we have not attempted to *define* what a particle is, we will not define a field. Particles and fields in fact are not necessarily mathematical concepts, but are physical notions. A field, generally speaking, can be thought of, or represented by, some function ϕ on some space-time (or some section of a bundle over some space-time). The dynamics of the field is determined by some action functional $S(\phi)$ given by integration of some Lagrangian density L involving ϕ and its derivatives. A variational principle yields the equation of motion of the field, similarly as one obtains the Euler–Lagrange equation of motion of a particle: One sets the variation of $S(\phi)$ with respect to ϕ (a Fréchet derivative) equal to zero; compare the remarks following equation (2.7.20). Solutions of these field equations can represent particles—*quantized oscillations* of the field, where various properties of these particles correspond to various modes of oscillation. The photons discussed in Chapter 1, for example, correspond to oscillations of the electromagnetic field, where as the *Higgs particle* (or Higgs boson, assuming it exists) corresponds to oscillation of a Higgs field (after Peter Higgs). Recall from Chapter 10 that bosons are particles with symmetric wave functions.

The simplest quantum field equation is the Klein–Gordon equation

$$(\Box + m^2)\phi = 0 \tag{E.1.1}$$

for a *scalar field* ϕ on Minkowski space-time $(\mathbb{R}^4, g)$ where $\Box$ is the Laplace–Beltrami operator of the Lorentzian metric

$$g = \begin{bmatrix} 1 & & & 0 \\ & -1 & & \\ & & -1 & \\ 0 & & & -1 \end{bmatrix} \tag{E.1.2}$$

on $\mathbb{R}^4$. If one labels the coordinates on $\mathbb{R}^4$ by $(x_0, x_1, x_2, x_3) = (t, x, y, z)$ (where t represents time), then by definition 16.1.4 one immediately computes that

$$\Box = \sum_{j=0}^{3} \frac{\partial}{\partial x_j} g^{jj} \frac{\partial}{\partial x_j} = \frac{\partial^2}{\partial t^2} - \nabla^2 \tag{E.1.3}$$

for

$$\nabla^2 = \frac{\partial^2}{\partial x^2} + \frac{\partial^2}{\partial y^2} + \frac{\partial^2}{\partial z^2}. \tag{E.1.4}$$

Equation (E.1.1) is therefore the equation

$$\left(\frac{\partial^2}{\partial t^2} - \nabla^2 + m^2\right)\phi = 0. \tag{E.1.5}$$

An appropriate Lagrangian density whose variation of the corresponding action with respect to ϕ yields equation (E.1.5) is

$$L(\phi) = \frac{1}{2}\sum_{j=0}^{3} \partial_j \phi \partial^j \phi - \frac{m^2}{2}\phi, \tag{E.1.6}$$

where $\partial_j = \frac{\partial}{\partial x_j}$, $\partial^j = \sum_i g^{ji}\partial_i$.

The parameter m in (E.1.1) can be thought of as the mass of some particle but, strictly speaking, only after the field ϕ has been *quantized* so that some particle interpretation of it can be considered. This quantization process, called *second quantization* (better terminology for this could have been invented) involves regarding ϕ as an *operator* subject to commutation rules analogous to the Heisenberg–Born–Jordan rules considered earlier, where it is necessary, first of all, to construct a field momentum operator conjugate to ϕ. The latter construction is done via of functional differentiation of L with respect to the time derivative of ϕ, which is the analogue of definition (2.7.13).

The Klein–Gordon equation was introduced originally as the analogue of Schrödinger's equation for the quantum mechanical study of relativistic particles. However, in this attempt two drawbacks emerge. One of them is that one loses the probabilistic meaning of the integral over a region of the modulus squared of a solution ϕ (in contrast to the initial remarks of this appendix). Here we refer to the relativistic version of the "modulus squared," which is no longer necessarily positive definite. A second drawback is that some ϕ's might yield negative relativistic energy values. For example, fix $\kappa = (\kappa_1, \kappa_2, \kappa_3) \in \mathbb{R}^3$ and $\omega \in \mathbb{R}$. For $v = (x_1, x_2, x_3) \in \mathbb{R}^3$ define ϕ on $\mathbb{R}^4$ by

$$\phi(t, v) = e^{i[\kappa \cdot v + \omega t]}. \tag{E.1.7}$$

Then for $|\kappa|^2 = \kappa \cdot \kappa = \kappa_1^2 + \kappa_2^2 + \kappa_3^2$,

$$\frac{\partial^2 \phi}{\partial t^2} = -\omega^2 \phi, \nabla^2 \phi = -|\kappa|^2 \phi \tag{E.1.8}$$

which means that ϕ is a solution of the Klein–Gordon equation (equation (E.1.5)) if and only if $\omega^2 = |\kappa|^2 + m^2$. On the other hand, the energy of the plane wave (E.1.7) turns out to be $-\omega = -(|\kappa|^2 + m^2)^{1/2}$. One would need to find some meaning of a negative energy particle. A proper interpretation of equation (E.1.1) is that it describes the field distribution of certain spin-zero particles. Following the consideration of the Klein–Gordon equation, for spin-0 fields, one may move on to consider Dirac's equation for spin-$\frac{1}{2}$ fields, Maxwell and Proca equations for spin-1 fields, etc. We note that fields are characterized by a continuum of degrees of freedom contrary to systems of particles.

It is of interest to consider quantum fields at some *finite temperature*, a situation that arose in Chapter 16, and in particular to consider quantum partition functions.

Recall from Chapter 14 the significance of such functions, and the Euclidean path integral representation given in (14.1.2) by "integration on the diagonal" of the density matrix ρ. Our remarks here provide for a further perspective on that chapter—a chapter which served in a crucial way as motivation for some of the more difficult ideas of Chapter 16. In Theorem 14.3, for example, the logarithm of the partition function (i.e., the free energy of a harmonic oscillator, up to a constant) was expressed in terms of a suitable zeta function. That theorem helps us in a crucial way to realize that indeed in quite more difficult situations, such as the consideration of partition functions Z associated to quantum fields over some curved space-time, one should expect a zeta function to play a key role—especially as such Zs are also given by suitable path integrals. By this line of thought one understands why we called definition (16.1.18) the main definition of Chapter 16.

For physical systems consisting of a great number of interacting atoms whose states therefore cannot be specified exactly, one attempts to employ for their study statistical mechanical methods. In a copper coin of mass 4 grams, for example, there are 4×10^{22} atoms. Clearly it would be impossible (even with a sophisticated computer) to keep track of the interactions of these atoms. Instead of one attempting to pin down the behavior of individual particles in a complex system, the better procedure is to calculate their average, or most probable behavior. Thus the basic idea is that physical properties of such a system are to be thought of as some kind of statistical average which one calculates over an *ensemble* of considerably smaller states, called *microstates*, of that system.

Consider a system at a given temperature T which interacts with some external heat bath or thermal reservoir. The energy of the total closed system consisting of the system plus the bath will have a constant value, although the energy of the system itself will vary as it assumes the various values E_j of its microstates. What is the probability p_j of finding our system in a specific microstate S_j with energy value E_j? The answer is known to be given by the equation

$$Zp_j = e^{-\beta E_j} \tag{E.1.9}$$

where $\beta = \frac{1}{\kappa T}$ for a suitable constant κ (Boltzmann's constant, as in (14.1.1)), and where

$$Z = Z(T) = \sum_j e^{-\beta E_j} \tag{E.1.10}$$

is Boltzamann's *canonical partition function*. The letter Z is commonly used to denote this function, Z being the first letter in the german word *Zustandssumme*— the sum in (E.1.10) being taking over all of the microstates S_j. From a given partition function Z one derives the other thermodynamic quantities. For example the free energy of the system is given by

$$F = F(T) = -\kappa T \log Z, \tag{E.1.11}$$

and the internal energy is given by

$$U = \frac{\kappa T^2}{Z}\frac{\partial Z}{\partial T} = \frac{1}{Z}\sum_j E_j e^{-\beta_j E_j} = \sum_j E_j p_j. \tag{E.1.12}$$

The last statement of equality follows from equation (E.1.9), and it shows that we can regard U as the *mean energy* of the system; compare this remark with equation (3.9.10) and compare equations (E.1.11) and (E.1.12) with equations (14.1.3) and (14.1.7). We also have the entropy

$$\begin{aligned} S &= -\frac{\partial F}{\partial T} = \kappa\left[T\frac{1}{Z}\frac{\partial Z}{\partial T} + \log Z\right] \\ &= \frac{U}{T} + \kappa \log Z = \kappa\beta U + \kappa \log Z \end{aligned} \tag{E.1.13}$$

(compare this with (14.1.5)), and we may conclude that

$$F = U - TS, \tag{E.1.14}$$

as in (14.1.6).

A quantum mechanical analogue of the Boltzmann partition function Z in (E.1.10) is the quantum partition function (at a fixed temperature)

$$Z = \sum_j e^{-\beta E_j} \tag{E.1.15}$$

where now the sum is not over microstates but over eigenstates, the E_j being eigenvalues of a quantum Hamiltonian H of a system—say $H\psi_j = E_j\psi_j$ for a complete orthonormal basis $\{\psi_j\}_{j=1}^{\infty}$ of a suitable Hilbert space. If we form the *density* operator $\rho \overset{\text{def}}{=} e^{-\beta H}$, then we can regard Z in (E.1.15) as its *trace.* Analogous to formula (E.1.9) one has the formula

$$p_j = \frac{e^{-\beta E_j}}{Z} \tag{E.1.16}$$

for the probability p_j that the system has energy E_j. It should be pointed out that for many systems, where there are infinite degrees of freedom, the density operator ρ need not be of trace class. Of course H also might have continuous contributions to its spectrum. It is clear therefore that our little excursion into some of the preceding matters is by nature informal. The notion of "microstates," for example, also requires a far more serious discussion. The reader may consult John von Neumann's treatise [84] which gives careful attention to the thermodynamics of quantum mechanical ensembles, among other matters. Theorem 14.2 shows that for a harmonic oscillator the Euclidean path integral representation of Z given in (14.1.2) agrees with the representation (E.1.15).

Boltzmann's name has come up a couple of times. This Austrian physicist (Ludwig Boltzmann (1844–1906)) was absolutely convinced of the reality of atoms. His brilliant ideas were met with sharp criticism during his lifetime. Despite the rejection of his molecular or atomistic kinetic theory of gases, for example, he continued to make outstanding discoveries that marked him as the creator of statistical thermodynamics. His formula $S = \kappa \log W$ which relates the entropy S and the probability W of a given macroscopic state was the starting point for Planck's quantum mechanics theory.

Work on Brownian motion by Einstein and Smoluchowski, and experiments of Jean Perrin, for example, eventually confirmed once and for all the existence of atoms. Boltzmann, however, never lived to see his work find vindication. The many attacks on him wounded him deeply to the point of suicide. Peace to his spirit. And to him and other great atomists, like Democritus and John Dalton, many salutations.

References

[1] N. Balazs and A. Voros. *Chaos on the Pseudosphere*. Physics Reports, 143, No. 3. North-Holland Amsterdam, 109–240, (1986).

[2] N. Bandyopadhyay. *Partition Functions for Quantized Fields in Curved Space-Times*. J. Math. Physics, 33, No. 7, 2562–2566, (1992).

[3] F. Berezin and M. Shubin. *The Schrödinger Equation*. Mathematics and its Applications (Soviet Series). Kluwer Academic Pub., 1991.

[4] S. Blau and M. Visser. *Zeta Functions and the Casimir Energy*. Nuclear Physics B, 310, No. 1, 163–180, (1988).

[5] D. Bohm. *Quantum Theory*. Prentice-Hall, 1951.

[6] N. Bohr. *On the Series of Spectrum of Hydrogen and the Structure of Atoms*. Phil. Magazine, 29, 332–335, (1915).

[7] M. Born. *Atomic Physics*. Eighth Edition, from the original translation of J. Dougall, revised by R. Blin-Stoyle and J. Radcliffe. Dover Pub., 1935.

[8] M. Born, W. Heisenberg, and P. Jordan. *On Quantum Mechanics* II. Zeits. f. Phys., 35, 557–615, (1926).

[9] M. Born and P. Jordan. *On Quantum Mechanics*. Zeits. f. Phys., 34, 858–888, (1925).

[10] I. Brevik, A. Bytsenko, A. Goncalves, and F. Williams. *Zeta Function Regularization and the Thermodynamic Potential for Quantum Fields in Symmetric Spaces*. J. Physics A, 31, No. 19, 4437–4448, (1998).

[11] A. Bytsenko, G. Cognola, L. Vanzo, and S. Zerbini. *Quantum Fields and Extended Objects in Space-Times with Constant Curvature Spatial Section.* Physics Reports, 266, No. 1 and 2, 1–126, (1996).

[12] A. Bytsenko, E. Elizalde, S. Odintsov, A. Romeo, and S. Zerbini. *Zeta Regularization Techniques with Applications.* World Scientific, 1994.

[13] A. Bytsenko, A. Goncalves, and F. Williams. *The Conformal Anomaly in General Rank* 1 *Symmetric Spaces and Associated Operator Product.* Modern Physics Letters A, 13, No. 2, 99–108, (1998).

[14] A. Bytsenko and Y. Goncharov. *Topological Casimir Effect for a Class of Hyperbolic Three-Dimensional Clifford–Klein Space-Times.* Classical and Quantum Gravity, 8, 2269–2275, (1991).

[15] A. Bytsenko, L. Vanzo, and S. Zerbini. *Zeta Function Regularization for Kaluza–Klein Finite Temperature Theories with Chemical Potentials.* Physics Letters B, 291, 26–31, (1992).

[16] A. Bytsenko and F. Williams. Editors of the Proceedings of the 1999 Londrina Winter School on *Mathematical Methods in Physics.* World Scientific, 2000.

[17] R. Camporesi. *Finite Temperature and Chemical Potentials in Higher Dimensions.* Classical and Quantum Gravity, 8, 529–549, (1991).

[18] G. Cognola, E. Elizalde, and S. Zerbini. *Applications in Physics of the Multiplicative Anomaly Formula Involving Some Basic Differential Operators.* Nuclear Physics B, 532, 407–428, (1998).

[19] G. Cognola and L. Vanzo. *Thermodynamic Potential for Scalar Fields in Space-Time with Hyperbolic Spatial Part.* Modern Physics Letters A7, No. 39, 3677–3688, (1992).

[20] A. Connes. *Noncommutative Geometry.* Academic Press, 1994.

[21] A. Connes and J. Lott. *Particle Models and Noncommutative Geometry.* Nuclear Physics B (Proc. Suppl.), B18, 29–47, (1990).

[22] H. Cordes, A. Jensen, S. Kuroda, G. Ponce, B. Simon, and M. Taylor. *Tosio Kato* (1917–1999). Notices of the Am. Math. Soc., 47, No. 6, pages 650–657, June/July (2000).

[23] P. Cox. *Introduction to Quantum Theory and Atomic Structure.* Oxford Chemistry Primers 37. Oxford University Press, 1996.

[24] L. de Broglie. *Comments on the New Wave Theory.* Comptes Rendus, 183, 272–274, (1926).

[25] L. Debnath and P. Mikusiński. *Introduction to Hilbert Spaces with Applications*. Academic Press, 1990.

[26] E. D'Hoker and D. Phong. *On Determinants of Laplacians on Riemann Surfaces*. Comm. Math. Physics, 104, 537–545, (1986).

[27] P. Dirac. *The Principles of Quantum Mechanics*. Clarendon Press, Oxford, Fourth Edition, 1958.

[28] E. Elizalde. *Ten Physical Applications of Spectral Zeta Functions*. Lecture Notes in Physics Series 35. Springer Verlag, 1995.

[29] E. Elizalde, L. Vanzo, and S. Zerbini. *Zeta-Function Regularization, the Multiplicative Anomaly and the Wodzicki Residue*. Comm. Math. Physics, 194, 613–630, (1998).

[30] G. Fano. *Mathematical Methods of Quantum Mechanics*. Translated by Scripta Technica, McGraw–Hill, 1967.

[31] R. Feynman. *The Principle of Least Action in Quantum Mechanics*. Ph.D. Thesis. Princeton University, 1942.

[32] R. Feynman. *Space-Time Approach to Nonrelativistic Quantum Mechanics*. Rev. Modern Physics, 20, 367–387, (1948).

[33] R. Feynman and A. Hibbs. *Quantum Mechanics and Path Integrals*. McGraw–Hill, 1965.

[34] G. Folland. *Fourier Analysis and its Applications*. Brooks/Cole Pub. I(T)P, 1992.

[35] P. Fong. *Elementary Quantum Mechanics*. Addison-Wesley, 1964.

[36] D. Freed. *Determinants, Torsion, and Strings*. Comm. Math. Physics, 107, 483–513, (1986).

[37] H. Friedrich and B. Eckhardt. Editors. *Classical, Semiclassical and Quantum Dynamics in Atoms*. Lecture Notes in Physics Series 485. Springer Verlag, 1997.

[38] R. Gangolli. *Zeta Functions of Selberg's Type for Compact Space Forms of Symmetric Spaces of Rank One*. Illinois J. Math., 21, 1–42, (1977).

[39] A. Goswami. *Quantum Mechanics*. Wm. C. Brown Pub., 1992.

[40] I. S. Gradshteyn and I. M. Ryzhik. *Table of Integrals, Series, and Products*. Collected and Enlarged Edition Prepared by A. Jeffrey. Academic Press, 1980.

[41] W. Greiner. *Quantum Mechanics: An Introduction*. Third Edition. Springer-Verlag, 1994.

[42] C. Grosche and F. Steiner. *Handbook of Feynman Path Integrals*. Springer Tracts in Modern Physics 145. Springer Verlag, 1998.

[43] H. Hameka. *Quantum Mechanics*. John Wiley and Sons, 1981.

[44] L. Harris and A. Loeb. *Introduction to Wave Mechanics*. McGraw–Hill, 1963.

[45] S. Hawking. *Zeta Function Regularization of Path Integrals in Curved Space-Time*. Comm. Math. Physics, 55, 133–148, (1977).

[46] W. Heisenberg. *The Physical Principles of the Quantum Theory*. Dover Pub., 1930.

[47] W. Heitler. *Elementary Wave Mechanics with Applications to Quantum Chemistry*. Second Edition, Oxford University Press, 1956.

[48] S. Helgason. *Differential Geometry and Symmetric Spaces*. Academic Press, 1962.

[49] E. Hille. *Lectures on Ordinary Differential Equations*. Addison-Wesley, 1969.

[50] S. Holland, Jr. *Applied Analysis by the Hilbert Space Method. An Introduction with Applications to the Wave, Heat, and Schrödinger Equations*. Pure and Applied Math. Series 137. Marcel Dekker, 1990.

[51] J. Humphreys. *Introduction to Lie Algebras and Representation Theory*. Graduate Texts in Mathematics 9. Springer-Verlag, 1972.

[52] N. Hurt. *Mathematical Physics of Quantum Wires and Devices. From Spectral Resonances to Anderson Localization*. Mathematics and its Applications 506. Kluwer Academic Pub., 2000.

[53] E. Ikenberry. *Quantum Mechanics for Mathematicians and Physicists*. Oxford University Press, 1962.

[54] D. Kastler. *A Detailed Account of Alain Connes' Version of the Standard Model*. Parts I and II, Rev. Math. Physics, 5, 477–523 (1993). Part III, Rev. Math. Physics 8, 103–165, (1996).

[55] D. Kastler and T. Schücker. *A Detailed Account of Alain Connes' Version of the Standard Model*. Part IV, Rev. Math. Physics, 8, 205–228, (1996). Also, *The Standard Model à la Connes-Lott*, hep-th /9412185.

[56] T. Kato. *Fundamental Properties of Hamiltonian Operators of Schrödinger Type*. Trans. Am. Math. Soc., 70, 190–211, (1951).

[57] T. Kato. *On the Existence of Solutions of the Helium Wave Equation*. Trans. Am. Math. Soc., 70, 212–218, (1951).

[58] E. Kemble. *The Fundamental Principles of Quantum Mechanics.* Dover Pub., 1958.

[59] M. Kontsevich and S. Vishik. *Geometry of Determinants of Elliptic Operators.* From Functional Analysis on the Eve of the 21st Century, Vol. 1, Progr. Math. 131, 173–197. Birkhäuser Boston, 1995.

[60] L. Landau and E. Lifshitz. *Quantum Mechanics, Non-relativistic Theory.* Translated from the Russian by J. Sykes and J. Bell. Pergamon Press LTD/ Addison-Wesley, 1958.

[61] R. Lindsay and H. Margenau. *Foundations of Physics.* Dover Pub. 1935, new edition, 1957.

[62] G. Mackey. *Mathematical Foundations of Quantum Mechanics.* W. A. Benjamin, 1963.

[63] A. Messiah. *Quantum Mechanics.* Vol.1 Translated from the French by G. Temmer. North-Holland Amsterdam, 1964.

[64] R. Miatello. *The Minakshisundaram-Pleijel Coefficients for the Vector-Valued Heat Kernel on Compact Locally Symmetric Spaces of Negative Curvature.* Trans. Am. Math. Soc., 260, 1–33, (1980).

[65] S. Minakshisundaram and A. Pleijel. *Some Properties of the Eigenfunctions of the Laplace Operator on Riemannian Manifolds.* Canadian J. Math., 1, 242–256, (1949).

[66] A. Nikiforov and V. Uvarov. *Special Functions of Mathematical Physics. A Unified Introduction with Applications.* Translated from the Russian by R. Boas. Birkhäuser, 1988.

[67] F.W.J. Olver. *Asymptotics and Special Functions.* Computer Science and Applied Math. Series. Academic Press, 1974.

[68] D. Park. *Introduction to the Quantum Theory.* Second Edition. McGraw–Hill, 1974.

[69] L. Pauling and E. Wilson, Jr. *Introduction to Quantum Mechanics with Applications to Chemistry.* Dover Pub., 1935.

[70] A. Polyakov. *Quantum Geometry of Bosonic Strings.* Physics Letters, 103 B, 207–210, (1981).

[71] E. Prugovečki. *Quantum Mechanics in Hilbert Space.* Second Edition, Pure and Applied Math. Series Vol. 92. Academic Press, 1981.

[72] B. Randol. *On the Analytic Continuation of the Minakshisundaram-Pleijel Zeta Function for Compact Riemann Surfaces.* Trans. Am. Math. Soc., 201, 241–246, (1975).

[73] V. Rojansky. *Introductory Quantum Mechanics.* Prentice-Hall, 1938.

[74] P. Sarnak. *Determinants of Laplacians.* Comm. Math. Physics, 110, 113–120, (1987).

[75] E. Schrödinger. *Quantization as an Eigenvalue Problem,* I, II, III, IV. Annalen der Physik, Volumes, 79, 79, 80, 81, 361–376, 489–527, 734–756, 437–490, 109–139, (1926a), (1926b), (1926d), (1926e).

[76] E. Schrödinger. *Collected Papers on Wave Mechanics.* Chelsa, New York, (1928).

[77] A. Selberg. *Harmonic Analysis on Discontinuous Groups in Weakly Symmetric Riemannian Spaces with Applications to Dirichlet Series.* J. Indian Math. Soc., 20, 47–87, (1956).

[78] J. Slater. *Quantum Theory of Atomic Structure.* McGraw–Hill, 1960.

[79] S. Sternberg. *Group Theory and Physics.* Cambridge University Press, 1994.

[80] A. Sudbery. *Quantum Mechanics and the Particles of Nature. An outline for Mathematicians.* Cambridge University Press, 1986.

[81] E. Titchmarsh. *Eigenfunction Expansions Associated with Second-Order Differential Equations.* Oxford University Press, 1946.

[82] B. L. van der Waerden. *Sources of Quantum Mechanics.* Classics of Science Series, Vol. 5, under the General Editorship of G. Holton. Dover Pub., 1966.

[83] B. L. van der Waerden. *Group Theory and Quantum Mechanics.* Die Grundlehren der mathematischen Wissenschaften in Einzeldarstellungen 214. Springer-Verlag, 1974.

[84] J. von Neumann. *Mathematical Foundations of Quantum Mechanics.* Translated from the German edition by R. Beyer. Princeton University Press, 1955.

[85] A. Voros. *Spectral Zeta Functions.* From Advanced Studies in Pure Math. 21, Zeta Functions in Geometry, 327–358. Academic Press, 1992.

[86] N. Wallach. *On the Selberg Trace Formula in the Case of Compact Quotient.* Bulletin of the Am. Math. Soc., 2, 171–195, (1976).

[87] H. Weyl. *The Theory of Groups and Quantum Mechanics.* Translated from the Second (Revised) German Edition by H. P. Robertson. Dover Pub., 1931.

[88] F. Williams. *Topological Casimir Energy for a General Class of Clifford–Klein Space-Times.* J. of Math. Physics, 38, 796–808, (1977).

[89] F. Williams. *Lectures on the Spectrum of* $L^2(\Gamma\backslash G)$. Pitman Research Notes in Math. Science 242. Longman/Addison-Wesley, 1991.

[90] F. Williams. *Meromorphic Continuation of Minakshisundaram-Pleijel Series for Semisimple Lie Groups.* Pacific J. of Math., 182, 137–156, (1998).

[91] F. Williams. *One-Loop Effective Potential for Quantum Fields on Hyperbolic Kaluza–Klein Space-Times.* From Proceedings of the Fourth Alexander Friedmann International Seminar on Gravitation and Cosmology, St. Petersburg, Russia, June 17-25, 1998. IMECC, 351–358, (1999).

[92] M. Wodzicki. *Local Invariants of Spectral Asymmetry.* Inventiones Math., 75, 143–177, (1984).

[93] M. Wodzicki. *Non-commutative Residue. From K-theory, Arithmetic and Geometry.* Lecture Notes in Mathematics, Chapter 1, Vol. 1289, Edited by Yu I. Manin. Springer-Verlag, 1987.

Index

Ampère's law, 218
azimuthal number, 124, 138, 260, 261
    quantum, 138, 148, 171

Balmer
    formula, 9, 11, 14, 150
    series, 11, 14
    value, 14
Bessel equation, 82
Bessel function
    K-, 312, 329, 330, 335
    modified, 335
Bohr
    –Wilson–Sommerfeld quantization rule, 65
    complementarity principle, 23
    frequencies, 195, 197
    frequency rule, 4, 7, 9, 12–14, 150, 200
    magneton, 6, 238
    radius, 5, 132, 148
    theory, 9, 13, 14, 51, 123
Bose–Einstein particles, 265
Bosonic string, 369, 371
bosons, 265
Brackett series, 11

Casimir energy, 317, 330, 331
center of mass coordinates, 49, 72
Chebyshev
    equation, 82
    polynomials, 98
Clifford–Klein space-time, 330
Coulomb's law, 5, 14, 60, 62, 225
    electrostatic, 60
Coulomb potential, 49, 123, 126, 218, 240, 243, 260

de Broglie wavelength, 7, 15, 17, 20, 22, 151, 253
Dedekind eta function, 370

Euclidean path integrals, 280, 299–301, 303, 307, 323
Euler's differential operator, 141
Euler's theorem, 144, 174
Euler–Lagrange equations, 42–44

Faraday's law, 218
Faraday magnetic induction laws, 218
fermions, 265

Feynman
  integrals, 280, 284, 323
  path integrals, 271, 281, 283, 286, 299, 301
Fourier analysis, 165
Fourier expansion, 194, 236
Fourier transform, 165, 166, 168, 267, 374, 377
Fresnel integrals, 271, 273
  multiple, 283, 294
Fubini's theorem, 337, 374–377

Gauge theory, 224, 341, 356
Gauss' law, 218
Gauss equation, 82
Gaussian curvature, 349
Gaussian system, 217
Gegenbauer polynomials, 98

Haar measure, 185, 187, 188, 209, 327, 328, 339, 375
Hamilton's equation of motion, 25, 38, 39
Hamilton's equations, 28, 38
Hamilton's principle, 356
Hamiltonian, 16, 28, 36, 38, 39, 42, 59, 79, 199, 203, 204, 217, 236, 237, 240, 258, 265, 280, 307
  atomic, 79
  function, 223
  operators, 319
  Pauli's magnetic, 240, 245
  quantized, 224, 240, 243, 255
  quantized operator, 51, 52, 59, 71
  quantum, 264, 280
  time-dependent, 191
Harish-Chandra–Plancherel density, 330
Hawking path integral, 280, 317
Heisenberg's uncertainty principle, 7, 20, 22, 23, 157, 162–165, 169, 170, 253
Heisenberg–Born–Jordan commutation relations, 17
Heisenberg inequality, 161, 168
Hermite
  equation, 82, 87, 95
  orthogonality, 107–109
  polynomials, 83, 92, 245, 304
Hilbert space, 51, 52, 77, 131, 178, 181, 183, 202, 235, 236
  infinite-dimensional, 209
hypergeometric equation(s), 53, 81, 82, 87, 88, 113, 128, 130
  Gauss, 82
hypergeometric function, 339
  Gauss, 337
Hölder's inequality, 159–161, 166

Jacobi, 49, 72
  differential equation, 82, 83, 97, 98, 111
  equation, 87, 128
  equation of motion, 25
  inversion formula, 311, 312, 314, 322, 324
  orthogonality, 102, 104
  polynomials, 83, 97, 98, 115
Jacobian determinant, 349
Joseph Polchinski, 371

Kaluza–Klein space-time, 280, 315, 317, 321, 329
Kronecker limit formula, 370

Lagrange's equation of motion, 25, 40, 41, 44, 222
Lagrangian, 40–42, 221, 222, 281
  classical mechanics, 280
Laguerre
  equation, 82, 87, 96
  function, 129
  orthogonality, 98, 100
  polynomials, 83, 94, 121, 128
Landau
  energies, 245
  spectra, 217

spectrum, 217
states, 245
Laplace
equation, 81
operator, 320, 333, 334, 339
non-Euclidean, 317
Laplace–Beltrami operator, 322, 333, 369
Laplace–Mellin transform, 335
Lebesgue
measurable function, 162
measure, 16, 52, 58, 131, 159, 188, 189, 288, 289, 373
Legendre duplicating formula, 336
Legendre equation, 82, 87, 96
Legendre functions, 83, 106, 112, 116, 119, 124, 128, 129, 336
Legendre orthogonality, 104, 106
Legendre polynomials, 82, 83, 98
Leibniz' rule, 87, 101
Levi–Civita
connection, 350
covariant derivative, 353
Lorentz
force, 223
formula, 219
invariance, 371
metric, 353
Lyman line, 150, 151
Lyman series, 11, 150

Maxwell's equations, 218, 341, 352, 355, 356
Maxwell's theory of electromagnetism, 321
measurement, 21
laws of, 165
units of, 3
Mehler formula, 304
Mellin transform, 324
Mendeléev periodic table, 263, 365
Minakshisundaram–Pleijel
spectral zeta function, 330

Oersted magnetic induction laws, 218
oscillator, 110
anharmonic, 255, 256, 319
harmonic, 53, 65, 71, 109, 169, 170, 245, 255, 291, 293, 300, 303, 308, 310, 317, 319, 323
quantized, 67, 87, 93
quantum, 152
quantum, 53

Pöschl–Teller potential, 112
Paschen series, 11, 151
Pauli
exclusion principle, 253, 261, 264, 265
spin matrices, 234
spin wave functions, 236
Pauli's equation, 225, 236–238, 240
Pfund series, 11
Plancherel
density, 330, 331, 338
measure, 328, 339
Plancherel's
formula, 166
theorem, 167
Planck's constant, 4
Planck quantization, 7–9, 12, 253
Pochhammer symbol, 337
Polyakov
integral, 369, 371
string theory, 317, 320

Quantum
angular momentum, 139, 233
condition, 93
conditions, 15
data, 124, 138
dynamics, 190
electrodynamics, 280, 299
energy, 15, 51, 55, 130, 170
fields, 311, 317, 321, 323, 329
scalar, 324
field theory, 310, 317, 332
Hamiltonian, 264, 280
harmonic oscillator, 152

level, 263
mean values, 164
mechanical effects, 7
mechanical principle, 51
Mechanical Selection Rule, 202
mechanical system, 202
mechanics, 7, 8, 10, 14, 17, 51, 58, 77, 79, 81, 126, 152, 163, 165, 171, 194, 200, 204, 233, 236, 253, 260, 280, 299, 332, 345
number, 124, 138, 148, 171, 233, 234, 242, 260–262
angular momentum, 260
azimuthal, 260
electron spin, 260
magnetic, 124, 138, 171, 260
orbital, 138
principal, 260
operator, 235
oscillator, 53
state, 149, 170, 171, 190
statistical mechanics, 280, 304, 307
statistical quantities, 321
system, 9, 51, 65, 77, 81, 123, 138, 190, 191, 201, 255, 280
energy of, 15
theory, 8, 15, 20, 132, 165, 248, 263, 310, 356
wave function, 345, 353

Riemann, 350
curvature tensor, 350, 351
surface, 323
zeta function, 72, 318, 371
Riemann–Lebesgue lemma, 166
Riemannian
canonical structure, 322
manifold, 350, 351
metrics, 369
Ritz–Rydberg formula, 12, 14, 150
Rodrigues form, 87
Rydberg constant, 13
Rydberg units, 259

Schrödinger's differential operator, 49
Schrödinger's equation, 7, 12, 15–17, 19, 51–54, 58, 59, 63–66, 72–74, 79, 81, 91, 109, 112, 123, 125, 130, 146, 170, 190, 191, 204, 243, 253–256, 264, 284, 319, 341, 342, 346, 348, 349, 353, 355
time dependent, 284
time independent, 62, 63, 76, 245
Schrödinger's wave function, 236
Selberg
trace formula, 271, 322, 325, 327, 328, 334
zeta function, 330–332
Stoner's rule, 261

Titchmarsh theorem, 67

Weierstrass' theorem, 71
Wiener integrals, 280, 299
Wodzicki
non-commutative residue, 339
Wodzicki residue, 339
Wronskian, 98

Zeeman effect, 124, 217, 233, 240, 242, 243
zeta function, 70–72, 271, 280, 310–312, 314, 317, 320, 323–325, 330, 335, 338
finite temperature, 325
Riemann, 72, 318, 371
Selberg, 330–332
spectral, 70, 318, 319, 332, 334, 370
Minakshisundaram–Pleijel type, 330
zero temperature, 325
zeta functions, 317
zeta regularization, 317, 318, 333, 369

Progress in Mathematical Physics

Progress in Mathematical Physics is a book series encompassing all areas of mathematical physics. It is intended for mathematicians, physicists and other scientists, as well as graduate students in the above related areas.

This distinguished collection of books includes authored monographs and textbooks, the latter primarily at the senior undergraduate and graduate levels. Edited collections of articles on important research developments or expositions of particular subject areas may also be included.

This series is reasonably priced and is easily accessible to all channels and individuals through international distribution facilities.

Preparation of manuscripts is preferable in LaTeX. The publisher will supply a macro package and examples of implementation for all types of manuscripts.

Proposals should be sent directly to the series editors:

Anne Boutet de Monvel
Mathématiques, case 7012
Université Paris VII Denis Diderot
2, place Jussieu
F-75251 Paris Cedex 05
France

Gerald Kaiser
The Virginia Center for Signals and Waves
1921 Kings Road
Glen Allen, VA 23059
U.S.A.

or to the Publisher:

Birkhäuser Boston
675 Massachusetts Avenue
Cambridge, MA 02139
U.S.A.
Attn: Ann Kostant

Birkhäuser Verlag
40-44 Viadukstrasse
CH-4010 Basel
Switzerland
Attn: Thomas Hempfling

1 COLLET/ECKMANN. Iterated Maps on the Interval as Dynamical Systems
ISBN 3-7643-3510-6
2 JAFFE/TAUBES. Vortices and Monopoles, Structure of Static Gauge Theories
ISBN 3-7643-3025-2
3 MANIN. Mathematics and Physics
ISBN 3-7643-3027-9
4 ATWOOD/BJORKEN/BRODSKY/STROYNOWSKI. Lectures on Lepton Nucleon Scattering and Quantum Chromodynamics
ISBN 3-7643-3079-1
5 DITA/GEORGESCU/PURICE. Gauge Theories: Fundamental Interactions and Rigorous Results
ISBN 3-7643-3095-3

6 FRAMPTON/GLASHOW/VAN DAM. Third Workshop on Grand Unification, 1982
ISBN 3-7643-3105-4

7 FRÖHLICH. Scaling and Self-Similarity in Physics: Renormalization in Statistical Mechanics and Dynamics
ISBN 3-7643-3168-2

8 MILTON/SAMUEL. Workshop on Non-Perturbative Quantum Chromodynamics
ISBN 3-7643-3127-5

9 LANGACKER/STEINHARDT/WELDON. Fourth Workshop on Grand Unification
ISBN 3-7643-3169-0

10 FRITZ/JAFFE/SZÁSZ. Statistical Physics and Dynamical Systems: Rigorous Results
ISBN 3-7643-3300-6

11 CEAUSESCU/COSTACHE/GEORGESCU. Critical Phenomena: 1983 Brasov School Conference
ISBN 3-7643-3289-1

12 PIGUET/SIBOLD. Renormalized Supersymmetry: The Perturbation Theory of N=1 Supersymmetric Theories in Flat Space-Time
ISBN 3-7643-3346-4

13 HABA/SOBCZYK. Functional Integration, Geometry and Strings: Proceedings of the XXV Karpacz Winter School of Theoretical Physics
ISBN 3-7643-2387-6

14 SMIRNOV. Renormalization and Asymptotic Expansions
ISBN 3-7643-2640-9

15 Leznov/Saveliev. Group-Theoretical Methods for Integration of Nonlinear Dynamical Systems
ISBN 3-7643-2615-8

16 MASLOV. The Complex WKB Method for Nonlinear Equations I: Linear Theory
ISBN 3-7643-5088-1

17 BAYLIS. Electrodynamics: A Modern Geometric Approach
ISBN 0-8176-4025-8

18 ABŁAMOWICZ/FAUSER. Clifford Algebras and their Applications in Mathematical Physics, Volume 1: Algebra and Physics
ISBN 0-8176-4182-3

19 RYAN/SPRÖßIG. Clifford Algebras and their Applications in Mathematical Physics, Volume 2: Clifford Analysis
ISBN 0-8176-4183-1

20 STOLLMANN. Caught by Disorder: Bound States in Random Media
ISBN 0-8176-4210-2

21 PETTERS/LEVINE/WAMBSGANSS. Singularity Theory and Gravitational Lensing
ISBN 0-8176-3668-4

22 CERCIGNANI. The Relativistic Boltzmann Equation: Theory and Applications
ISBN 3-7643-6693-1

23 KASHIWARA/MIWA. MathPhys Odyssey 2001: Integrable Models and Beyond—*In Honor of Barry M. McCoy*
ISBN 0-8176-4260-9

24 CNOPS. An Introduction to Dirac Operators on Manifolds
ISBN 0-8176-4298-6

25 KLAINERMAN/NICOLÒ. The Evolution Problem in General Relativity
ISBN 0-8176-4254-4

26 BLANCHARD/BRÜNING. Mathematical Methods in Physics
ISBN 0-8176-4228-5

27 WILLIAMS. Topics in Quantum Mechanics
ISBN 0-8176-4311-7

Lightning Source UK Ltd.
Milton Keynes UK
UKOW01n0727080218
317556UK00012B/780/P

9 780817 643119